Fundamental Electrical and Electronic Principles

Fundamental Electrical and Electronic Principles

Christopher R Robertson

Second Edition

AMSTERDAM • BOSTON • HEIDELBERG • LONDON • NEW YORK • OXFORD
PARIS • SAN DIEGO • SAN FRANCISCO • SINGAPORE • SYDNEY • TOKYO
Newnes is an imprint of Elsevier

Newnes is an imprint of Elsevier
Linacre House, Jordan Hill, Oxford OX2 8DP, UK
30 Corporate Drive, Suite 400, Burlington, MA 01803, USA

First published as *Electrical and Electronic Principles I* 1993
Second edition 2001
Reprinted 2002, 2004, 2005, 2006

British Library Cataloguing in Publication Data
A catalogue record for this book is available from the British Library

Library of Congress Cataloging-in-Publication Data
A catalog record for this book is available from the Library of Congress

ISBN–13: 978-0-7506-5145-5
ISBN–10: 0-7506-5145-8

For information on all Newnes publications
visit our website at www.newnespress.com

Printed and bound in *The Netherlands*

06 07 08 09 10 10 9 8 7 6 5

Contents

Preface

This Textbook supersedes the title *Electrical and Electronic Principles 1* with the addition of two new chapters on semiconductors and semiconductor circuits. The original Chapter 2 has also been expanded with the inclusion of further network theorems. Also, in response to comments from colleges, even more worked examples have been included in the form of supplementary worked examples, which appear at the ends of the relevant chapters.

This book continues with the philosophy of the original in that it may be used as a complete set of student course notes for students undertaking the study of Electrical and Electronic Principles in the first year of a BTEC National Diploma/Certificate course. It also provides coverage of the following:

(i) Electrical Principles content of the BTEC Science unit.
(ii) Electrical Applications of the VCE unit Applied Science in Engineering.
(iii) AVCE unit Electrical and Electronic Principles.
(iv) Any other course of study or foundation/bridging course leading to a qualification in Electrical and Electronic Engineering.

Fundamental Electrical and Electronic Principles contains over 440 illustrations, 180 worked examples, 25 suggested practical assignments and 225 assignment questions, the answers to which are to be found at the end of the book.

The order of the chapters does not necessarily follow the order set out in any syllabus, but rather follows a logical step-by-step sequence through the subject. Some topic areas may well extend beyond current syllabus requirements, but do so for the sake of completeness.

Coverage of the second year syllabus for the BTEC National Diploma/Certificate unit in Electrical and Electronic Principles, plus certain AVCE units appears in the companion volume entitled *Further Electrical and Electronic Principles*. This title supersedes the original *Electrical and Electronic Principles 2*.

C. Robertson

Tonbridge, 2001

Introduction

The chapters follow a sequence that I consider to be a logical progression through the subject matter, and in the main follow the order of objectives stated in the BTEC module. The major exception to this sequence is that the topics of instrumentation and measurements do not appear in a specific chapter of that title. Instead, the various instruments and measurement methods are integrated within those chapters where the relevant electrical theory is dealt with.

Occasionally you will see a word or phrase in **bold type**, and close by a box with an explanation. These emphasised words or terms may be ones that are not familiar to you. Within the box will be an explanation of the words used in the text.

Throughout the book, Worked Examples appear as **Q questions in bold type**, followed by **A** answers. In most chapters Assignment Questions are provided for students to solve.

The first chapter deals with the basic concepts of electricity; the use of standard form and its adaptation to scientific notation; SI and derived units; and the plotting of graphs. This chapter is intended to provide a means of ensuring that all students on a given course start with the same background knowledge. Also included in this chapter are notes regarding communication. In particular, the way in which the solution of Assignment Questions, and Practical Assignment reports are presented.

1 Fundamentals

1.1 Units

Wherever measurements are performed there is a need for a coherent and practical system of units. In science and engineering the International System of units (SI units) form the basis of all units used. There are seven 'base' units from which all the other units are derived, called derived units.

Table 1.1 *The SI base units*

Quantity	*Unit*	*Unit symbol*
Mass	kilogram	kg
Length	metre	m
Time	second	s
Electric current	ampere	A
Temperature	kelvin	K
Luminous intensity	candela	cd
Amount of substance	mole	mol

A few examples of derived units are shown in Table 1.2, and it is worth noting that different symbols are used to represent the quantity and its associated unit in each case.

Table 1.2 *Some SI derived units*

Quantity		*Unit*	
Name	*Symbol*	*Name*	*Symbol*
Force	F	Newton	N
Power	P	Watt	W
Energy	W	Joule	J
Resistance	R	Ohm	Ω

For a more comprehensive list of SI units see Appendix A at the back of the book.

1.2 Standard Form Notation

Standard form is a method of writing large and small numbers in a form that is more convenient than writing a large number of trailing or leading zeroes.

For example the speed of light is approximately 300 000 000 m/s. When written in standard form this figure would appears as

3.0×10^8 m/s, where 10^8 represents 100 000 000

Similarly, if the wavelength of 'red' light is approximately 0.000 000 767 m, it is more convenient to write it in standard form as

7.67×10^{-7} m, where $10^{-7} = 1/10\,000\,000$

It should be noted that whenever a 'multiplying' factor is required, the base 10 is raised to a positive power. When a 'dividing' factor is required, a negative power is used. This is illustrated below:

$$\begin{array}{rr} 10 = 10^1 & 1/10 = 0.1 \quad = 10^{-1} \\ 100 = 10^2 & 1/100 = 0.01 \quad = 10^{-2} \\ 1000 = 10^3 & 1/1000 = 0.001 \quad = 10^{-3} \\ \text{etc.} & \text{etc.} \end{array}$$

One restriction that is applied when using standard form is that only the first non-zero digit must appear before the decimal point.

Thus, 46 500 is written as

4.65×10^4 and *not* as 46.5×10^3

Similarly, 0.002 69 is written as

2.69×10^{-3} and *not* as 26.9×10^{-4} or 269×10^{-5}

1.3 'Scientific' Notation

This notation has the advantage of using the base 10 raised to a power but it is not restricted to the placement of the decimal point. It has the added advantage that the base 10 raised to certain powers have unique symbols assigned.

For example if a body has a mass $m = 500\,000$ g.

In standard form this would be written as

$m = 5.0 \times 10^5$ g.

Using scientific notation it would appear as

$m = 500$ kg (500 kilogram)

where the 'k' in front of the g for gram represents 10^3.

Not only is the latter notation much neater but it gives a better 'feel' to the meaning and relevance of the quantity.

See Table 1.3 for the symbols (prefixes) used to represent the various powers of 10. It should be noted that these prefixes are arranged in multiples of 10^3. It is also a general rule that the positive powers of 10 are represented by capital letters, with the negative powers being represented by lower case (small) letters. The exception to this rule is the 'k' used for kilo.

Table 1.3 *Unit prefixes used in 'scientific' notation*

Multiplying factor	*Prefix name*	*Symbol*
10^{12}	tera	T
10^{9}	giga	G
10^{6}	mega	M
10^{3}	kilo	k
10^{-3}	milli	m
10^{-6}	micro	μ
10^{-9}	nano	n
10^{-12}	pico	p

Worked Example 1.1

Question

Write the following quantities in a concise form using (a) standard form, and (b) scientific notation (i) 0.000 018 A (ii) 15 000 V (iii) 250 000 000 W

Answer

(a) (i) 0.000 018 A = 1.8×10^{-5} A

(ii) 15 000 V = 1.5×10^{4} V

(iii) 250 000 000 W = 2.5×10^{8} W

(b) (i) 0.000 018 A = 18 μA

(ii) 15 000 V = 15 kV

(iii) 250 000 000 W = 250 MW **Ans**

The above example illustrates the neatness and convenience of the scientific or engineering notation.

1.4 Conversion of Areas and Volumes

Consider a square having sides of 1 m as shown in Fig. 1.1. In this case each side can also be said to be 100 cm or 1000 mm. Hence the area A enclosed could be stated as:

$$A = 1 \times 1 = 1 \text{ m}^2$$

$$\text{or } A = 100 \times 100 = 10^2 \times 10^2 = 10^4 \text{ cm}^2$$

$$\text{or } A = 1000 \times 1000 = 10^3 \times 10^3 = 10^6 \text{ mm}^2.$$

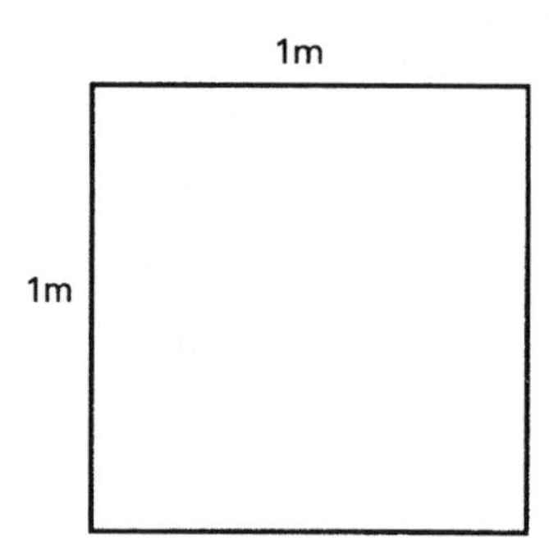

Fig. 1.1

From the above it may be seen that

$$1 \text{ m}^2 = 10^4 \text{ cm}^2$$

$$\text{and that } 1 \text{ m}^2 = 10^6 \text{ mm}^2.$$

Similarly, if the square had sides of 1 cm the area would be

$$A = 1 \text{ cm}^2 = 10^{-2} \times 10^{-2} = 10^{-4} \text{ m}^2.$$

Again if the sides were of length 1 mm the area would be

$$A = 1 \text{ mm}^2 = 10^{-3} \times 10^{-3} = 10^{-6} \text{ m}^2.$$

Thus $1 \text{ cm}^2 = 10^{-4} \text{ m}^2$ and $1 \text{ mm}^2 = 10^{-6} \text{ m}^2$.

Since the basic unit for area is m^2, then areas quoted in other units should firstly be converted into square metres before calculations proceed. This procedure applies to all the derived units, and it is good practice to convert all quantities into their 'basic' units before proceeding with calculations.

It is left to the reader to confirm that the following conversions for volumes are correct:

$$1 \text{ mm}^3 = 10^{-9} \text{ m}^3$$

$$1 \text{ cm}^3 = 10^{-6} \text{ m}^3.$$

Worked Example 1.2

Question

A mass m of 750 g is acted upon by a force F of 2 N. Calculate the resulting acceleration given that the three quantities are related by the equation

$F = ma$ newton

Answer

$m = 750\,\text{g} = 0.75\,\text{kg};\quad F = 2\,\text{N}$

Since $F = ma$ newton, then

$$a = \frac{F}{m}\ \text{metre/second}^2$$

$$= \frac{2}{0.75}$$

so $a = 2.667\ \text{m/s}^2$ **Ans**

1.5 Graphs

A graph is simply a pictorial representation of how one quantity or variable relates to another. One of these is known as the dependent variable and the other as the independent variable. It is general practice to plot the dependent variable along the vertical axis and the independent variable along the horizontal axis of the graph. To illustrate the difference between these two types of variable consider the case of a vehicle that is travelling between two points. If a graph of the distance travelled versus the time elapsed is plotted, then the distance travelled would be the dependent variable. This is because the distance travelled depends on the time that has elapsed. But the time is independent of the distance travelled, since the time will continue to increase regardless of whether the vehicle is moving or not.

Such a graph is shown in Fig. 1.2, from which it can be seen that over the first three hours the distance travelled was 30 km. Over the next two hours a further 10 km was travelled, and subsequently no further distance was travelled. Since distance travelled divided by the time taken is velocity, then the graph may be used to determine the speed of the vehicle at any time. Another point to note about this graph is that it consists of straight lines. This tells us that the vehicle was travelling at different but constant velocities at different times. It should be apparent that the steepest part of the graph occurs when the vehicle was travelling fastest. To be more precise, we refer to the slope or gradient of the graph. In order to calculate the velocity over the first three hours, the slope can be determined as follows:

Change in time, $\delta t_1 = 3 - 0 = 3$ h

change in distance $\delta s_1 = 30 - 0 = 30$ km

slope or gradient $\equiv$ velocity $v = \dfrac{\delta s_1}{\delta t_1} = \dfrac{30}{3} = 10$ km/h

Similarly, for the second section of the graph:

$$v = \frac{\delta s_2}{\delta t_2} = \frac{(40 - 30)}{(5 - 3)} = 5 \text{ km/h}$$

Considering the final section of the graph, it can be seen that there is no change in distance (the vehicle is stationary). This is confirmed, since if δs is zero then the velocity must be zero.

In some ways this last example is a special case, since it involved a straight line graph. In this case we can say that the distance is directly proportional to time. In many cases a non-linear graph is produced, but the technique for determining the slope at any given point is similar. Such a graph is shown in Fig. 1.3, which represents the displace-

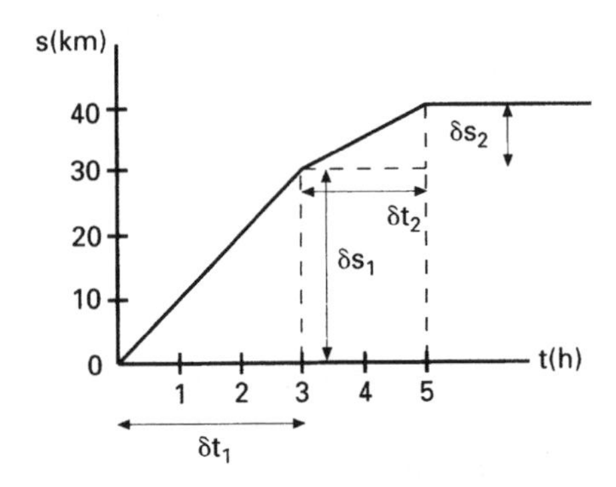

Fig. 1.2

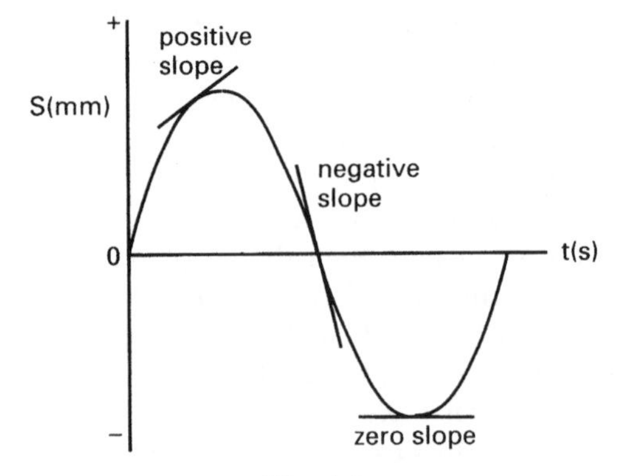

Fig. 1.3

ment of a mass when subjected to simple harmonic motion. The resulting graph is a sinewave. To determine the slope at any given instant in time we would have to determine the slope of the tangent to the curve at that point on the graph. If this is done then the figure obtained in each case would be the velocity of the mass at that instant. Notice that the slope is steepest at the instants that the curve passes through the zero displacement axis (maximum velocity). It is zero at the 'peaks' of the graph (zero velocity). Also note that if the graph is sloping upwards as you trace its path from left to right it is called a positive slope. If it slopes downwards it is called a negative slope.

1.6 Basic Electrical Concepts

All matter is made up of atoms, and there are a number of 'models' used to explain physical effects that have been both predicted and subsequently observed. One of the oldest and simplest of these is the Bohr model. This describes the atom as consisting of a central nucleus containing minute particles called protons and neutrons. Surrounding the nucleus are a number of electrons in various orbits. This model is illustrated in Fig. 1.4. The possible presence of neutrons in the nucleus has been ignored, since these particles play no part in the electrical concepts to be described. It should be noted that this atomic model is greatly over-simplified. It is this very simplicity that makes it ideal for the beginner to achieve an understanding of what electricity is and how many electrical devices operate.

The model shown in Fig. 1.4 is not drawn to scale since a proton is approximately 2000 times more massive than an electron. Due to this *relatively* large mass the proton does not play an active part in electrical current flow. It is the behaviour of the electrons that is more important. However, protons and electrons do share one thing in common; they both possess a property known as electric charge. The unit of charge is called the coulomb (C). Since charge is considered as the *quantity* of electricity it is given the symbol Q. An electron and proton have exactly the same amount of charge. The electron has a negative charge, whereas the proton has a positive charge. Any atom in its 'normal' state is electrically neutral (has no net charge). So, in this state the atom must possess as many orbiting electrons as there are protons in its nucleus. If one or more of the orbiting electrons can somehow be persuaded to leave the parent atom then this charge balance is upset. In this case the atom acquires a net positive charge, and is then known as a positive ion. On the other hand, if 'extra' electrons can be made to orbit the nucleus then the atom acquires a net negative charge. It then becomes a negative ion.

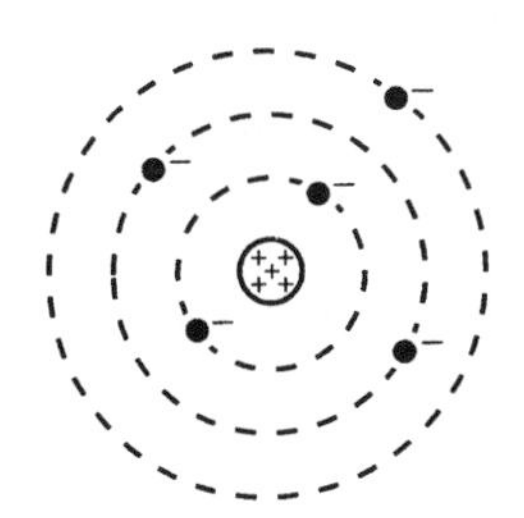

Fig. 1.4

> An **analogy** is a technique where the **behaviour** of one system is compared to the **behaviour** of another system. The system chosen for this comparison will be one that is more familiar and so more easily understood. HOWEVER, it must be borne in mind that an analogy should not be extended too far. Since the two systems are usually very different physically there will come a point where comparisons are no longer valid

You may now be wondering why the electrons remain in orbit around the nucleus anyway. This can best be explained by considering an **analogy**. Thus, an electron orbiting the nucleus may be compared to a satellite orbiting the Earth. The satellite remains in orbit due to a balance of forces. The gravitational force of attraction towards the Earth is balanced by the centrifugal force on the satellite due to its high velocity. This high velocity means that the satellite has high kinetic energy. If the satellite is required to move into a higher orbit, then its motor must be fired to speed it up. This will increase its energy. Indeed, if its velocity is increased sufficiently, it can be made to leave Earth orbit and travel out into space. In the case of the electron there is also a balance of forces involved. Since both electrons and protons have mass, there will be a gravitational force of attraction between them. However the masses involved are so minute that the gravitational force is negligible. So, what force of attraction does apply here? Remember that electrons and protons are oppositely charged particles, and oppositely charged bodies experience a force of attraction. Compare this to two simple magnets, whereby opposite polarities attract and like (the same) polarities repel each other. The same rule applies to charged bodies. Thus it is the balance between this force of electrostatic attraction and the kinetic energy of the electron that maintains the orbit. It may now occur to you to wonder why the nucleus remains intact, since the protons within it are all positively charged particles! It is beyond the scope of this book (and of the course of study on which you are now embarked) to give a comprehensive answer. Suffice to say that there is a force within the nucleus far stronger than the electrostatic repulsion between the protons that binds the nucleus together.

All materials may be classified into one of three major groups—conductors, insulators and semiconductors. In simple terms, the group into which a material falls depends on how many 'free' electrons it has. The term 'free' refers to those electrons that have acquired sufficient energy to leave their orbits around their parent atoms. In general we can say that conductors have many free electrons which will be drifting in a *random* manner within the material. Insulators have very few free electrons (ideally none), and semiconductors fall somewhere between these two extremes.

Electric current This is the rate at which free electrons can be made to drift through a material *in a particular direction*. In other words, it is the rate at which charge is moved around a circuit. Since charge is measured in coulombs and time in seconds then logically the unit for electric current would be the coulomb/second. In fact, the amount of current flowing through a circuit may be calculated by dividing the amount of charge passing a given point by the time taken. The unit however is given a special name, the ampere (often abbreviated to amp). This is fairly common practice with SI units, whereby the names chosen are those of famous scientists whose pioneering work is thus commemorated. The relationship between current, charge and time can be expressed as a mathematical equation as follows:

$$I = \frac{Q}{t} \text{ amp, or } Q = It \text{ coulomb} \qquad (1.1)$$

Worked Example 1.3

Question

A charge of 35 mC is transferred between two points in a circuit in a time of 20 ms. Calculate the value of current flowing.

Answer

$Q = 35 \times 10^{-3}$ C; $t = 20 \times 10^{-3}$ s

$$I = \frac{Q}{t} \text{ amp}$$

$$= \frac{35 \times 10^{-3}}{20 \times 10^{-3}}$$

$I = 1.75$ A **Ans**

Worked Example 1.4

Question

If a current of 120 μA flows for a time of 15 s, determine the amount of charge transferred.

Answer

$I = 120 \times 10^{-6}$A; $t = 15$ s

$Q = It$ coulomb

$= 120 \times 10^{-6} \times 15$

$Q = 1.8$ mC **Ans**

Worked Example 1.5

Question

80 coulombs of charge was transferred by a current of 0.5 A. Calculate the time for which the current flowed.

Continued on p. 10

Worked Example 1.5 (*Continued*)

Answer

$Q = 80\,\text{C}$; $I = 0.5\,\text{A}$

$$t = \frac{Q}{I} \text{ seconds}$$

$$= \frac{80}{0.5}$$

$t = 160\,\text{s}$ **Ans**

Electromotive Force (emf) The *random* movement of electrons within a material does not constitute an electrical current. This is because it does not result in a drift in one particular direction. In order to cause the 'free' electrons to drift in a given direction an electromotive force must be applied. Thus the emf is the 'driving' force in an electrical circuit. The symbol for emf is E and the unit of measurement is the volt (V). Typical sources of emf are cells, batteries and generators.

The amount of current that will flow through a circuit is directly proportional to the size of the emf applied to it. The circuit diagram symbols for a cell and a battery are shown in Figs 1.5(a) and (b) respectively. Note that the positively charged plate (the long line) usually does not have a plus sign written alongside it. Neither does the negative plate normally have a minus sign written by it. These signs have been included here merely to indicate (for the first time) the symbol used for each plate.

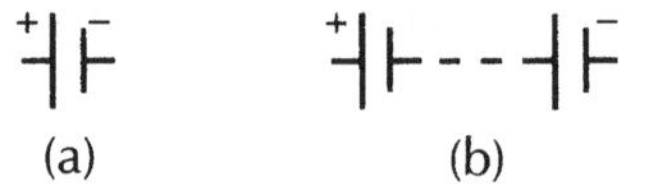

(a) (b)

Fig. 1.5

Resistance (R) Although the amount of electrical current that will flow through a circuit is directly proportional to the applied emf, the other property of the circuit (or material) that determines the resulting current is the opposition offered to the flow. This opposition is known as the electrical resistance, which is measured in ohms (Ω). Thus conductors, which have many 'free' electrons available for current carrying, have a low value of resistance. On the other hand, since insulators have very few 'free' charge carriers then insulators have a very high resistance. Pure semiconductors tend to behave more like insulators in this respect. However, in practice, semiconductors tend to be used in an impure form, where the added impurities improve the conductivity of the material. An electrical device that is designed to have a specified value of resistance is called a resistor. The circuit diagram symbol for a resistor is shown in Fig. 1.6.

Fig. 1.6

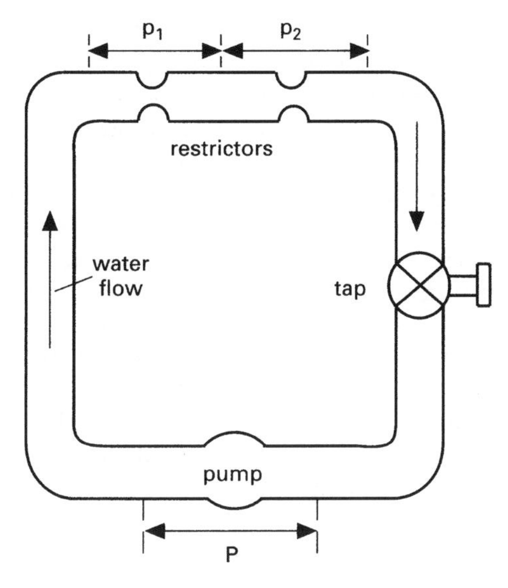

Fig. 1.7

Potential Difference (p.d.) Whenever current flows through a resistor there will be a p.d. developed across it. The p.d. is measured in volts, and is quite literally the difference in voltage levels between two points in a circuit. Although both p.d. and emf are measured in volts they are not the same quantity. Essentially, emf (being the driving force) causes current to flow; whilst a p.d. is the result of current flowing through a resistor. Thus emf is a *cause* and p.d. is an *effect*. It is a general rule that the symbol for a quantity is different to the symbol used for the unit in which it is measured. One of the few exceptions to this rule is that the quantity symbol for p.d. happens to be the same as its unit symbol, namely V. In order to explain the difference between emf and p.d. we shall consider another analogy.

Figure 1.7 represents a simple hydraulic system consisting of a pump, the connecting pipework and two restrictors in the pipe. The latter will have the effect of limiting the rate at which the water flows around the circuit. Also included is a tap that can be used to interrupt the flow completely. Figure 1.8 shows the equivalent electrical circuit, comprising a battery, the connecting conductors (cables or leads) and two resistors. The latter will limit the amount of current flow. Also included is a switch that can be

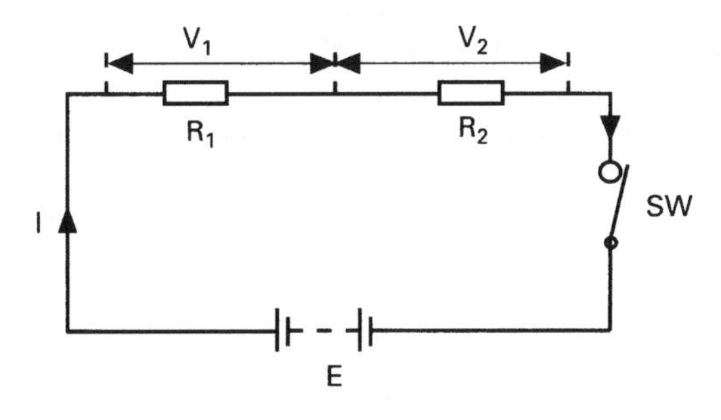

Fig. 1.8

used to 'break' the circuit and so prevent any current flow. As far as each of the two systems is concerned we are going to make some assumptions.

For the water system we will assume that the connecting pipework has no slowing down effect on the flow, and so will not cause any pressure drop. Provided that the pipework is relatively short then this is a reasonable assumption. The similar assumption in the electrical circuit is that the connecting wires have such a low resistance that they will cause no p.d. If anything, this is probably a more legitimate assumption to make. Considering the water system, the pump will provide the total system pressure (P) that circulates the water through it. Using some form of pressure measuring device it would be possible to measure this pressure together with the pressure drops (p_1 and p_2) that would occur across the two restrictors. Having noted these pressure readings it would be found that the total system pressure is equal to the sum of the two pressure drops. Using a similar technique for the electrical circuit, it would be found that the sum of the two p.d.s (V_1 and V_2) is equal to the total applied emf E volts. These relationships may be expressed in mathematical form as:

$$\mathrm{P} = \mathrm{p}_1 + \mathrm{p}_2 \text{ pascal}$$

$$\text{and} \quad E = V_1 + V_2 \text{ volt} \tag{1.2}$$

When the potential at some point in a circuit is quoted as having a particular value (say 10 V) then this implies that it is 10 V above some reference level or datum. Compare this with altitudes. If a mountain is said to be 5000 m high it does not necessarily mean that it rises 5000 m from its base to its peak. The figure of 5000 m refers to the height of its peak above mean sea level. Thus, mean sea level is the reference point or datum from which altitudes are measured. In the case of electrical potentials the datum is taken to be the potential of the Earth which is 0 V. Similarly, −10 V means 10 V below or less than 0 V.

Conventional current and electron flow You will notice in Fig. 1.8 that the arrows used to show the direction of current flow indicate that this is from the positive plate of the battery, through the circuit, returning to the negative battery plate. This is called conventional current flow. However, since electrons are negatively charged particles, then these must be moving in the opposite direction. The latter is called electron flow. Now, this poses the problem of which one to use. It so happens that before science was sufficiently advanced to have knowledge of the electron, it was assumed that the positive plate represented the 'high' potential and the negative the 'low' potential. So the convention was adopted that the current flowed around the circuit from the high potential to the low potential. This compares with water which can naturally only flow from a high level to a lower level. Thus the concept of conventional current flow was adopted. All the subsequent 'rules' and conventions were based on this direction of current flow. On the discovery of the nature of the electron, it was decided to retain the concept of conventional current flow. Had this not been the case then all the other rules and conventions would have needed to be changed! Hence, true electron flow is used only when it is necessary to explain certain effects (as in semiconductor devices such as diodes and transistors). Whenever we are considering basic electrical circuits and devices we shall use *conventional current flow* i.e. current flowing around the circuit *from* the *positive* terminal of the source of emf *to* the *negative* terminal.

Ohm's Law This states that the p.d. developed between the two ends of a resistor is directly proportional to the value of current flowing through it, provided that all other factors (e.g. temperature) remain constant. Writing this in mathematical form we have:

$$V \propto I$$

However, this expression is of limited use since we need an equation. This can only be achieved by introducing a constant of proportionality; in this case the resistance value of the resistor.

$$\text{Thus} \quad V = IR \text{ volt} \qquad (1.3)$$

$$\text{or} \quad I = \frac{V}{R} \text{ amp} \qquad (1.4)$$

$$\text{and} \quad R = \frac{V}{I} \text{ ohm} \qquad (1.5)$$

Worked Example 1.6

Question

A current of 5.5 mA flows through a 33 kΩ resistor. Calculate the p.d. thus developed across it.

Answer

$I = 5.5 \times 10^{-3}$ A; $R = 33 \times 10^{3}\ \Omega$

$$V = IR \text{ volt}$$

$$= 5.5 \times 10^{-3} \times 33 \times 10^{3}$$

$V = 181.5$ V **Ans**

Worked Example 1.7

Question

If a p.d. of 24 V exists across a 15 Ω resistor then what must be the current flowing through it?

Continued on p. 14

Worked Example 1.7 (*Continued*)

Answer

$V = 24$ V; $R = 15\,\Omega$

$$I = \frac{V}{R} \text{ amp}$$

$$= \frac{24}{15}$$

$$I = 1.6 \text{ A}$$ **Ans**

Internal Resistance (r) So far we have considered that the emf E volts of a source is available at its terminals when supplying current to a circuit. If this were so then we would have an ideal source of emf. Unfortunately this is not the case in practice. This is due to the internal resistance of the source. As an example consider a typical 12 V car battery. This consists of a number of oppositely charged plates, appropriately inter-connected to the terminals, immersed in an **electrolyte**. The plates themselves, the internal connections and the electrolyte all combine to produce a small but **finite** resistance, and it is this that forms the battery internal resistance.

An **electrolyte** is the chemical 'cocktail' in which the plates are immersed. In the case of a car battery, this is an acid/water mixture.

In this context, **finite** simply means measurable.

Figure 1.9 shows such a battery with its terminals on open circuit (no external circuit connected). Since the circuit is incomplete no current can flow. Thus there will be no p.d. developed across the battery's internal resistance r. Since the term p.d. quite literally means a difference in potential between the two ends of r, then the terminal A must be at a potential of 12 V, and terminal B must be at a potential of 0 V. Hence, under these conditions, the full emf 12 V is available at the battery terminals.

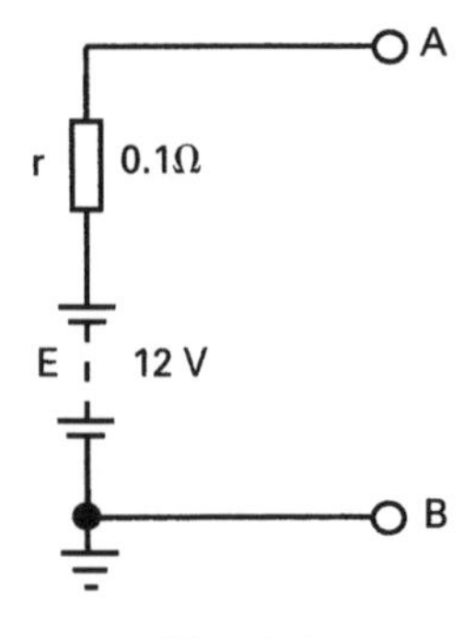

Fig. 1.9

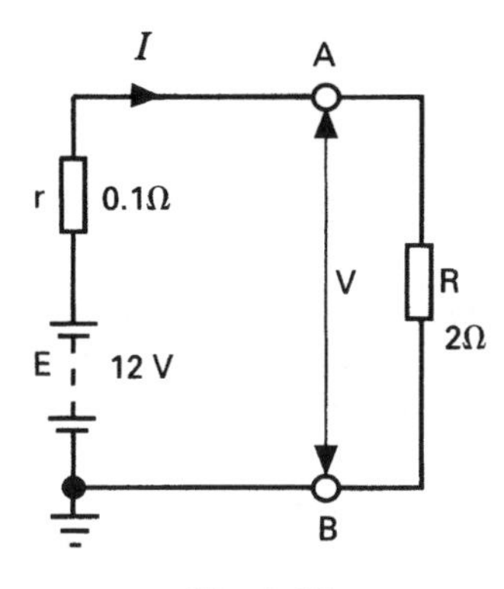

Fig. 1.10

Figure 1.10 shows an external circuit, in the form of a 2 Ω resistor, connected across the terminals. Since we now have a complete circuit then current I will flow as shown. The value of this current will be 5.71 A (the method of calculating this current will be dealt with early in the next chapter). This current will cause a p.d. across r and also a p.d. across R. These calculations and the consequences for the complete circuit now follow:

$$\begin{aligned}\text{p.d. across } r &= Ir \text{ volt} \quad \text{(Ohms law applied)}\\ &= 5.71 \times 0.1\\ &= 0.571\,\text{V}\\ \text{p.d. across } R &= IR \text{ volt}\\ &= 5.71 \times 2\\ &= 11.42\,\text{V}\end{aligned}$$

Note: $0.571 + 11.42 = 11.991$ V but this figure *should* be 12 V. The very small difference is simply due to 'rounding' the figures obtained from the calculator.

The p.d. across R is the battery terminal p.d. V. Thus it may be seen that when a source is supplying current, the terminal p.d. will always be less than its emf. To emphasise this point let us assume that the external resistor is changed to one of 1.5 Ω resistance. The current now drawn from the battery will be 7.5 A. Hence:

$$\begin{aligned}\text{p.d. across } r &= 7.5 \times 0.1 = 0.75\,\text{V}\\ \text{and p.d. across } R &= 7.5 \times 1.5 = 11.25\,\text{V}\end{aligned}$$

Note that $11.25 + 0.75 = 12$ V (rounding error not involved). Hence, the battery terminal p.d. has fallen still further as the current drawn has increased. This example brings out the following points.

1 Assuming that the battery's charge is maintained, then its emf remains constant. But its terminal p.d. varies as the current drawn is varied, such that

$$V = E - Ir \text{ volt} \tag{1.6}$$

2 Rather than having to write the words 'p.d. across R′ it is more convenient to write this as V_{AB}, which translated, means the potential difference between points A and B.
3 In future, if no mention is made of the internal resistance of a source, then for calculation purposes you may assume that it is zero, i.e. an ideal source.

Energy (W) This is the property of a system that enables it to do work. Whenever work is done energy is transferred from that system to another one. The most common form into which energy is transformed is heat. Thus one of the effects of an electric current is to produce heat (e.g. an electric kettle). J. P. Joule carried out an investigation into this effect. He reached the conclusion that the amount of heat so produced was proportional to the value of the square of the current flowing and the time for which it flowed. Once more a constant of proportionality is required, and again it is the resistance of the circuit that is used. Thus the heat produced (or energy dissipated) is given by the equation

$$W = I^2Rt \text{ joule} \qquad (1.7)$$

and applying Ohm's law as shown in equations (1.3) to (1.5)

$$W = \frac{V^2t}{R} \text{ joule} \qquad (1.8)$$

$$\text{or } W = VIt \text{ joule} \qquad (1.9)$$

Worked Example 1.8

Question

A current of 200 mA flows through a resistance of 750 Ω for a time of 5 minutes. Calculate (a) the p.d. developed, and (b) the energy dissipated.

Answer

$I = 200\text{ mA} = 0.2\text{ A};\ t = 5 \times 60 = 300\text{ s};\ R = 750\ \Omega$

(a) $V = IR$ volt

$= 0.2 \times 750$

$V = 150$ V **Ans**

(b) $W = I^2Rt$ joule

$= 0.2 \times 750 \times 300$

$W = 9000$ J or 9 kJ **Ans**

Note: It would have been possible to use either equation (1.8) or (1.9) to calculate W. However, this would have involved using the calculated value for V. If this value had been miscalculated, then the answers to both parts of the question would have been incorrect. So, whenever possible, make use of data that are given in the question in preference to values that you have calculated. Please also note that the time has been converted to its basic unit, the second.

Power (P) This is the rate at which work is done, or at which energy is dissipated. The unit in which power is measured is the watt (W).

Warning: Do not confuse this unit symbol with the quantity symbol for energy.
In general terms we can say that power is energy divided by time.

i.e. $P = \frac{W}{t}$ watt

Thus, by dividing each of equations (1.7), (1.8) and (1.9), in turn, by t, the following equations for power result:

$$P = I^2R \text{ watt} \quad (1.10)$$

$$P = \frac{V^2}{R} \text{ watt} \quad (1.11)$$

$$\text{or } P = VI \text{ watt} \quad (1.12)$$

Worked Example 1.9

Question

A current of 1.4 A when flowing through a circuit for 15 minutes dissipates 200 KJ of energy. Calculate (a) the p.d., (b) power dissipated and (c) the resistance of the circuit.

Answer

$I = 1.4$ A; $t = 15 \times 60$ s; $W = 2 \times 10^5$ J

(a) $W = VIt$ joule

so $V = \frac{W}{It}$ volt

$$= \frac{2 \times 10^5}{1.4 \times 15 \times 60}$$

$V = 158.7$ V **Ans**

Continued on p. 18

Worked Example 1.9 (*Continued*)

(b)

$$P = VI \text{ watt}$$
$$= 158.7 \times 1.4$$
$$P = 222.2 \text{ W } \textbf{Ans}$$

(a)

$$R = \frac{V}{I} \text{ ohm}$$
$$= \frac{158.7}{1.4}$$
$$R = 113.4\,\Omega \textbf{ Ans}$$

The Commercial Unit of Energy (kWh) Although the joule is the SI unit of energy, it is too small a unit for some practical uses. e.g. where large amounts of power are used over long periods of time. The electricity meter in your home actually measures the energy consumption. So, if a 3 kW heater was in use for 12 hours the amount of energy used would be 129.6 MJ. In order to record this the meter would require at least ten dials to indicate this very large number. In addition to which, many of them would have to rotate at an impossible rate. Hence the commercial unit of energy is the kilowatt-hour (kWh). Kilowatt-hours are the 'units' that appear on electricity bills. The number of units consumed can be calculated by multiplying the power (in kW) by the time interval (in hours). So, for the heater mentioned above, the number of 'units' consumed would be written as 36 kWh. It should be apparent from this that to record this particular figure, fewer dials are required, and their speed of rotation is perfectly acceptable.

Worked Example 1.10

Question

Calculate the cost of operating a 12.5 kW machine continuously for a period of 8.5 h if the cost per unit is 7.902 p.

Answer

$$W = 12.5 \times 8.5 \text{ kWh}$$
$$\text{so } W = 106.25 \text{ kWh}$$
$$\text{and cost} = 106.25 \times 7.902$$
$$= \pounds 8.40 \textbf{ Ans}$$

Note: When calculating energy in kWh the power must be expressed in kW, and the time in hours respectively, rather than in their basic units of watts and seconds respectively.

Alternating and Direct Quantities The sources of emf and resulting current flow so far considered are called d.c. quantities. This is because a battery or cell once connected to a circuit is capable of driving current around the circuit in one direction only. If it is required to reverse the current it is necessary to reverse the battery connections. The term d.c., strictly speaking, means 'direct current'. However, it is also used to describe unidirectional voltages. Thus a d.c. voltage refers to a unidirectional voltage that may only be reversed as stated above.

However, the other commonly used form of electrical supply is that obtained from the electrical mains. This is the supply that is generated and distributed by the power companies. This is an alternating or a.c. supply in which the current flows alternately in opposite directions around a circuit. Again, the term strictly means 'alternating current', but the emfs and p.d.s associated with this system are referred to as a.c. voltages. Thus, an a.c. generator (or alternator) produces an alternating voltage. Most a.c. supplies provide a sinusoidal waveform (a sinewave shape). Both d.c. and a.c. waveforms are illustrated in Fig. 1.11. The treatment of a.c. quantities and circuits is dealt with in Chapters 6 and 7, and need not concern you any further at this stage.

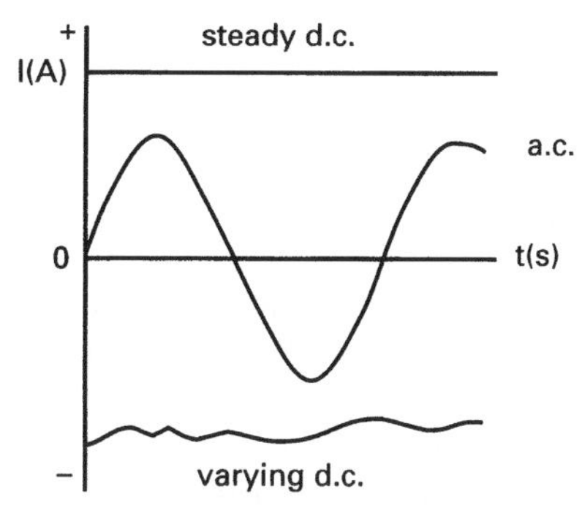

Fig. 1.11

Factors affecting Resistance The resistance of a sample of material depends upon four factors

(i) its length
(ii) its cross-sectional area (csa)
(iii) the actual material used
(iv) its temperature

Simple experiments can show that the resistance is directly proportional to the length and inversely proportional to the csa. Combining these two statements we can write:

$$R \propto \frac{l}{A} \quad \text{where } l = \text{length (in metres)} \quad \text{and } A = \text{csa (in square metres)}$$

The constant of proportionality in this case concerns the third factor listed above, and is known as the resistivity of the material. This is defined as the resistance that exists between the opposite faces of a 1 m cube of that material, measured at a defined temperature. The symbol for resistivity is ρ. The unit of measurement is the ohm-metre

(Ωm). Thus the equation for resistance using the above factor is

$$R = \rho\frac{l}{A} \text{ ohm} \quad (1.13)$$

Worked Example 1.11

Question

A coil of copper wire 200 m long and of csa 0.8 mm^2 has a resistivity of 0.02 $\mu\Omega$m at normal working temperature. Calculate the resistance of the coil.

Answer

$\ell = 200$ m; $\rho = 2 \times 10^{-8}$ Ωm; $A = 8 \times 10^{-7}$ m^2

$$R = \frac{\rho\ell}{A} \text{ ohm}$$

$$= \frac{2 \times 10^{-8} \times 200}{8 \times 10^{-7}}$$

$R = 5\,\Omega$ **Ans**

The resistance of a material also depends on its temperature and has a property known as its temperature coefficient of resistance. The resistance of all pure metals increases with increase of temperature. The resistance of carbon, insulators, semiconductors and electrolytes decreases with increase of temperature. For these reasons, conductors (metals) are said to have a positive temperature coefficient of resistance. Insulators etc. are said to have a negative temperature coefficient of resistance. Apart from this there is another major difference. Over a moderate range of temperature, the change of resistance for conductors is relatively small and is a very close approximation to a straight line. Semiconductors on the other hand tend to have very much larger changes of resistance over the same range of temperatures, and follow an exponential law. These differences are illustrated in Fig. 1.12.

Temperature coefficient of resistance is defined as the ratio of the change of resistance per degree change of temperature, to the resistance at some specified temperature. The quantity symbol is α and the unit of measurement is per degree, e.g. /°C. The reference temperature usually quoted is 0°C, and the resistance at this temperature is referred to as R_0. Thus the resistance at some other temperature θ°C can be obtained from:

$$R_1 = R_0(1 + \alpha\theta_1) \text{ ohm} \quad (1.14)$$

In general, if a material having a resistance R_0 at 0°C has a resistance R_1 at θ_1°C and R_2 at θ_2°C, and if α is the temperature coefficient at 0°C, then

$$R_1 = R_0(1 + \alpha\theta_1) \text{ and } R_2 = R_0(1 + \alpha\theta_2)$$

$$\text{so } \frac{R_1}{R_2} = \frac{1 + \alpha\theta_1}{1 + \alpha\theta_2} \quad (1.15)$$

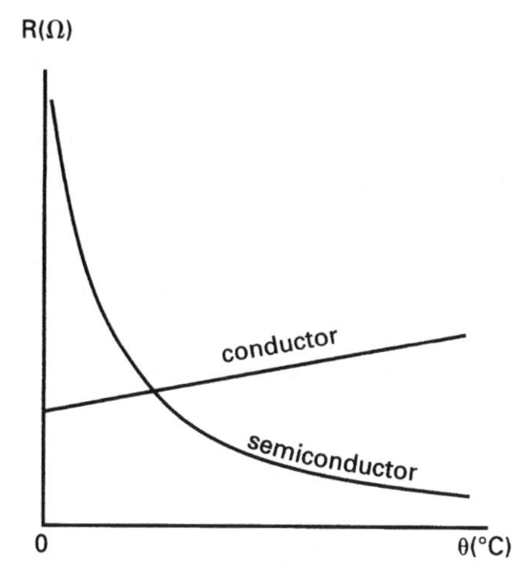

Fig. 1.12

Worked Example 1.12

Question

The field coil of an electric motor has a resistance of 250 Ω at 15°C. Calculate the resistance if the motor attains a temperature of 45°C when running. Assume that $\alpha = 0.00428/°C$ referred to 0°C.

Answer

$R_1 = 250\,\Omega$; $\theta_1 = 15°C$; $\theta_2 = 45°C$; $\alpha = 4.28 \times 10^{-3}$

Using equation (1.15):

$$\frac{250}{R_2} = \frac{1 + (4.28 \times 10^{-3} \times 15)}{1 + (4.28 \times 10^{-3} \times 45)}$$

$$\frac{250}{R_2} = 0.8923$$

$R_2 = 280.2\ \Omega$ **Ans**

Use of meters The measurement of electrical quantities is an essential part of engineering, so you need to be proficient in the use of the various types of measuring instrument. In this chapter we will consider only the use of the basic current and voltage measuring instruments, namely the ammeter and voltmeter respectively.

An ammeter is a current measuring instrument. It has to be connected into the circuit in such a way that the current to be measured is forced to flow through it. If you need to measure the current flowing in a section of a circuit that is already connected together, you will need to 'break' the circuit at the appropriate point and connect the ammeter in the 'break'. If you are connecting a circuit (as you will frequently have to do when carrying out practical assignments), then insert the ammeter as the circuit connections are being made. Most ammeters will have their terminals colour coded: red for the positive and black for the negative. PLEASE NOTE that these polarities refer to *conventional current flow*, so the current should enter the meter at the red terminal and leave via the black terminal. The ammeter circuit symbol is shown in Fig. 1.13.

Fig. 1.13

As you would expect, a voltmeter is used for measuring voltages; in particular, p.d.s. Since a p.d. is a voltage between two points in a circuit, then this meter is NOT connected into the circuit in the same way as an ammeter. In this sense it is a simpler instrument to use, since it need only be connected across (between the two ends of) the component whose p.d. is to be measured. The terminals will usually be colour coded in the same way as an ammeter, so the red terminal should be connected to the more positive end of the component, i.e. follow the same principle as with the ammeter. The voltmeter symbol is shown in Fig. 1.14.

Fig. 1.14

It is most probable that you will have to make use of meters that are capable of combining the functions of an ammeter, a voltmeter and an ohmmeter. These instruments are known as multimeters. One of the most common examples is the AVO. This meter is an example of the type known as analogue instruments, whereby the 'readings' are indicated by the position of a pointer along a graduated scale. The other type of multimeter is of the digital type (often referred to as a DMM). In this case, the 'readings' are in the form of a numerical display, using either light emitting diodes or a liquid crystal, as on calculator displays. Although the digital instruments are easier to read, it does not necessarily mean that they give more accurate results. The choice of the type of meter to use involves many considerations. At this stage it is better to rely on advice from your teacher as to which ones to use for a particular measurement.

All multimeters have switches, either rotary or pushbutton, that are used to select between a.c. or d.c. measurements. There is also a facility for selecting a number of current and voltage ranges. To gain a proper understanding of the use of these meters you really need to have the instrument in front of you. This is a practical exercise that your teacher will carry out with you. I will conclude this section by outlining some important general points that you should observe when carrying out practical measurements.

All measuring instruments are quite fragile, not only mechanically (please handle them carefully) but even more so electrically. If an instrument becomes damaged it is very inconvenient, but more importantly, it is expensive to repair and/or replace. So whenever you use them please observe the following rules:

1 Do not switch on (or connect) the power supply to a circuit until your connections have been checked by the teacher or laboratory technician.
2 Starting with all meters switched to the OFF position, select the highest possible range, and then carefully select lower ranges until a suitable deflection (analogue instrument) or figure is displayed (DMM).
3 When taking a series of readings try to select a range that will accommodate the whole series. This is not always possible. However, if the range(s) *are* changed and the results are used to plot a graph, then a sudden unexpected gap or 'jump' in the plotted curve may well occur.
4 When finished, turn off and disconnect all power supplies, and turn all meters to their OFF position.

1.7 Communication

It is most important that an engineer is a good communicator. He or she must be capable of transmitting information orally, by the written word and by means of sketches and drawings. He or she must also be able to receive and translate information in all of these forms. Most of these skills can be perfected only with guidance and practice. Thus, an engineering student should at every opportunity strive to improve these skills. The art of good communication is a specialised area, and this book does not pretend to be authoritative on the subject. However, there are a number of points, given below, regarding the presentation of written work, that may assist you.

The assignment questions at the end of each chapter are intended to fulfil three main functions. To reinforce your knowledge of the subject matter by repeated application of the underlying principles. To give you the opportunity to develop a logical and methodical approach to the solution of problems. To use these same skills in the presentation of technical information. Therefore when you complete each assignment, treat it as a vehicle for demonstrating your understanding of the subject. This means that your method and presentation of the solution, are more important than always obtaining the 'correct' numerical answer. To help you to achieve this use the following procedures:

1 Read the question carefully from beginning to end in order to ensure that you understand fully what is required.
2 Extract the numerical data from the question and list this at the top of the page, using the relevant quantity symbols and units. This is particularly important when values are given for a number of quantities. In this case, if you try to extract the data in the midst of calculations, it is all too easy to pick out the wrong figure amongst all the words. At the same time, convert all values into their basic units. Another advantage of using this technique is that the resulting list, with the quantity symbols, is likely to jog your memory as to the appropriate equation(s) that will be required.

3 Whenever appropriate, sketch the relevant circuit diagram, clearly identifying all components, currents, p.d.s etc. If the circuit is one in which there are a number of junctions then labelling as shown in Fig. 1.15 makes the presentation of your solution very much simpler. For example

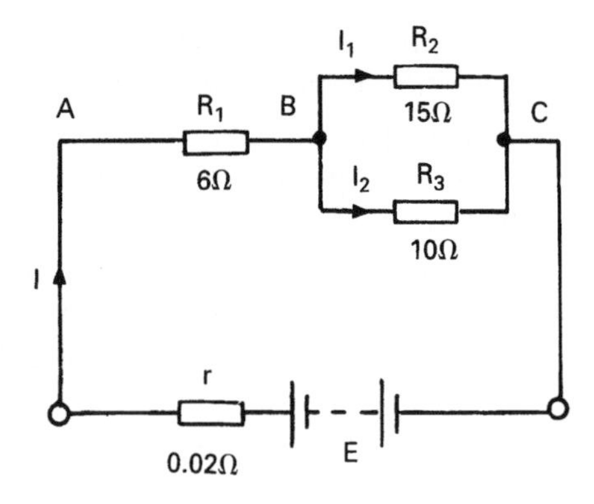

Fig. 1.15

If the diagram had not been labelled, and you wished to refer to the effective resistance between points B and C, then you would have to write out 'the effective resistance of R_2 and R_3 in parallel = ...'. However, with the diagram labelled you need simply write 'R_{BC} = ...'. Similarly, instead of having to write 'the current through the 15 Ω resistor = ...', all that is required is 'I_1 = ...'.

4 Before writing down any figures quote the equation being used, together with the appropriate unit of measurement. This serves to indicate your method of solution. Also note that the units should be written in words. The unit symbols should only be used when preceded by a number thus. Thus, $V = IR$ volt for equation, and 24 V for actual value.

5 Avoid the temptation to save space by having numerous '=' signs along one line. If the line is particularly short then a maximum of two equals signs per line is acceptable.

6 Show ALL figures used in the calculations including any sub-answers obtained.

7 Clearly identify your answer(s) by either underlining or by writing 'Ans' alongside.

You will notice that the above procedures have been followed in all of the worked examples throughout this book.

In addition to written assignments you will be required to undertake others. Although these may not entail using exactly the same procedures as outlined above, they will still require a logical and methodical approach to the presentation. Practical assignments will also be a major feature of your course of study. These will normally involve the use of equipment and measuring instruments in order that you may discover certain technical facts or to verify your theories, etc. This form of exercise also needs to be well documented so that, if necessary, another person can repeat your procedure to confirm (or refute!) your findings. In these cases a more formalised form of written report is required, and the following format is generally acceptable:

'Title'

Objective: This needs to be a concise and clear statement as to what it is that you are trying to achieve.

Apparatus: This will be a list of all equipment and instruments used, quoting serial numbers where appropriate.

Diagram(s): The circuit diagram, clearly labelled.

Method: This should consist of a series of numbered paragraphs that describe each step of the procedure carried out. This section of the report (as with the rest) should be impersonal and in the past tense. Words such as 'I', 'we', 'they', 'us' should not be used. Thus, instead of writing a phrase such as 'We set the voltage to 0 V and I adjusted it in 0.1 V steps ...' you should write 'The voltage was set to 0 V and then adjusted in 0.1 V steps...'

Results: All measurements and settings should be neatly tabulated. To avoid writing unit symbols alongside figures in the body of the table the appropriate units need to be stated at the top of each column. If there is more than one table of results then each one should be clearly identified. These points are illustrated below:

Table 1

p.d. (V)	*I* (mA)	*R*(kΩ)
0.0	0.00	—
1.0	1.25	0.80
2.0	1.75	1.14
	etc	

Calculations: This section may not always be required, but if you have carried out any calculations using the measured data then these should be shown here.

Graphs: In most cases the tabulated results form the basis for a graph or graphs which must be carefully plotted on approved graph paper. Simple lined or squared paper is NOT adequate. Try to use as much of the page as possible, but at the same time choose sensible scales. For example let the graduations on the graph paper represent increments such as 0.1, 0.2, 0.5, 10 000, etc. and not 0.3, 0.4, 0.6, 0.7, 0.8, 0.9 etc. By doing this you will make it much simpler to plot the graph in the first place, and more importantly, very much easier to take readings from it subsequently.

Conclusions: Having gone through the above procedure you need to complete the assignment by drawing some conclusions based on all the data gathered. Any such conclusions must be justified. For example, you may have taken measurements of some variable y for corresponding values of a second variable x. If your plotted graph happens to be a straight line that passes through the origin then your conclusion would be as follows:

'Since the graph of y versus x is a straight line passing through the origin then it can be concluded that y is directly proportional to x'.

Assignment Questions

1 Convert the following into standard form.
(a) 456.3 (b) 902 344 (c) 0.000 285 (d) 8 000 (e) 0.047 12 (f) 180 μA
(g) 38 mV (h) 80 GN (i) 2 000 μF

2 Write the following quantities in scientific notation.
(a) 1 500 Ω (b) 0.0033 Ω (c) 0.000 025 A (d) 750 V (e) 800 000 V
(f) 0.000 000 047 F

3 Calculate the charge transferred in 25 minutes by a current of 500 mA.

4 A current of 3.6 A transfers a charge of 375 mC. How long would this take?

5 Determine the value of charging current required to transfer a charge of 0.55 mC in a time of 600 μs.

6 Calculate the p.d. developed across a 750 Ω resistor when the current flowing through it is (a) 3 A, (b) 25 mA.

7 An emf of 50 V is applied in turn to the following resistors: (a) 22 Ω, (b) 820 Ω, (c) 2.7 MΩ, (d) 330 kΩ. Calculate the current flow in each case.

8 The current flowing through a 470 Ω resistance is 4 A. Determine the energy dissipated in a time of 2 h. Express your answer in both joules and in commercial units.

9 A small business operates three pieces of equipment for nine hours continuously per day for six days a week. If the three equipments consume 10 kW, 2.5 kW and 600 W respectively, calculate the weekly cost if the charge per unit is 7.9 pence.

10 A charge of 500 μC is passed through a 560 Ω resistor in a time of 1 ms. Under these conditions determine (a) the current flowing, (b) the p.d. developed, (c) the power dissipated, and (d) the energy consumed in 5 min.

11 A battery of emf 50 V and internal resistance 0.2 Ω supplies a current of 1.8 A to an external load. Under these conditions determine (a) the terminal p.d. and (b) the resistance of the external load.

12 The terminal p.d. of a d.c. source is 22.5 V when supplying a load current of 10 A. If the emf is 24 V calculate (a) the internal resistance and (b) the resistance of the external load.

13 For the circuit arrangement specified in Q12 above determine the power and energy dissipated by the external load resistor in 5 minutes.

14 A circuit of resistance 4 Ω dissipates a power of 16 W. Calculate (a) the current flowing through it, (b) the p.d. developed across it, and (c) the charge displaced in a time of 20 minutes.

15 In a test the velocity of a body was measured over a period of time, yielding the results shown in the table below. Plot the corresponding graph and use it to determine the acceleration of the body at times $t = 0$, $t = 5$ s and $t = 9$ s. You may assume that the graph consists of a series of straight lines.

v(m/s)	0.0	3.0	6.0	10.0	14.0	15.0	16.0
t(s)	0.0	1.5	3.0	4.5	6.0	8.0	10.0

16 The insulation resistance between a conductor and earth is 30 MΩ. Calculate the leakage current if the supply voltage is 240 V.

17 A 3 kW immersion heater is designed to operate from a 240 V supply. Determine its resistance and the current drawn from the supply.

18 A 110 V d.c. generator supplies a lighting load of forty 100 W bulbs, a heating load of 10 kW and other loads which consume a current of 15 A. Calculate the power output of the generator under these conditions.

19 The field winding of a d.c. motor is connected to a 110 V supply. At a temperature of 18°C, the current drawn is 0.575 A. After running the machine for some time the current has fallen to 0.475 A, the voltage remaining unchanged. Calculate the temperature of the winding under the new conditions, assuming that the temperature coefficient of resistance of copper is 0.004 26/°C at 0°C.

20 A coil consists of 1500 turns of aluminium wire having a cross-sectional area of 0.75 mm^2. The mean length per turn is 60 cm and the resistivity of aluminium at the working temperature is 0.028 $\mu\Omega$ m. Calculate the resistance of the coil.

Supplementary Worked Examples

Example 1: Write the following quantities in scientific (engineering) notation. (a) 25×10^{-5} A, (b) 3×10^{4} W, (c) 850 000 J, (d) 0.0016 V.

Solution: **(a)**

$$25 \times 10^{-5} = 25 \times 10 \times 10^{-6}$$
$$= 250 \times 10^{-6}$$

and since 10^{-6} is represented by μ (micro)

then $25 \times 10^{-5}\,\text{A} = 250\,\mu\text{A}$ **Ans**

Alternatively, $25 \times 10^{-5} = 25 \times 10^{-2} \times 10^{-3}$

$$= 0.25 \times 10^{-3}$$

so $25 \times 10^{-5}\,\text{A} = 0.25\,\text{mA}$ **Ans**

(b)

$$3 \times 10^{-4} = 3 \times 10^{-1} \times 10^{-3} \text{ or } 300 \times 10^{-6}$$

so $3 \times 10^{-4}\,\text{W} = 0.3\,\text{mW}$ or $300\,\mu\text{W}$ **Ans**

(c)

$$850\,000 = 850 \times 10^{3} \text{ or } 0.85 \times 10^{6}$$

so $850\,000\,\text{J} = 850\,\text{kJ}$ or $0.85\,\text{MJ}$ **Ans**

(d)

$$0.0016 = 1.6 \times 10^{-3}$$

so $0.0016\,\text{V} = 1.6\,\text{mV}$ **Ans**

Example 2: Calculate the current flowing through a circuit when a charge of 50 μC is transferred in a time of 250 milliseconds.

Solution: $Q = 50 \times 10^{-6}\,\text{C};\ t = 250 \times 10^{-3}\,\text{s}$

(a) $Q = It$ coulomb

$$\text{so, } I = \frac{Q}{t} \text{ amp}$$

$$= \frac{50 \times 10^{-6}}{250 \times 10^{-3}} = 2 \times 10^{-4}$$

hence, $I = 0.2\,\text{mA}$ **Ans**

Example 3: Calculate the charge transferred by a current of 9.5 A when it flows for a time of 15 minutes.

Solution: $I = 9.45\,\text{A};\ t = 15 \times 60\,\text{s}$

$$Q = It \text{ coulomb}$$

$$= 9.5 \times 15 \times 60 = 8550$$

so, $Q = 8.55\,\text{kJ}$ **Ans**

Example 4: A current of 50 mA flows through a 20 Ω resistor for a time of 4 minutes. Calculate (a) the p.d. developed across the resistor, (b) the charge transferred, (c) the energy dissipated, and (d) the power dissipated.

Solution: $I = 50 \times 10^{-3}\,\text{A};\ R = 20\,\Omega;\ t = 4 \times 60\,\text{s}$

(a) $V = IR$ volt

$$= 50 \times 10^{-3} \times 20$$

so, $V = 1\,\text{V}$ **Ans**

(b) $Q = It$ coulomb

$$= 50 \times 10^{-3} \times 4 \times 60$$

so, $Q = 12\,\text{C}$ **Ans**

(c) $W = I^2Rt = \dfrac{V^2t}{R} = VIt$ joule

Now, the second and third of the above equations require the use of the p.d. calculated in part (a). However, if this calculation was incorrect, then the error involved would cause the answer to part (b) to be wrong also. For this reason the first equation will be used, since all the data are given in the question.

Hence, $W = I^2Rt$ watt

$$= (50 \times 10^{-3})^2 \times 20 \times 240$$

$$= 2500 \times 10^{-6} \times 20 \times 240$$

so, $W = 12\,\text{J}$ **Ans**

(d) $P = \frac{W}{t}$ watt $= \frac{12}{240}$

so, $P = 50\,\text{mW}$ **Ans**

Example 5: A 1.5 kΩ resistor is connected to a variable voltage supply. Calculate the current flowing through it when the p.d. across the resistor is (a) 5 mV, (b) 0.35 V and (c) 4 V.

Solution: Applying Ohm's law, $I = \frac{V}{R}$ amp

(a) $I = \frac{5 \times 10^{-3}}{1.5 \times 10^3} = 3.33 \times 10^{-6}$

so, $I = 3.33\,\mu\text{A}$ **Ans**

(b) $I = \frac{0.35}{1.5 \times 10^3} = 2.33 \times 10^{-4}$

so, $I = 233\,\mu\text{A}$ or $0.233\,\text{mA}$ **Ans**

(c) $I = \frac{4}{1.5 \times 10^3} = 2.67 \times 10^{-3}$

so, $I = 2.67\,\text{mA}$ **Ans**

Example 6: A resistor of 680 Ω, when connected in a circuit, dissipates a power of 85 mW. Calculate (a) the p.d. developed across it, and (b) the current flowing through it.

Solution: $R = 680\,\Omega$; $P = 85 \times 10^{-3}\,\text{W}$

(a) $P = \frac{V^2}{R}$ watt

so, $V^2 = PR$

and $V = \sqrt{PR}$ volt

$= \sqrt{85 \times 10^{-3} \times 680} = \sqrt{57.8}$

so, $V = 7.6\,\text{V}$ **Ans**

(b) $P = I^2R$ watt

so, $I^2 = \frac{P}{R}$

and $I = \sqrt{\frac{P}{R}}$ amp

$= \sqrt{\frac{85 \times 10^{-3}}{680}} = \sqrt{1.25 \times 10^{-4}}$

so, $I = 11.18\,\text{mA}$ **Ans**

Note: Since $P = VI$ watt, the calculations may be checked as follows
$P = 7.6 \times 11.18 \times 10^{-3}$
so, $P = 84.97$ mW, which when rounded up to one decimal place gives 85.0 mW—the value given in the question

Example 7: An electricity bill totalled £78.75, which included a standing charge of £15.00. The number of units charged for was 750. Calculate (a) the charge per unit, and (b) the total bill if the charge/unit had been 9p, and the standing charge remained unchanged.

Solution: Total bill = £78.75; standing charge = £15.00; units used = 750 = 750 kWh

(a) Cost of the energy (units) used = total − standing charge

$$= £78.75 - £15.00 = £63.75$$

$$\text{Cost/unit} = \frac{£63.75}{750} = £0.085$$

so, cost/unit = 8.5 p **Ans**

(b) If the cost/unit is raised to 9p, then

cost of energy used = £0.09 × 750 = £67.50

total bill = cost of units used + standing charge = £67.50 + £15.00

so, total bill = £82.50 **Ans**

Example 8: A coil of wire has a resistance value of 350 Ω when its temperature is 0°C. Given that the temperature coefficient of resistance of the wire is $4.26 \times 10^{-3}/°C$ referred to 0°C, calculate its resistance at a temperature of 60°C.

Solution: $R_0 = 350\,\Omega$; $\alpha = 4.26 \times 10^{-3}/°C$; $\theta_1 = 60°C$

$R_1 = R_0(1 + \alpha\theta_1)$ ohm; where R_1 is the resistance at 60°C

$$= 350\ \{1 + (4.26 \times 10^{-3} \times 60)\}$$

$$= 350\ \{1 + 0.22556\}$$

$$= 350 \times 1.2556$$

so, $R_1 = 439.6\,\Omega$ **Ans**

Example 9: A carbon resistor has a resistance value of 120 Ω at a room temperature of 16°C. When it is connected as part of a circuit, with current flowing through it, its temperature rises to 32°C. If the temperature coefficient of resistance of carbon is −0.000 48/°C referred to 0°C, calculate its resistance under these operating conditions.

Solution: $\theta_1 = 16°C$; $\theta_2 = 32°C$; $R_1 = 120\,\Omega$; $\alpha = -0.000\,48/°C$

$$\frac{R_1}{R_2} = \frac{1 + \alpha\theta_1}{1 + \alpha\theta_2}$$

$$\frac{120}{R_2} = \frac{1 + (-0.000\,48 \times 16)}{1 + (-0.000\,48 \times 32)}$$

$$\frac{120}{R_2} = 1.0078$$

$$R_2 = \frac{120}{1.0078}$$

so, $R_2 = 119.1\,\Omega$ **Ans**

Example 10: A wirewound resistor is made from a 250 metre length of copper wire having a circular cross-section of diameter 0.5 mm. Given that the wire has a resistivity of 0.018 $\mu\Omega$m, calculate its resistance value.

Solution: $\ell = 250\,\text{m}$; $d = 5 \times 10^{-4}\,\text{m}$; $\rho = 1.8 \times 10^{-8}\,\Omega\,\text{m}$

$$R = \frac{\rho\ell}{A}\ \text{ohm, where cross-sectional area, } A = \frac{\pi d^2}{4}\ \text{metre}^2$$

$$\text{hence, } A = \frac{\pi \times (5 \times 10^{-4})^2}{4} = 1.9635 \times 10^{-7}\,\text{m}^2$$

$$\text{hence, } R = \frac{1.8 \times 10^{-8} \times 250}{1.9635 \times 10^{-7}}$$

so, $R = 22.92\,\Omega$ **Ans**

Example 11: A battery of emf 6 V has an internal resistance of 0.15 Ω. Calculate its terminal p.d. when delivering a current of (a) 0.5 A, (b) 2 A, and (c) 10 A.

Solution: $E = 6\,\text{V}$; $r = 0.15\,\Omega$

(a) $V = E - Ir$ volt

$= 6 - (0.5 \times 0.15) = 6 - 0.075$

so, $V = 5.925\,\text{V}$ **Ans**

(b) $V = 6 - (2 \times 0.15) = 6 - 0.3$

so, $V = 5.7\,\text{V}$ **Ans**

(c) $V = 6 - (10 \times 0.15) = 6 - 1.5$

so, $V = 4.5\,\text{V}$ **Ans**

Note: This example verifies that the terminal p.d. of a source of emf decreases as the load on it (the current drawn from it) is increased.

Example 12: A battery of emf 12 V supplies a circuit with a current of 5 A. If, under these conditions, the terminal p.d. is 11.5 V, determine (a) the battery internal resistance, (b) the resistance of the external circuit, and (c) the power dissipated in the external circuit.

Solution: $E = 12\,\text{V}$; $I = 5\,\text{A}$; $V = 11.5\,\text{V}$

As with the vast majority of electrical problems, a simple sketch of the circuit diagram will help you to visualise the problem. For the above problem the circuit diagram would be as shown in Fig. 1.16.

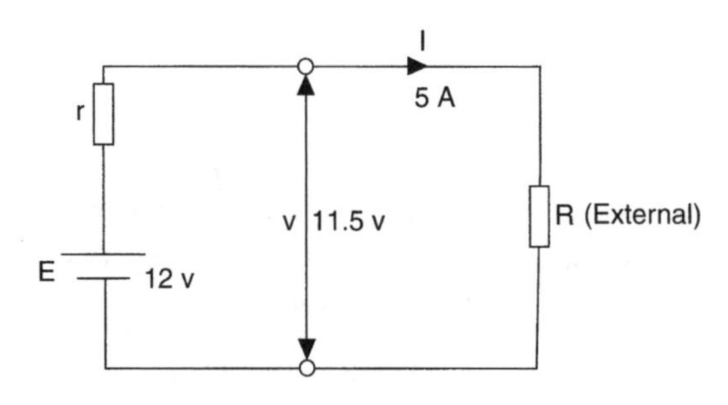

Fig. 1.16

(a)

$$E = V + Ir\,\text{volt}$$
$$E - V = Ir\,\text{volt}$$
$$\text{so, } r = \frac{E - V}{I}\,\text{ohm}$$
$$= \frac{12 - 11.5}{5}$$
$$\text{hence, } r = 0.1\,\Omega\ \textbf{Ans}$$

(b)

$$R = \frac{V}{I}\,\text{ohm}$$
$$= \frac{11.5}{5}$$
$$\text{so, } R = 2.3\,\Omega\ \textbf{Ans}$$

(c)

$$P = I^2R\,\text{watt}$$
$$= 5^2 \times 2.3$$
$$\text{so, } P = 57.5\,\text{W}\ \textbf{Ans}$$

$$\text{Alternatively, } P = VI\,\text{watt}$$
$$= 11.5 \times 5$$
$$\text{so, } P = 57.5\,\text{W}\ \textbf{Ans}$$

Note: The use of the second equation for part (c) is preferable. Explain why.

Summary of Equations

Charge: $Q = It$ coulomb

Ohm's law: $V = IR$ volt

Terminal p.d.: $V = E - Ir$ volt

Energy: $W = VIt = I^2Rt = \frac{V^2t}{R}$ joule

Power: $P = \frac{W}{t} = VI = I^2R = \frac{V^2}{R}$ watt

Resistance: $R = \frac{\rho \ell}{A}$ ohm

Resistance at specified temp.: $R_1 = R_0(1 + \alpha\theta)$ ohm

or $\frac{R_1}{R_2} = \frac{1 + \alpha\theta_1}{1 + \alpha\theta_2}$

2 D.C. Circuits

This chapter explains how to apply circuit theory to the solution of simple circuits and networks by the application of Ohm's law, Kirchhoff's laws, other network theorems and the concepts of potential and current dividers.

This means that on completion of the chapter you should be able to:

1 Calculate current flows, potential differences, power and energy dissipations for circuit components and simple circuits, by applying Ohm's law.
2 Carry out the above calculations for more complex networks using Kirchhoff's and other network theorems.
3 Use decibel notation for power, voltage and current gain/attenuation.
4 Calculate circuit p.d.s using the potential divider technique, and branch currents using the current divider technique.
5 Understand the principles and use of a Wheatstone Bridge.
6 Understand the principles and use of a slidewire potentiometer.

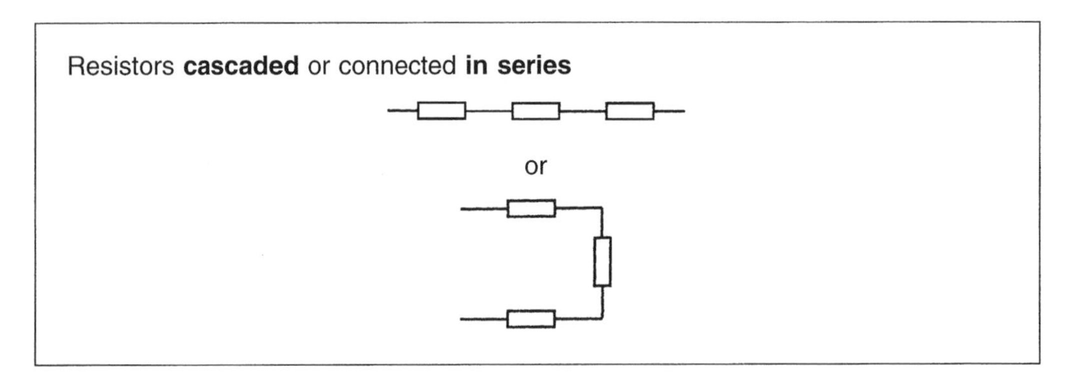

2.1 Resistors in Series

When resistors are connected 'end-to-end' so that the same current flows through them all they are said to be **cascaded** or **connected in series**. Such a circuit is shown in Fig. 2.1. Note that, for the sake of simplicity, an ideal source of emf has been used

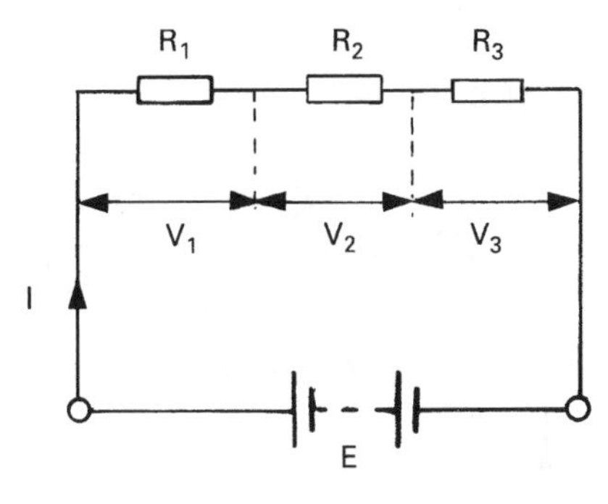

Fig. 2.1

(no internal resistance). From the previous chapter we know that the current flowing through the resistors will result in p.d.s being developed across them. We also know that the sum of these p.d.s must equal the value of the applied emf. Thus

$V_1 = IR_1$ volt; $V_2 = IR_2$ volt; and $V_3 = IR_3$ volt

However, the circuit current I depends ultimately on the applied emf E and the total resistance R offered by the circuit. Hence

$$E = IR \text{ volt.}$$

Also, $E = V_1 + V_2 + V_3$ volt

and substituting for E, V_1, V_2 and V_3 in this last equation

we have $IR = IR_1 + IR_2 + IR_3$ volt

and dividing this last equation by the common factor I

$$R = R_1 + R_2 + R_3 \text{ ohm} \qquad (2.1)$$

where R is the total circuit resistance. From this result it may be seen that when resistors are connected in series the total resistance is found simply by adding together the resistor values.

Worked Example 2.1

Question

For the circuit shown in Fig. 2.2 calculate (a) the circuit resistance, (b) the circuit current, (c) the p.d. developed across each resistor and (d) the power dissipated by the complete circuit.

Answer

$E = 24$ V; $R_1 = 330\,\Omega$; $R_2 = 1500\,\Omega$; $R_3 = 470\,\Omega$

Continued on p. 36

Worked Example 2.1 (*Continued*)

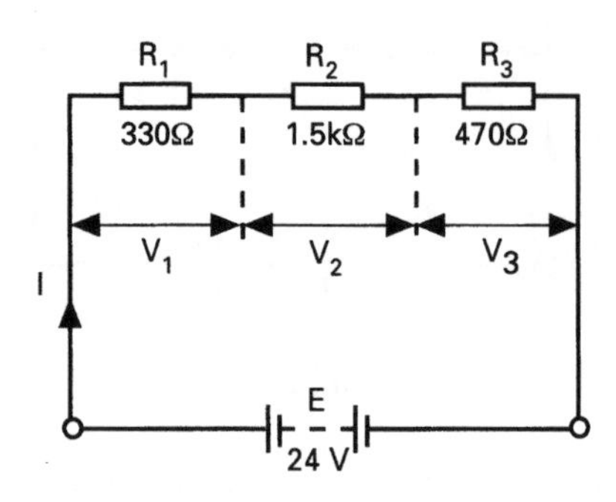

Fig. 2.2

(a) $R = R_1 + R_2 + R_3$ ohm

$= 330 + 1500 + 470$

$R = 2300\,\Omega$ or $2.3\,\text{k}\Omega$ **Ans**

(b) $I = \dfrac{E}{R}$ amp

$= \dfrac{24}{2300}$

$I = 10.43$ mA **Ans**

(c) $V_1 = IR_1$ volt

$= 10.43 \times 10^{-3} \times 330$

$V_1 = 3.44$ V **Ans**

$V_2 = IR_2$ volt

$= 10.43 \times 10^{-3} \times 1500$

$V_2 = 15.65$ volts **Ans**

$V_3 = IR_3$ volt

$= 10.43 \times 10^{-3} \times 470$

$V_3 = 4.90$ V **Ans**

Note: The sum of the above p.d.s is 23.99 V instead of 24 V due to the rounding errors in the calculation. It should also be noted that the value quoted for the current was 10.43 mA whereas the calculator answer is 10.4347 mA. This latter value was then stored in the calculator memory and used in the calculations for part (c), thus reducing the rounding errors to an acceptable minimum.

(d) $P = EI$ watt

$= 24 \times 10.43 \times 10^{-3}$

$P = 0.25$ W or 250 mW **Ans**

It should be noted that the power is dissipated by the three resistors in the circuit. Hence, the circuit power could have been determined by calculating the power dissipated by each of these and adding these values to give the total. This is shown below, and serves, as a check for the last answer.

$$P_1 = I^2 R_1 \text{ watt}$$
$$= (10.43 \times 10^{-3})^2 \times 330$$
$$P_1 = 35.93 \text{ mW}$$
$$P_2 = (10.43 \times 10^{-3})^2 \times 1500$$
$$= 163.33 \text{ mW}$$
$$P_3 = (10.43 \times 10^{-3})^2 \times 470$$
$$= 51.18 \text{ mW}$$

total power: $P = P_1 + P_2 + P_3$ watt

so $P = 250.44$ mW

(Note the worsening effect of continuous rounding error)

Worked Example 2.2

Question

Two resistors are connected in series across a battery of emf 12 V. If one of the resistors has a value of 16 Ω and it dissipates a power of 4 W, then calculate (a) the circuit current and (b) the value of the other resistor.

Answer

Since the only two pieces of data that are directly related to each other concern the 16 Ω resistor and the power that it dissipates, then this information must form the starting point for the solution of the problem. Using these data we can determine either the current through or the p.d. across the 16 Ω resistor (and it is not important which of these is calculated first). To illustrate this point both methods will be demonstrated. The appropriate circuit diagram, which forms an integral part of the solution, is shown in Fig. 2.3.

$E = 12$ V; $R_{BC} = 16\ \Omega$; $P_{BC} = 4$ W

Continued on p. 38

Worked Example 2.2 (*Continued*)

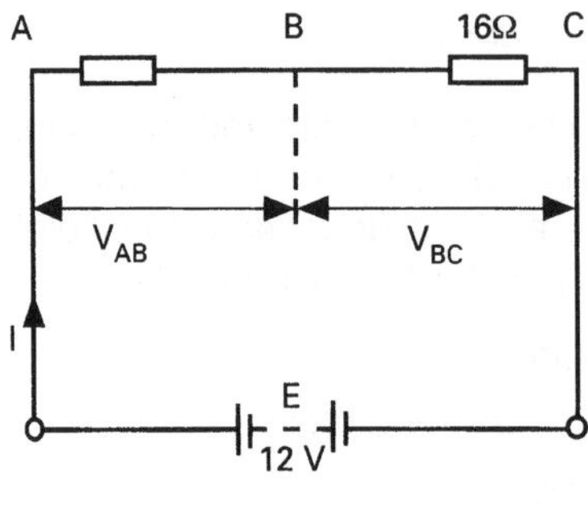

Fig. 2.3

(a) $I^2 R_{BC} = P_{BC}$ watt

$$I^2 = \frac{P_{BC}}{R_{BC}}$$

$$= \frac{4}{16} = 0.25$$

so $I = 0.5$ A **Ans**

(b) total resistance, $R = \frac{E}{I}$ ohm

$$= \frac{12}{0.5} = 24\,\Omega$$

$$R_{AB} = R - R_{BC}$$

$$= 24 - 16$$

so $R_{AB} = 8\,\Omega$ **Ans**

Alternatively, the problem can be solved thus:

(a) $\frac{V_{BC}^2}{R_{BC}} = P_{BC}$ watt

$$V_{BC}^2 = P_{BC} \cdot R_{BC} = 4 \times 16$$

$$= 64$$

so $V_{BC} = 8$ V

$$I = \frac{V_{BC}}{R_{BC}} \text{ amp}$$

$$= \frac{8}{16}$$

so $I = 0.5$ A **Ans**

(b)

$$V_{AB} = E - V_{BC} \text{ volt}$$
$$= 12 - 8$$
$$V_{AB} = 4\,\text{V}$$
$$R_{AB} = \frac{V_{AB}}{I}$$
$$= \frac{4}{0.5}$$
so $R_{AB} = 8\,\Omega$ **Ans**

2.2 Resistors in Parallel

When resistors are joined 'side-by-side' so that their corresponding ends are connected together they are said to be connected in **parallel**. Using this form of connection means that there will be a number of paths through which the current can flow. Such a circuit consisting of three resistors is shown in Fig. 2.4, and the circuit may be analysed as follows:

Since all three resistors are connected directly across the battery terminals then they all have the same voltage developed across them. In other words the voltage is the

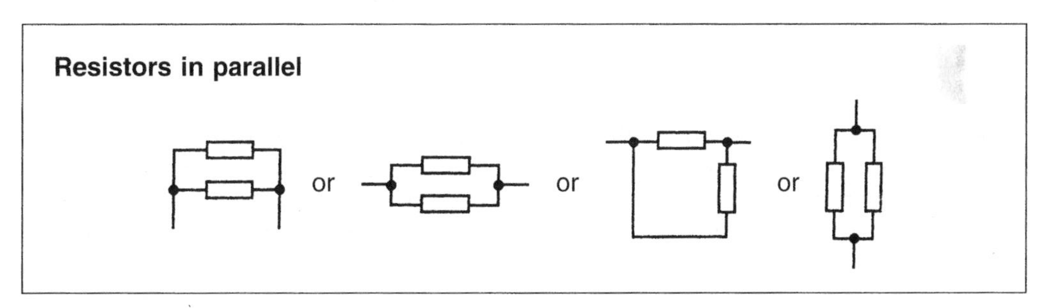

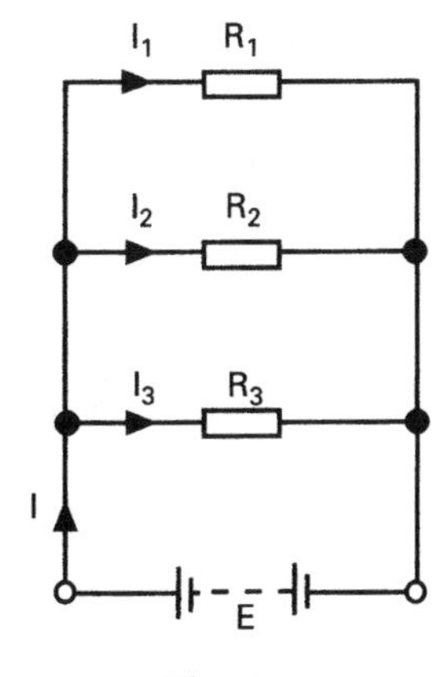

Fig. 2.4

common factor in this arrangement of resistors. Now, each resistor will allow a certain value of current to flow through it, depending upon its resistance value. Thus

$$I_1 = \frac{E}{R_1}\text{ amp; } I_2 = \frac{E}{R_2}\text{ amp; and } I_3 = \frac{E}{R_3}\text{ amp}$$

The total circuit current I is determined by the applied emf and the total circuit resistance R,

$$\text{so } I = \frac{E}{R}\text{ amp}$$

Also, since all three branch currents originate from the battery, then the total circuit current must be the sum of the three branch currents

$$\text{so } I = I_1 + I_2 + I_3$$

and substituting the above expression for the currents:

$$\frac{E}{R} = \frac{E}{R_1} + \frac{E}{R_2} + \frac{E}{R_3}$$

and dividing the above equation by the common factor E:

$$\frac{1}{R} = \frac{1}{R_1} + \frac{1}{R_2} + \frac{1}{R_3}\text{ siemen} \qquad (2.2)$$

Note: The above equation does NOT give the total resistance of the circuit, but does give the total circuit **conductance** (G) which is measured in siemens (S). Thus, conductance is the reciprocal of resistance, so to obtain the circuit resistance you must then take the reciprocal of the answer obtained from an equation of the form of equation (2.2).

> **Conductance** is a measure of the 'willingness' of a material or circuit to allow current to flow through it

$$\text{That is } \frac{1}{R} = G\text{ siemen; and } \frac{1}{G} = R\text{ ohm} \qquad (2.3)$$

However, when only two resistors are in parallel the combined resistance may be obtained directly by using the following equation:

$$R = \frac{R_1 \times R_2}{R_1 + R_2}\text{ ohm} \qquad (2.4)$$

If there are 'x' **identical** resistors in parallel the total resistance is simply R/x ohms.

In this context, the word **identical** means having the same value of resistance

Worked Example 2.3

Question

Considering the circuit of Fig. 2.5, calculate (a) the total resistance of the circuit, (b) the three branch currents and (c) the current drawn from the battery.

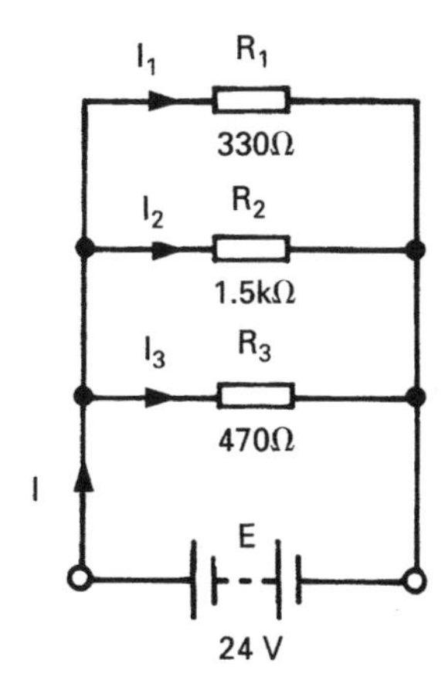

Fig. 2.5

Answer

$E = 24\,\text{V}$; $R_1 = 330\,\Omega$; $R_2 = 1500\,\Omega$; $R_3 = 470\,\Omega$

(a)
$$\frac{1}{R} = \frac{1}{R_1} + \frac{1}{R_2} + \frac{1}{R_3} \text{ siemen}$$
$$= \frac{1}{330} + \frac{1}{1500} + \frac{1}{470}$$
$$= 0.00303 + 0.000667 + 0.00213$$
$$= 0.005825\,\text{S}$$

so $R = 171.68\,\Omega$ **Ans** (reciprocal of 0.005825)

(b)
$$I_1 = \frac{E}{R_1} \text{ amp}$$
$$= \frac{24}{330}$$
$$I_1 = 72.73\,\text{mA}$$ **Ans**

Continued on p. 42

Worked Example 2.3 (*Continued*)

$$I_2 = \frac{E}{R_2} \text{ amp}$$
$$= \frac{24}{1500}$$
$$I_2 = 16\,\text{mA}$$ **Ans**

$$I_3 = \frac{E}{R_3} \text{ amp}$$
$$= \frac{24}{470}$$
$$I_3 = 51.06\,\text{mA}$$ **Ans**

(c) $I = I_1 + I_2 + I_3$ amp
$= 72.73 + 16 + 51.06\,\text{mA}$
so $I = 139.8$ mA **Ans**

Alternatively, the circuit current could have been determined by using the values for E and R as follows

$$I = \frac{E}{R} \text{ amp}$$
$$= \frac{24}{171.68}$$
$$I = 139.8\,\text{mA}$$ **Ans**

Compare this example with worked example 2.1 (the same values for the resistors and the emf have been used). From this it should be obvious that when resistors are connected in parallel the total resistance of the circuit is reduced. This results in a corresponding increase of current drawn from the source. This is simply because the parallel arrangement provides more paths for current flow.

Worked Example 2.4

Question

Two resistors, one of 6 Ω and the other of 3 Ω resistance, are connected in parallel across a source of emf of 12 V. Determine (a) the effective resistance of the combination, (b) the current drawn from the source and (c) the current through each resistor.

Answer

The corresponding circuit diagram, suitably labelled is shown in Fig. 2.6

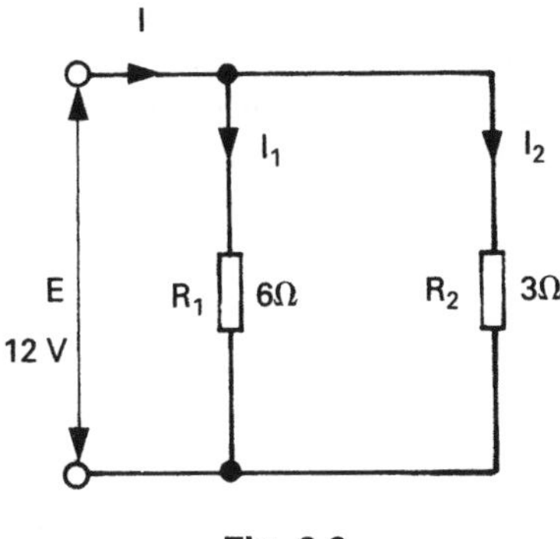

Fig. 2.6

$E = 12$ V; $R_1 = 6\ \Omega$; $R_2 = 3\ \Omega$

(a) $R = \dfrac{R_1 R_2}{R_1 + R_2}$ ohm

$= \dfrac{6 \times 3}{6 + 3} = \dfrac{18}{9}$

so $R = 2\,\Omega$ **Ans**

(b) $I = \dfrac{E}{R}$ amp

$= \dfrac{12}{2}$

so $I = 6$ A **Ans**

(c) $I_1 = \dfrac{E}{R_1}$ amp

$= \dfrac{12}{6}$

$I_1 = 2$ A **Ans**

$I_2 = \dfrac{E}{R_2}$

$= \dfrac{12}{3}$

$I_2 = 4$ A **Ans**

2.3 Potential Divider

When resistors are connected in series the p.d. developed across each resistor will be in direct proportion to its resistance value. This is a useful fact to bear in mind, since it means it is possible to calculate the p.d.s without first having to determine the circuit current. Consider two resistors connected across a 50 V supply as shown Fig. 2.7. In order to demonstrate the potential divider effect we will in this case firstly calculate circuit current and hence the two p.d.s by applying Ohms law:

$$R = R_1 + R_2 \text{ ohm}$$
$$R = 75 + 25 = 100\,\Omega$$
$$I = \frac{E}{R} \text{ amp}$$
$$I = \frac{50}{100} = 0.5\,\text{A}$$
$$V_1 = IR_1 \text{ volt}$$
$$= 0.5 \times 75$$
$$V_1 = 37.5\,\text{V}\ \textbf{Ans}$$
$$V_2 = IR_2 \text{ volt}$$
$$= 0.5 \times 25$$
$$V_2 = 12.5\,\text{V}\ \textbf{Ans}$$

Applying the potential divider technique, the two p.d.s may be obtained by using the fact that the p.d. across a resistor is given by the ratio of its resistance value to the total resistance of the circuit, expressed as a proportion of the applied voltage. Although this sounds complicated it is very simple to put into practice. Expressed in the form of an equation it means

$$V_1 = \frac{R_1}{R_1 + R_2} \times E \text{ volt} \tag{2.5}$$

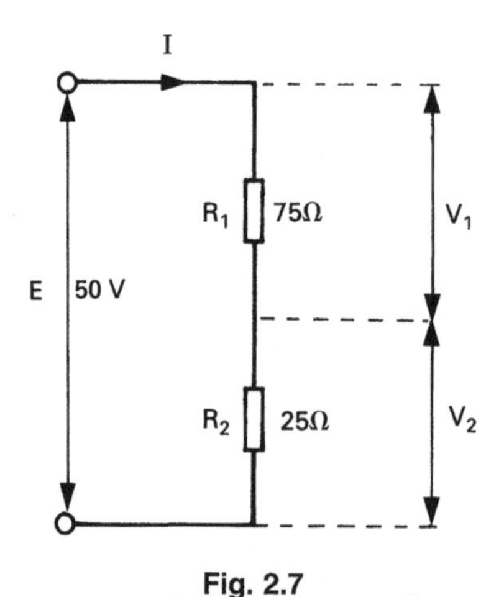

Fig. 2.7

and $$V_2 = \frac{R_2}{R_1 + R_2} \times E \text{ volt} \qquad (2.6)$$

and using the above equations the p.d.s can more simply be calculated as follows:

$$V_1 = \frac{75}{100} \times 50 = 37.5\,\text{V} \textbf{ Ans}$$

and $$V_2 = \frac{25}{100} \times 50 = 12.5\,\text{V} \textbf{ Ans}$$

This technique is not restricted to only two resistors in series, but may be applied to any number. For example, if there were three resistors in series, then the p.d. across each may be found from

$$V_1 = \frac{R_1}{R_1 + R_2 + R_3} \times E$$

$$V_2 = \frac{R_2}{R_1 + R_2 + R_3} \times E$$

and $$V_3 = \frac{R_3}{R_1 + R_2 + R_3} \times E \text{ volt}$$

2.4 Current Divider

It has been shown that when resistors are connected in parallel the total circuit current divides between the alternative paths available. So far we have determined the branch currents by calculating the common p.d. across a parallel branch and dividing this by the respective resistance values. However, these currents can be found directly, without the need to calculate the branch p.d., by using the current divider theory. Consider two resistors connected in parallel across a source of emf 48 V as shown in Fig. 2.8. Using the p.d. method we can calculate the two currents as follows:

$$I_1 = \frac{E}{R_1} \quad \text{and} \quad I_2 = \frac{E}{R_2} \text{ amp}$$

$$= \frac{48}{12} \qquad\qquad = \frac{48}{24}$$

$$I_1 = 4\,\text{A} \quad \text{and} \quad I_2 = 2\,\text{A}$$

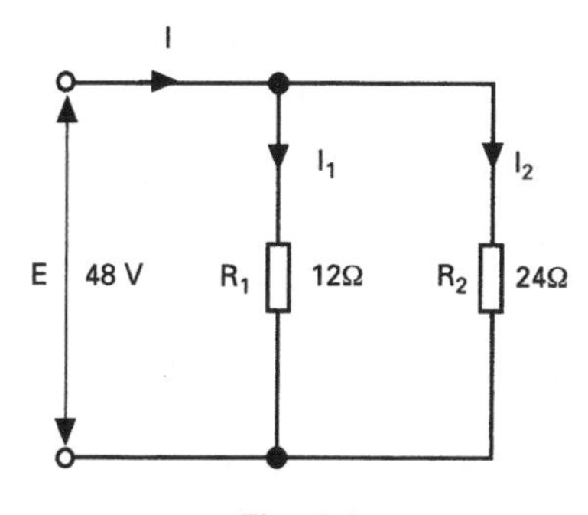

Fig. 2.8

It is now worth noting the values of the resistors and the corresponding currents. It is clear that R_1 is half the value of R_2. So, from the calculation we obtain the quite logical result that I_1 is twice the value of I_2. That is, a ratio of 2:1 applies in each case. Thus, the smaller resistor carries the greater proportion of the total current. By stating the ratio as 2:1 we can say that the current is split into three equal 'parts'. Two 'parts' are flowing through one resistor and the remaining 'part' through the other resistor.

Thus $\frac{2}{3} \times I$ flows through R_1

and $\frac{1}{3} \times I$ flows through R_2

Since $I = 6\,\text{A}$ then

$$I_1 = \frac{2}{3} \times 6 = 4\,\text{A}$$

$$I_2 = \frac{1}{3} \times 6 = 2\,\text{A}$$

In general we can say that

$$I_1 = \frac{R_2}{R_1 + R_2} \times I \qquad (2.7)$$

and

$$I_2 = \frac{R_1}{R_1 + R_2} \times I \qquad (2.8)$$

Note: This is NOT the same ratio as for the potential divider. If you compare (2.5) with (2.7) you will find that the numerator in (2.5) is R_1 whereas in (2.7) the numerator is R_2. There is a similar 'cross-over' when (2.6) and (2.8) are compared.

Again, the current divider theory is not limited to only two resistors in parallel. Any number can be accommodated. However, with three or more parallel resistors the current division method can be cumbersome to use, and it is much easier for mistakes to be made. For this reason it is recommended that where more than two resistors exist in parallel the 'p.d. method' is used. This will be illustrated in the next section, but for completeness the application to three resistors is shown below.

Consider the arrangement shown in Fig. 2.9:

$$\frac{1}{R} = \frac{1}{R_1} + \frac{1}{R_2} + \frac{1}{R_3} = \frac{1}{3} + \frac{1}{4} + \frac{1}{6} = \frac{4+3+2}{12}$$

and examining the numerator, we have 4+3+2 = 9 'parts'.

Thus, the current ratios will be 4/9, 3/9 and 2/9 respectively for the three resistors.

So, $I_1 = \frac{4}{9} \times 18 = 8\,\text{A};\; I_2 = \frac{3}{9} \times 18 = 6\,\text{A};\; I_3 = \frac{2}{9} \times 18 = 4\,\text{A}$

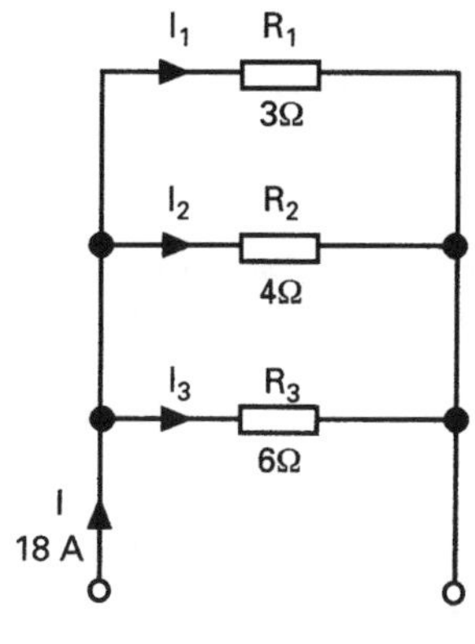

Fig. 2.9

2.5 Series/Parallel Combinations

Most practical circuits consist of resistors which are interconnected in both series and parallel forms. The simplest method of solving such a circuit is to reduce the parallel branches to their equivalent resistance values and hence reduce the circuit to a simple series arrangement. This is best illustrated by means of a worked example.

Worked Example 2.5

Question

For the circuit of Fig. 2.10 calculate (a) the current drawn from the source, (b) the p.d. across each resistor, (c) the current through each resistor and (d) the power dissipated by the 5 Ω resistor.

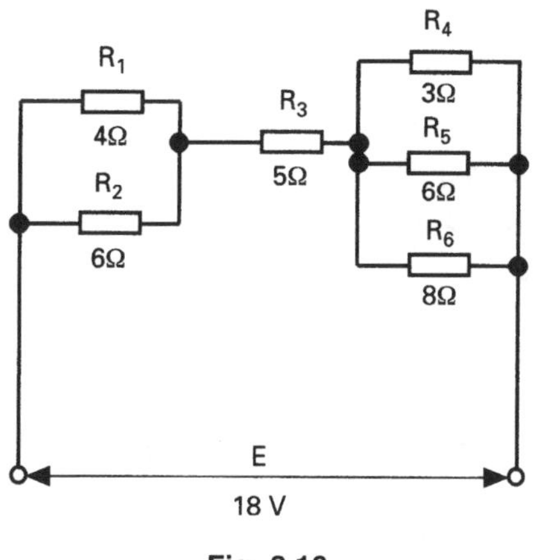

Fig. 2.10

Continued on p. 48

Worked Example 2.5 (*Continued*)

Answer

The first step in the solution is to label the diagram clearly with letters at the junctions and identifying p.d.s and branch currents. This is shown in Fig. 2.11.

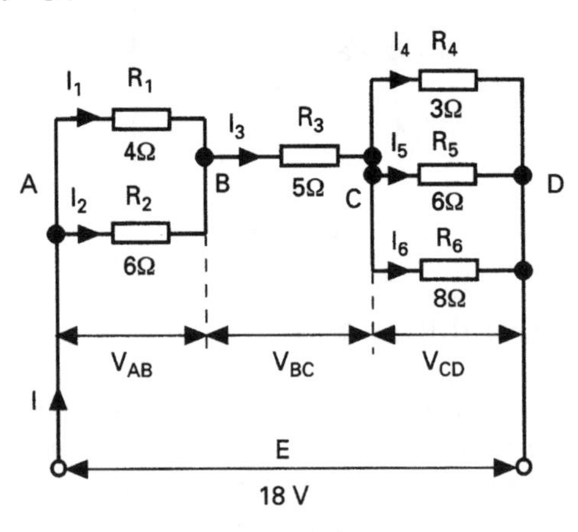

Fig. 2.11

(a) $R_{AB} = \dfrac{R_1 R_2}{R_1 + R_2}$ ohm $= \dfrac{4 \times 6}{4 + 6} = 2.4\,\Omega$

$R_{BC} = 5\,\Omega$

$$\frac{1}{R_{CD}} = \frac{1}{R_4} + \frac{1}{R_5} + \frac{1}{R_6} = \frac{1}{3} + \frac{1}{6} + \frac{1}{8}$$

$$= \frac{8 + 4 + 3}{24} = \frac{15}{24}\,\text{S}$$

$$R_{CD} = \frac{24}{15} = 1.6\,\Omega$$

$$R = R_{AB} + R_{BC} + R_{CD} \text{ ohm}$$

$$R = 2.4 + 5 + 1.6 = 9\,\Omega$$

$$I = \frac{E}{R} \text{ amp} = \frac{18}{9}$$

$I = 2$ A **Ans**

(b) The circuit has been reduced to its series equivalent as shown in Fig. 2.12. Using this equivalent circuit it is now a simple matter to calculate the p.d. across each section of the circuit.

$V_{AB} = IR_{AB}$ volt $= 2 \times 2.4$

$V_{AB} = 4.8$ V **Ans**

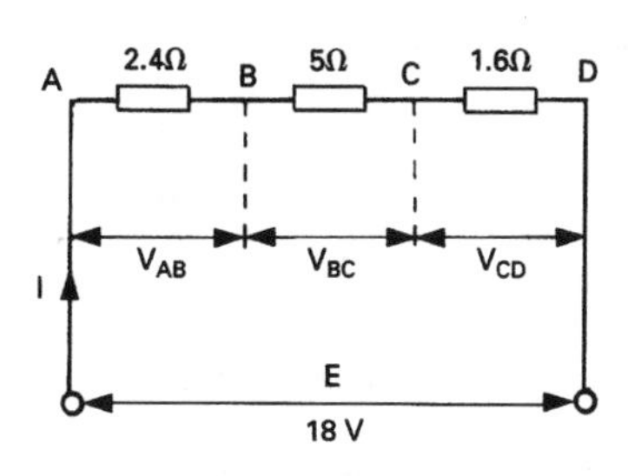

Fig. 2.12

(this p.d. is common to both R_1 and R_2)

$V_{BC} = IR_{BC}$ volt $= 2 \times 5$

$V_{BC} = 10\,\text{V}$ **Ans**

$V_{CD} = IR_{CD}$ volt $= 2 \times 1.6$

$V_{CD} = 3.2\,\text{V}$ **Ans**

(this p.d. is common to R_4, R_5 and R_6)

(c) $I_1 = \dfrac{V_{AB}}{R_1} = \dfrac{4.8}{4}$ or $I_1 = \dfrac{R_2}{R_1 + R_2} \times I = \dfrac{6}{10} \times 2$

$I_1 = 1.2\,\text{A}$ **Ans** $\qquad I_1 = 1.2\,\text{A}$ **Ans**

$I_2 = \dfrac{V_{AB}}{R_2} = \dfrac{4.8}{6}$ or $I_2 = \dfrac{R_1}{R_1 + R_2} \times I = \dfrac{4}{10} \times 2$

$I_2 = 0.8\,\text{A}$ **Ans** $\qquad I_2 = 0.8\,\text{A}$ **Ans**

$I_3 = I = 2\,\text{A}$ **Ans**

$I_4 = \dfrac{V_{CD}}{R_4} = \dfrac{3.2}{3}$ or $\dfrac{1}{R_{CD}} = \dfrac{1}{R_4} + \dfrac{1}{R_5} + \dfrac{1}{R_6}$

$I_4 = 1.067\,\text{A}$ **Ans** $\qquad = \dfrac{1}{3} + \dfrac{1}{6} + \dfrac{1}{8}$

$I_5 = \dfrac{V_{CD}}{R_5} = \dfrac{3.2}{6}$ $\qquad = \dfrac{8 + 4 + 3}{24} = \dfrac{15}{24}$

$I_5 = 0.533\,\text{A}$ **Ans** $\qquad$ so $I_4 = \dfrac{8}{15} \times 2$

$I_6 = \dfrac{V_{CD}}{R_6} = \dfrac{3.2}{8}$ $\qquad I_4 = 1.067\,\text{A}$ **Ans**

$I_6 = 0.4\,\text{A}$ **Ans** $\qquad I_5 = \dfrac{4}{15} \times 2$

Continued on p. 50

Worked Example 2.5 (*Continued*)

$$I_5 = 0.533\text{ A } \textbf{Ans}$$

$$I_6 = \frac{3}{15} \times 2$$

$$I_6 = 0.4\text{ A } \textbf{Ans}$$

Notice that the p.d. method is an easier and less cumbersome one than current division when more than two resistors are connected in parallel.

(d) $P_3 = I_3^2 R_3$ watt or $V_{BC} I_3$ watt

or $\frac{V_{BC}^2}{R_3}$ watt

and using the first of these alternative equations:

$$P_3 = 2^2 \times 5$$

$$P_3 = 20\text{ W } \textbf{Ans}$$

It is left to the reader to confirm that the other two power equations above yield the same answer.

2.6 Kirchhoff's Current Law

We have already put this law into practice, though without stating it explicitly. The law states that the algebraic sum of the currents at any junction of a circuit is zero. Another, and perhaps simpler, way of stating this is to say that the sum of the currents arriving at a junction is equal to the sum of the currents leaving that junction. Thus we have applied the law with parallel circuits, where the assumption has been made that the sum of the branch currents equals the current drawn from the source. Expressing the law in the form of an equation we have:

$$\Sigma I = 0 \qquad (2.9)$$

where the symbol Σ means 'the sum of'.

Figure 2.13 illustrates a junction within a circuit with a number of currents arriving and leaving the junction. Applying Kirchhoff's current law yields:

$$I_1 - I_2 + I_3 + I_4 - I_5 = 0$$

where '+' signs have been used to denote currents arriving and '−' signs for currents leaving the junction. This equation can be transposed to comply with the alternative statement for the law, thus:

$$I_1 + I_3 + I_4 = I_2 + I_5$$

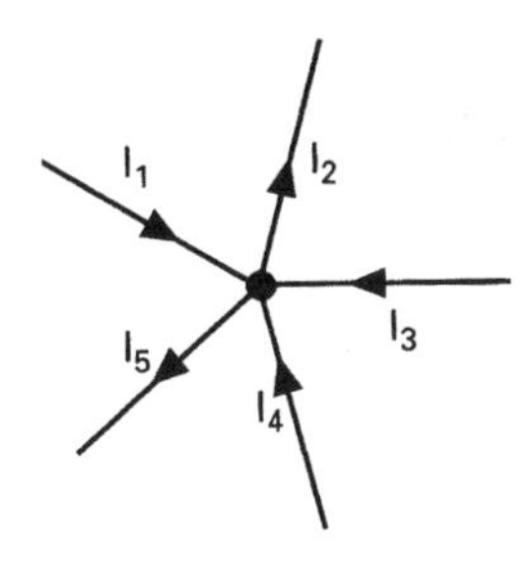

Fig. 2.13

Worked Example 2.6

Question

For the network shown in Fig. 2.14 calculate the values of the marked currents.

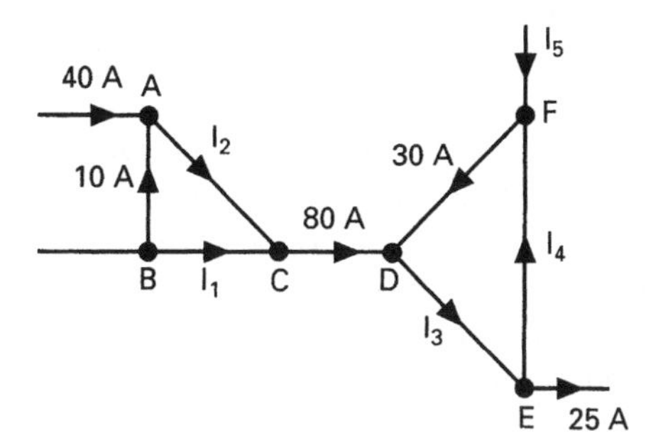

Fig. 2.14

Answer

Junction A: $I_2 = 40 + 10 = 50\,\text{A}$ **Ans**

Junction C: $I_1 + I_2 = 80$

$I_1 + 50 = 80$

so $I_1 = 30\,\text{A}$ **Ans**

Junction D: $I_3 = 80 + 30 = 110\,\text{A}$ **Ans**

Junction E: $I_4 + 25 = I_3$

$I_4 = 110 - 25$

so $I_4 = 85\,\text{A}$ **Ans**

Continued on p. 52

Worked Example 2.6 (*Continued*)

Junction F: $I_5 + I_4 = 30$

$$I_5 + 85 = 30$$
$$I_5 = 30 - 85$$
$$\text{so } I_5 = -55\text{ A} \quad \textbf{Ans}$$

Note: The minus sign in the last answer tells us that the current I_5 is actually flowing away from the junction rather than towards it as shown.

2.7 Kirchhoff's Voltage Law

This law also has already been used—in the explanation of p.d. and in the series and series/parallel circuits. This law states that in any closed network the algebraic sum of the emfs is equal to the algebraic sum of the p.d.s taken in order about the network. Once again, the law sounds very complicated, but it is really only common sense, and is simple to apply. So far, it has been applied only to very simple circuits, such as resistors connected in series across a source of emf. In this case we have said that the sum of the p.d.s is equal to the applied emf (e.g. $V_1 + V_2 = E$). However, these simple circuits have had only one source of emf, and could be solved using simple Ohm's law techniques. When more than one source of emf is involved, or the network is more complex, then a network analysis method must be used. Kirchhoff's is one of these methods.

Expressing the law in mathematical form:

$$\Sigma E = \Sigma IR \tag{2.10}$$

A generalised circuit requiring the application of Kirchhoff's laws is shown in Fig. 2.15. Note the following:

1 The circuit has been labelled with letters so that it is easy to refer to a particular loop and the direction around the loop that is being considered. Thus, if the left-hand loop is considered, and you wish to trace a path around it in a clockwise direction, this would be referred to as ABEFA. If a counterclockwise path was required, it would be referred to as FEBAF or AFEBA.

2 Current directions have been assumed and marked on the diagram. As was found in the previous worked example (2.6), it may well turn out that one or more of these currents actually flows in the opposite direction to that marked. This result would be indicated by a negative value obtained from the calculation. However, to ensure consistency, make the initial assumption that all sources of emf are discharging current into the circuit; i.e. current leaves the positive terminal of each battery and enters at its negative terminal. The current law is also applied at this stage, which is why the current flowing through R_3 is marked as $(I_1 + I_2)$ and not as I_3. This is an important point since the solution involves the use of simultaneous equations, and

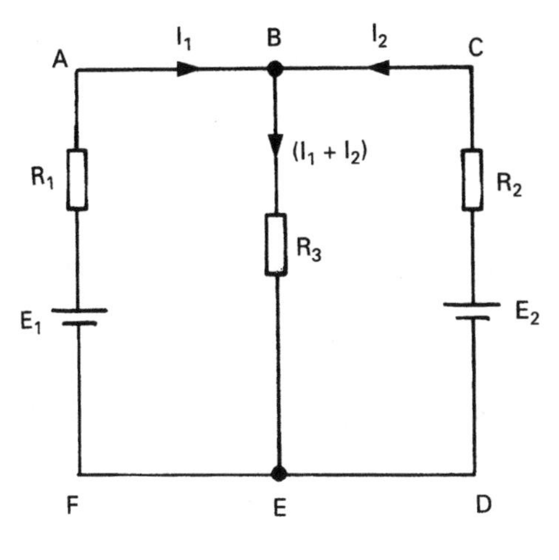

Fig. 2.15

the fewer the number of 'unknowns' the simpler the solution. Thus marking the third-branch current in this way means that there are only two 'unknowns' to find, namely I_1 and I_2. The value for the third branch current, I_3, is then simply found by using the values obtained for I_1 and I_2.

3 If a negative value is obtained for a current then the minus sign MUST be retained in any subsequent calculations. However, when you quote the answer for such a current, make a note to the effect that it is flowing in the opposite direction to that marked, e.g. from C to D.

4 When tracing the path around a loop, concentrate solely on that loop and ignore the remainder of the circuit. Also note that if you are following the marked direction of current then the resulting p.d.(s) are assigned positive values. If the direction of 'travel' is opposite to the current arrow then the p.d. is assigned a negative value.

Let us now apply these techniques to the circuit of Fig. 2.15.

Consider first the left-hand loop in a clockwise direction. Tracing around the loop it can be seen that there is only one source of emf within it (namely E_1). Thus the sum of the emfs is simply E_1 volt. Also, within the loop there are only two resistors (R_1 and R_2) which will result in two p.d.s, $I_1 R_1$ and $(I_1 + I_2)R_3$ volt. The resulting loop equation will therefore be:

<u>ABEFA</u>: $E_1 = I_1R_1 + (I_1 + I_2)R_3$ [1]

Now taking the right-hand loop in a counterclockwise direction it can be seen that again there is only one source of emf and two resistors. This results in the following loop equation:

<u>CBEDC</u>: $E_2 = I_2R_2 + (I_1 + I_2)R_3$ [2]

Finally, let us consider the loop around the edges of the diagram in a clockwise direction. This follows the 'normal' direction for E_1 but is opposite to that for E_2, so the sum of the emfs is $E_1 - E_2$ volt. The loop equation is therefore

$$\underline{\text{ABCDEFA}}: E_1 - E_2 = I_1R_1 - I_2R_2 \qquad [3]$$

Since there are only two unknowns then only two simultaneous equations are required, and three have been written. However it is a useful practice to do this as the 'extra' equation may contain more convenient numerical values for the coefficients of the 'unknown' currents.

The complete technique for the applications of Kirchhoff's laws becomes clearer by the consideration of a worked example containing numerical values.

Worked Example 2.7

Question

For the circuit of Fig. 2.16 determine the value and direction of the current in each branch, and the p.d. across the 10 Ω resistor.

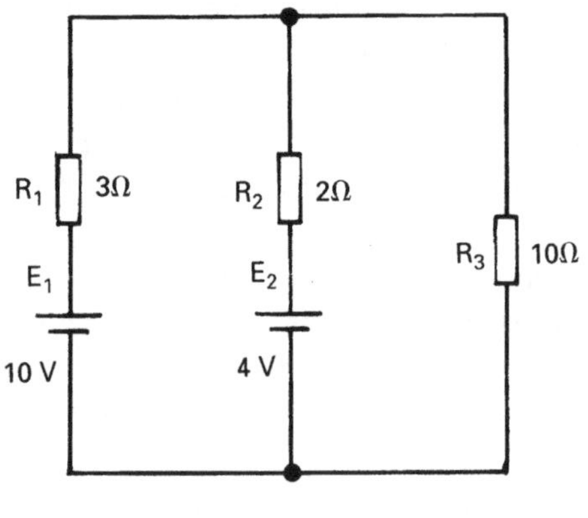

Fig. 2.16

Answer

The circuit is first labelled and current flows identified and marked by applying the current law. This is shown in Fig. 2.17.

<u>ABEFA</u>: $10 - 4 = 3I_1 - 2I_2$

so $6 = 3I_1 - 2I_2$ .... [1]

<u>ABCDEFA</u>: $10 = 3I_2 + 10(I_1 + I_2)$

$= 3I_2 + 10I_1 + 10I_2$

so $10 = 13I_1 + 10I_2$ .... [2]

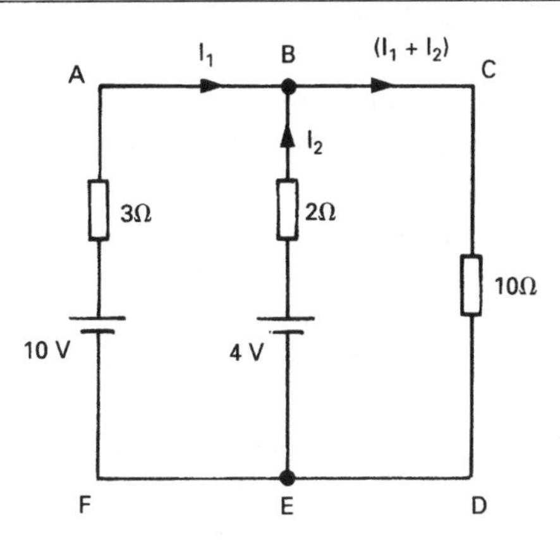

Fig. 2.17

BCDEB: $4 = 2I_2 + 10(I_1 + I_2)$

$= 2I_2 + 10I_1 + 10I_2$

so $4 = 10I_1 + 12I_2$ .... [3]

Inspection of equations [1] and [2] shows that if equation [1] is multiplied by 5 then the coefficient of I_2 will be the same in both equations. Thus, if the two are now added then the term containing I_2 will be eliminated, and hence a value can be obtained for I_1.

$30 = 15I_1 - 10I_2$ .... [1] × 5

$10 = 13I_1 + 10I_2$ .... [2]

$40 = 28I_1$

so $I_1 = \dfrac{40}{28} = 1.429$ A **Ans**

Substituting this value for I_1 into equation [3] yields;

$4 = 14.29 + 12I_2$

$12I_2 = 4 - 14.29$

so $I_2 = \dfrac{-10.29}{12} = -0.857$ A (charge) **Ans**

$(I_1 + I_2) = 1.429 - 0.857 = 0.572$ A **Ans**

$V_{CD} = (I_1 + I_2) \cdot R_{CD}$ volt

$= 0.572 \times 10$

so $V_{CD} = 5.72$ V **Ans**

A **network theorem** is a set of rules or laws which enable the calculation of currents, p.d.s etc. in a network where the application of Ohm's law alone cannot yield these results. Examples are: Thévenin's, Norton's, and Superposition theorems, and Kirchhoff's laws

2.8 Other Network Theorems

These theorems provide an alternative to the application of Kirchhoff's laws to the solution of networks containing one or more sources of emf.

2.9 Superposition Theorem

The superposition principle allows us to break up a complex circuit into separate sections, and then apply simple Ohm's law techniques to each section in turn. The theorem is defined as follows:

In any network consisting of resistors, the current flowing in any branch is the algebraic sum of the currents that would flow in that branch, if each emf source was considered separately, all others being replaced by resistors equal to their internal resistance.

In the above definition the term resistance has been used. This is because you are required to deal only with resistive circuits. However, the theorem applies equally to a.c. circuits, which may also include reactive elements. In this case, the word *impedance* should be substituted for resistance. This comment applies also to the other network theorems that follow shortly.

As with most theorems, it sounds most complicated. However, putting the principle into practice is relatively simple, and is best illustrated by means of an example.

Worked Example 2.8

Question

A circuit containing two sources of emf is shown in Fig. 2.18. Using the principle of Superposition, determine the current flowing in the 5 Ω resistor.

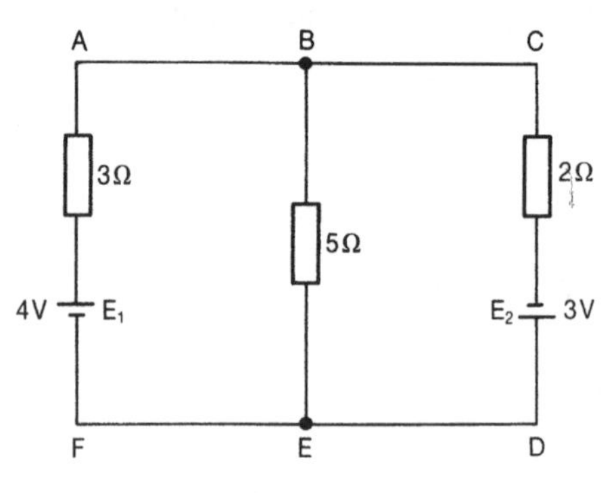

Fig. 2.18

Answer

The theorem allows us to split the original circuit into two sub-circuits (shown in Figs. 2.19 and 2.20), each of which contains only one battery, the other battery being replaced by its internal resistance. Note that it is not clear, from the diagram, whether the 3 Ω and 2 Ω resistors are the internal resistances of E_1 and E_2 respectively; or that E_1 and E_2 are ideal sources (having zero internal resistance), with the two resistors being external resistors in the circuit. In this case it makes no difference, so the two sub-circuits shown apply in both circumstances.

Considering Fig. 2.19, currents I_1 and I_2 may be calculated as follows:

Total resistance of the circuit, R consists of the 3 Ω in series with the parallel combination of the 5 Ω and the 2 Ω.

$$\text{Hence, } R = 3 + \frac{2 \times 5}{2 + 5} = 4.429\,\Omega$$

$$I_1 = \frac{E_1}{R}\text{ amp} = \frac{4}{4.429}$$

$$\text{so, } I_1 = 0.903\text{ A}$$

Using current divider technique, current I_2 is:

$$I_2 = \frac{2}{7} \times 0.903 = 0.258\text{ A from B to E} \ldots\ldots\ldots\ldots [1]$$

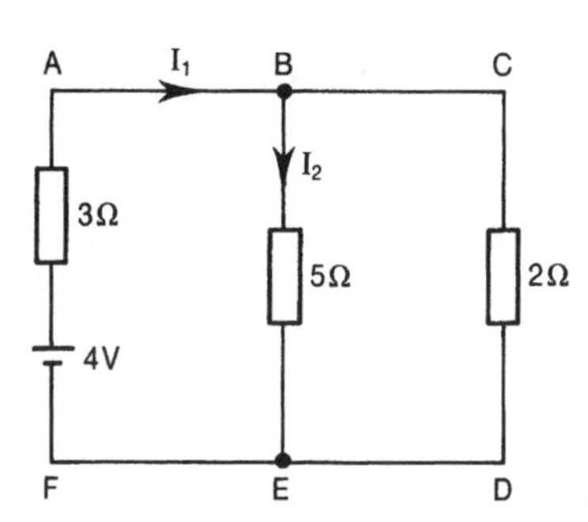

Fig. 2.19

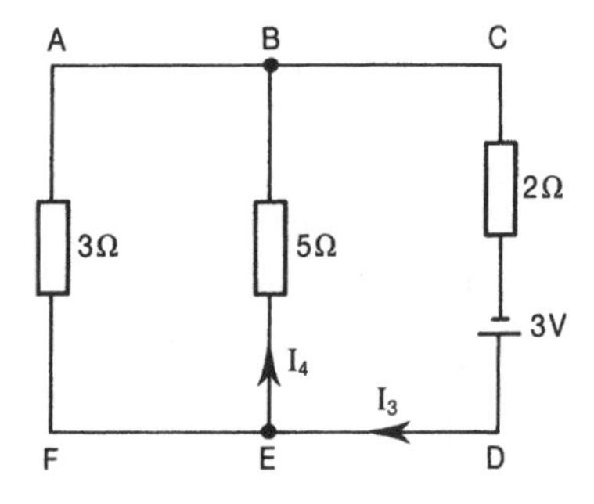

Fig. 2.20

Continued on p. 58

Worked Example 2.8 (*Continued*)

Considering Fig. 2.20, we can apply the same techniques to evaluate currents I_3 and I_4:

$$R = 2 + \frac{5 \times 3}{5 + 3} = 3.875\,\Omega$$

$$I_3 = \frac{3}{3.875} = 0.774\,\text{A}$$

$$I_4 = \frac{3}{8} \times 0.774 = 0.29\,\text{A from E to B} \qquad [2]$$

The current through the $5\,\Omega$ resistor is the algebraic sum of the currents I_2 and I_4 as in [1] and [2] above. However, notice that these two currents are in opposite directions. The algebraic sum is therefore the *difference* between them.

Hence, current through the $5\,\Omega = I_4 - I_2$ amp

$= 0.032\,\text{A}$ from E to B **Ans**

2.10 Constant Voltage and Constant Current Sources and their Equivalence

Thus far, we have considered only voltage sources, in the form of a constant emf in series with its internal resistance. In some circuits it is more convenient to consider the source as a constant current generator, with its internal resistance in parallel with it. This is often the case when dealing with the equivalent circuit for a transistor.

Consider a constant voltage source of emf E volt and internal resistance r ohm, supplying a load resistor R, as shown in Fig. 2.21.

$$\text{load current, } I = \frac{E}{r + R} \text{ amp} \ldots\ldots\ldots\ldots\ldots [1]$$

$$\text{and terminal p.d., } V = IR \text{ volt} \ldots\ldots\ldots\ldots\ldots\ldots [2]$$

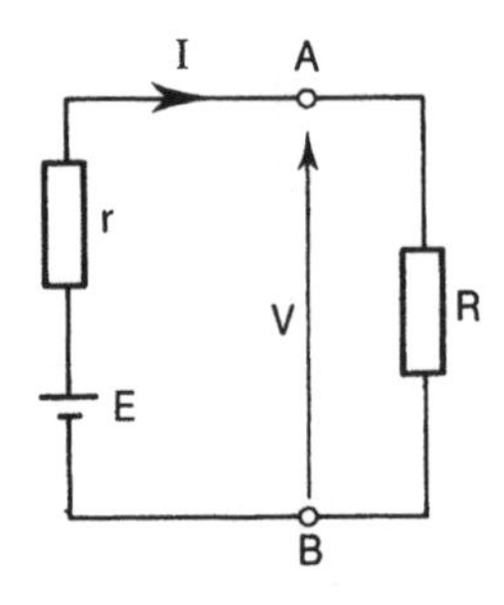

Fig. 2.21

Let the load R now be replaced by a short-circuit between terminals A and B, as shown in Fig. 2.22.

$$I_{sc} = \frac{E}{r} \text{ amp} \qquad [3]$$

Let us now replace the original voltage source by a constant current source, of value I_{sc} and internal resistance of r ohm in parallel with it. This is shown in Fig. 2.23. The linked circles form the circuit symbol for a current generator.

Considering this figure, the current through the load, and the terminal p.d. may be determined by applying the current divider technique as follows:

$$I = \frac{r}{r+R} \times I_{sc} \text{ amp}$$

and substituting [3] for I_{sc}:

$$I = \frac{r}{r+R} \times \frac{E}{r}$$

$$\text{so } I = \frac{E}{r+R} \text{ amp} \qquad [4]$$

$$\text{and } V = IR \text{ volt} \qquad [5]$$

Since the expressions for the load current and terminal voltage in [1] and [2] are identical to those in [4] and [5] respectively, then it follows that the constant voltage generator of Fig. 2.21 is equivalent to the constant current generator of Fig. 2.23, and vice versa.

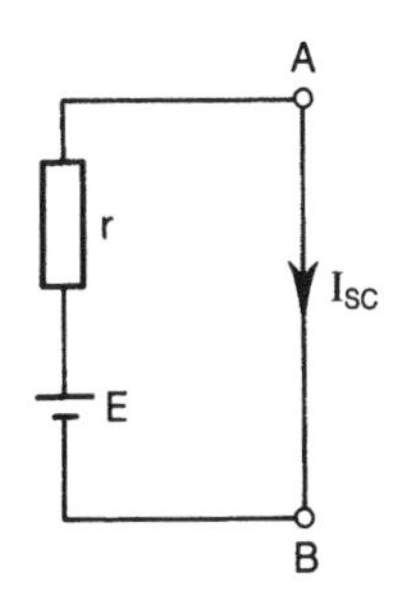

Fig. 2.22

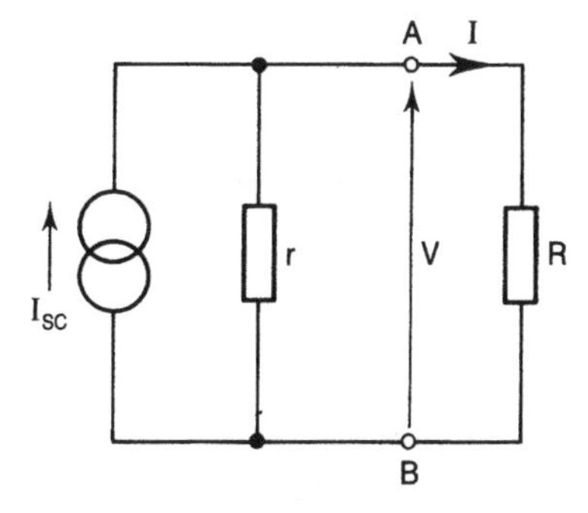

Fig. 2.23

Worked Example 2.9

Question

(a) Convert the voltage source shown in Fig. 2.24 into its equivalent constant current source, and (b) convert the current source of Fig. 2.25 into its equivalent voltage source.

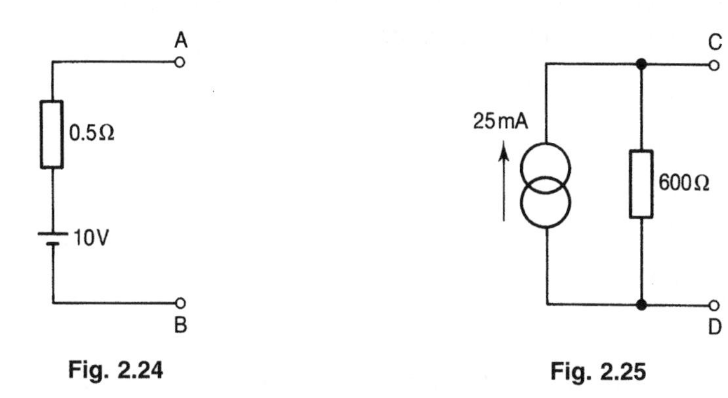

Fig. 2.24 **Fig. 2.25**

Answer

(a) Placing a short-circuit between the terminals of the voltage generator will produce a current of:

$$I_{sc} = \frac{10}{0.5} \text{ amp} = 20\,\text{A}$$

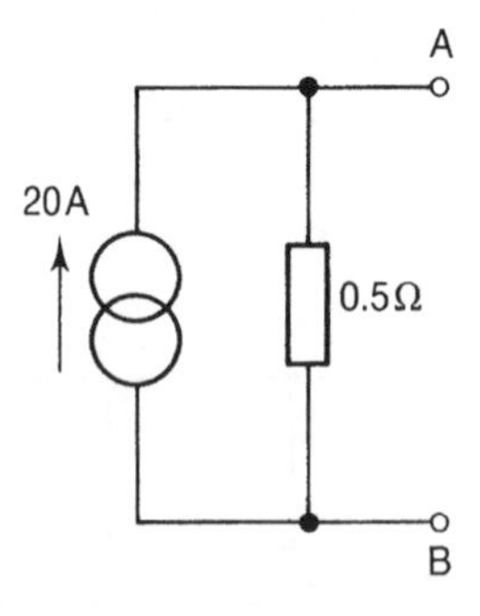

Fig. 2.26

so the constant current generator will produce a current of 20 A, with an internal resistance of 0.5 Ω in parallel, as shown in Fig. 2.26.

(b) From Fig. 2.25, the open-circuit terminal voltage is:

$$V = 25 \times 10^{-3} \times 600 = 15\,\text{V}$$

so the constant voltage generator will have an emf of 15 V, with an internal resistance of 600 Ω in series, as shown in Fig. 2.27.

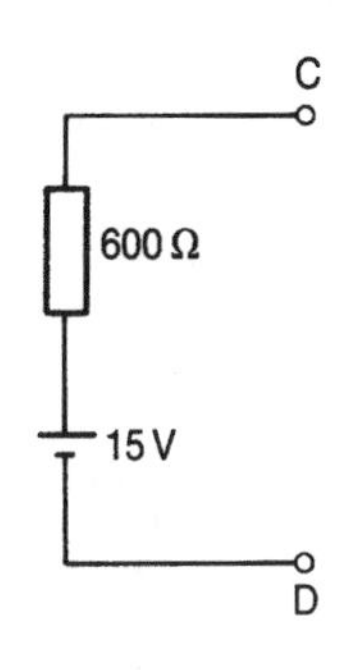

Fig. 2.27

2.11 Thévenin's Theorem

Thévenin's theorem states that any network containing sources and resistors can be replaced by a *single* voltage source of emf E_o and series internal resistance R_o, as shown in Fig. 2.28. The value of E_o is equal to the open-circuit terminal voltage of the network, measured between terminals A and B. The resistance R_o is the resistance of the network measured between terminals A and B, with all network sources replaced by their internal resistances.

Note: If any of the sources in the original network are shown as being ideal (zero internal resistance), these will be replaced by a short-circuit.

The application of Thévenin's theorem to the solution of network is illustrated in the following example.

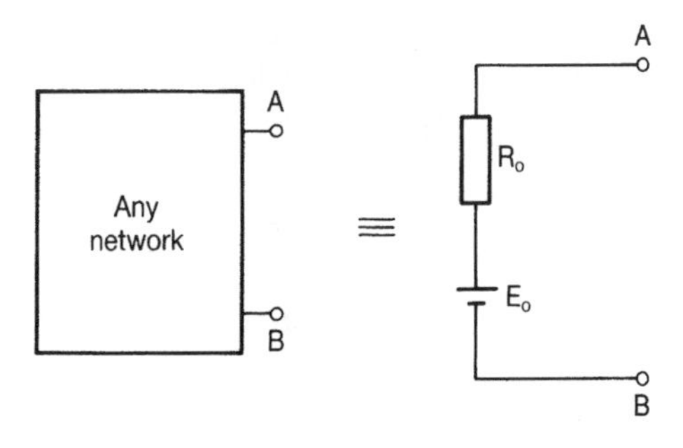

Fig. 2.28

Worked Example 2.10

Question

For the network shown in Fig. 2.29, (a) determine the Thévenin equivalent generator for the circuit to the left of terminals A and B, (b) hence calculate the p.d. across and the current flowing through the 5 Ω load resistor, and (c) calculate the p.d. across and the current through the load if this resistance is changed to 3 Ω.

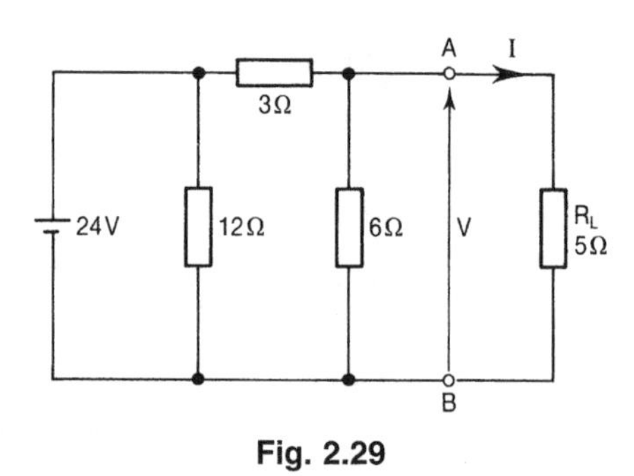

Fig. 2.29

Answer

(a) With the 5 Ω resistor removed, terminals A and B will be open-circuit. The full 24 V will be developed across the 12 Ω resistor, so the 6 Ω and 3 Ω resistors form a simple potential divider.

Thus, $E_o = \frac{6}{9} \times 24 = 16\,\text{V}$

Since the battery is shown as an ideal source, then it is replaced by a short-circuit. Thus, 'looking in' at the terminals, the circuit appears as in Fig. 2.30. As the 12 Ω resistor is now short-circuited, then the 6 Ω and the 3 Ω resistors are effectively connected in parallel.

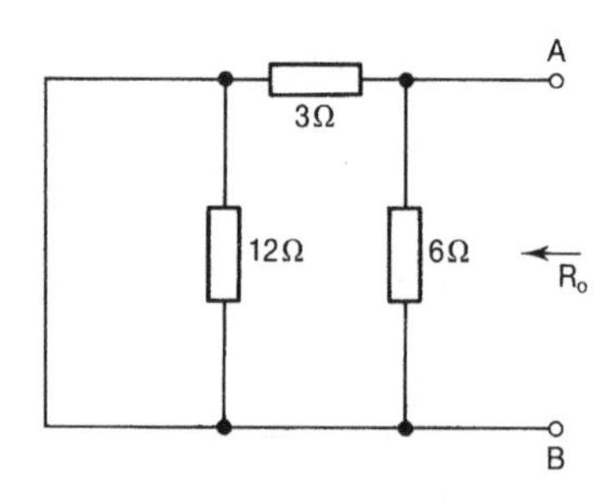

Fig. 2.30

Hence, $R_o = \frac{3 \times 6}{9} = 2\,\Omega$

The Thévenin equivalent generator therefore has an emf of 16 V and internal resistance of 2 Ω. The complete equivalent circuit is shown in Fig. 2.31.

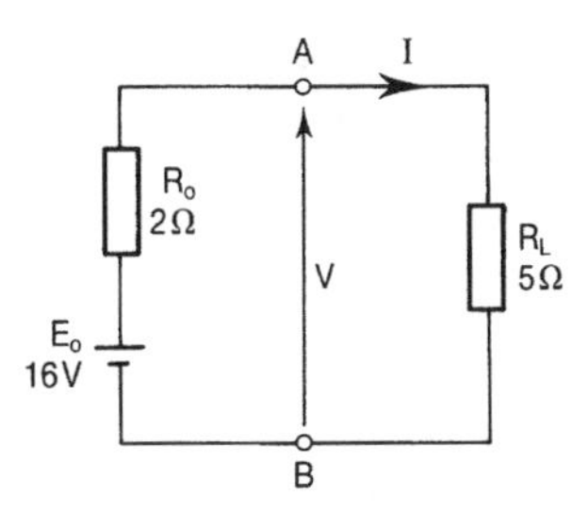

Fig. 2.31

(b) From Fig. 2.31:

$$V = \frac{5}{7} \times 16 = 11.43\,\text{V}\ \textbf{Ans}$$

$$\text{and } I = \frac{V}{R_L}\ \text{amp} = \frac{11.43}{5}$$

$$\text{so, } I = 2.29\,\text{A}\ \textbf{Ans}$$

(c)

$$V = \frac{3}{5} \times 16 = 9.6\,\text{V}\ \textbf{Ans}$$

$$\text{and } I = \frac{9.6}{3} = 3.2\,\text{A}\ \textbf{Ans}$$

Note: The original circuit could have been solved using normal Ohm's law techniques, but the calculations would have been more extensive. The real advantage of deriving the Thévenin equivalent generator is that the load value may be changed to any value, as many times as is wished, but the subsequent calculations for V and I would be very quick and simple. Using normal Ohm's law techniques, *all* the circuit currents etc. would have to be recalculated each time.

Worked Example 2.11

Question

For the circuit of Fig. 2.32 (a) determine the Thévenin equivalent generator for the network to the left of terminals A and B, and hence calculate the current flowing through the 5 Ω load resistor, and (b) determine the load current if the load resistor is changed to 10 Ω.

Continued on p. 64

Worked Example 2.11 (*Continued*)

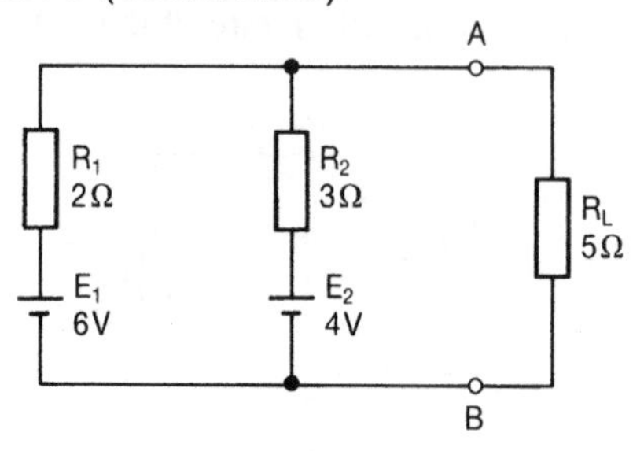

Fig. 2.32

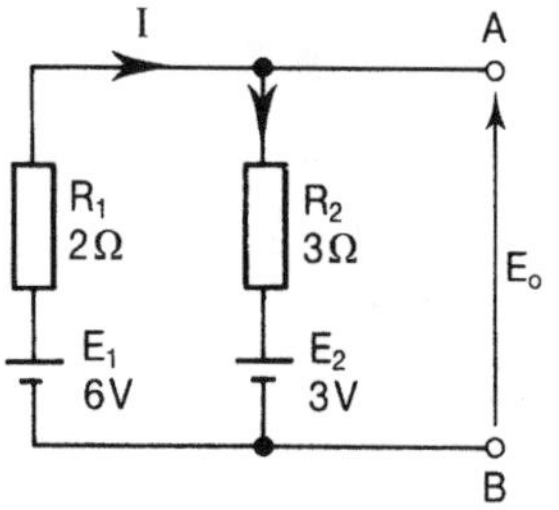

Fig. 2.33

Answer

(a) Removing the load resistor, the circuit is as shown in Fig. 2.33, and the terminal p.d. (E_o) calculated as follows.

$$I = \frac{E_1 - E_2}{R_1 + R_2} \text{ amp} = \frac{6-4}{2+3}$$

therefore, $I = 0.4\,\text{A}$

$$E_o = E_1 - IR_1 \quad \text{or} \quad E_2 + IR_2 \text{ volt}$$
$$= 6 - (0.4 \times 2) \quad \text{or} \quad 4 + (0.4 \times 3)$$

so, $E_o = 5.2\,\text{V}$

From Fig. 2.34, where the two sources have been replaced by their internal resistances:

$$R_o = \frac{R_1 R_2}{R_1 + R_2} \text{ ohm} = \frac{6}{5} = 1.2\,\Omega$$

Hence, the Thévenin equivalent circuit will be as shown in Fig. 2.35, and from this circuit:

$$I = \frac{E_o}{R_o + R_L} \text{ amp} = \frac{5.2}{6.2}$$

so, $I = 0.839\,\text{A}$ **Ans**

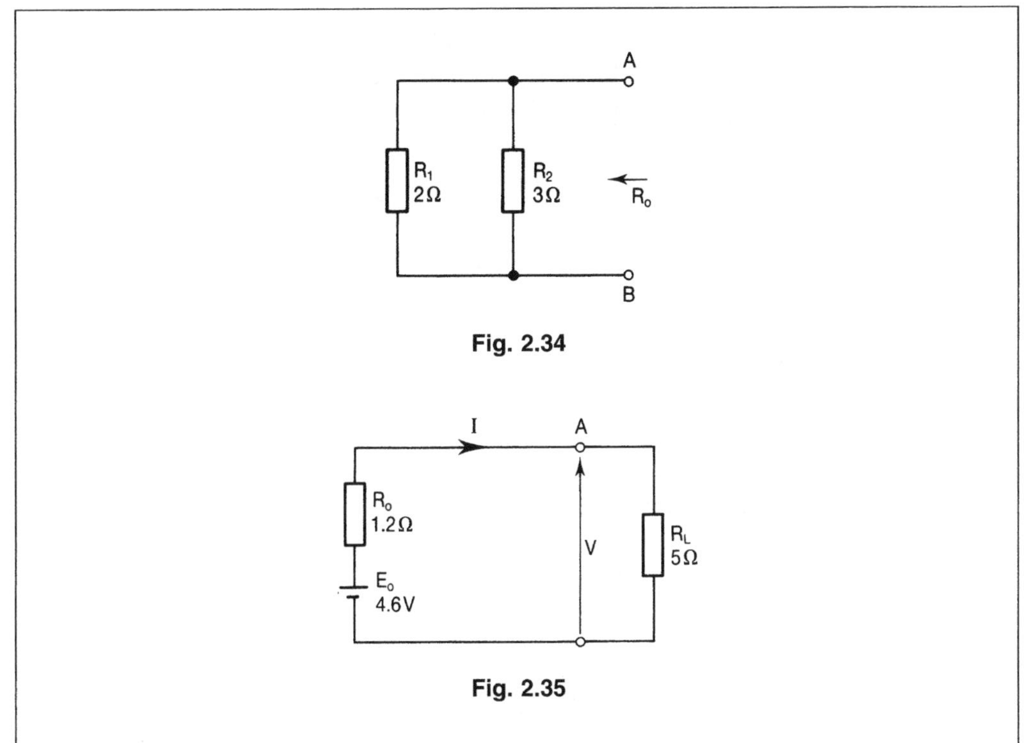

Fig. 2.34

Fig. 2.35

(b) With R_L changed to $10\,\Omega$, and using the Thévenin circuit, the load current is simply obtained thus:

$$I = \frac{E_o}{R_o + R_L}\ \text{amp} = \frac{5.2}{11.2} = 0.464\,\text{A}\ \textbf{Ans}$$

Note: This problem could have been solved using Kirchhoff's laws (or superposition theorem). However, in order to complete part (b), all of the loop equations would have to be derived again to provide a new set of simultaneous equations for solution. This clearly illustrates the advantage of using the Thévenin generator.

2.12 Norton's Theorem

This theorem complements Thévenin's in that the network is replaced by a constant current generator of I_{sc} amp, and parallel internal resistance R_o ohm. The current, I_{sc}, is the current that would flow between the terminals when they are short-circuited. The resistance R_o is defined in exactly the same way as it is for the Thévenin circuit. Thus R_o will be the same for both a Norton and a Thévenin equivalent circuit. The Norton constant current generator is illustrated in Fig. 2.36.

The application of Norton's theorem is illustrated in the following example.

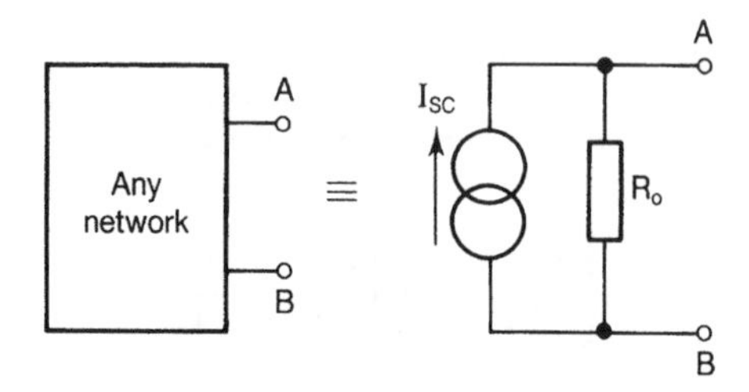

Fig. 2.36

Worked Example 2.12

Question

Derive the Norton equivalent circuit for the network to the left of terminals A and B in Fig. 2.37, and hence determine the current through the 10 Ω load resistor.

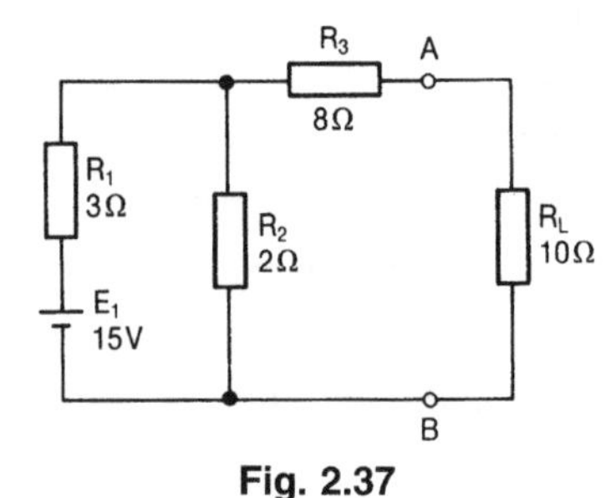

Fig. 2.37

Answer

With terminals A and B short-circuited, the circuit will be as in Fig. 2.38. Using this circuit, the short-circuit current I_{sc} is found as follows:

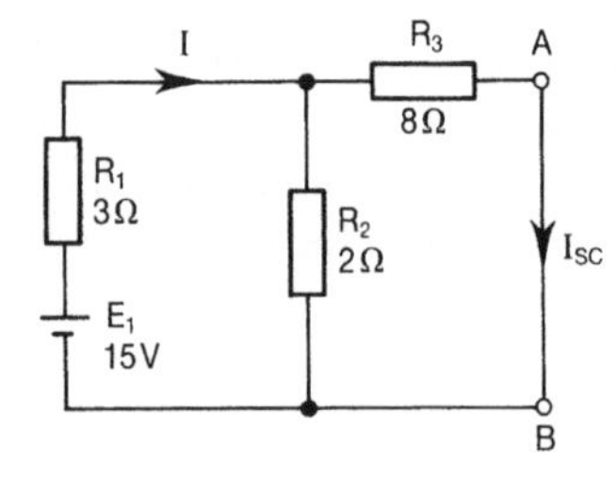

Fig. 2.38

$$\text{Total resistance, } R = R_1 + \frac{R_2 R_3}{R_2 + R_3} \text{ ohm} = 3 + \frac{16}{10}$$

$$\text{so, } R = 4.6\,\Omega$$

$$I = \frac{E_1}{R} \text{ amp} = \frac{15}{4.6} = 3.261\text{ A}$$

Using current division,

$$I_{sc} = \frac{R_2}{R_2 + R_3} \times I \text{ amp} = \frac{2}{10} \times 3.261$$

therefore, $I_{sc} = 0.652\,\text{A}$

From Fig. 2.39, the resistance R_o is obtained thus:

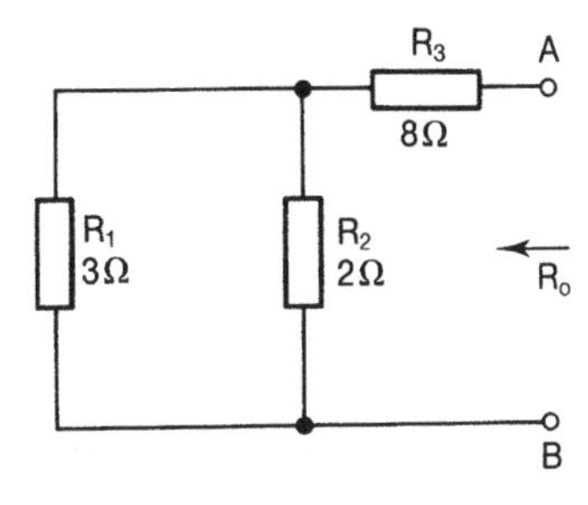

Fig. 2.39

$$R_o = R_3 + \frac{R_1 R_2}{R_1 + R_2} \text{ ohm} = 8 + \frac{6}{5}$$

so, $R_o = 9.2\,\Omega$

The complete Norton equivalent circuit will be as shown in Fig. 2.40, and from this figure:

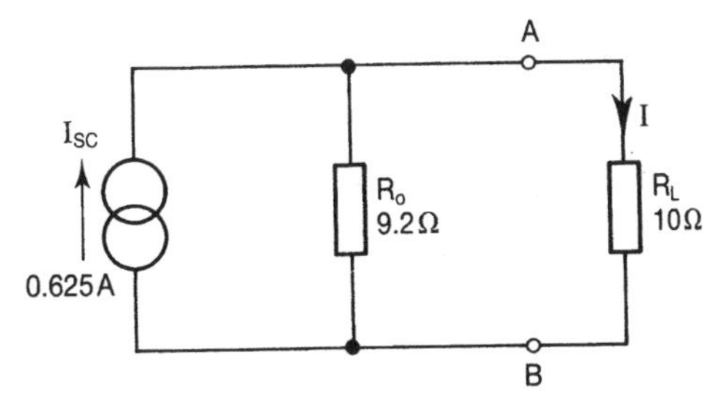

Fig. 2.40

$$I = \frac{R_o}{R_o + R_L} \times I_{sc} \text{ amp} = \frac{9.2}{19.2} \times 0.652$$

hence, $I = 0.312\,\text{A}$ **Ans**

The advantage of the Norton equivalent circuit is the same as that for the Thévenin circuit; i.e. if the load resistance value is repeatedly changed, the new load current can be very simply and rapidly calculated.

2.13 The Maximum Power Transfer Theorem

This theorem states that maximum power is transferred from a source to a load when the load resistance is of the same value as the internal resistance of the source.

Again, note that this theorem also applies to a.c. circuits; in which case the word *resistance* should be replaced by the word *impedance*. However, the generator may also possess internal reactance. In this case, for maximum power transfer, the load must also possess an equal value of reactance, but of the opposite type. In other words, if the source has inductive reactance, then the load must have an equivalent capacitive reactance, and vice versa.

Worked Example 2.13

Question

For the circuit shown in Fig. 2.37 (Example 2.12), determine the value of load resistor R_L that would result in maximum power dissipation in this load, and calculate the value of this power.

Answer

In order to solve this problem, the Norton (or Thévenin) equivalent circuit would be derived, as in Example 2.12. Thus, for this example, we know that the equivalent Norton generator has a current of 0.652 A, and internal resistance of 9.2 Ω. Hence the value of R_L that will result in maximum power transfer will be 9.2 Ω **Ans**.

The generator current I_{sc} will divide equally between R_o and R_L, hence:

$$\text{the load current, } I = \frac{0.652}{2} = 0.326 \text{ A}$$

$$P_L = I^2 R_L \text{ watt} = 0.326^2 \times 9.2$$

therefore $P_L = 0.978$ W **Ans**

In order to ensure that maximum power is transferred from a source to a load, the load has to be matched to the source. This effect can be achieved in a.c. circuits by means of an impedance matching transformer. Using this technique, the source is effectively connected to a load of impedance value equal to its own internal resistance. To illustrate the principle, consider a single-phase transformer, connected to a resistive load, as shown in Fig. 2.41.

The primary and secondary windings will be designed to have only a very small resistance (i.e. we may consider them to be perfect inductors). Thus the power developed in the secondary circuit must be due only to the secondary current flowing through the load resistor, R_L. However, the secondary of the transformer must derive its power from the primary circuit. The power in the primary circuit must therefore be due to the primary current flowing through some *effective* resistance. This effective resistance is known as 'the secondary resistance, referred to the primary'. In other words, the supply connected to the primary winding 'sees' an effective resistance

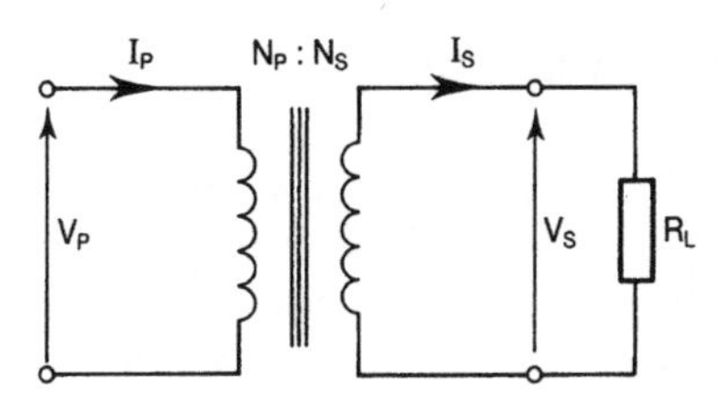

Fig. 2.41

connected between the primary terminals. The value of this referred resistance is obtained as follows.

$$R_L = \frac{V_s}{I_s} \text{ ohm; and } R'_p = \frac{V_p}{I_p} \text{ ohm}$$

$$\text{so, } \frac{R'_p}{R_L} = \frac{V_p}{I_p} \times \frac{I_s}{V_s} = \frac{V_p}{V_s} \times \frac{I_s}{I_p}$$

$$\text{but, } \frac{V_p}{V_s} = \frac{N_p}{N_s} \text{ and } \frac{I_s}{I_p} = \frac{N_p}{N_s}$$

$$\text{therefore, } \frac{R'_p}{R_L} = \left(\frac{N_p}{N_s}\right)^2 \qquad (2.11)$$

$$\text{hence, } R'_p = \left(\frac{N_p}{N_s}\right)^2 \times R_L$$

i.e. the resistance referred to the primary circuit is equal to the load resistance multiplied by the *square* of the turns ratio.

Worked Example 2.14

Question

A loudspeaker of impedance 8 Ω is fed from an amplifier of output impedance 1.8 kΩ. Determine the turns ratio of a suitable matching transformer.

Answer

$R_p = 1800\,\Omega$; $R_L = 8\,\Omega$

$$R'_p = \left(\frac{N_p}{N_s}\right)^2 R_L \text{ ohm}$$

$$\text{so, } \left(\frac{N_p}{N_s}\right)^2 = \frac{R'_p}{R_L} = \frac{1800}{8}$$

$$\text{therefore, } \left(\frac{N_p}{N_s}\right)^2 = 225$$

$$\text{hence, } \frac{N_p}{N_s} = 15 \text{ i.e. turns ratio of 15:1 } \textbf{Ans}$$

2.14 The Decibel and its Usage

Consider a two-stage amplifier system, as shown in Fig. 2.42. Each amplifier provides an increase of the *signal* power. This effect is referred to as the power gain, A_p, of the amplifier. This means that the signal output power from an amplifier is greater than its signal input power. This power gain may be expressed as:

$$A_p = \frac{P_{out}}{P_{in}} \quad \text{or} \quad \frac{P_o}{P_i}$$

For the system shown, if the input power P_1 is 1 mW, and the gains of the two amplifier stages are 10 times and 100 times respectively, then the final output power, P_3, may be determined as follows.

$$A_{p1} = \frac{P_2}{P_1}; \quad \text{so} \quad P_2 = A_{p1} \times P_1 \text{ watt}$$

$$\text{therefore,} \quad P_2 = 10 \times 10^{-3} = 10\,\text{mW}$$

$$A_{p2} = \frac{P_3}{P_2}; \quad \text{so} \quad P_3 = A_{p2} \times P_2 \text{ watt}$$

$$\text{therefore,} \quad P_3 = 100 \times 10 \times 10^{-3} = 1\,\text{W}$$

From these results it may be seen that the output power of the system is 1000 times that of the input.

In other words, overall signal power gain,

$$A_p = \frac{P_3}{P_1} = \frac{1}{10^{-3}} = 1000$$

or, overall signal power gain,

$$A_p = A_{p1}A_{p2}$$

In general, when amplifiers (or other devices) are cascaded in this way, the overall gain (or loss) is given by the product of the individual stage gains (or losses).

Note: The efficiency of any machine or device is defined as the ratio of its output power to its input power. However, this does NOT mean that an amplifier is more than 100% efficient! The reason is that only the *signal* input and output powers are considered when quoting the power gain. No account is taken of the comparatively large amount of power injected from the d.c. power supply, without which the amplifier cannot function. In practice, small signal voltage amplifiers will have an *efficiency* figure of less than 25%. Power amplifiers may have an efficiency in the order of 70%.

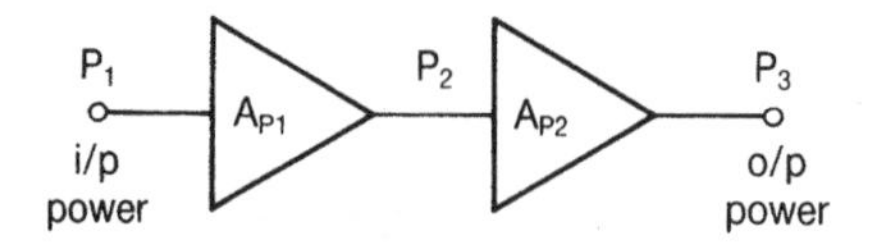

Fig. 2.42

It is often more convenient to express power gain ratios in a logarithmic form, known as the Bel (named after Alexander Graham Bell). Thus a power gain expressed in this way is:

$$A_p = \log \frac{P_o}{P_i} \text{ Bel}$$

where P_i = input power; P_o = output power; and logarithms to the base 10 are used.

The Bel is an inconveniently large unit for practical purposes, so the decibel (one tenth of a Bel) is used.

$$\text{Hence, } A_p = 10 \log \frac{P_o}{P_i} \text{ decibel} \tag{2.12}$$

The unit symbol for the decibel is dB. For the two-stage amplifier system considered, the power gains would be expressed as follows:

$$A_{p1} = 10 \log 10 = 10 \text{ dB}$$
$$A_{p2} = 10 \log 100 = 20 \text{ dB}$$
$$\text{and } A_p = 10 \log 1000 = 30 \text{ dB}$$

Note that the overall system gain, A_p, when expressed in dB is simply the sum of the individual stage gains, also expressed in dB.

Worked Example 2.15

Question

A communications system, involving transmission lines and amplifiers, is illustrated in Fig. 2.43. Each section of transmission line attenuates (reduces) the signal power by a factor of 35.5%, and each amplifier has a gain ratio of 5 times. Calculate the overall power gain of the system as (a) a power ratio, and (b) in decibels.

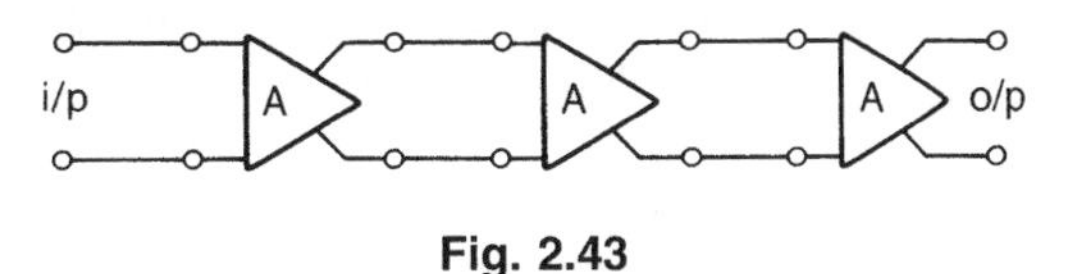

Fig. 2.43

Answer

(a) For each line, $\frac{P_o}{P_i} = 0.355$

so, total loss $= 0.355 \times 0.355 \times 0.355 \times 0.355$
$= 0.0159$

Continued on p. 72

Worked Example 2.15 (*Continued*)

For each amplifier, $\frac{P_o}{P_i} = 5$

$$\text{so, total gain} = 5 \times 5 \times 5 = 125$$

$$\text{therefore, overall gain} = 125 \times 0.0159$$

$$= 2 \text{ times } \textbf{Ans}$$

(b) For each line, attenuation is

$$10 \log 0.355 = -4.5 \text{ dB}^*$$

$$\text{so total attenuation} = -4.5 \times 4$$

$$= -18 \text{ dB}$$

$$\text{For each amplifier, gain} = 10 \log 5$$

$$= +7 \text{ dB}$$

$$\text{so total gain} = 7 \times 3 = 21 \text{ dB}$$

Hence, overall gain of the system = 21 − 18 = 3 dB **Ans**

*Note that a loss or attenuation expressed in dB has a negative value; whereas a gain has a positive value. A further point to note is that if the gains and attenuations of the system had originally been expressed in dB, then the calculation would simply have been as follows:

Overall system gain = (3 × 7) − (4 × 4.5) = 3 dB **Ans**

The above example illustrates the convenience of using decibel notation, since it involves only simple addition and subtraction to determine the overall gain or attenuation of a system. It has also been shown that a gain of 2 times is equivalent to a power gain of 3 dB (more precisely, 3.01 dB). It is left to the reader to confirm, by using a calculator, that an attenuation of 2 times (i.e. a gain of 0.5) is equal to −3 dB. This figure of −3 dB will frequently be met when dealing with the frequency response curves for amplifiers and other frequency-dependent circuits, such as series and parallel tuned circuits.

In order to gain a 'feel' for power gains and losses expressed in dB, the following list shows the corresponding power gain ratios.

$$\text{gains} \begin{cases} 10\,000 \text{ times} = 40 \text{ dB} \\ 1\,000 \text{ times} = 30 \text{ dB} \\ 100 \text{ times} = 20 \text{ dB} \\ 10 \text{ times} = 10 \text{ dB} \\ 2 \text{ times} = 3 \text{ dB} \\ 1 \text{ times} = 0 \text{ dB} \end{cases}$$

$$\text{losses}\begin{cases}0.5\ \text{times} = -3\,\text{dB}\\ 0.1\ \text{times} = -10\,\text{dB}\\ 0.01\ \text{times} = -20\,\text{dB}\\ 0.001\ \text{times} = -30\,\text{dB}\\ 0.0001\ \text{times} = -40\,\text{dB}\end{cases}$$

Worked Example 2.16

Question

(a) Convert the following gain ratios into dB.

(*i*) 250, (*ii*) 50, (*iii*) 0.4

(b) Convert the following gains and losses into ratios.

(*i*) 25 dB, (*ii*) 8 dB, (*iii*) −15 dB

Answer

(a) (i) $A_p = 10\log 250 = 24\,\text{dB}$ **Ans**

(ii) $A_p = 10\log 50 = 17\,\text{dB}$ **Ans**

(iii) $A_p = 10\log 0.4 = -4\,\text{dB}$ **Ans**

(b)

(i) $25 = 10\log(\text{ratio})\,\text{dB}$

so $2.5 = \log(\text{ratio})$

therefore, $(\text{ratio}) = \text{antilog}\ 2.5 = 316$ times **Ans**

(ii) $8 = 10\log(\text{ratio})\,\text{dB}$

$0.8 = \log(\text{ratio})$

therefore, $(\text{ratio}) = \text{antilog}\ 0.8 = 6.3$ times **Ans**

(iii) $-15 = 10\log(\text{ratio})\,\text{dB}$

$-1.5 = \log(\text{ratio})$

therefore, $(\text{ratio}) = \text{antilog}\ (-1.5) = 0.032$ **Ans**

Although the decibel is defined in terms of a power ratio, it may also be used to express both voltage and current ratios, provided that certain conditions are met. These conditions are that the resistance of the load is the same as that of the source, i.e. the conditions for maximum power transfer. Consider such a system, whereby the two resistance values are R ohm, the input voltage is V_1 volt, and the output voltage is V_2 volt. Let the corresponding currents be I_1 and I_2 ampere.

$$P_1 = \frac{V_1^2}{R} \text{ watt and } P_2 = \frac{V_2^2}{R} \text{ watt}$$

$$\text{gain} = 10\log\left(\frac{V_2^2}{R} \times \frac{R}{V_1^2}\right)$$

$$= 10\log\left(\frac{V_2}{V_1}\right)^2$$

hence, voltage gain, $A_v = 20\log \frac{V_2}{V_1}$ dB

Also, using the fact that $P_1 = I_1^2R$ watt and $P_2 = I_2^2R$ watt it is left to the reader to verify that:

$$\text{current gain, } A_i = 20\log \frac{I_2}{I_1} \text{ dB} \tag{2.14}$$

It is shown in Volume 2, that a tuned circuit used as a pass-band or stop-band filter has a bandwidth. The same concept also applies to a.c. amplifiers. In the latter case, the frequency response curve for a voltage amplifier would be similar to that shown in Fig. 2.44. The bandwidth is defined as that range of frequencies over which the voltage gain is greater than, or equal to, $A_{vm}/\sqrt{2}$, where A_{vm} is the mid-frequency gain. A similar response curve for the amplifier current gain could also be plotted.

Now, power gain = voltage gain × current gain

$$\text{or, } A_p = A_v \times A_i$$

The cut-off frequencies, f_1 and f_2, define the bandwidth, and at these frequencies, the current and voltage gains will be:

$$A_v = \frac{A_{vm}}{\sqrt{2}} \quad \text{and} \quad A_i = \frac{A_{im}}{\sqrt{2}} \text{ respectively}$$

$$\text{thus } A_p = \frac{A_{vm}}{\sqrt{2}} \times \frac{A_{im}}{\sqrt{2}} = \frac{A_{pm}}{2}$$

These points on the response curve are therefore referred to as either the cut-off points, the half-power points, or the −3 dB points.

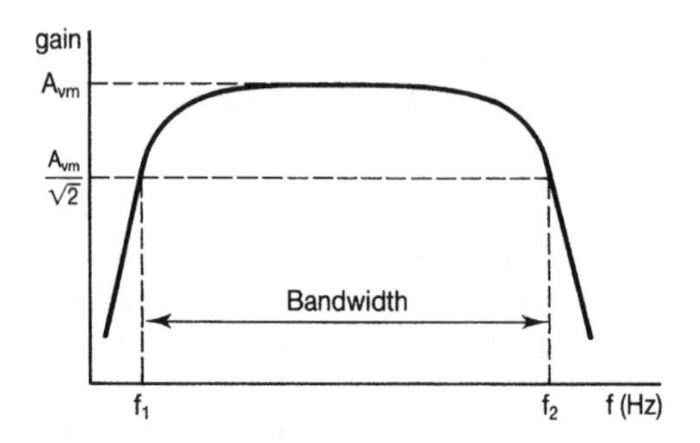

Fig. 2.44

Worked Example 2.17

Question

An amplifier is fed with a 50 mV, 200 μA signal. The amplifier has a voltage gain of 75 times and a current gain of 150 times. Determine (a) the voltage, current and power gains, expressed in dB, and (b) the output voltage, current and power. You may assume that input and output resistances are the same value.

Answer

$V_1 = 50 \times 10^{-3}$ V; $I_1 = 200 \times 10^{-6}$ A; $A_v = 75$; $A_i = 150$

(a) $A_v = 20 \log\ 75\,\text{dB} = 37.5\,\text{dB}$ **Ans**

$A_i = 20 \log\ 150\,\text{dB} = 43.5\,\text{dB}$ **Ans**

$A_p = A_v \times A_i = 75 \times 150 = 11\,250$ times

so, $A_p = 10 \log\ 11\,250\,\text{dB} = 40.5\,\text{dB}$ **Ans**

(b) $V_2 = V_1 \times A_v$ volt $= 50 \times 10^{-3} \times 75 = 3.75$ V **Ans**

$I_2 = I_1 \times A_i$ amp $= 200 \times 10^{-6} \times 150 = 30$ mA **Ans**

$P_2 = V_2 I_2$ watt $= 3.75 \times 0.03 = 112.5$ mW **Ans**

Worked Example 2.18

Question

An amplifier has a voltage gain of 10 times at a frequency of 150 Hz; 60 times between 2 kHz and 12 kHz; and 15 times at 35 kHz. Determine (a) the voltage gain, expressed in decibel, for each case, and (b) the voltage gain at the limits of its bandwidth.

Answer

(a) at 150 Hz: $A_v = 20 \log\ 10 = 20\,\text{dB}$ **Ans**

at mid-frequencies: $A_v = 20 \log\ 60 = 35.6\,\text{dB}$ **Ans**

at 35 kHz: $A_v = 20 \log\ 15 = 23.5\,\text{dB}$ **Ans**

Continued on p. 76

Worked Example 2.18 (*Continued*)

(b) The limits of the bandwidth occur when A_v is 3 dB down on the mid-frequency value, or $A_v = A_{vm}/\sqrt{2}$.

Therefore, $A_v = A_{vm} - 3 = 35.6 - 3 = 32.6\,\text{dB}$ **Ans**

$$\text{or, } A_v = \frac{A_{vm}}{\sqrt{2}} = \frac{60}{\sqrt{2}} = 42.426 \text{ times}$$

hence, $A_v = 20\log 42.426 = 32.6\,\text{dB}$ **Ans**

The decibel is defined in terms of a power *ratio*. Thus, the power output of a device cannot be quoted directly in dB. For example, to say that an amplifier has an output power of 20 dB is completely meaningless. In this case it only makes sense to say that the amplifier has a power *gain* of 20 dB. However, actual values of output power may be expressed in decibel form *provided* a known reference power input is used. In electronic work, this reference power is one milliwatt (1 mW), and powers expressed in this way are designated as dBm.

Considering such an amplifier, it could be said to have a power output of 20 dBm. This would mean that its actual output power was 0.1 W. This figure is obtained thus:

$$10\log\frac{P_2}{P_1} = 20\,\text{dBm}$$

$$\log\frac{P_2}{P_1} = 2$$

$$\frac{P_2}{P_1} = \text{antilog } 2 = 100$$

and since $P_1 = 1\,\text{mW}$, then $P_2\,100 \times 10^{-3} = 0.1\,\text{W}$ or 100 mW.

2.15 The Wheatstone Bridge Network

This is a network of interconnected resistors or other components, depending on the application. Although the circuit contains only one source of emf, it requires the application of a network theorem such as the Kirchhoff's method for its solution. A typical network, suitably labelled and with current flows identified is shown in Fig. 2.45.

Notice that although there are five resistors, the current law has been applied so as to minimise the number of 'unknown' currents to three. Thus only three simultaneous equations will be required for the solution, though there are seven possible loops to choose from. These seven loops are:

ABCDA; ADCA; ABDCA; ADBCA; ABDA; BCDB; and ABCDA

If you trace around these loops you will find that the last three do not include the source of emf, so for each of these loops the sum of the emfs will be ZERO! Up to a point it doesn't matter which three loops are chosen provided that at least one of them

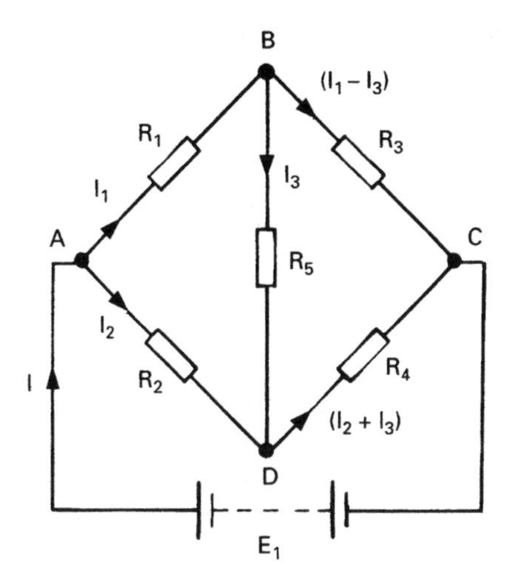

Fig. 2.45

includes the source. If you chose to use only the last three 'zero emf' loops you would succeed only in proving that zero equals zero!

The present level of study does not require you to solve simultaneous equations containing three unknowns. It is nevertheless good practice in the use of Kirchhoff's laws, and the seven equations for the above loops are listed below. In order for you to gain this practice it is suggested that you attempt this exercise before reading further, and compare your results with those shown below.

ABCA: $E_1 = I_1R_1 + (I_1 - I_3)R_3$

ADCA: $E_1 = I_2R_2 + (I_2 + I_3)R_4$

ABDCA: $E_1 = I_1R_1 + I_3R_5 + (I_2 + I_3)R_4$

ADBCA: $E_1 = I_2R_2 - I_3R_5 + (I_1 - I_3)R_3$

ABDA: $0 = I_1R_1 + I_3R_5 - I_2R_2$

BCDB: $0 = (I_1 - I_3)R_3 - (I_2 + I_3)R_4 - I_3R_5$

ABCDA: $0 = I_1R_1 + (I_1 - I_3)R_3 - (I_2 + I_3)R_4 - I_2R_2$

As a check that the current law has been correctly applied, consider junctions B and C:

current arriving at B $= I$

total current leaving $= I_1 + I_2$

so $I = I_1 + I_2$

current arriving at C $= (I_1 - I_3) + (I_2 + I_3)$

$= I_1 - I_3 + I_2 + I_3$

$= I_1 + I_2$

$= I$

Hence, current leaving battery = current returning to battery.

Worked Example 2.19

Question

For the bridge network shown in Fig. 2.46 calculate the current through each resistor, and the current drawn from the supply.

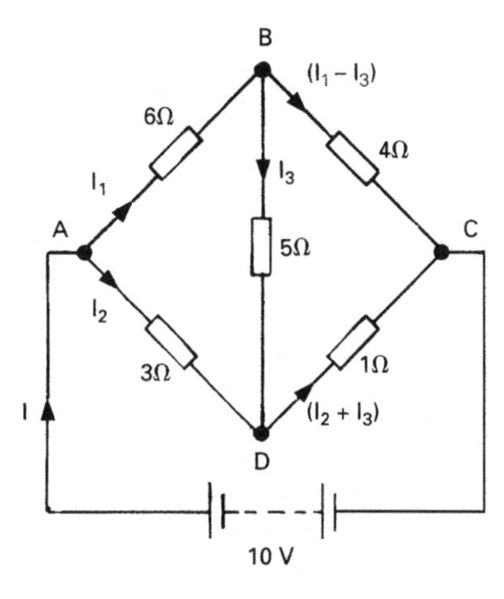

Fig. 2.46

Answer

The circuit is first labelled and the currents identified using the current law.

$$\underline{\text{ABDA}}: 0 = 6I_1 + 5I_3 - 3I_2$$
$$0 = 6I_1 - 3I_2 + 5I_3 \ldots\ldots\ldots\ldots\ldots\ldots [1]$$

$$\underline{\text{BDCB}}: 0 = 5I_3 + 1(I_2 + I_3) - 4(I_1 - I_3)$$
$$= 5I_3 + I_2 + I_3 - 4I_1 + 4I_3$$
$$0 = -4I_1 + I_2 + 10I_3 \ldots\ldots\ldots\ldots\ldots\ldots [2]$$

$$\underline{\text{ADCA}}: 10 = 3I_2 + 1(I_2 + I_3)$$
$$= 3I_2 + I_2 + I_3$$
$$10 = 4I_2 + I_3 \ldots\ldots\ldots\ldots\ldots\ldots [3]$$

Multiplying equation [1] by 2, equation [2] by 3 and then adding them

$$0 = 12I_1 - 6I_2 + 10I_3 \ldots\ldots\ldots\ldots\ldots\ldots [1] \times 2$$
$$0 = -12I_1 + 3I_2 + 30I_3 \ldots\ldots\ldots\ldots\ldots\ldots [2] \times 3$$
$$0 = -3I_2 + 40I_3 \ldots\ldots\ldots\ldots\ldots\ldots [4]$$

Multiplying equation [3] by 3, equation [4] by 4 and then adding them

$$30 = 12I_2 + 3I_3 \ldots\ldots [3] \times 3$$
$$0 = -12I_2 + 160I_3 \ldots\ldots [4] \times 4$$
$$30 = 163I_3$$
$$I_3 = \frac{30}{163} = 0.184 \text{ A } \textbf{Ans}$$

Substituting for I_3 in equation [3]

$$10 = 4I_2 + 0.184$$
$$4I_2 = 9.816$$
$$I_2 = \frac{9.816}{4} = 2.454 \text{ A } \textbf{Ans}$$

Substituting for I_3 and I_2 in equation [2]

$$0 = -4I_1 + 2.454 + 1.84$$
$$4I_1 = 4.294$$
$$I_1 = \frac{4.294}{4} = 1.074 \text{ A } \textbf{Ans}$$
$$I = I_1 + I_2$$
$$= 1.074 + 2.454$$
$$I = 3.529 \text{ A } \textbf{Ans}$$

Since all of the answers obtained are positive values then the currents will flow in the directions marked on the circuit diagram.

Worked Example 2.20

Question

If the circuit of Fig. 2.46 is now amended by simply changing the value of R_{DC} from 1 Ω to 2 Ω, calculate the current flowing through the 5 Ω resistor in the central limb.

Answer

The amended circuit diagram is shown in Fig 2.47.

ABDA: $0 = 6I_1 + 5I_3 - 3I_2$

$$0 = 6I_1 - 3I_2 + 5I_3 \ldots\ldots [1]$$

BDCB: $0 = 5I_3 + 2(I_2 + I_3) - 4(I_1 - I_3)$

$$= 5I_3 + 2I_2 + 2I_3 - 4I_1 + 4I_3$$
$$0 = -4I_1 + 2I_2 + 11I_3 \ldots\ldots [2]$$

Continued on p. 80

Worked Example 2.20 (*Continued*)

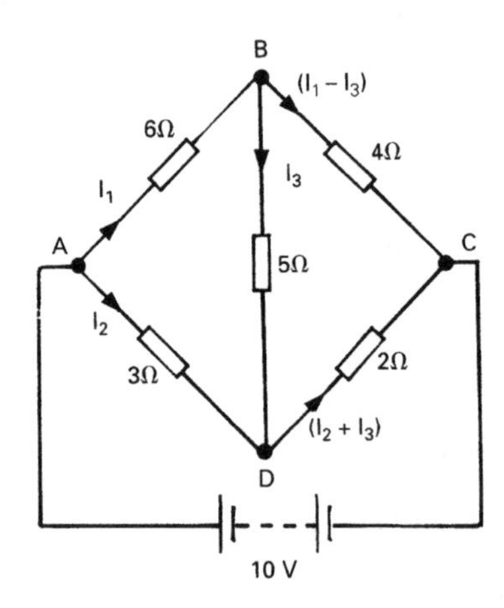

Fig. 2.47

Multiplying equation [1] by 2, equation [2] by 3 and adding them

$$0 = 12I_1 - 6I_2 + 10I_3 \ldots\ldots\ldots\ldots [1] \times 2$$

$$0 = -12I_1 + 6I_2 + 33I_3 \ldots\ldots\ldots\ldots [2] \times 3$$

$$0 = 43I_3$$

so $I_3 = 0\,\text{A}$ **Ans**

At first sight this would seem to be a very odd result. Here we have a resistor in the middle of a circuit with current being drawn from the source, yet no current flows through this particular resistor! Now, in any circuit, current will flow between two points only if there is a difference of potential between the two points. So we must conclude that the potentials at junctions B and D must be the same. Since junction A is a common point for both the 6 Ω and 3 Ω resistors, then the p.d. across the 6 Ω must be the same as that across the 3 Ω resistor. Similarly, since point C is common to the 4 Ω and 2 Ω resistors, then the p.d. across each of these must also be equal. This may be verified as follows.

Since I_3 is zero then the 5 Ω resistor plays no part in the circuit. In this case we can ignore its presence and re-draw the circuit as in Fig. 2.48.

Thus the circuit is reduced to a simple series/parallel arrangement that can be analysed using simple Ohm's law techniques.

$$R_{ABC} = R_1 + R_3 \text{ ohm} = 6 + 4$$

$$\text{so } R_{ABC} = 10\,\Omega$$

$$I_1 = \frac{E}{R_{ABC}} \text{ amp}$$

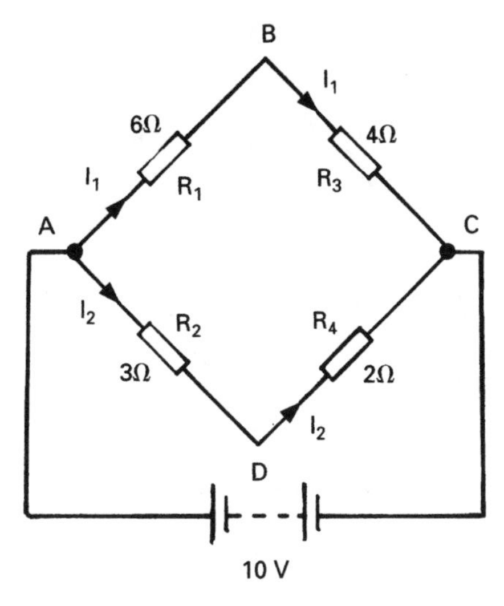

Fig. 2.48

$$\text{so } I_1 = \frac{10}{10} = 1\,\text{A}$$
$$V_{AB} = I_1 R_1 \text{ volt} = 1 \times 6 = 6\,\text{V}$$

$$\text{Similarly, } R_{ADC} = 3 + 2 = 5\,\Omega$$
$$I_2 = \frac{10}{5} = 2\,\text{A}$$
$$\text{and } V_{AD} = 2 \times 3 = 6\,\text{V}$$

Thus $V_{AB} = V_{AD} = 6\,\text{V}$, so the potentials at B and D are equal. In this last example, the values of R_1, R_2, R_3 and R_4 are such to produce what is known as the **balance condition** for the bridge. Being able to produce this condition is what makes the bridge circuit such a useful one for many applications in measurement systems. The value of resistance in the central limb has no effect on the balance conditions. This is because, at balance, zero current flows through it. In addition, the value of the emf also has no effect on the balance conditions, but will of course affect the values for I_1 and I_2. Consider the general case of a bridge circuit as shown in Fig. 2.49, where the values of resistors R_1 to R_4 are adjusted so that I_3 is zero.

> **Balance condition** refers to that condition when zero current flows through the central arm of the bridge circuit, due to a particular combination of resistor values in the four 'outer' arms of the bridge

$$V_{AB} = I_1 R_1 \text{ and } V_{AD} = I_2 R_2$$

but under the balance condition $V_{AB} = V_{AD}$

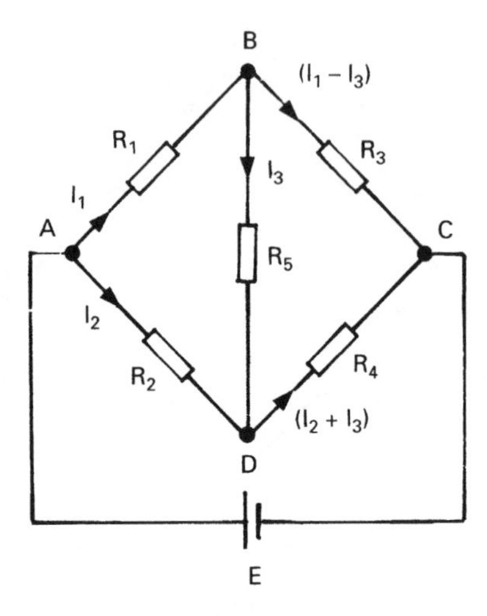

Fig. 2.49

so $I_1 R_1 = I_2 R_2$ .... [1]

Similarly, $V_{BC} = V_{DC}$

so $(I_1 - I_3) \cdot R_3 = (I_2 + I_3)R_4$

but, $I_3 = 0$, so current through $R_3 = I_1$
and current through $R_4 = I_2$, therefore

$I_1 R_3 = I_2 R_4$ .... [2]

Dividing equation [1] by equation [2]:

$$\frac{I_1 R_1}{I_1 R_3} = \frac{I_2 R_2}{I_2 R_4}$$

$$\text{so } \frac{R_1}{R_3} = \frac{R_2}{R_4}$$

This last equation may be verified by considering the values used in the previous example where $R_1 = 6\,\Omega$, $R_2 = 3\,\Omega$, $R_3 = 4\,\Omega$ and $R_4 = 2\,\Omega$.

$$\text{i.e. } \frac{6}{3} = \frac{4}{2}$$

So for balance, the ratio of the two resistors on the left-hand side of the bridge equals the ratio of the two on the right-hand side.

However, a better way to express the balance condition in terms of the resistor values is as follows. If the product of two diagonally opposite resistors equals the product of the other pair of diagonally opposite resistors, then the bridge is balanced, and zero current flows through the central limb

$$\text{i.e. } R_1 R_4 = R_2 R_3 \qquad (2.16)$$

and transposing equation (2.16) to make R_4 the subject we have

$$R_4 = \frac{R_2}{R_1} R_3 \tag{2.17}$$

Thus if resistors R_1, R_2 and R_3 can be set to known values, and adjusted until a sensitive current measuring device inserted in the central limb indicates zero current, then we have the basis for a sensitive resistance measuring device.

2.16 The Wheatstone Bridge Instrument

This is an instrument used for the accurate measurement of resistance over a wide range of resistance values. It comprises three arms, the resistances of which can be adjusted to known values. A fourth arm contains the 'unknown' resistance, and a central limb contains a sensitive microammeter (a galvanometer or 'galvo'). The general arrangement is shown in Fig. 2.50. Comparing this circuit with that of Fig. 2.49 and using equation (2.17), the value of the resistance to be measured (R_x) is given by

$$R_x = \frac{R_m}{R_d} R_v \text{ ohm}$$

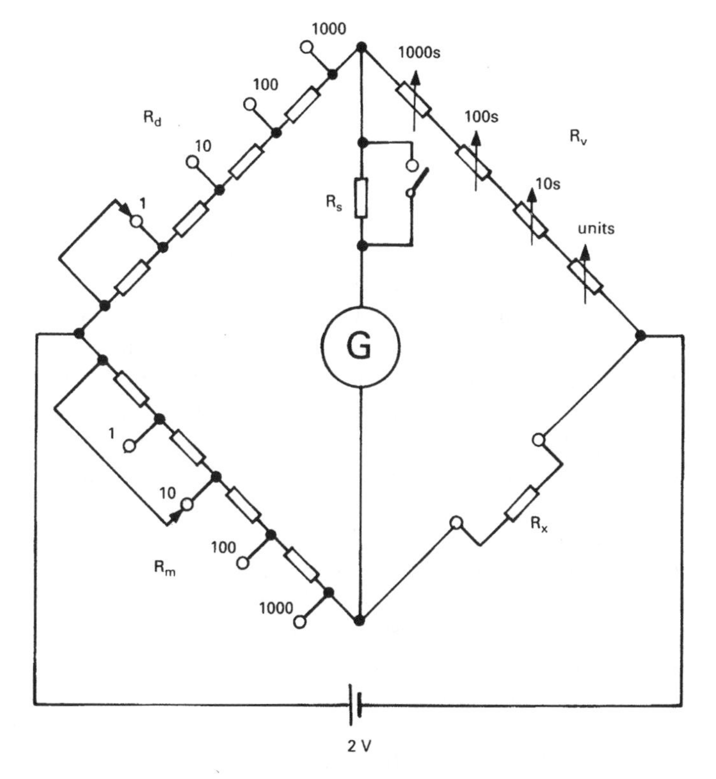

Fig. 2.50

R_m and R_d are known collectively as the ratio arms, where R_m is the multiplier and R_d is the divider arm. Both of these arms are variable in decade steps (i.e. 1, 10, 100, 1000). This does not mean that these figures represent actual resistance values, but they indicate the appropriate ratio between these two arms. Thus, if R_d is set to 10 whilst R_m is set to 1000, then the resistance value selected by the variable arm R_v is 'multiplied' by the ratio $1000/10 = 100$.

Worked Example 2.21

Question

Two resistors were measured using a Wheatstone Bridge, and the following results were obtained.

(a) $R_m = 1000$; $R_d = 1$; $R_v = 3502\,\Omega$

(b) $R_m = 1$; $R_d = 1000$; $R_v = 296\,\Omega$

For each case determine the value of the resistance being measured.

Answer

(a) $R_x = \dfrac{1000}{1} \times 3502 = 3.502\,\mathrm{M\Omega}$ **Ans**

(b) $R_x = \dfrac{1}{1000} \times 296 = 0.296\,\Omega$ **Ans**

From the above example it may be appreciated that due to the ratio arms, the Wheatstone Bridge is capable of measuring a very wide range of resistance values. The instrument is also very **accurate** because it is what is known as a null method of measurement. This term is used because no settings on the three arms are used to determine the value of R_x until the galvo (G) in the central limb indicates zero (null reading). Since the galvo is a very sensitive microammeter it is capable of indicating fractions of a microamp. Hence, the slightest imbalance of the bridge can be detected. Also, since the bridge is adjusted until the galvo indicates zero, then this condition can be obtained with maximum accuracy. The reason for this accuracy is that before any measurements are made (no current through the galvo) it is a simple matter to ensure that the galvo pointer indicates zero. Thus, only the sensitivity of the galvo is utilised, and its accuracy over the remainder of its scale is unimportant. Included in the central limb are a resistor and a switch. These are used to limit the galvo current to a value that will not cause damage to the galvo when the bridge is well off balance. When the ratio arms and the variable arm have been adjusted to give only a small deflection of the galvo pointer, the switch is then closed to bypass the swamp resistor R_s. This will revert the galvo to its maximum sensitivity for the final balancing using R_v. The bridge supply is normally provided by a 2 V cell as shown.

Do not confuse **accuracy** with sensitivity. For an instrument to be accurate it must also be sensitive. However, a sensitive instrument is not necessarily accurate. Sensitivity is the ability to react to small changes of the quantity being measured. Accuracy is to do with the closeness of the indicated value to the true value

2.17 The Slidewire Potentiometer

This instrument is used for the accurate measurement of small voltages. Like the Wheatstone Bridge, it is a null method of measurement since it also utilises the fact that no current can flow between points of equal potential. In its simplest form it comprises a metre length of wire held between two brass or copper blocks on a base board, with a graduated metre scale beneath the wire. Connected to one end of the wire is a contact, the other end of which can be placed at any point along the wire. A 2 V cell causes current to flow along the wire. This arrangement, including a voltmeter, is shown in Fig. 2.51. The wire between the blocks A and B must be of uniform cross-section and resistivity throughout its length, so that each millimetre of its length has the same resistance as the next. Thus it may be considered as a number of equal resistors connected in series between points A and B. In other words it is a continuous potential divider.

Let us now conduct an imaginary experiment. If the movable contact is placed at point A then both terminals of the voltmeter will be at the same potential, and it will indicate zero volts. If the contact is now moved to point B then the voltmeter will indicate 2 V. Consider now the contact placed at point C which is midway between A and B. In this case it is exactly halfway along our 'potential divider', so it will indicate 1 V. Finally, placing the contact at a point D (say 70 cm from A), the voltmeter will indicate 1.4 V. These results can be summarised by the statement that there is a uniform potential gradient along the wire. Therefore, the p.d. 'tapped off' by the moving contact, is in direct proportion to the distance travelled along the wire from point A. Since the source has an emf of 2 V and the wire is of 1 metre length, then the potential gradient must be 2 V/m. In general we can say that

$$V_{AC} = \frac{AC}{AB} E \text{ volt} \qquad (2.18)$$

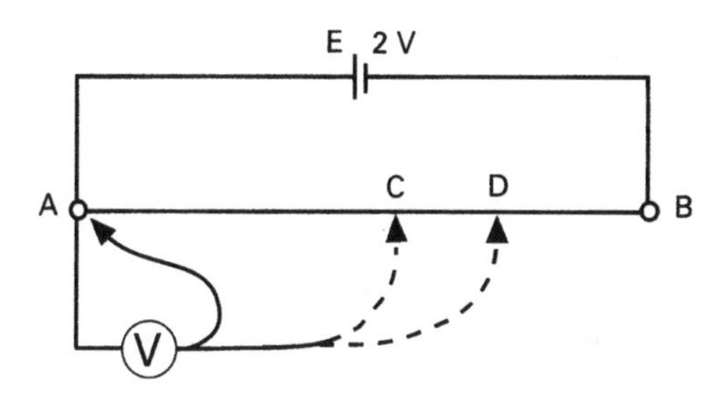

Fig. 2.51

where AC = distance travelled along wire

AB = total length of the wire

and E = the source voltage

Utilising these facts the simple circuit can be modified to become a measuring instrument, as shown in Fig. 2.52. In this case the voltmeter has been replaced by a galvo. The movable contact can be connected either to the cell to be measured or the standard cell, via a switch. Using this system the procedure would be as follows:

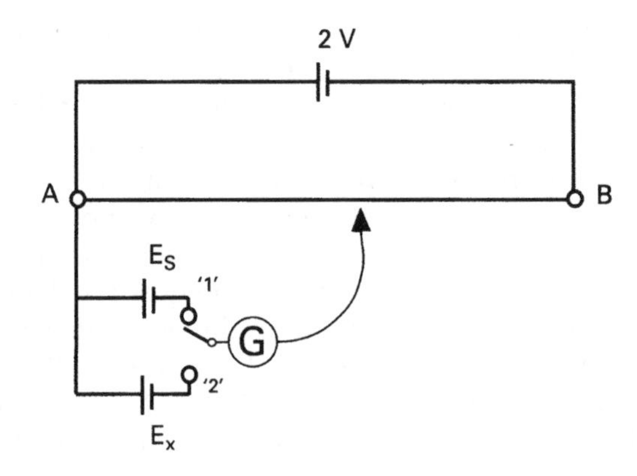

Fig. 2.52

1 The switch is moved to position '1' and the slider moved along the wire until the galvo indicates zero current. The position of the slider on the scale beneath the wire is then noted. This distance from A represents the emf E_s of the standard cell.

2 With the switch in position '2', the above procedure is repeated, whereby the distance along the scale represents the emf E_x of the cell to be measured.

3 The value of E_x may now be calculated from

$$E_x = \frac{\text{AD}}{\text{AC}} \times E_s$$

where AC represents the scale reading obtained for the standard cell and AD the scale reading for the unknown cell.

It should be noted that this instrument will measure the true emf of the cell since the readings are taken when the galvo carries zero current (i.e. no current is being drawn from the cell under test), hence there will be no p.d. due to its internal resistance.

Worked Example 2.22

Question

A slidewire potentiometer when used to measure the emfs of two cells provided balance conditions at scale settings of (a) 600 mm and (b) 745 mm. If the standard cell has an emf of 1.0186 V and a scale reading of 509.3 mm then determine the values for the two cell emfs.

Answer

Let l_s, l_1 and l_2 represent the scale readings for the standard cell and cells 1 and 2 respectively. Hence:

$l_s = 509.3$ mm $l_1 = 600$ mm; $l_2 = 745$ mm $E_s = 1.0186$ V

$$E_1 = \frac{l_1}{l_s} \times E_s \text{ volt}$$

$$= \frac{600}{509.3} \times 1.0186$$

$$E_1 = 1.2 \text{ V } \textbf{Ans}$$

$$E_2 = \frac{l_2}{l_s} \times E_s \text{ volt}$$

$$= \frac{745}{509.3} \times 1.0186$$

$$E_2 = 1.49 \text{ V } \textbf{Ans}$$

It is obviously inconvenient to have an instrument that needs to be one metre in length and requires the measurements of lengths along a scale. In the commercial version of the instrument the long wire is replaced by a series of precision resistors plus a small section of wire with a movable contact. The standard cell and galvo would also be built-in features. Also, to avoid the necessity for separate calculations, there would be provision for standardising the potentiometer. This means that the emf values can be read directly from dials on the front of the instrument.

Assignment Questions

1 Two 560 Ω resistors are placed in series across a 400 V supply. Calculate the current drawn.

2 When four identical hotplates on a cooker are all in use, the current drawn from a 240 V supply is 33 A. Calculate (a) the resistance of each hotplate, (b) the current drawn when only three plates are switched on. The hotplates are connected in parallel.

3 Calculate the total current when six 120 Ω torch bulbs are connected in parallel across a 9 V supply.

4 Two 20 Ω resistors are connected in parallel and this group is connected in series with a 4 Ω resistor. What is the total resistance of the circuit?

5 A 12 Ω resistor is connected in parallel with a 15 Ω resistor and the combination is connected in series with a 9 Ω resistor. If this circuit is supplied at 12 V, calculate (a) the total resistance, (b) the current through the 9 Ω resistor and (c) the current through the 12 Ω resistor.

6 For the circuit shown in Fig. 2.53 calculate the values for (a) the current through each resistor, (b) the p.d. across each resistor and (c) the power dissipated by the 20 Ω resistor.

7 Determine the p.d. between terminals E and F of the circuit in Fig. 2.54.

8 For the circuit of Fig. 2.55 calculate (a) the p.d. across the 8 Ω resistor, (b) the current through the 10 Ω resistor and (c) the current through the 12 Ω resistor.

9 Three resistors of 5 Ω, 6 Ω and 7 Ω respectively are connected in parallel. This combination is connected in series with another parallel combination of 3 Ω and 4 Ω. If the complete circuit is supplied from a 20 V source, calculate (a) the total resistance, (b) the total current, (c) the p.d. across the 3 Ω resistor and (d) the current through the 4 Ω resistor.

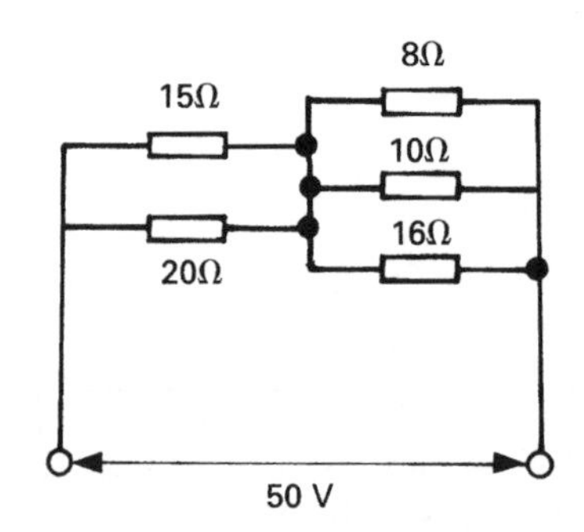

Fig. 2.53

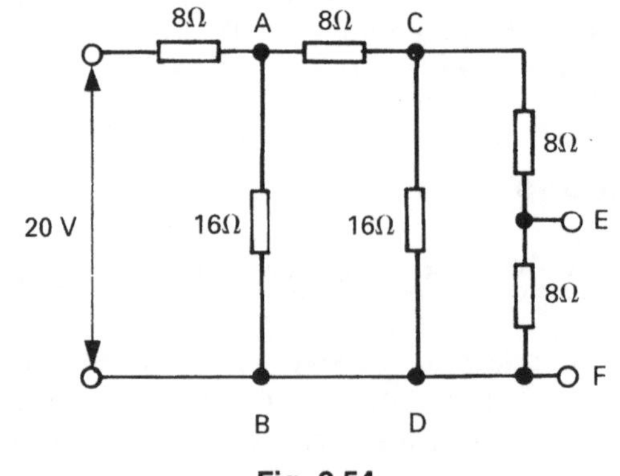

Fig. 2.54

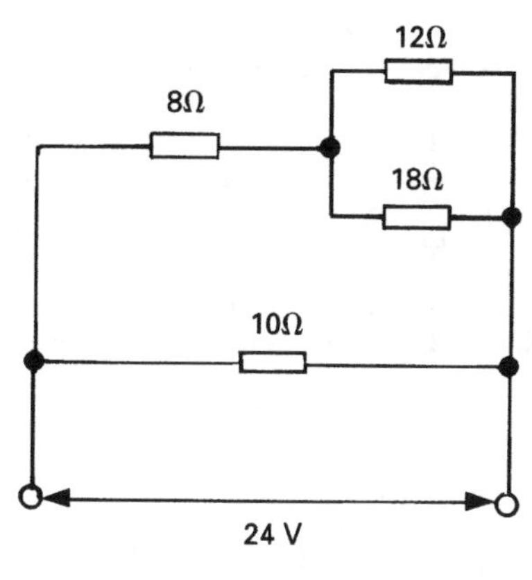

Fig. 2.55

10 Two resistors of 18 Ω and 12 Ω are connected in parallel and this combination is connected in series with an unknown resistor R_x. Determine the value of R_x if the complete circuit draws a current of 0.6 A from a 12 V supply.

11 Three loads, of 24 A, 8 A, and 12 A are supplied from a 200 V source. If a motor of resistance 2.4 Ω is also connected across the supply, calculate (a) the total resistance and (b) the total current drawn from the supply.

12 Two resistors of 15 Ω and 5 Ω connected in series with a resistor R_x and the combination is supplied from a 240 volt source. If the p.d. across the 5 Ω resistor is 20 V calculate the value of R_x.

13 A 200 V, 0.5 A lamp is to be connected in series with a resistor across a 240 V supply. Determine the resistor value required for the lamp to operate at its correct voltage.

14 A 12 Ω and a 6 Ω resistor are connected in parallel across the terminals of a battery of emf 6 V and internal resistance 0.6 Ω. Sketch the circuit diagram and calculate (a) the current drawn from the battery, (b) the terminal p.d. and (c) the current through the 6 Ω resistor.

15 An electric cooker element consists of two parts, each having a resistance of 18 Ω, which can be connected (a) in series, (b) in parallel, or (c) using one part only. Calculate the current drawn from a 240 V supply for each connection.

16 A cell of emf 2 V has an internal resistance 0.1 Ω. Calculate the terminal p.d. when (a) there is no load connected and (b) a 2.9 Ω resistor is connected across the terminals. Explain why these two answers are different.

17 A battery has a terminal voltage of 1.8 V when supplying a current of 9 A. This voltage rises to 2.02 V when the load is removed. Calculate the internal resistance.

18 Four resistors of values 10 Ω, 20 Ω, 40 Ω, and 40 Ω are connected in parallel across the terminals of a generator having an emf of 48 V and internal resistance 0.5 Ω. Sketch the circuit diagram and calculate (a) the current drawn from the generator, (b) the p.d. across each resistor and (c) the current flowing through each resistor.

19 Calculate the p.d. across the 3 Ω resistor shown in Fig. 2.56 given that V_{AB} is 11 V.

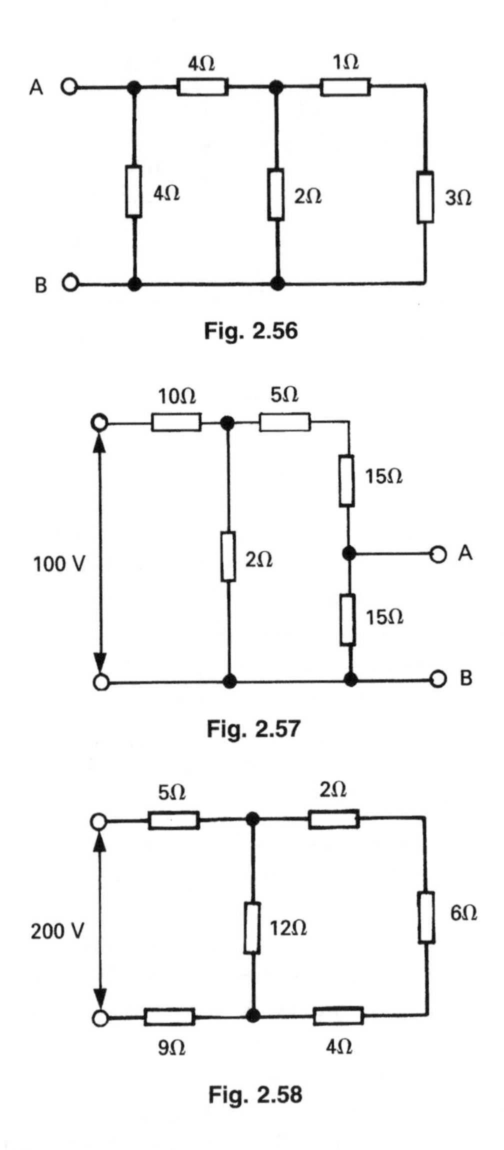

Fig. 2.56

Fig. 2.57

Fig. 2.58

20 Calculate the p.d. V_{AB} in Fig. 2.57.

21 For the network shown in Fig. 2.58, calculate (a) the total circuit resistance, (b) the supply current, (c) the p.d. across the 12 Ω resistor, (d) the total power dissipated in the whole circuit and (e) the power dissipated by the 12 Ω resistor.

22 A circuit consists of a 15 Ω and a 30 Ω resistor connected in parallel across a battery of internal resistance 2 Ω. If 60 W is dissipated by the 15 Ω resistor, calculate (a) the current in the 30 Ω resistor, (b) the terminal p.d. and emf of the battery, (c) the total energy dissipated in the external circuit in one minute and (d) the quantity of electricity through the battery in one minute.

23 Use Kirchhoff's laws to determine the three branch currents and the p.d. across the 5 Ω resistor in the network of Fig. 2.59.

24 Determine the value and direction of current in each branch of the network of Fig. 2.60, and the power dissipated by the 4 Ω load resistor.

25 Two batteries A and B are connected in parallel (positive to positive) with each other and this combination is connected in parallel with a battery C; this is in series with a 25 Ω resistor, the negative terminal of C being connected to the positive terminals of A and B. Battery A has an emf of 108 V and internal resistance 3 Ω, and the corresponding values for B are 120 V and 2 Ω. Battery C has an emf of 30 V and negligible internal resistance. Sketch the circuit and calculate (a) the value and direction of current in each battery and (b) the terminal p.d. of A.

26 For the circuit of Fig. 2.61 determine (a) the current supplied by each battery, (b) the current through the 15 Ω resistor and (c) the p.d. across the 10 Ω resistor.

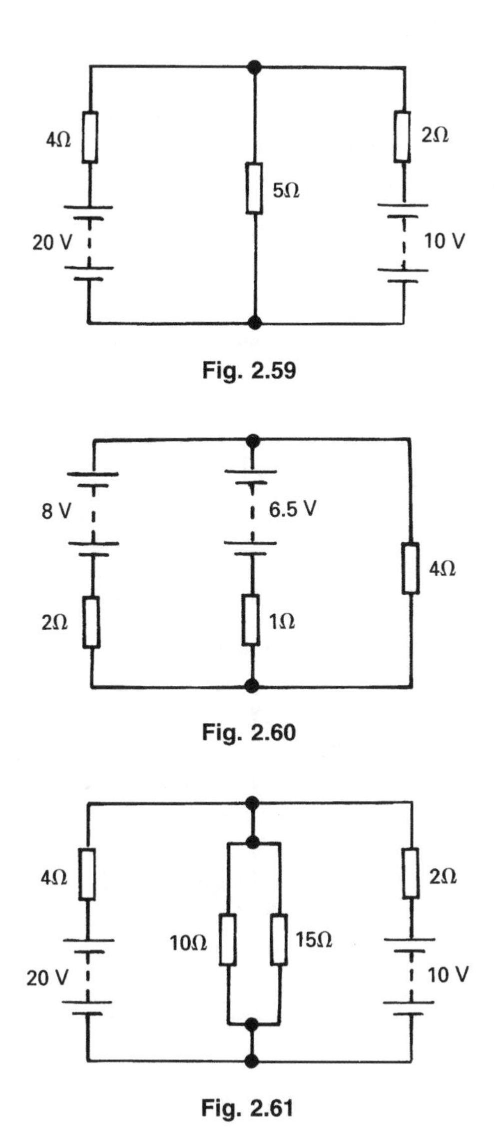

Fig. 2.59

Fig. 2.60

Fig. 2.61

27 For the network of Fig. 2.62, calculate the value and direction of all the branch currents and the p.d. across the 80 Ω load resistor.

28 Figure 2.63 shows a Wheatstone Bridge network. (a) For this network, write down (but do not solve) the loop equations for loops ABDA, ABCDA, ADCA, and CBDC, (b) to what value must the 2 Ω resistor be changed to ensure zero current through the 8 Ω resistor? (c) Under this condition, calculate the currents through and p.d.s across the other four resistors.

29 Three resistances were measured using a commercial Wheatstone Bridge, yielding the following results for the settings on the multiplying, dividing and variable arms. Determine the resistance value in each case.

R_d	1000	10	100
R_m	10	100	100
$R_v(\Omega)$	349.8	1685	22.5

30 The slidewire potentiometer instrument shown in Fig. 2.64 when used to measure the emf of cell E_x yielded the following results:

(a) galvo current was zero when connected to the standard cell and the movable contact was 552 mm from A;

(b) galvo current was zero when connected to E_s and the movable contact was 647 mm from A.

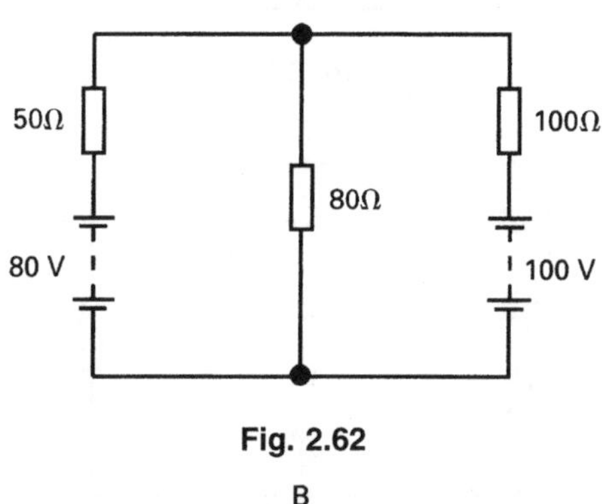

Fig. 2.62

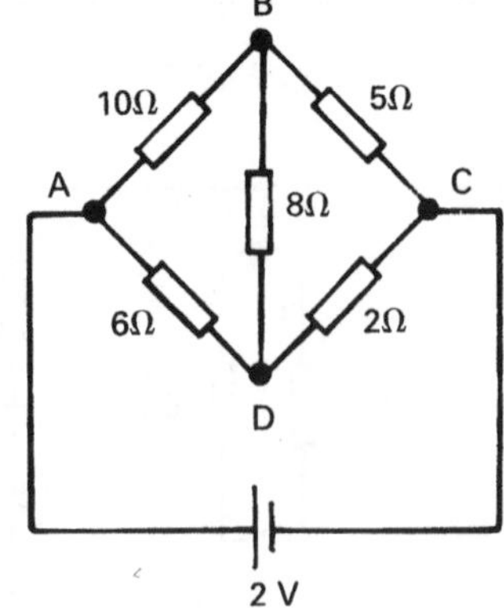

Fig. 2.63

Calculate the value of E_x, given $E_s = 1.0183\,\text{V}$.

It was found initially that E_x was connected the opposite way round and a balance could not be obtained. Explain this result.

31 Using the principle of superposition, determine the value of current flowing through the $6\,\Omega$ resistor of the circuit shown in Fig. 2.65

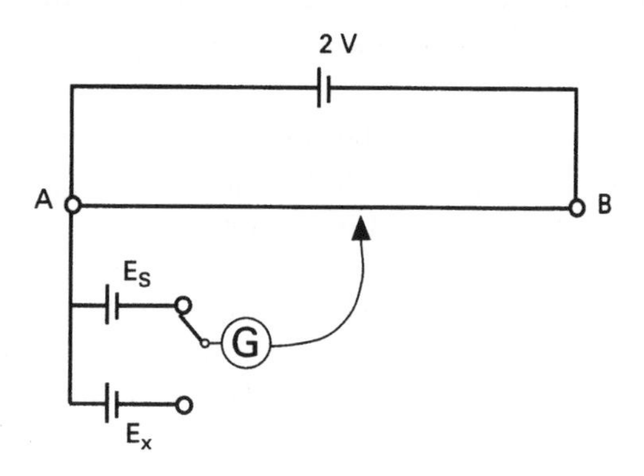

Fig. 2.64

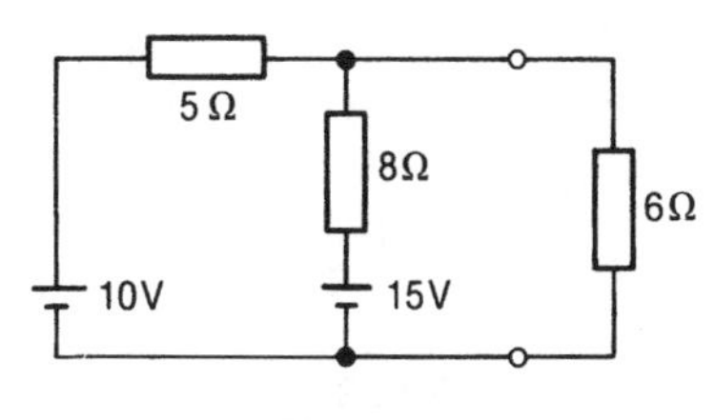

Fig. 2.65

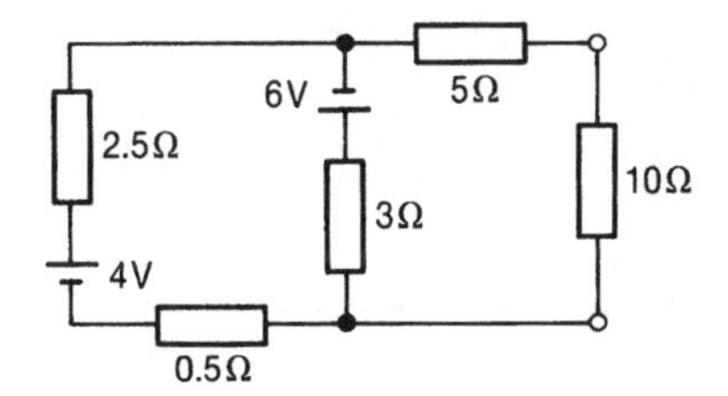

Fig. 2.66

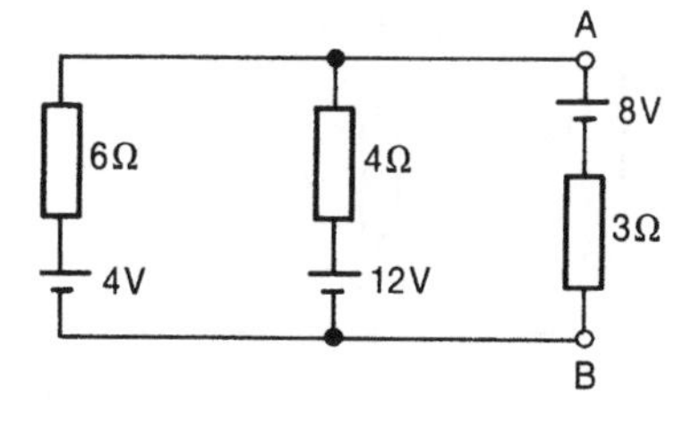

Fig. 2.67

32 By means of the superposition theorem, determine the value and direction of the current flow through the 10 Ω resistor in Fig. 2.66

33 Use the superposition theorem to calculate the p.d. and polarity between terminals A and B of the circuit shown in Fig. 2.67.

34 Determine the Thévenin equivalent circuit for the network to the left of terminals A and B in Fig. 2.68. Hence calculate the current flowing through the 6 Ω resistor.

35 Use Norton's theorem to calculate the current through the 6 Ω resistor of the circuit shown in Fig. 2.65.

36 Use Thévenin's theorem to solve Question 32 above.

37 For the network shown in Fig. 2.69, derive the Norton equivalent generator for the circuit to the left of terminals A and B, and hence calculate the p.d. across the 0.8 Ω resistor.

38 For the network of Fig. 2.70, determine (a) the Thévenin equivalent circuit, (b) the Norton equivalent circuit, and (c) the value of resistor, connected between terminals X and Y, which will result in the transfer of maximum power from the source.

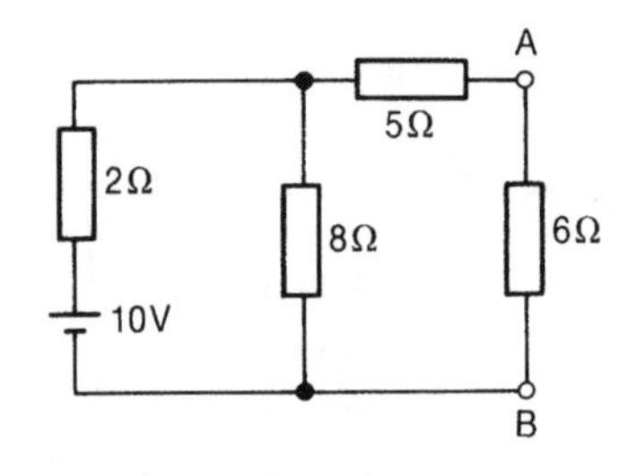

Fig. 2.68

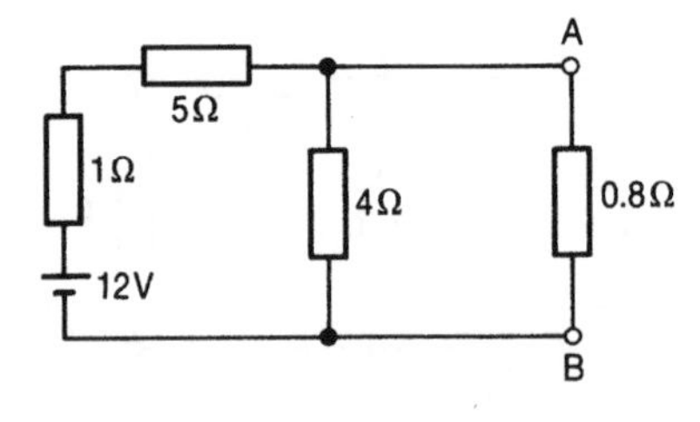

Fig. 2.69

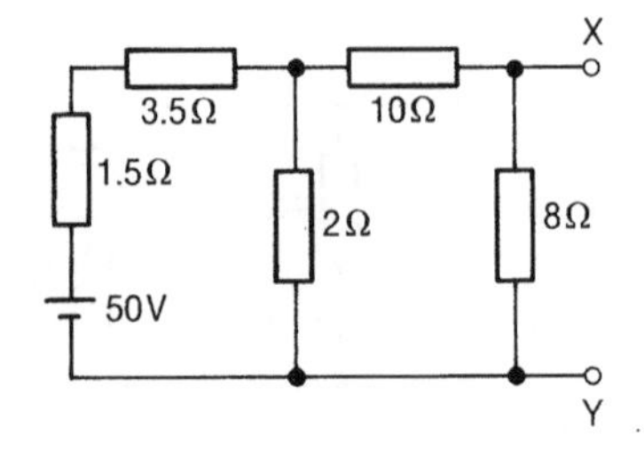

Fig. 2.70

39 Calculate the p.d. across, and current through the 3.9 Ω resistor in Fig. 2.71.

40 For the circuit of Fig. 2.72, (a) using either Thévenin's or Norton's theorem, determine the p.d. across, and current through R_L, (b) the value to which R_L must be changed to ensure maximum power transfer into the load, and (c) the value of this maximum power.

41 An a.c. source of internal impedance 600 Ω supplies a load of impedance 150 Ω. Determine the turns ratio of the matching transformer required for maximum power transfer.

42 Express the following power gains/losses in decibels:

(a) 25, (b) 0.36, (c) 255, (d) 10^6, (e) 0.022, (f) 85

43 An amplifier has a voltage gain of 25 dB. If its output voltage is 6 V, determine the corresponding input voltage.

44 An attenuator provides a voltage attenuation of −15 dB. Calculate the resulting output voltage if a 20 V signal is applied to its input terminals.

45 Express the following powers in dBm:

(a) 100 mW, (b) 2.5 W, (c) 150 μW, (d) 0.25 mW, (e) 10 W

46 Four circuit elements are connected in series. If the power gains/losses of the individual elements are −8 dB, +5 dB, +12.5 dB, and −2.4 dB, calculate the output power if 0 dBm is applied at the input.

47 An amplifier has a voltage gain of 8 at a frequency of 100 Hz, and a voltage gain of 45 at 5 kHz. Determine the relative gain at 5 kHz to that at 100 Hz, expressing your answer in dB.

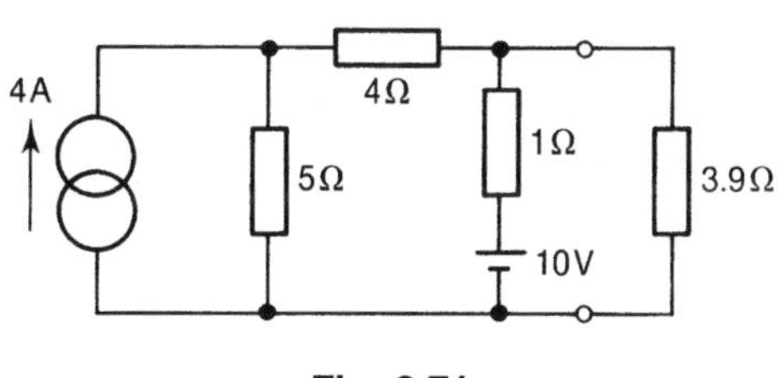

Fig. 2.71

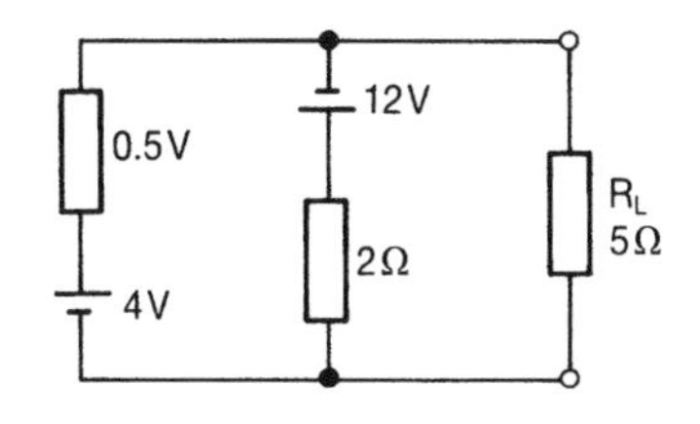

Fig. 2.72

Suggested Practical Assignments

Note: Component values and specific items of equipment when quoted here are only suggestions. Those used in practice will of course depend upon availability within a given institution.

Assignment 1

To investigate Ohm's law and Kirchhoff's laws as applied to series and parallel circuits.

Apparatus:

Three resistors of different values
1 × variable d.c. power supply unit (psu)
1 × ammeter
1 × voltmeter (DMM)

Method:

1 Connect the three resistors in series across the terminals of the psu with the ammeter connected in the same circuit. Adjust the current (as measured with the ammeter) to a suitable value. Measure the applied voltage and the p.d. across each resistor. Note these values and compare the p.d.s to the theoretical (calculated) values.

2 Reconnect your circuit so that the resistors are now connected in parallel across the psu. Adjust the psu to a suitable voltage and measure, in turn, the current drawn from the psu and the three resistor currents. Note these values and compare to the theoretical values.

3 Write an assignment report and in your conclusions justify whether the assignment confirms Ohm's law and Kirchhoff's laws, allowing for experimental error and resistor tolerances.

Assignment 2

To investigate the application of Kirchhoff's laws to a network containing more than one source of emf.

Apparatus:

2 × variable d.c. psu
3 × different value resistors
1 × ammeter
1 × voltmeter (DMM)

Method:

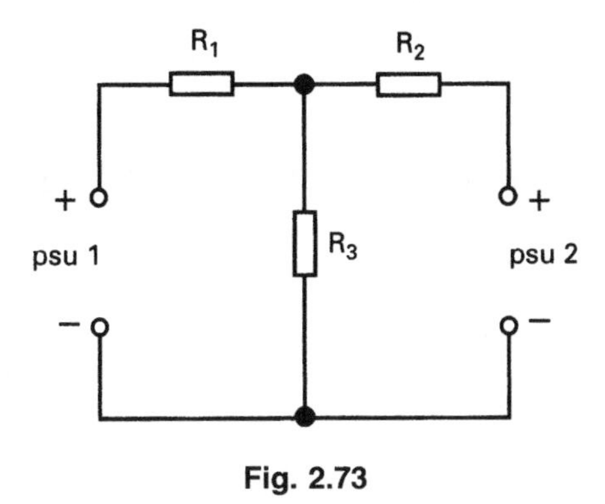

Fig. 2.73

1 Connect the circuit as shown in Fig. 2.73. Set psu 1 to 2 V and psu 2 to 4 V. Measure, in turn, the current in each limb of the circuit, and the p.d. across each resistor. For each of the three possible loops in the circuit compare the sum of the p.d.s measured with the sum of the emfs. Carry out a similar exercise regarding the three currents.
2 Reverse the polarity of psu 2 and repeat the above.
3 Write the assignment report and in your conclusions justify whether or not Kirchhoff's laws have been verified for the network.

Assignment 3

To investigate potential and current dividers.

Apparatus:

2 × decade resistance boxes
1 × ammeter
1 × voltmeter (DMM)
1 × d.c. psu

Method:

1 Connect the resistance boxes in series across the psu. Adjust one of them (R_1) to 3 kΩ and the other (R_2) to 7 kΩ. Set the psu to 10 V and measure the p.d. across each resistor. Compare the measured values with those predicted by the voltage divider theory.
2 Reset both R_1 and R_2 to two or more different values and repeat the above procedure.
3 Reconnect the two resistance boxes in parallel across the psu and adjust the current drawn from the psu to 10 mA. Measure the current flowing through each resistance and compare to those values predicted by the current division theory.

4 Repeat the procedure of 3 above for two more settings of R_1 and R_2, but let one of these settings be such that $R_1 = R_2$.

Assignment 4

To make resistance measurements using a Wheatstone Bridge.

Apparatus:

1 × commercial form of Wheatstone Bridge
3 × decade resistance boxes
10 Ω, 6.8 kΩ, and 470 kΩ resistors
1 × centre-zero galvo
1 × d. c. psu

Method:

1 Using the decade boxes, galvo and psu (set to 2 V) connect your own Wheatstone Bridge circuit and measure the three resistor values.
2 Use the commercial bridge to re-measure the resistors and compare the results obtained from both methods.

Assignment 5

Use a slidewire potentiometer to measure the emf of a number of primary cells (nominal emf no more than 1.5 V).

Assignment 6

To investigate the maximum power theorem.

Apparatus:

1 × a.c. signal generator (impedance 600 Ω)
1 × decade resistance box
1 × wattmeter

Method:

1 Connect the resistance box to the 600 Ω output terminals of the signal generator, with the wattmeter connected to measure the power dissipated.
2 Set the generator output to 10 V, at a frequency of 1 kHz.
3 Set the resistance to 100 Ω, and record the power dissipated.
4 Increase the resistance, in 50 Ω steps, up to 1 kΩ, noting the power dissipation at each step.

5 Plot a graph of power dissipation versus load resistance, and from your graph, determine the value of load resistance resulting in maximum load power.

6 Submit an assignment report.

Supplementary Worked Examples

Example 1: A 10 Ω resistor, a 20 Ω resistor and a 30 Ω resistor are connected (a) in series, and then (b) in parallel with each other. Calculate the total resistance for each of the two connections.

Solution: $R_1 = 10\,\Omega;\ R_2 = 20\,\Omega;\ R_3 = 30\,\Omega$

(a)
$$R = R_1 + R_2 + R_3 \text{ ohm}$$
$$= 10 + 20 + 30$$
$$\text{so, } R = 60\,\Omega \textbf{ Ans}$$

(b)
$$\frac{1}{R} = \frac{1}{R_1} + \frac{1}{R_2} + \frac{1}{R_3} \text{ siemen}$$
$$= \frac{1}{10} + \frac{1}{20} + \frac{1}{30} = 0.1 + 0.05 + 0.033$$
$$\text{so, } R = \frac{1}{0.183} = 5.46\,\Omega \textbf{ Ans}$$

Alternatively,

$$\frac{1}{R} = \frac{1}{10} + \frac{1}{20} + \frac{1}{30} = \frac{6+3+2}{60}$$
$$= \frac{11}{60}\text{ S}$$
$$\text{so, } R = \frac{60}{11} = 5.46\,\Omega \textbf{ Ans}$$

Example 2: For the circuit shown in Fig. 2.74, calculate (a) the current drawn from the supply, (b) the current through the 6 Ω resistor, and (c) the power dissipated by the 5.6 Ω resistor.

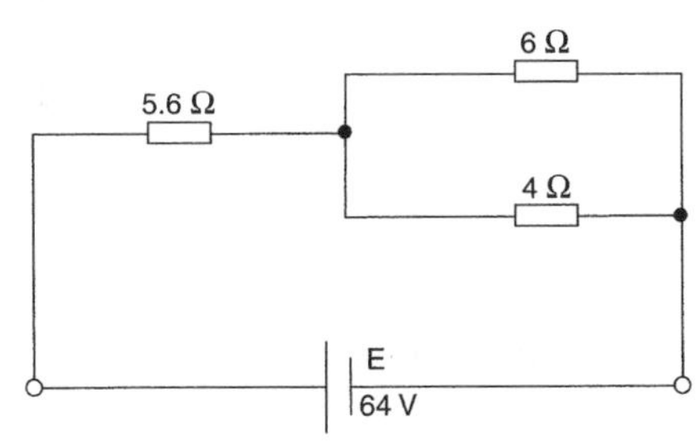

Fig. 2.74

The first step in the solution is to sketch and label the circuit diagram, clearly showing all currents flowing and identifying each part of the circuit as shown in Fig. 2.75. Also note that since there is no mention of internal resistance it may be assumed that the source of emf is ideal.

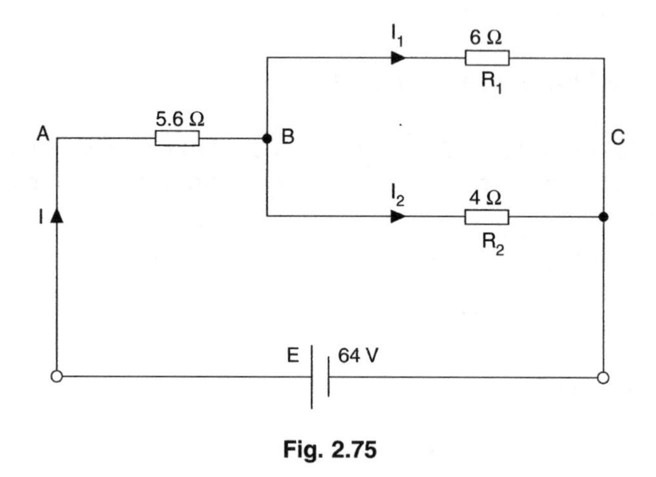

Fig. 2.75

(a) To determine the current I drawn from the battery we need to know the total resistance R_{AC} of the circuit.

$$R_{BC} = \frac{6 \times 4}{6 + 4} \text{ (using } \frac{\text{product}}{\text{sum}} \text{ for two resistors in parallel)}$$

$$= \frac{24}{10}$$

so $R_{BC} = 2.4\,\Omega$

The original circuit may now be redrawn as in Fig 2.76.

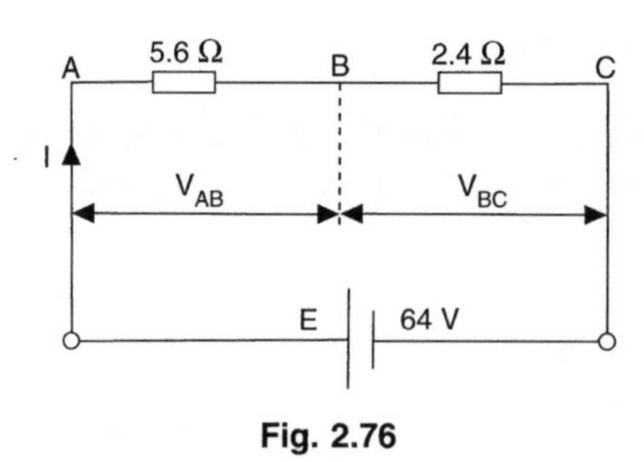

Fig. 2.76

$$R_{AC} = R_{AB} + R_{BC} \text{ ohm (resistors in series)}$$

$$= 5.6 + 2.4$$

so $R_{AC} = 8\,\Omega$

$$I = \frac{E}{R_{AC}} \text{ amp} = \frac{64}{8}$$

so $I = 8\,\text{A}$ **Ans**

(b) To find the current I_1 through the $6\,\Omega$ resistor we may use either of two methods. Both of these are now demonstrated.

p.d. method:

$$V_{BC} = IR_{BC}\text{ volt (Fig. 2.76)}$$
$$= 8 \times 2.4$$
$$\text{so, } V_{BC} = 19.2\,\text{V}$$
$$I_1 = \frac{V_{BC}}{R_1}\text{ amp (Fig. 2.75)}$$
$$= \frac{19.2}{6}$$
$$\text{so, } I_1 = 3.2\,\text{A } \textbf{Ans}$$

This answer may be checked as follows:

$$I_2 = \frac{V_{BC}}{R_2}\text{ amp}$$
$$= \frac{19.2}{4} = 4.8\,\text{A}$$
$$\text{and since } I = I_1 + I_2 = 3.2 + 4.8 = 8\,\text{A}$$

which agrees with the value found in (a).

current division method:

Considering Fig. 2.75, the current I splits into the components I_1 and I_2 according to the ratio of the resistor values. *However*, you must bear in mind that the *larger* resistor carries the *smaller* proportion of the total current.

$$I_1 = \frac{R_2}{R_1 + R_2} \times I\text{ amp}$$
$$= \frac{4}{6+4} \times 8$$
$$\text{so, } I_1 = 3.2\,\text{A } \textbf{Ans}$$

(c)

$$P_{AB} = I^2 R_{AB}\text{ watt}$$
$$= 8^2 \times 5.6$$
$$\text{so, } P_{AB} = 358.4\,\text{W } \textbf{Ans}$$

Alternatively, $P_{AB} = V_{AB}\, I$ watt

where $V_{AB} = E - V_{BC}$ volt $= 64 - 19.2 = 44.8\,\text{V}$

$$P_{AB} = 44.8 \times 8$$
$$\text{so, } P_{AB} = 358.4\,\text{W } \textbf{Ans}$$

Example 3: A source of emf 24 V and internal resistance 2 Ω supplies an external circuit which consists of a 5 Ω resistor connected in series with three paralleled resistors of 12 Ω, 10 Ω, and 30 Ω respectively. Calculate (a) the current drawn from the battery, (b) the p.d. across the 30 Ω resistor, (c) the current through the 12 Ω resistor, and (d) the power dissipated by the 10 Ω resistor.

Solution: A labelled circuit diagram is sketched as shown in Fig. 2.77.

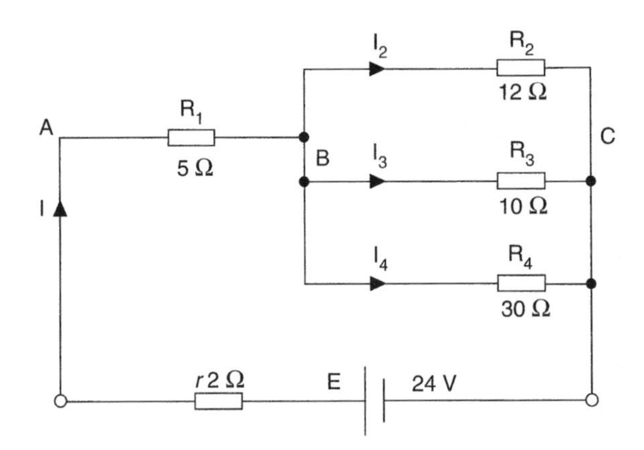

Fig. 2.77

(a)
$$\frac{1}{R_{BC}} = \frac{1}{R_1} + \frac{1}{R_2} + \frac{1}{R_3} \text{ siemen}$$
$$= \frac{1}{12} + \frac{1}{10} + \frac{1}{30} = \frac{5+6+2}{60}$$
$$\text{so, } R_{BC} = \frac{60}{13} = 4.615\,\Omega$$

The original circuit may now be redrawn as in Fig. 2.78.

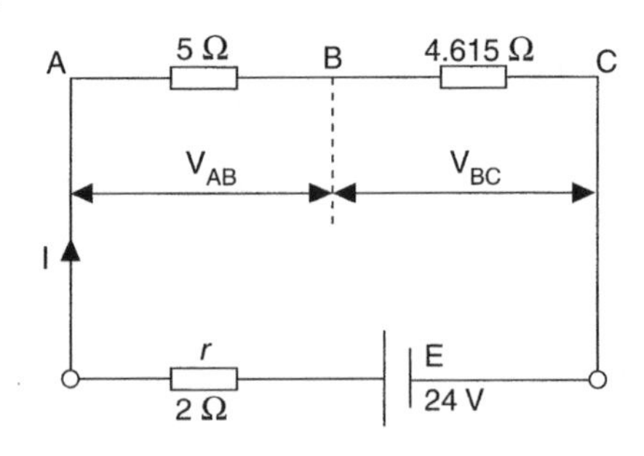

Fig. 2.78

$$\text{Total resistance, } R = r + R_{AB} + R_{BC} \text{ ohm}$$
$$= 2 + 5 + 4.615$$
$$\text{so, } R = 11.615\,\Omega$$

$$I = \frac{E}{R} \text{ amp}$$

$$= \frac{24}{11.615}$$

so, $I = 2.066\,\text{A}$ **Ans**

(b) The p.d. across R_2, R_3, and R_4 is the *same*, namely V_{BC}

$$V_{BC} = IR_{BC} \text{ volt}$$

$$= 2.066 \times 4.615$$

so, $V_{BC} = \text{p.d. across } 30\,\Omega = 9.54\,\text{V}$ **Ans**

(c) Current through $12\,\Omega$ resistor is I_2, where

$$I_2 = \frac{V_{BC}}{R_2} \text{ amp}$$

$$= \frac{9.54}{12}$$

so, $I_2 = 0.795\,\text{A}$ **Ans**

(d) $$P_{10\,\Omega} = \frac{V_{BC}^2}{R_3} \text{ watt}$$

$$= \frac{9.54^2}{10}$$

so, $P_{10\,\Omega} = 9.1\,\text{W}$ **Ans**

Example 4: For the circuit shown in Fig. 2.79, determine (a) the current through the 12 Ω resistor, (b) the p.d. across the 30 Ω resistor, and (c) the total power taken from the supply.

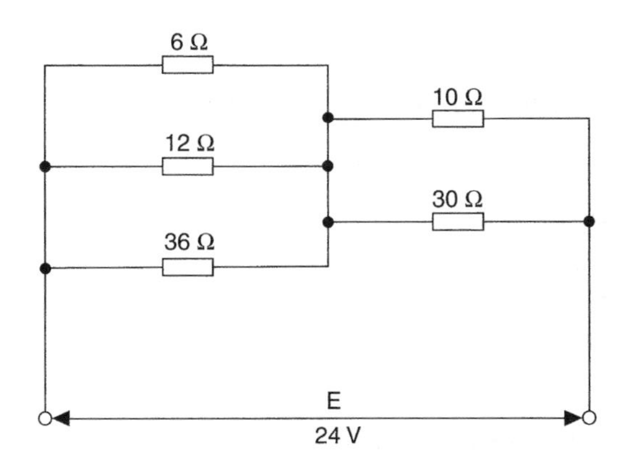

Fig. 2.79

Solution: The circuit diagram is sketched and labelled as shown in Fig. 2.80.

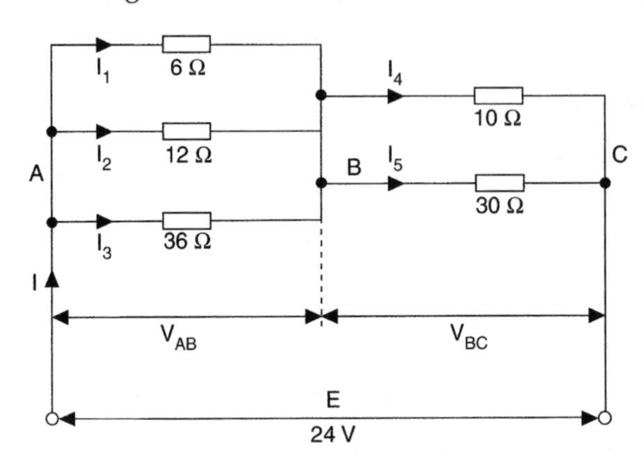

Fig. 2.80

$$\frac{1}{R_{AB}} = \frac{1}{6} + \frac{1}{12} + \frac{1}{36}\ \text{S}$$

$$= \frac{6+3+1}{36}$$

$$\text{so, } R_{AB} = \frac{36}{10} = 3.6\,\Omega$$

$$R_{BC} = \frac{10 \times 30}{10 + 30} = \frac{300}{40}$$

$$\text{so, } R_{BC} = 7.5\,\Omega$$

$$R_{AC} = R_{AB} + R_{BC}\ \text{ohm} = 7.5 + 3.6$$

$$\text{so, } R_{AC} = 11.1\,\Omega$$

$$I = \frac{E}{R_{AC}}\ \text{amp} = \frac{24}{11.1}$$

$$\text{so, } I = 2.162\ \text{A}$$

Current through $12\,\Omega$ resistor is I_2 and the p.d. across it is V_{AB}.

$$\text{Now, } V_{AB} = IR_{AB}\ \text{volt} = 2.162 \times 3.6$$

$$\text{so, } V_{AB} = 7.784\ \text{V}$$

$$\text{and } I_2 = \frac{V_{AB}}{R_2}\ \text{amp} = \frac{7.78}{12}$$

$$\text{so, } I_2 = 0.65\ \text{A}\ \textbf{Ans}$$

(b) p.d. across $30\,\Omega$ resistor is V_{BC} volt, where

$$V_{BC} = E - V_{AB}\ \text{volt} = 24 - 7.784$$

$$\text{so, } V_{BC} = 16.22\ \text{V}\ \textbf{Ans}$$

(c) Circuit power, $P = VI$ watt $= 24 \times 2.162$

so, $P = 51.9\,\text{W}$ **Ans**

Example 5: For the circuit shown in Fig 2.81, use Kirchhoff's Laws to calculate (a) the current flowing in each branch of the circuit, and (b) the p.d. across the 5Ω resistor.

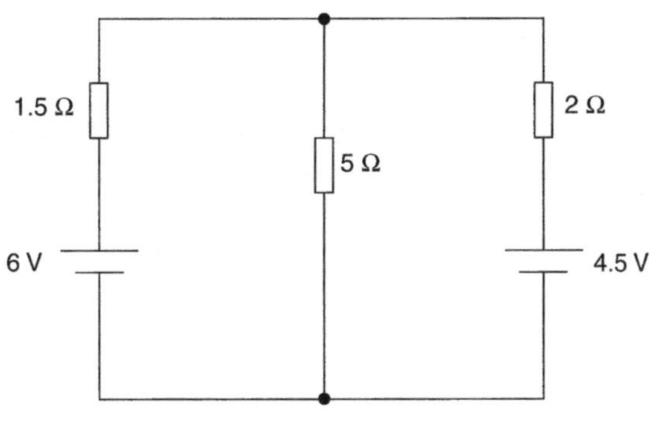

Fig. 2.81

Solution: Firstly the circuit is sketched and labelled and currents identified using Kirchhoff's current law. This is shown in Fig. 2.82.

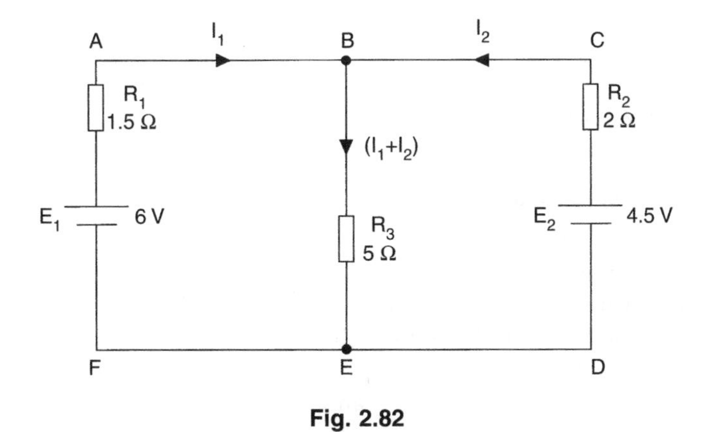

Fig. 2.82

(a) We can now consider three loops in the circuit and write down the corresponding equations using Kirchhoff's voltage law:

ABEFA: $E_1 = I_1R_1 + (I_1 + I_2)R_3$ volt

$$6 = 1.5I_1 + 5(I_1 + I_2) = 1.5I_1 + 5I_1 + 5I_2$$

so, $6 = 6.5\,I_1 + 5I_2$ .... [1]

CBEDC: $E_2 = I_2R_2 + (I_1 + I_2)R_3$ volt

$$4.5 = 2I_2 + 5(I_1 + I_2) = 2I_2 + 5I_1 + 5I_2$$

so, $4.5 = 5I_1 + 7I_2 \ldots\ldots\ldots\ldots [2]$

ABCDEFA: $E_1 - E_2 = I_1R_1 - I_2R_2$ volt

$$6 - 4.5 = 1.5I_1 - 2I_2$$

so, $1.5 = 1.5I_1 - 2I_2 \ldots\ldots\ldots\ldots [3]$

Now, any pair of these three equations may be used to solve the problem, using the technique of simultaneous equations. We shall use equations [1] and [3] to eliminate the unknown current I_2, and hence obtain a value for current I_1. To do this we can multiply [1] by 2 and [3] by 5, and then add the two modified equations together, thus:

$12 = 13I_1 + 10I_2 \ldots\ldots\ldots\ldots [1] \times 2$

$7.5 = 7.5I_1 - 10I_2 \ldots\ldots\ldots\ldots [3] \times 5$

$19.5 = 20.5I_1$

hence, $I_1 = \dfrac{19.5}{20.5} = 0.951$ A **Ans**

Substituting this value for I_1 into equation [3] gives:

$$1.5 = (1.5 \times 0.951) - 2I_2$$

$$1.5 = 1.427 - 2I_2$$

hence, $2I_2 = 1.427 - 1.5 = -0.0732$

and $I_2 = -0.0366$ A **Ans**

Note: The minus sign in the answer for I_2 indicates that this current is actually flowing in the opposite direction to that marked in Fig. 2.82. This means that battery E_1 is both supplying current to the $5\,\Omega$ resistor *and charging* battery E_2.

Current through $5\,\Omega$ resistor $= I_1 + I_2$ amp $= 0.951 + (-0.0366)$

so current through $5\,\Omega$ resistor $= 0.951 - 0.0366 = 0.914$ A **Ans**

(b) To obtain the p.d. across the $5\,\Omega$ resistor we can either subtract the p.d. (voltage drop) across R_1 from the emf E_1, or *add* the p.d. across R_2 to emf E_2, because E_2 is being *charged*. A third alternative is to multiply R_3 by the current flowing through it. All three methods will be shown here, and, provided that the same answer is obtained each time, the correctness of the answers obtained in part (a) will be confirmed.

$$V_{BE} = E_1 - I_1R_1 \text{ volt} = 6 - (0.951 \times 1.5)$$
$$= 6 - 1.4265$$
$$\text{so, } V_{BE} = 4.574\,\text{V } \textbf{Ans}$$
$$\text{OR: } V_{BE} = E_2 + I_2R_2 \text{ volt} = 4.5 + (0.0366 \times 2)$$
$$= 4.5 + 0.0732$$
$$\text{so, } V_{BE} = 4.573\,\text{V } \textbf{Ans}$$
$$\text{OR: } V_{BE} = (I_1 + I_2)R_3 \text{ volt} = 0.914 \times 5$$
$$\text{so, } V_{BE} = 4.57\,\text{V } \textbf{Ans}$$

The very small differences between these three answers is due simply to rounding errors, and so the answers to part (a) are verified as correct.

Example 6: A Wheatstone Bridge type circuit is shown in Fig. 2.83. Determine (a) the p.d. between terminals B and D, and (b) the value to which R_4 must be adjusted in order to reduce the current through R_3 to zero (balance the bridge).

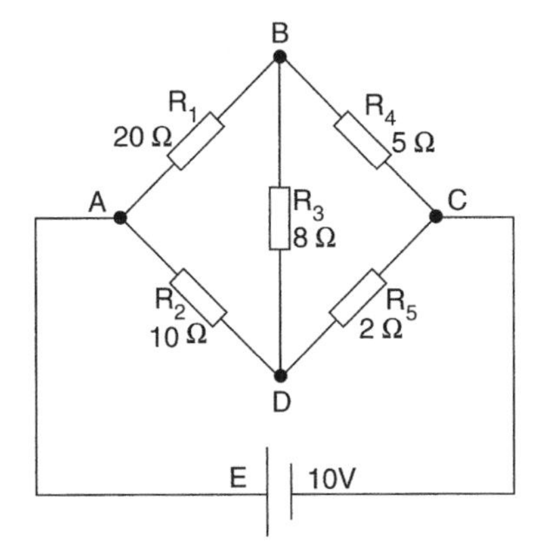

Fig. 2.83

Solution: The circuit is sketched and currents marked, applying Kirchhoff's current law, as shown in Fig. 2.84.

Kirchhoff's voltage law is now applied to any three loops. Note that as in this case there are three unknowns (I_1, I_2, and I_3) then we must have at least three equations in order to solve the problem.

$$\textbf{ABDA: } 0 = 20I_1 + 8I_3 - 10I_2$$
$$\text{so, } 0 = 20I_1 - 10I_2 + 8I_3 \ldots\ldots\ldots\ldots\ldots\ldots [1]$$

$$\textbf{BDCB: } 0 = 8I_3 + 2(I_2 + I_3) - 5(I_1 - I_3)$$
$$= 8I_3 + 2I_2 + 2I_3 - 5I_1 + 5I_3$$
$$\text{so, } 0 = -5I_1 + 2I_2 + 15I_3 \ldots\ldots\ldots\ldots\ldots\ldots [2]$$

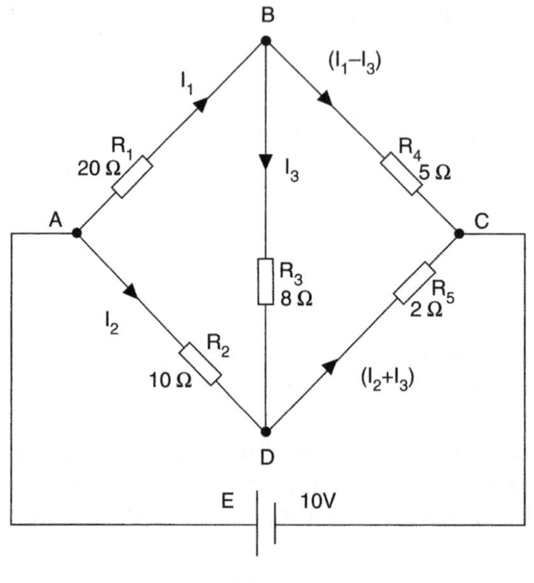

Fig. 2.84

ADCA: $10 = 10I_2 + 2(I_2 + I_3)$

$$= 10I_2 + 2I_2 + 2I_3$$

so, $10 = 12I_2 + 2I_3$ .... [3]

Using equations [1] and [2] to eliminate I_1 we have:

$$0 = 20I_1 - 10I_2 + 8I_3 \quad \text{.......... [1]}$$

$$0 = -20I_1 + 8I_2 + 60I_3 \quad \text{.......... [2]} \times 4$$

and adding, $0 = -2I_2 + 68I_3$ .... [4]

and now using equations [3] and [4] we can eliminate I_2 as follows:

$$10 = 12I_2 + 2I_3 \quad \text{.......... [3]}$$

$$0 = -12I_2 + 408I_3 \quad \text{.......... [4]} \times 6$$

$$10 = 410I_3$$

and $I_3 = \frac{10}{410} = 0.0244\,\text{A}$

$$V_{BD} = I_3R_3 \text{ volt}$$

$$= 0.0244 \times 8$$

so, $V_{BD} = 0.195\,\text{V}$ **Ans**

(b) For balance conditions

$$R_2R_4 = R_1R_5$$
$$R_4 = \frac{R_1R_5}{R_2}\text{ ohm}$$
$$= \frac{20 \times 2}{10}$$
$$\text{so, } R_4 = 4\,\Omega \textbf{ Ans}$$

Example 7: A potential divider circuit consists of a 2.5 kΩ resistor in series with a 7.5 kΩ resistor, this combination being connected to a 100 V supply. *Without* calculating the circuit current, determine the p.d. across each resistor.

Solution: The relevant circuit diagram is shown in Fig 2.85.

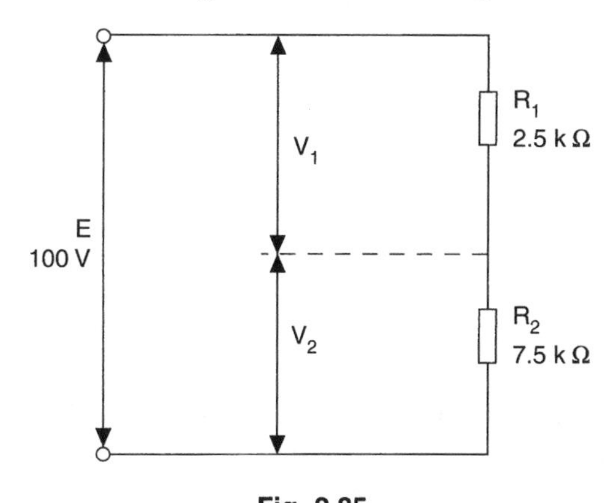

Fig. 2.85

$$V_1 = \frac{R_1}{R_1 + R_2}E\text{ volt}$$
$$= \frac{2.5 \times 10^3}{10 \times 10^3} \times 100$$
$$\text{so, } V_1 = 25\text{ V} \textbf{ Ans}$$
$$V_2 = \frac{R_2}{R_1 + R_2}E\text{ volt}$$
$$= \frac{7.5 \times 10^3}{10 \times 10^3} \times 100$$
$$\text{so, } V_2 = 75\text{ V} \textbf{ Ans}$$

Example 8: A 270 Ω resistor is connected in parallel with a 330 Ω resistor and this combination is connected to a source which supplies a current of 2.5 A. Calculate the current through each resistor.

Solution: The circuit diagram is shown in Fig. 2.86.

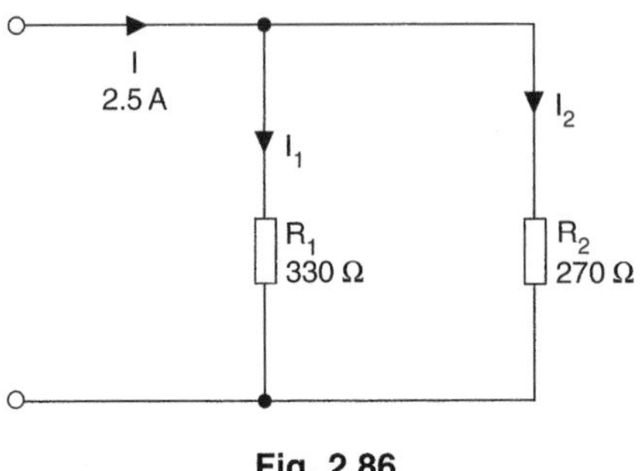

Fig. 2.86

Using current division theory:

$$I_1 = \frac{R_2}{R_1 + R_2} \times I \text{ amp}$$
$$= \frac{270}{600} \times 2.5$$

so, $I_1 = 1.125$ A **Ans**

$$I_2 = \frac{R_1}{R_1 + R_2} \times I \text{ amp}$$
$$= \frac{330}{600} \times 2.5$$

so, $I_2 = 1.375$ A **Ans**

Example 9: For the network shown in Fig. 2.87, determine (a) the Norton equivalent circuit, and (b) the Thévenin equivalent circuit.

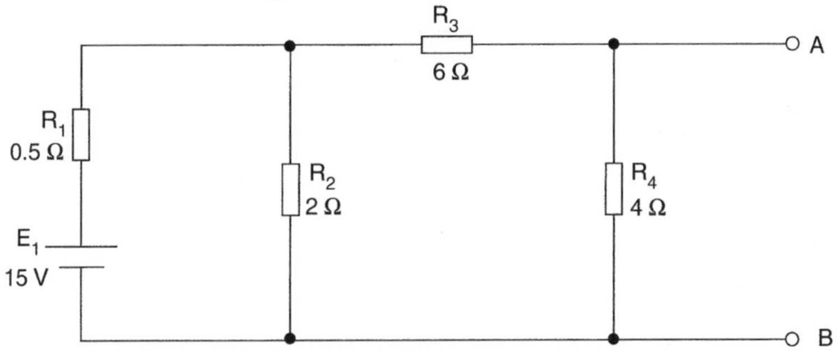

Fig. 2.87

Solution:

(a) With terminals A and B short-circuited, R_4 is also short-circuited so the circuit becomes as shown in Fig. 2.88.

Current from E_1 flows through R_1 and then through R_2 and R_3 in parallel, thus total resistance, $R = R_1 + \frac{R_2 R_3}{R_2 + R_3}$ ohm $= 0.5 + \frac{6 \times 2}{6 + 2} = 2\,\Omega$

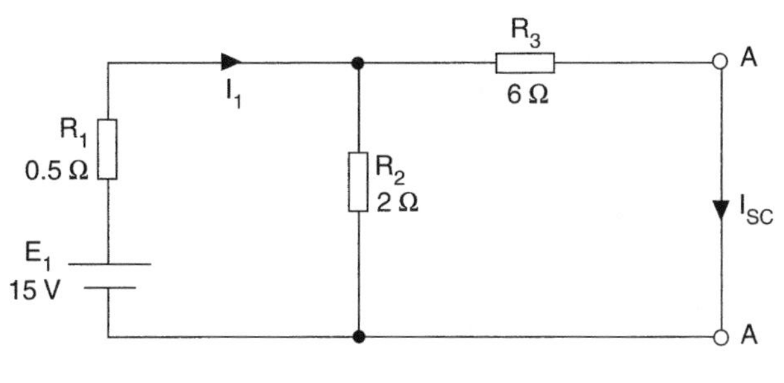

Fig. 2.88

$$I_1 = \frac{E_1}{R} = \frac{15}{2} = 7.5\,\text{A}$$

Using current division, $I_{SC} = \dfrac{R_2}{R_2 + R_3} \times I_1$ amp

$$I_{SC} = \frac{2}{8} \times 7.5 = 1.875\,\text{A}$$

With terminals A and B open-circuited and with E_1 replaced by its internal resistance, R_1, the circuit is as shown in Fig. 2.89

$$R_{CD} = \frac{R_1 R_2}{R_1 + R_2} = \frac{0.5 \times 2}{2.5} = 0.4\,\Omega$$

$$R_{ACD} = R_3 + R_{CD} = 6.4\,\Omega$$

$$R_O = \frac{R_4 R_{ACD}}{R_4 + R_{ACD}} = \frac{4 \times 6.4}{10.64}$$

$$R_O = 2.462\,\Omega$$

The Norton equivalent circuit will therefore be as shown in Fig. 2.90 **Ans**

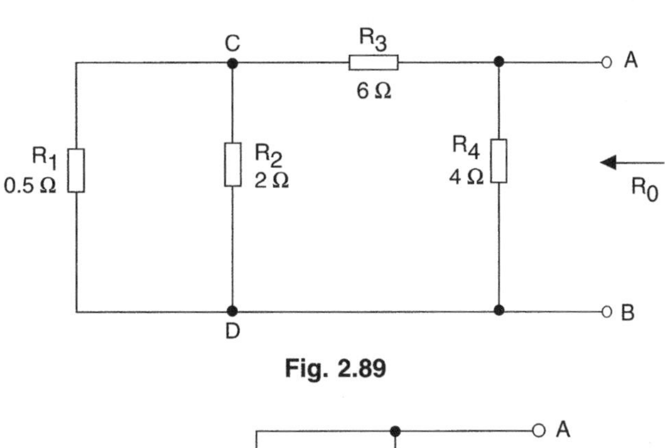

Fig. 2.89

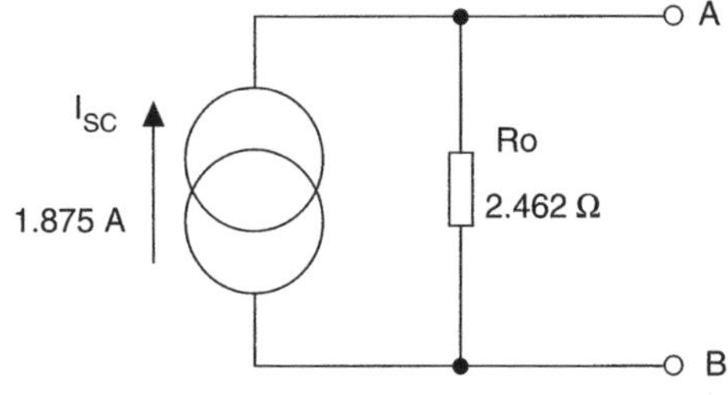

Fig. 2.90

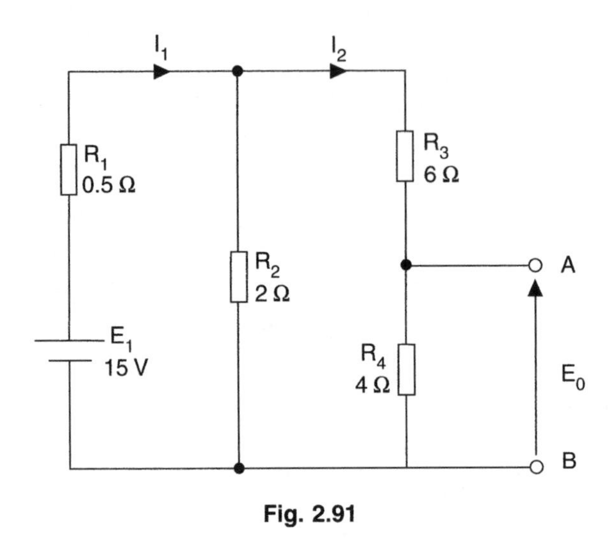

Fig. 2.91

(b) The circuit with terminals A and B open-circuited is shown in Fig. 2.91, and in this case current from E_1 will flow through R_1 and then split between R_2 and the series combination of R_3 and R_4.

$$\text{total resistance, } R = R_1 + \frac{R_2(R_3 + R_4)}{R_2 + R_3 + R_4} \text{ ohm}$$

$$R = 0.5 + \frac{2 \times 10}{12} = 2.1667\,\Omega$$

$$\text{current drawn from the battery, } I_1 = \frac{E_1}{R} \text{ amp} = \frac{15}{2.1667}$$

$$I_1 = 6.923\text{ A}$$

$$I_2 = \frac{R_2}{R_2 + R_3 + R_4} \times I_1 \text{ amp} = \frac{2}{12} \times 6.923$$

$$I_2 = 1.1538\text{ A}$$

$$E_O = I_2 R_4 \text{ volt} = 1.1538 \times 4$$

$$E_O = 4.615\text{ V}$$

The resistance of the Thévenin generator, R_O, is obtained in exactly the same manner as already done for the Norton equivalent, so the Thévenin equivalent circuit will be as shown in Fig. 2.92.

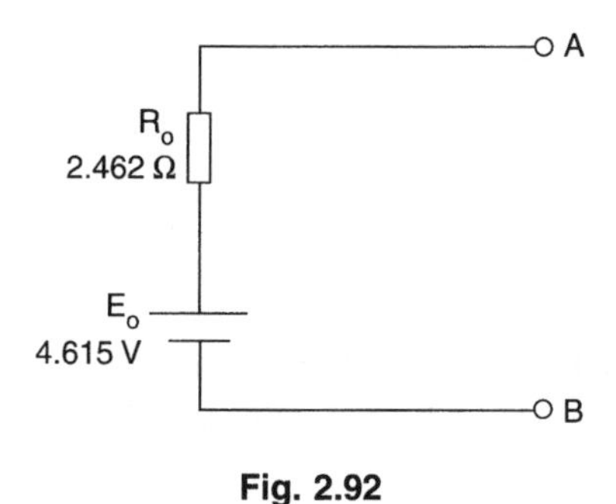

Fig. 2.92

Example 10: For the network shown in Fig. 2.93, (a) using Thévenin's theorem determine the current flowing through the 6 Ω resistor, and (b) verify your answer by using Norton's theorem.

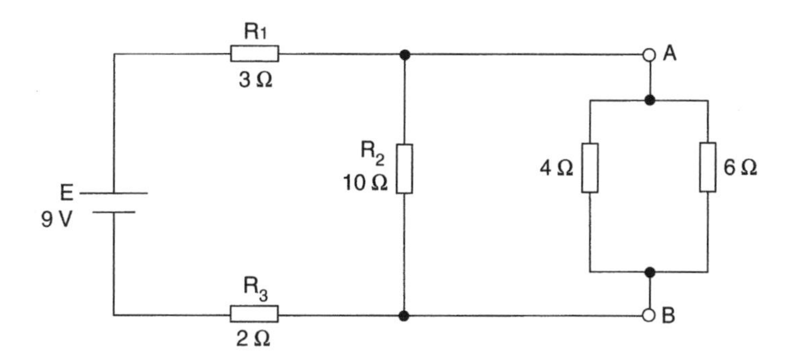

Fig. 2.93

Solution:

(a) With terminals A and B open-circuited the circuit appears as in Fig. 2.94, and the total resistance R is:

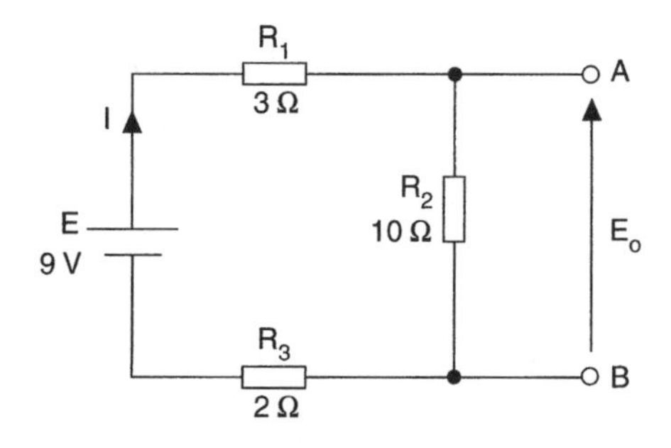

Fig. 2.94

$$R = R_1 + R_2 + R_3 \text{ ohm} = 3 + 2 + 10 = 15\,\Omega$$

$$I = \frac{E}{R} \text{ amp} = \frac{9}{15} = 0.6\text{ A}$$

$$E_O = V_{AB} = IR_2 \text{ volt} = 0.6 \times 10$$

$$E_O = 6\text{ V}$$

Replacing E with its internal resistance ($0\,\Omega$) the circuit is as in Fig. 2.95, and looking in at terminals A and B, it can be seen that R_2 is in parallel with the series combination of R_1 and R_3.

$$R_O = \frac{R_2(R_1 + R_3)}{R_2 + R_1 + R_3} \text{ ohm} = \frac{10 \times 5}{10 + 5}$$

$$R_O = 3.33\,\Omega$$

The Thévenin equivalent circuit is shown in Fig. 2.96

$$R_{AB} = \frac{6 \times 4}{6 + 4} = 2.4\,\Omega$$

$$V_{AB} = \frac{R_{AB}}{R_{AB} + R_O} \times E_O = \frac{2.4}{2.4 + 3.33} \times 6$$

$$V_{AB} = 2.512\text{ V}$$

$$I_{6\Omega} = \frac{V_{AB}}{6} \text{ amp} = \frac{2.512}{6}$$

$$I_{6\Omega} = 0.419\text{ A } \textbf{Ans}$$

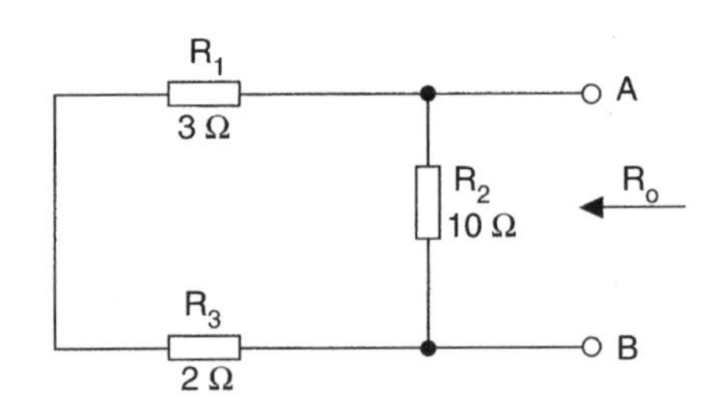

Fig. 2.95

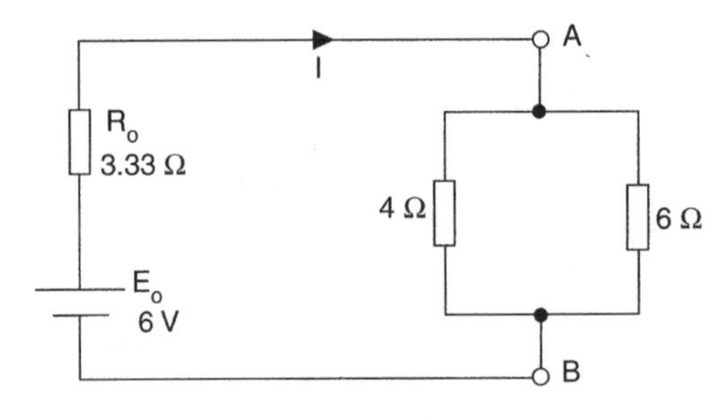

Fig. 2.96

(b) With terminals A and B short-circuited, the circuit will be as in Fig. 2.97, from which it may be seen that R_2 is also short-circuited. Thus the only opposition to the current flow will be the effect of R_1 and R_3 in series.

$$I_{SC} = \frac{E}{R_1 + R_3}\text{ amp} = \frac{9}{5} = 1.8\text{ A}$$

The internal resistance of the Norton constant current generator, R_O, may be found in exactly the same manner as in part (a), so $R_O = 3.33\,\Omega$, and the Norton equivalent circuit will be as in Fig. 2.98.

Also from part (a) we know that $R_{AB} = 2.4\,\Omega$, so

$$I = \frac{R_O}{R_O + R_{AB}} \times I_{SC}\text{ amp} = \frac{3.33}{5.73} \times 1.8$$

$$I = 1.0465\text{ A}$$

$$I_{6\Omega} = \frac{4}{10} \times 1.0465$$

$I_{6\Omega} = 0.419\text{ A}$ **Ans** Which verifies the answer in part (a)

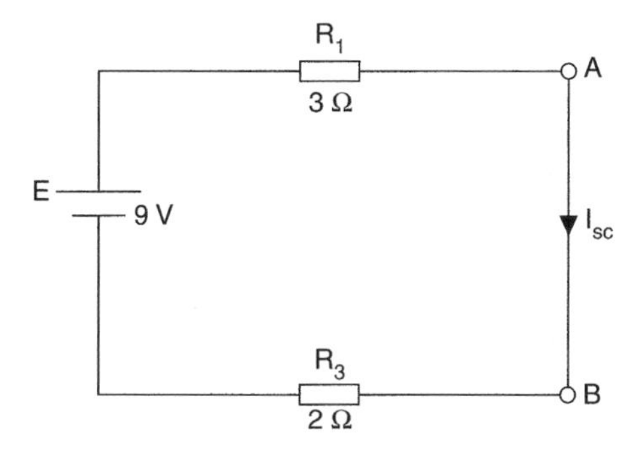

Fig. 2.97

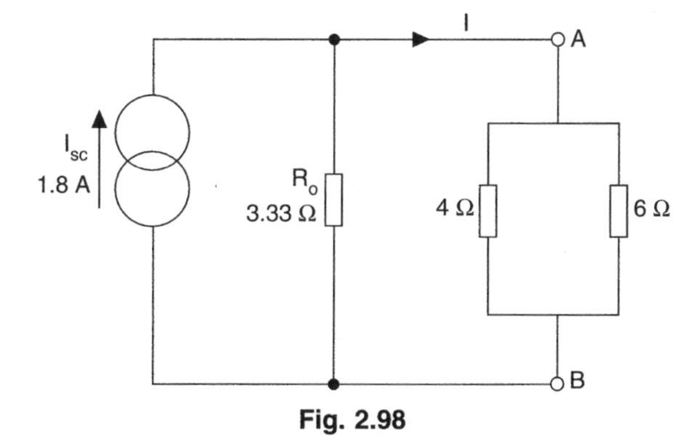

Fig. 2.98

Example 11: For the circuit of Fig. 2.99 determine the current through the 10 Ω resistor using (a) Thévenin's theorem, and (b) Norton's theorem.

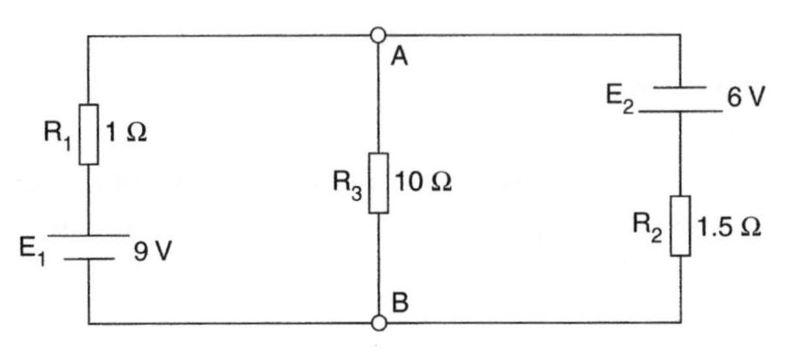

Fig. 2.99

Solution:

(a) With the 10 Ω resistor removed the circuit is as in Fig. 2.100.

Since the batteries are in series aiding, then

$$E = E_1 + E_2 \text{ volt} = 9 + 6 = 15\,\text{V}$$
$$R = R_1 + R_2 \text{ ohm} = 1 + 1.5 = 2.5\,\Omega$$
$$I = \frac{E}{R} = \text{amp} = \frac{15}{2.5} = 6\,\text{A}$$
$$E_O = E_1 - IR_1 \text{ volt} = 9 - (6 \times 1)$$
$$E_O = 3\,\text{V}$$

From Fig. 2.101, $R_O = \dfrac{R_1 R_2}{R_1 + R_2}$ ohm $= \dfrac{1.5}{2.5}$

$$R_O = 0.6\,\Omega$$

The Thévenin equivalent circuit is shown in Fig. 2.102.

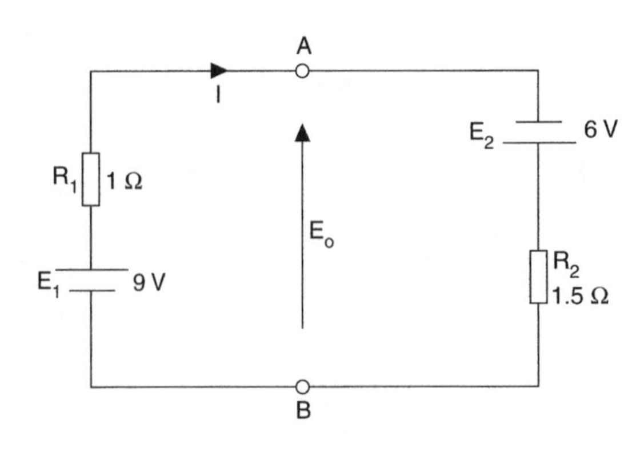

Fig. 2.100

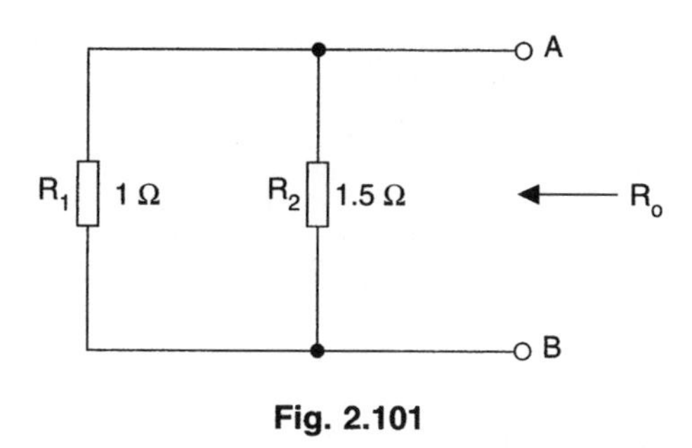

Fig. 2.101

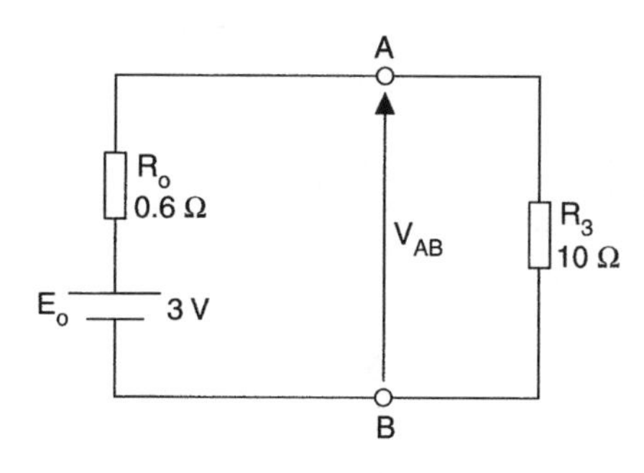

Fig. 2.102

R_O and R_3 act as a simple potential divider, so

$$V_{AB} = \frac{E_O R_3}{R_O + R_3} \text{ volt} = \frac{3 \times 10}{10.6}$$

$$V_{AB} = 2.83\,\text{V}$$

$$I_{10\Omega} = \frac{V_{AB}}{R_3} = \frac{2.83}{10}$$

$$I_{10\Omega} = 0.283\,\text{A } \textbf{Ans}$$

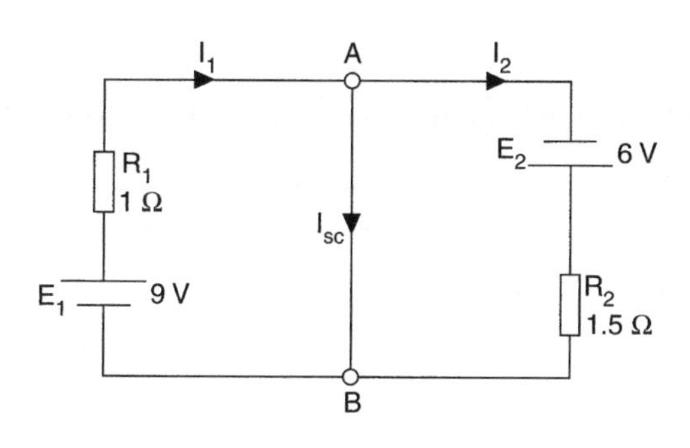

Fig. 2.103

(b) $I_1 = \dfrac{E_1}{R_1} = \dfrac{9}{1} = 9\,\text{A}$ (From Fig. 2.103)

$$I_2 = \frac{E_2}{R_2} = \frac{6}{1.5} = 4\,\text{A}$$

$$I_{SC} = I_1 - I_2\,\text{amp} = 9 - 4$$

$$I_{SC} = 5\,\text{A}$$

and from part (a), $R_O = 0.6\,\Omega$

From the Norton equivalent circuit of Fig. 2.104:

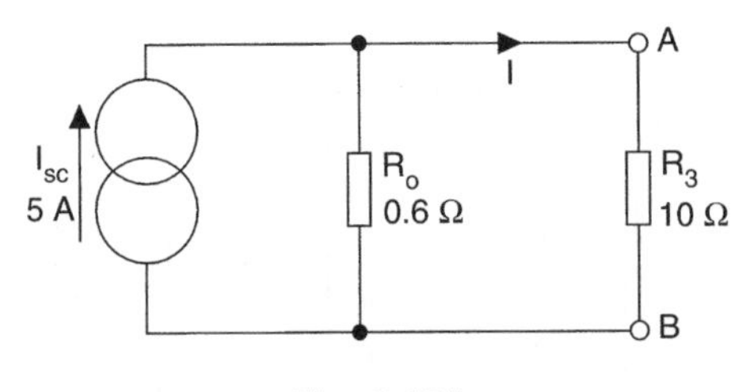

Fig. 2.104

$$I = \frac{R_O}{R_O + R_3} \times I_{SC} = \frac{0.6}{10.6} \times 5$$

$I = 0.283\,\text{A}$ **Ans**

Example 12: An attenuator circuit has input and output voltages of 8.5 V and 2.4 V respectively. Determine the attenuation, expressed in decibels.

Solution:

$V_i = 8.5\,\text{V};\ V_O = 2.4\,\text{V}$

$$\text{attenuation} = 20\log\frac{V_O}{V_i}\ \text{decibel}$$

$$= 20\log\frac{2.4}{8.5} = 20\log 0.2824$$

so, attenuation $= -11\,\text{dB}$ **Ans**

Example 13: An amplifier has a power gain of 40 dB. Express this gain as a power ratio.

Solution:

$$40 = 10\log(\text{ratio})\ \text{decibel}$$

$$4 = \log(\text{ratio})$$

antilog 4 = ratio

so, ratio $= 10^4$ **Ans**

Example 14: For the attenuator pad shown between terminals AB and CD in Fig. 2.105, calculate (a) the output (V_O) and input (V_i) voltages, (b) the voltage attenuation in decibels, and (c) the output power.

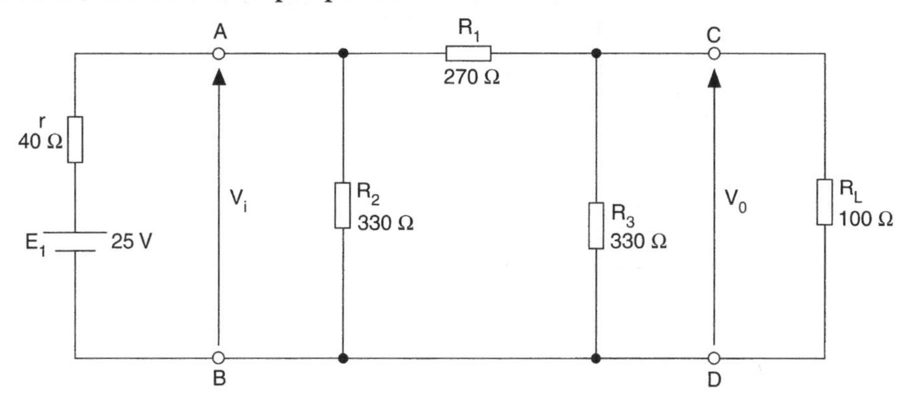

Fig. 2.105

Solution:

(a) Using Thévenin's theorem and looking in at terminals CD the circuit will appear as shown in Fig. 2.106.

$$R_{EF} = R_1 + \frac{rR_2}{r + R_2} \text{ ohm}$$

$$= 270 + \frac{40 \times 330}{40 + 330}$$

$$R_{EF} = 305.68\,\Omega$$

$$R_O = \frac{R_{EF}R_3}{R_{EF} + R_3} \text{ ohm} = \frac{305.68 \times 330}{305.68 + 330}$$

$$R_O = 158.7\,\Omega$$

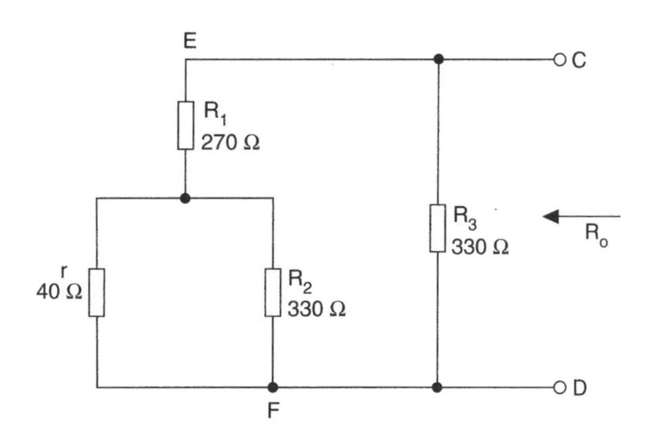

Fig. 2.106

To obtain the Thévenin generator emf E_O, consider Fig. 2.107.

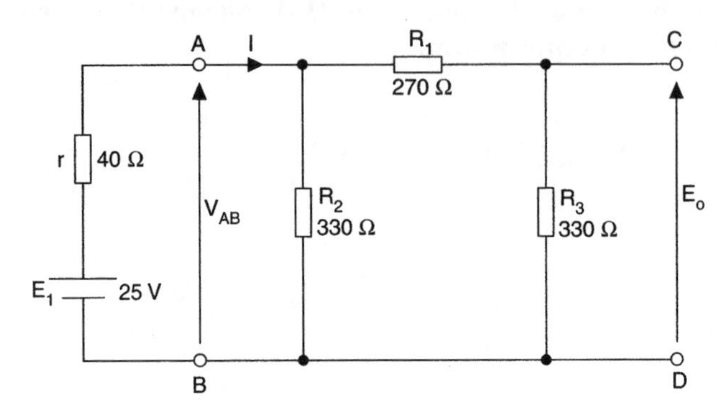

Fig. 2.107

$$\text{Total resistance, } R = r + \frac{R_2(R_1 + R_3)}{R_2 + (R_1 + R_3)} \text{ ohm}$$

$$= 40 + \frac{330 \times 600}{330 + 600}$$

$$R = 252.9\,\Omega$$

$$I = \frac{E_1}{R} \text{ amp} = \frac{25}{252.9} = 0.0989\text{ A}$$

$$V_{AB} = E_1 - Ir\,\text{volt} = 25 - (0.0989 \times 40)$$

$$V_{AB} = 21.05\text{ V}$$

Resistors R_1 and R_3 act as a potential divider with V_{AB} applied across them, so

$$E_O = \frac{R_3}{R_3 + R_1} \times V_{AB}\,\text{volt} = \frac{330}{600} \times 21.05$$

$E_O = 11.575\text{ V}$, and the Thévenin equivalent circuit is as Fig.2.108.

$$V_O = \frac{R_L}{R_L + R_O} \times E_O\,\text{volt} = \frac{100}{258.7} \times 11.575$$

$V_O = 4.474\text{ V}$ **Ans**

Looking in at terminals AB of the original circuit, the resistance 'seen' R_i is illustrated in Fig. 2.109, and may be obtained as follows:

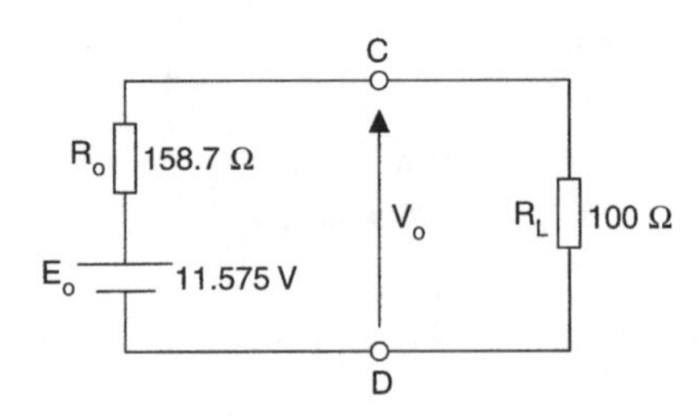

Fig. 2.108

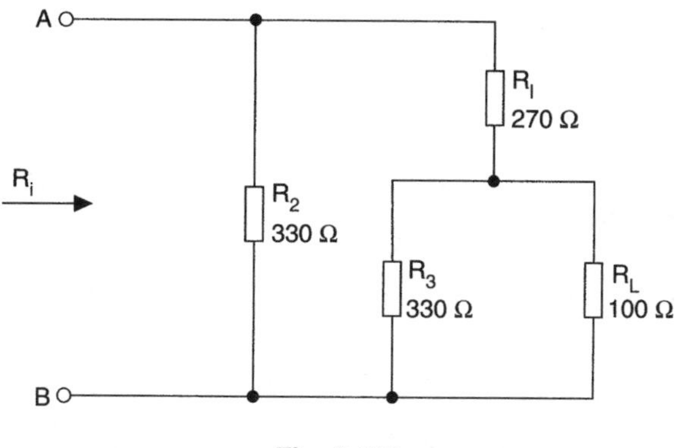

Fig. 2.109

$$R_{GH} = R_1 + \frac{R_3 R_L}{R_3 + R_L} \text{ ohm}$$
$$= 270 + \frac{330 \times 100}{330 + 100}$$
$$R_{GH} = 346.744\,\Omega$$
$$R_i = \frac{R_{GH} R_2}{R_{GH} + R_2} \text{ ohm} = \frac{346.7 \times 330}{346.7 + 330}$$
$$R_i = 169.08\,\Omega$$

So the circuit is equivalent to that shown in Fig. 2.110, from which:

$$V_i = \frac{R_i}{R_i + r} \times E_1 \text{ volt} = \frac{169.08}{209.08} \times 25$$

so, $V_i = 20.217\,\text{V}$ **Ans**

(b) attenuation $= 20 \log \frac{V_O}{V_i}$ decibel

$$= 20 \log \frac{4.474}{20.217}$$

attenuation $= -13\,\text{dB}$ **Ans**

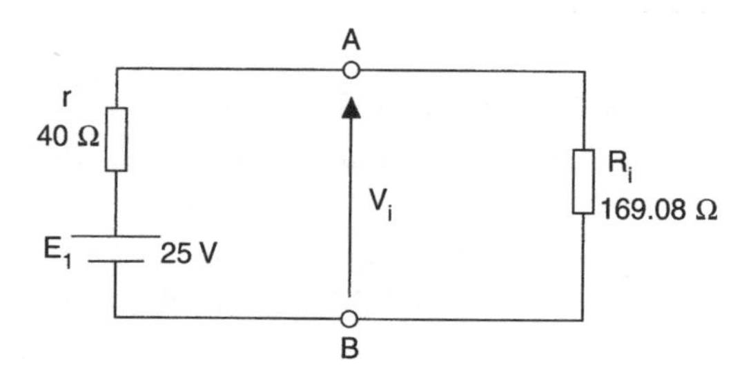

Fig. 2.110

(c) $P_O = \dfrac{V_O^2}{R_L}$ watt $= \dfrac{4.474^2}{100}$

$P_O = 0.2\,\text{W}$ **Ans**

Summary of Equations

Resistors in series: $R = R_1 + R_2 + R_3 + \ldots$ ohm

Resistors in parallel: $\dfrac{1}{R} = \dfrac{1}{R_1} + \dfrac{1}{R_2} + \dfrac{1}{R_3} + \ldots$ siemen

and for ONLY two resistors in parallel, $R = \dfrac{R_1R_2}{R_1 + R_2}$ ohm $\left(\dfrac{\text{product}}{\text{sum}}\right)$

Potential divider: $V_1 = \dfrac{R_1}{R_1 + R_2} \times E$ volt

Current divider: $I_1 = \dfrac{R_2}{R_1 + R_2} \times I$ amp

Kirchhoff's laws: $\Sigma I = 0$ (sum of the currents at a junction $= 0$)

$\Sigma E = \Sigma IR$ (sum of the emfs = sum of the p.d.s, in order)

Thévenin's and Norton's Theorems:

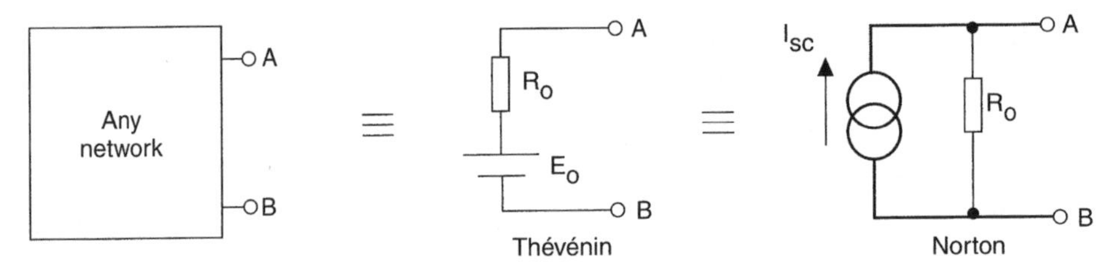

Decibel usage: Power gain, $A_p = 10 \log \dfrac{P_2}{P_1}$ decibel

Voltage gain, $A_v = 20 \log \dfrac{V_2}{V_1}$ decibel

Current gain, $A_i = 20 \log \dfrac{I_2}{I_1}$ decibel

Bandwidth: $B = (f_2 - f_1)$ hertz, where f_1 and f_2 occur at $\dfrac{A_{vm}}{\sqrt{2}}$

f_1 and f_2 are also known as the half-power or $-3\,\text{dB}$ points

Wheatstone Bridge: Balance condition when current through centre limb = 0
or, $R_1R_4 = R_2R_3$ (multiply diagonally across bridge circuit)

Slidewire potentiometer: $V_{AC} = \dfrac{AC}{AB} \times E$ volt

3 Electric Fields and Capacitors

This chapter deals with the laws and properties of electric fields and their application to electric components known as capacitors.

On completion of this chapter you should be able to:

1 Understand the properties of electric fields and insulating materials.
2 Carry out simple calculations involving these properties.
3 Carry out simple calculations concerning capacitors, and capacitors connected in series, parallel and series/parallel combinations.
4 Describe the construction and electrical properties of the different types of capacitor.
5 Understand the concept of energy storage in an electric field, and perform simple related calculations.

3.1 Coulomb's Law

A force exists between charged bodies. A force of attraction exists between opposite charges and a force of repulsion between like polarity charges. Coulomb's law states that the force is directly proportional to the product of the charges and is inversely proportional to the square of the distance between their centres. So for the two bodies shown in Fig. 3.1, this would be expressed as

$$F \propto \frac{Q_1 Q_2}{d^2}$$

In order to obtain a value for the force, a constant of proportionality must be introduced. In this case it is the permittivity of free space, ε_0. This concept of permittivity is

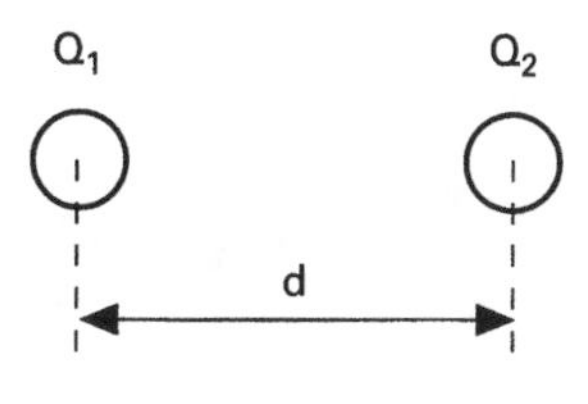

Fig. 3.1

dealt with later in this chapter, and need not concern you for the time being. The expression for the force in newtons becomes

$$F = \frac{Q_1 Q_2}{\epsilon_0 d^2} \text{ newton} \qquad (3.1)$$

This type of relationship is said to follow an inverse square law because $F \propto 1/d^2$. The consequence of this is that if the distance of separation is doubled then the force will be reduced by a factor of four times. If the distance is increased by a factor of four times, the force will be reduced by a factor of sixteen times, etc. The practical consequence is that although the force can never be reduced to zero, it diminishes very rapidly as the distance of separation is increased. This will continue until a point is reached where the force is negligible relative to other forces acting within the system.

3.2 Electric Fields

You are probably more familiar with the concepts and effects of magnetic and gravitational fields. For example, you have probably conducted simple experiments using bar magnets and iron filings to discover the shape of magnetic fields, and you are aware that forces exist between magnetised bodies. You also experience the effects of gravitational forces constantly, even though you probably do not consciously think about them.

Both of these fields are simply a means of transmitting the forces involved, from one body to another. However, the fields themselves cannot be detected by the human senses, since you cannot see, touch, hear or smell them. This tends to make it more difficult to understand their nature. An electric field behaves in the same way as these other two examples, except that it is the method by which forces are transmitted between charged bodies. In all three cases we can represent the appropriate field by means of arrowed lines. These lines are usually referred to as the lines of force.

To illustrate these points, consider Fig. 3.2 which shows two oppositely charged spheres with a small positively charged particle placed on the surface of Q_1. Since like charges repel and unlike charges attract each other, then the small charged particle will experience a force of repulsion from Q_1 and one of attraction from Q_2.

The force of repulsion from Q_1 will be very much stronger than the force of attraction from Q_2 because of the relative distance involved. The other feature of the forces is that they will act so as to be at right angles to the charged surfaces. Hence there will be a resultant force acting on the particle. Assuming that it is free to move, then it will start to move in the direction of this resultant force. For the sake of clarity, the distance

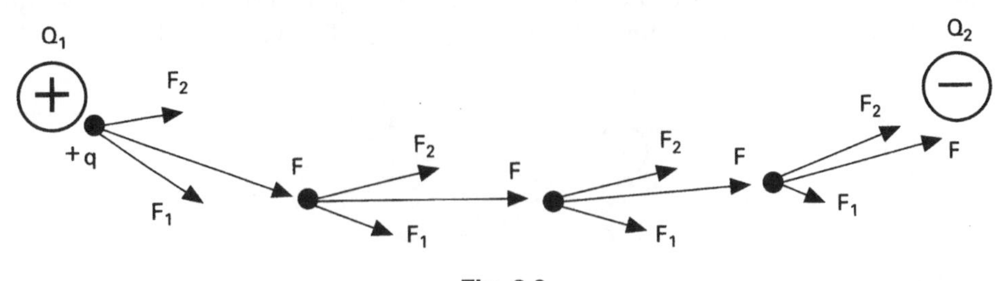

Fig. 3.2

moved in this direction is greatly exaggerated in the diagram. However, when the particle moves to the new position, force F_1 will have decreased and F_2 will have increased. In addition, the direction of action of each force will have changed. Thus the direction of action of the resultant will have changed, but its magnitude will have remained constant. The particle will now respond to the new resultant force F. This is a continuous process and the particle will trace out a curved path until it reaches the surface of Q_2. If this 'experiment' was carried out for a number of starting points at the surface of Q_1, the paths taken by the particle would be as shown in Fig. 3.3.

The following points should be noted:

1 The lines shown represent the possible paths taken by the positively charged particle in response to the force acting on it. Thus they are called the lines of electric force. They may also be referred to as the lines of electric flux, ψ.
2 The total electric flux makes up the whole electric field existing between and around the two charged bodies.
3 The lines themselves are imaginary and the field is three dimensional. The whole of the space surrounding the charged bodies is occupied by the electric flux, so there are no 'gaps' in which a charged particle would not be affected.
4 The lines of force (flux) radiate outwards from the surface of a positive charge and terminate at the surface of a negative charge.
5 The lines always leave (or terminate) at right angles to a charged surface.
6 Although the lines drawn on a diagram do not actually exist as such, they are a very convenient way to represent the existence of the electric field. They therefore aid the understanding of its properties and effects.
7 Since force is a vector quantity any line representing it must be arrowed. The convention used here is that the arrows point from the positive to the negative charge.

It is evident from Fig. 3.3 that the spacing between the lines of flux varies depending upon which part of the field you consider. This means that the field shown is non-uniform. A uniform electric field may be obtained between two parallel charged plates as shown in Fig. 3.4. Note that the electric field will exist in *all* of the space surrounding the two plates, but the uniform section exists only in the space between them. Some non-uniformity is shown by the curved lines at the edges (fringing effect). At this stage

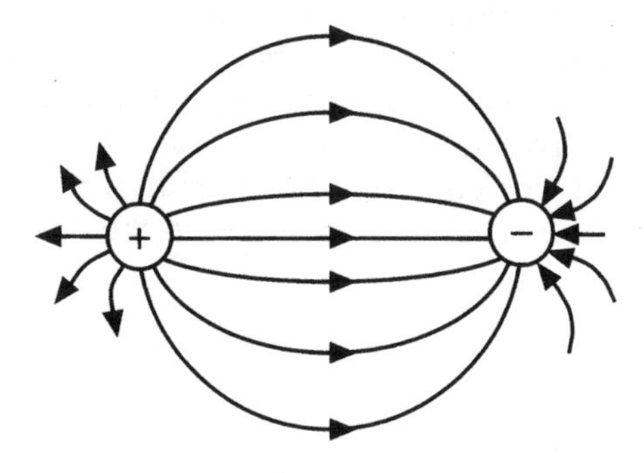

Fig. 3.3

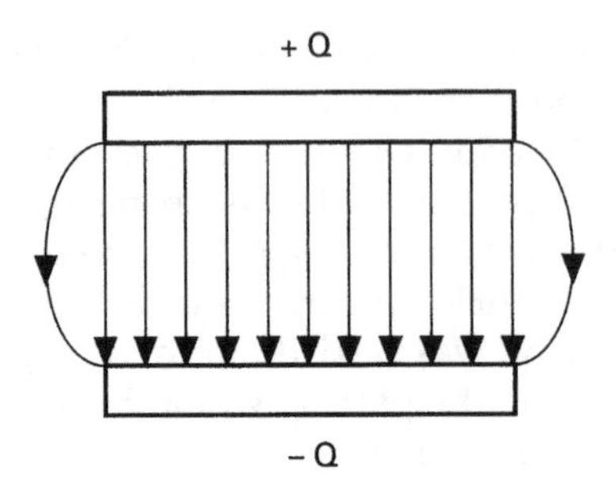

Fig. 3.4

we are concerned only with the uniform field between the plates. If a positively charged particle was placed between the plates it would experience a force that would cause it to move from the positive to the negative plate. The value of force acting on the particle depends upon what is known as the electric field strength.

3.3 Electric Field Strength (E)

This is defined as the force per unit charge exerted on a test charge placed inside the electric field. (An outdated name for this property is 'electric force').

$$\text{Hence, field strength} = \frac{\text{force}}{\text{charge}}$$

$$\text{So, } \mathbf{E} = \frac{F}{q} \text{ newton/coulomb} \qquad (3.2)$$

$$\text{or, } F = \mathbf{E}\, q \text{ newton} \qquad (3.3)$$

where q is the charge on the *particle*, and not the plates.

3.4 Electric Flux (ψ) and Flux Density (D)

In the SI system one 'line' of flux is assumed to radiate from the surface of a positive charge of one coulomb and terminate at the surface of a negative charge of one coulomb. Hence the electric flux has the same numerical value as the charge that produces it. Therefore the coulomb is used as the unit of electric flux. In addition, the Greek letter psi is usually replaced by the symbol for charge, namely Q.

The electric flux density D is defined as the amount of flux per square metre of the electric field. This area is measured at right angles to the lines of force. This gives the following equation

$$D = \frac{\psi}{A}$$

$$\text{or, } D = \frac{Q}{A} \text{ coulomb/metre}^2 \qquad (3.4)$$

Worked Example 3.1

Question

Two parallel plates of dimensions 30 mm by 20 mm are oppositely charged to a value of 50 mC. Calculate the density of the electric field existing between them.

Answer

$Q = 50 \times 10^{-3}\text{C};\ A = 30 \times 20 \times 10^{-6}\,\text{m}^2$

$$D = \frac{Q}{A}\ \text{coulomb/metre}^2$$

$$= \frac{50 \times 10^{-3}}{600 \times 10^{-6}}$$

so $D = 83.3\,\text{C/m}^2$ **Ans**

3.5 The Charging Process and Potential Gradient

We have already met the concept of a potential gradient when considering a uniform conductor (wire) carrying a current. This concept formed the basis of the slidewire potentiometer discussed in Chapter 2. However, we are now dealing with static charges that have been induced on to plates (the branch of science known as electrostatics). Current flow is only applicable during the charging process. The material between the plates is some form of insulator (a dielectric) which could be vacuum, air, rubber, glass, mica, pvc, etc. So under ideal conditions there will be no current flow from one plate to the other via the dielectric. None-the-less there will be a potential gradient throughout the dielectric.

Consider a pair of parallel plates (initially uncharged) that can be connected to a battery via a switch, as shown in Fig. 3.5. Note that the number of electrons and protons shown for each plate are in no way representative of the actual numbers involved. They are shown to aid the explanation of the charging process that will take place when the switch is closed. On closing the switch, some electrons from plate A will be attracted to the positive terminal of the battery. In this case, since plate A has lost electrons it will acquire a positive charge. This results in an electric field radiating out from plate A. The effect of this field is to induce a negative charge on the top surface of plate B, by attracting electrons in the plate towards this surface. Consequently, the lower surface of plate B must have a positive charge. This in turn will attract electrons from the negative terminal of the battery. Thus for every electron that is removed from plate A one is transferred to plate B. The two plates will therefore become equally but oppositely charged.

This charging process will not carry on indefinitely (in fact it will last for only a very short space of time). This is because as the charge on the plates increases so too does the voltage developed between them. Thus the charging process continues only until the

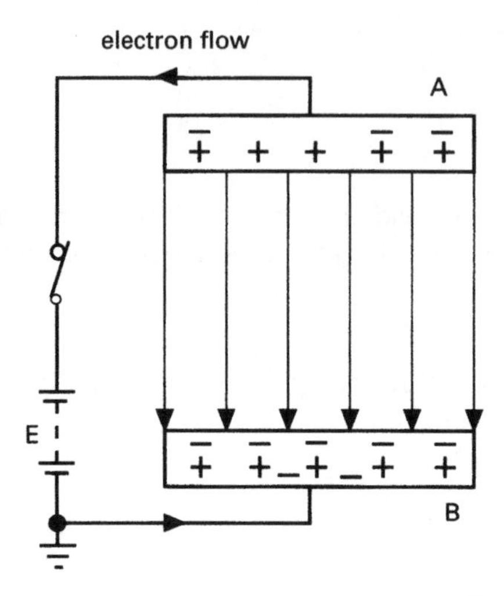

Fig. 3.5

p.d. between the plates, V is equal to the emf, E of the battery. The charging current at this time will become zero because plates A and B are positive and negative respectively. Thus, this circuit is equivalent to two batteries of equal emf connected in parallel as shown in Fig. 3.6. In this case each battery would be trying to drive an equal value of current around the circuit, but in opposite directions. Hence the two batteries 'balance out' each other, and no current will flow.

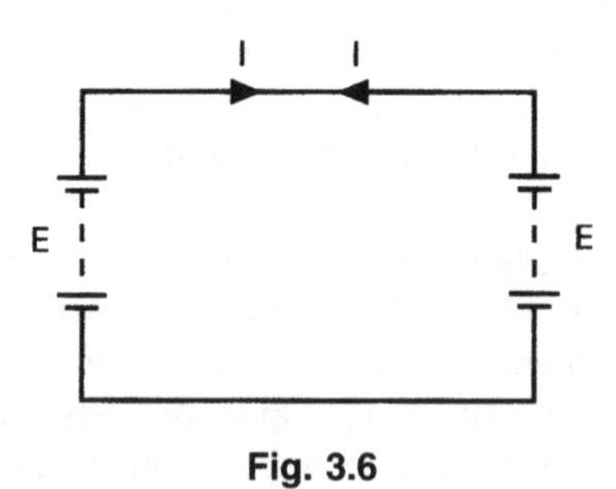

Fig. 3.6

With suitable instrumentation it would be possible to measure the p.d. between plate B and any point in the dielectric. If this was done, then a graph of the voltage versus distance from B would look like that in Fig. 3.7. The slope of this graph is uniform and has units of volts/metre i.e. the units of potential gradient

$$\text{so} \quad \text{potential gradient} = \frac{V}{d} \text{ volt/metre} \qquad (3.5.)$$

Now, energy $= VIt$ joule, and $I = \dfrac{Q}{t}$ amp

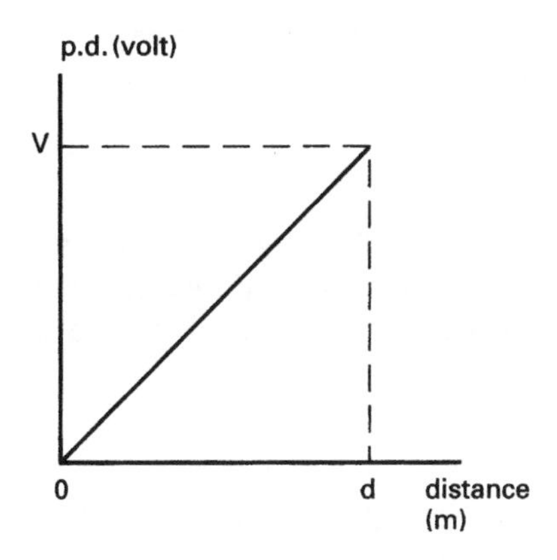

Fig. 3.7

therefore energy $= \frac{VQt}{t} = VQ$ joule, and transposing this

$$V = \frac{\text{energy}}{Q} \text{ joule/coulomb}$$

i.e. 1 volt $= 1\,\frac{\text{J}}{\text{C}}$ ....[1]

but the joule is the unit used for work done, and work is force × distance, i.e. newton metre,

so 1 J = 1 Nm ....[2]

Substituting [2] into [1]:

1 volt = 1 Nm/C or $V \equiv \frac{\text{Nm}}{\text{C}}$ ....[3]

Dividing both sides of [3] by distance of separation d:

$$\frac{V}{d} \equiv \frac{\text{Nm}}{\text{Cm}} = \frac{\text{N}}{\text{C}}$$

Referring back to equation (3.2), we know that electric field strength **E** is measured in N/C. So potential gradient and electric field strength must be one and the same thing. Now, electric field strength is defined in terms of the ratio of the force exerted on a charge to the value of the charge. This is actually an extremely difficult thing to measure. However, it is a very simple matter to measure the p.d. and distance between the charged plates. Hence, for practical purposes, electric field strength is always quoted in the units volt/metre

$$\text{i.e. } \mathbf{E} = \frac{V}{d} \text{ volt/metre} \qquad (3.6)$$

Notice that the symbol **E** has been used for electric field strength. This is in order to avoid confusion with the symbol E used for emf.

Worked Example 3.2

Question

Two parallel plates separated by a dielectric of thickness 3 mm acquire a charge of 35 mC when connected to a 150 V source. If the effective csa of the field between the plates is 144 mm^2, calculate (a) the electric field strength and (b) the flux density.

Answer

$d = 3 \times 10^{-3}$m; $Q = 35 \times 10^{-3}$C; $V = 150$ V; $A = 144 \times 10^{-6}$ m^2

(a) $$\mathbf{E} = \frac{V}{d} \text{ volt/metre} = \frac{150}{3 \times 10^{-6}}$$

so $\mathbf{E} = 50$ kV/m **Ans**

(b) $$D = \frac{Q}{A} \text{ coulomb/metre}^2 = \frac{35 \times 10^{-3}}{144 \times 10^{-6}}$$

so $D = 243.1$ C/m^2 **Ans**

3.6 Capacitance (C)

We have seen that in order for one plate to be at a different potential to the other one there is a need for a charge. This requirement is known as the capacity of the system. For a given system the ratio of the charge required to achieve a given p.d. is a constant for that system. This is called the *capacitance* (C) of the system

i.e. $$C = \frac{Q}{V} \text{ farad} \tag{3.7}$$

or $$Q = VC \text{ coulomb} \tag{3.8}$$

From equation (3.7) it may be seen that the unit for capacitance is the farad (F). This is defined as the capacitance of a system that requires a charge of one coulomb in order to raise its potential by one volt.

The farad is a very large unit, so in practice it is more usual to express capacitance values in microfarads (μF), nanofarads (nF), or picofarads (pF).

3.7 Capacitors

A capacitor is an electrical component that is designed to have a specified value of capacitance. In its simplest form it consists of two parallel plates separated by a dielectric; i.e. exactly the system we have been dealing with so far.

In order to be able to design a capacitor we need to know what dimensions are required for the plates, the thickness of the dielectric (the distance of separation d), and the other properties of the dielectric material chosen. Let us consider first the properties associated with the dielectric.

3.8 Permittivity of Free Space (ϵ_o)

When an electric field exists in a vacuum then the ratio of the electric flux density to the electric field strength is a constant, known as the permittivity of free space.

The value for $\epsilon_o = 8.854 \times 10^{-12}$ F/m.

Since a vacuum is a well defined condition, the permittivity of free space is chosen as the reference or datum value from which the permittivity of all other dielectrics are measured. This is a similar principle to using Earth potential as the datum for measuring voltages.

3.9 Relative Permittivity (ϵ_r)

The capacitance of two plates will be increased if, instead of a vacuum between the plates, some other dielectric is used. This difference in capacitance for different dielectrics is accounted for by the relative permittivity of each dielectric. Thus relative permittivity is defined as the ratio of the capacitance with that dielectric to the capacitance with a vacuum dielectric

$$\text{i.e.} \quad \epsilon_r = \frac{C_2}{C_1} \tag{3.9}$$

where C_1 is with a vacuum and C_2 is with the other dielectric.
Note: Dry air has the same effect as a vacuum so the relative permittivity for air dielectrics = 1.

3.10 Absolute Permittivity (ϵ)

For a given system the ratio of the electric flux density to the electric field strength is a constant, known as the absolute permittivity of the dielectric being used.

$$\text{so } \epsilon = \frac{D}{\mathbf{E}} \text{ farad/metre} \tag{3.10}$$

but we have just seen that a dielectric (other than air) is more effective than a vacuum by a factor of ϵ_r times, so the absolute permittivity is given by:

$$\epsilon = \epsilon_o \epsilon_r \text{ farad/metre} \tag{3.11}$$

3.11 Calculating Capacitor Values

From the equation (3.10):

$$\epsilon = \frac{D}{\mathbf{E}}$$

but $D = Q/A$ and $\mathbf{E} = V/d$

$$\text{so } \epsilon = \frac{Q}{V}\frac{d}{A}$$

also, $Q/V =$ capacitance C

$$\text{therefore } \epsilon = C\frac{d}{A}$$

and transposing this for C we have

$$C = \frac{\epsilon A}{d} = \frac{\epsilon_o \epsilon_r A}{d} \text{ farad} \tag{3.12}$$

Worked Example 3.3

Question

A capacitor is made from two parallel plates of dimensions 3 cm by 2 cm, separated by a sheet of mica 0.5 mm thick and of relative permittivity 5.8. Calculate (a) the capacitance and (b) the electric field strength if the capacitor is charged to a p.d. of 200 V.

Answer

$A = 3 \times 2 \times 10^{-4}\ \text{m}^2$; $d = 5 \times 10^{-4}\ \text{m}$; $\epsilon_r = 5.8$; $V = 200\ \text{V}$

(a)

$$C = \frac{\epsilon_o \epsilon_r A}{d} \text{ farad}$$

$$= \frac{8.854 \times 10^{-12} \times 5.8 \times 6 \times 10^{-4}}{5 \times 10^{-4}}$$

so $C = 61.62\ \text{pF}$ **Ans**

(b)

$$\mathbf{E} = \frac{V}{d} \text{ volt/metre}$$

$$= \frac{200}{5 \times 10^{-4}}$$

so $\mathbf{E} = 400\ \text{kV/m}$ **Ans**

Worked Example 3.4

Question

A capacitor of value 0.244 nF is to be made from two plates each 75 mm by 75 mm, using a waxed paper dielectric of relative permittivity 2.5. Determine the thickness of paper required.

Answer

$C = 0.224 \times 10^{-9}$ F; $A = 75 \times 75 \times 10^{-6}$ m^2; $\epsilon_r = 2.5$

$$C = \frac{\epsilon_0 \epsilon_r A}{d} \text{ farad}$$

$$\text{so } d = \frac{\epsilon_0 \epsilon_r A}{C} \text{ metre} = \frac{8.854 \times 10^{-12} \times 2.5 \times 75 \times 75 \times 10^{-6}}{0.224 \times 10^{-9}}$$

so $d = 5.6 \times 10^{-4}$ m $= 0.56$ mm **Ans**

Worked Example 3.5

Question

A capacitor of value 47 nF is made from two plates having an effective csa of 4 cm^2 and separated by a ceramic dielectric 0.1 mm thick. Calculate the relative permittivity.

Answer

$C = 4.7 \times 10^{-8}$ F; $A = 4 \times 10^{-4}$ m^2; $d = 1 \times 10^{-4}$ m

$$C = \frac{\epsilon_0 \epsilon_r A}{d} \text{ farad}$$

$$\text{so } \epsilon_r = \frac{Cd}{\epsilon_0 A} = \frac{4.7 \times 10^{-8} \times 10^{-4}}{8.854 \times 10^{-12} \times 4 \times 10^{-4}}$$

so $\epsilon_r = 1327$ **Ans**

Worked Example 3.6

Question

A p.d. of 180 V creates an electric field in a dielectric of relative permittivity 3.5, thickness 3 mm and of effective csa 4.2 cm^2. Calculate the flux and flux density thus produced.

Continued on p. 134

Worked Example 3.6 (*Continued*)

Answer

$V = 180\,\text{V}$; $d = 3 \times 10^{-3}\,\text{m}$; $\epsilon_r = 3.5$; $A = 4.2 \times 10^{-4}\,\text{m}^2$

There are two possible methods of solving this problem; either determine the capacitance and the use $Q = VC$ or determine the electric field strength and use $D = \epsilon_0\epsilon_r\mathbf{E}$. Both solutions will be shown.

$$C = \frac{\epsilon_0\epsilon_r A}{d}\ \text{farad} = \frac{8.854 \times 10^{-12} \times 3.5 \times 4.2 \times 10^{-4}}{3 \times 10^{-3}}$$

so $C = 4.34\,\text{pF}$ and since $Q = VC$ coulomb then

$$Q = 180 \times 4.34 \times 10^{-12}$$

hence, $Q = 0.78\ \text{nC}$ **Ans**

$$D = \frac{Q}{A}\ \text{coulomb/metre}^2 = \frac{7.8 \times 10^{-10}}{4.2 \times 10^{-4}}$$

so $D = 1.86\,\mu\text{C/m}^2$ **Ans**

Alternatively:

$$\mathbf{E} = \frac{V}{d}\ \text{volt/metre} = \frac{180}{3 \times 10^{-3}}$$

so $\mathbf{E} = 60\,\text{kV/m}$ and using $D = \epsilon_0\epsilon_r\mathbf{E}$ coulomb/metre^2

$$D = 8.854 \times 10^{-12} \times 3.5 \times 60 \times 10^3$$

so $D = 1.86\,\mu\text{C/m}^2$ **Ans**

$$Q = DA\ \text{coulomb} = 1.86 \times 10^{-6} \times 4.2 \times 10^{-4}$$

so $Q = 0.78\,\text{nC}$ **Ans**

3.12 Capacitors in Parallel

Consider two capacitors that are identical in every way (same plate dimensions, same dielectric material and same distance of separation between plates) as shown in Fig. 3.8. Let them now be moved vertically until the top and bottom edges respectively of their plates make contact. We will now effectively have a single capacitor of twice the csa of one of the original capacitors, but all other properties will remain unchanged.

Since $C = \epsilon A/d$ farad, then the 'new' capacitor so formed will have twice the capacitance of one of the original capacitors. The same effect could have been achieved if we had simply connected the appropriate plates together by means of a simple electrical connection. In other words connect them in parallel with each other. Both of the original capacitors have the same capacitance, and this figure is doubled when they are connected in parallel. Thus we can draw the conclusion that with this connection the total capacitance of the combination is given simply by adding the capacitance values. However, this might be considered as a special case. Let us verify this conclusion by considering the general case of three different value capacitors connected in parallel to a d.c. supply of V volts as in Fig. 3.9.

Each capacitor will take a charge from the supply according to its capacitance:

$$Q_1 = VC_1;\ Q_2 = VC_2;\ Q_3 = VC_3 \text{ coulomb}$$

but the total charge drawn from the supply must be:

$$Q = Q_1 + Q_2 + Q_3$$

also, total charge, $Q = VC$

where C is the total circuit capacitance.

Thus, $VC = VC_1 + VC_2 + VC_3$, and dividing through by V

$$C = C_1 + C_2 + C_3 \text{ farad} \qquad (3.13)$$

Note: This result is exactly the *opposite in form* to that for *resistors* in parallel.

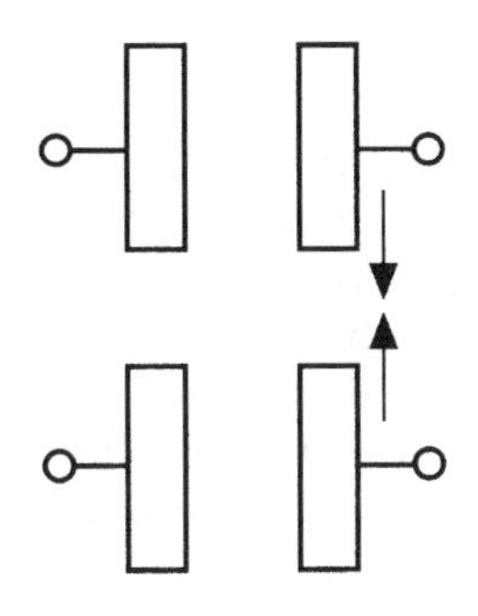

Fig. 3.8

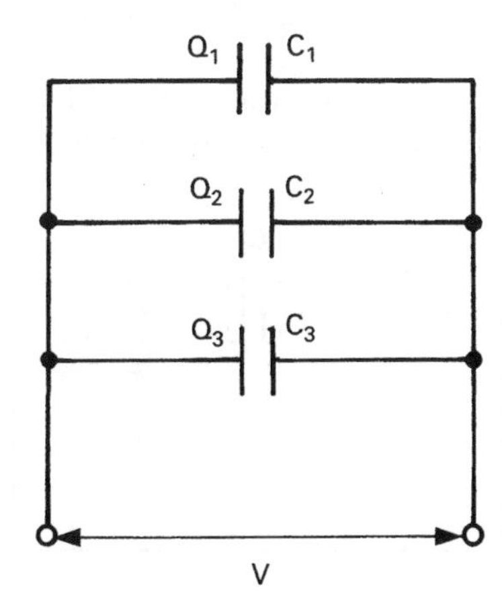

Fig. 3.9

Worked Example 3.7

Question

Three capacitors of value 4.7 μF, 3.9 μF and 2.2 μF are connected in parallel. Calculate the resulting capacitance of this combination.

Answer

In this case, since all of the capacitor values are in μF then it is not necessary to show the 10^{-6} multiplier in each case, since the answer is best quoted in μF. However it must be made clear that this is what has been done.

$$C = C_1 + C_2 + C_3 \text{ microfarad}$$
$$C = 4.7 + 3.9 + 2.2$$
$$\text{so } C = 10.8\,\mu\text{F} \quad \textbf{Ans}$$

3.13 Capacitors in Series

Three parallel plate capacitors are shown connected in series in Fig. 3.10. Each capacitor will receive a charge. However, you may wonder how capacitor C_2 can receive any charging current since it is sandwiched between the other two, and of course the charging current cannot flow through the dielectrics of these. The answer lies in the explanation of the charging process described in section 3.6 earlier. To assist the explanation, the plates of capacitors C_1 to C_3 have been labelled with letters.

Plate A will lose electrons to the positive terminal of the supply, and so acquires a positive charge. This creates an electric field in the dielectric of C_1 which will cause plate *B* to attract electrons from plate C of C_2. The resulting electric field in C_2 in turn causes plate D to attract electrons from plate E. Finally, plate F attracts electrons from the negative terminal of the supply. Thus all three capacitors become charged to the same value.

Having established that all three capacitors will receive the same amount of charge, let us now determine the total capacitance of the arrangement. Since the capacitors are

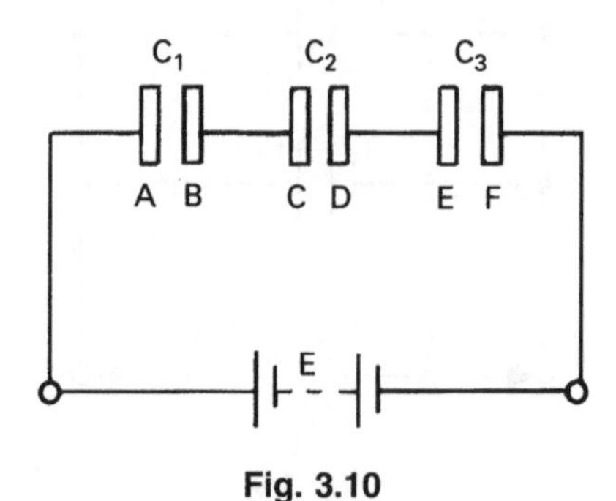

Fig. 3.10

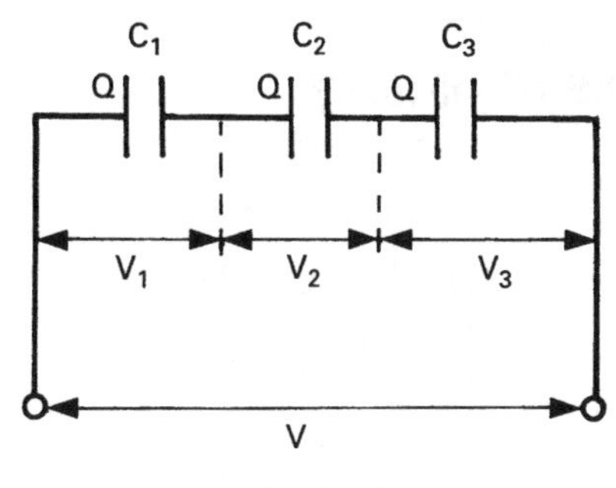

Fig. 3.11

of different values then each will acquire a different p.d. between its plates. This is illustrated in Fig. 3.11.

$$V_1 = \frac{Q}{C_1};\ V_2 = \frac{Q}{C_2};\ V_3 = \frac{Q}{C_3} \text{ volt}$$

and $V = V_1 + V_2 + V_3$ (Kirchhoff's voltage law)

also, $V = \frac{Q}{C}$ volt, where C is the total circuit capacitance.

Hence, $\frac{Q}{C} = \frac{Q}{C_1} + \frac{Q}{C_2} + \frac{Q}{C_3}$, and dividing by Q:

$$\frac{1}{C} = \frac{1}{C_1} + \frac{1}{C_2} + \frac{1}{C_3} \qquad (3.14)$$

Note: The above equation does not give the total capacitance directly. To obtain the value for C the reciprocal of the answer obtained from the equation (3.14) must be found. However, if ONLY TWO capacitors are connected in series the total capacitance may be obtained directly by using the 'product/sum' form

$$\text{i.e. } C = \frac{C_1 C_2}{C_1 + C_2} \text{ farad} \qquad (3.15)$$

Worked Example 3.8

Question

A 6 μF and a 4 μF capacitor are connected in series across a 150 V supply. Calculate (a) the total capacitance (b) the charge on each capacitor and (c) the p.d. developed across each.

Answer

Figure 3.12 shows the appropriate circuit diagram.

$C_1 = 6\,\mu\text{F}$; $C_2 = 4\,\mu\text{F}$; $V = 150\,\text{V}$

Continued on p. 138

Worked Example 3.8 (*Continued*)

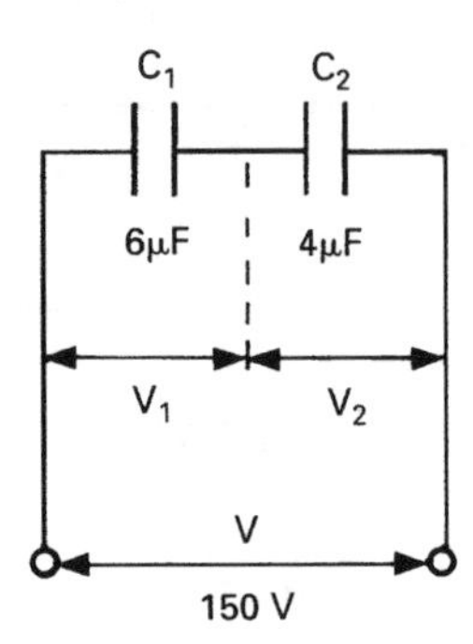

Fig. 3.12

(a) $C = \frac{C_1 C_2}{C_1 + C_2}$ farad

$= \frac{6 \times 4}{6 + 4}$ microfarad $= \frac{24}{10}$

so $C = 2.4\,\mu$F **Ans**

(b) $Q = VC$ coulomb $= 150 \times 2.4 \times 10^{-6}$

so $Q = 360\,\mu$C **Ans**

(same charge on both)

> Since capacitors in series all receive the same value of charge, then this must be the total charge drawn from the supply,
>
> $Q = \frac{V}{C}$
>
> This is equivalent to a series resistor circuit where the current drawn from the supply is common to all the resistors

(c) $V_1 = \frac{Q}{C_1}$ volt $= \frac{360 \times 10^{-6}}{6 \times 10^{-6}}$

so $V_1 = 60$ V **Ans**

Similarly, $V_2 = \frac{Q}{C_2}$ volt $= \frac{360 \times 10^{-6}}{4 \times 10^{-6}}$

so $V_2 = 90$ V **Ans**

Note that $V_1 + V_2 = 150\text{ V} = V$

Worked Example 3.9

Question

Capacitors of 3 μF, 6 μF and 12 μF are connected in series across a 400 V supply. Determine the p.d. across each capacitor.

Answer

Figure 3.13 shows the relevant circuit diagram.

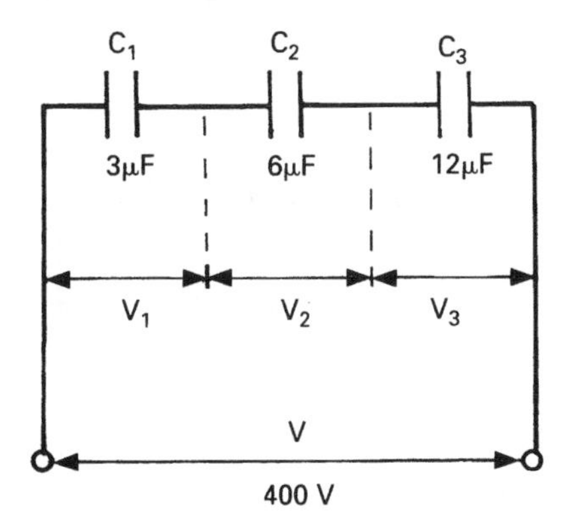

Fig. 3.13

$C_1 = 3\,\mu\text{F};\ C_2 = 6\mu\text{F};\ C_3 = 12\mu\text{F};\ V = 400\,\text{V}$

$$\frac{1}{C} = \frac{1}{C_1} + \frac{1}{C_2} + \frac{1}{C_3} = \frac{1}{3} + \frac{1}{6} + \frac{1}{2}$$

$$= \frac{4+2+1}{12} = \frac{7}{12}$$

$$\text{so } C = \frac{12}{7} = 1.714\,\mu\text{F}$$

$$Q = VC \text{ coulomb} = 400 \times 1.714 \times 10^{-6}$$

$$Q = 685.7\,\mu\text{C}$$

$$V_1 = \frac{Q}{C_1} \text{ volt} = \frac{685.7 \times 10^{-6}}{3 \times 10^{-6}}$$

so $V_1 = 228.6\,\text{V}$ **Ans**

Similarly, $V_2 = \dfrac{685.7 \times 10^{-6}}{6 \times 10^{-6}} = 114.3\,\text{V}$ **Ans**

and $V_3 = \dfrac{685.7 \times 10^{-6}}{12 \times 10^{-6}} = 57.1\,\text{V}$ **Ans**

3.14 Series/Parallel Combinations

The techniques required for the solution of this type of circuit are again best demonstrated by means of a worked example.

Worked Example 3.10

Question

For the circuit shown in Fig. 3.14, determine (a) the charge drawn from the supply, (b) the charge on the 8 μF capacitor, (c) the p.d. across the 4 μF capacitor, and (d) the p.d. across the 3 μF capacitor.

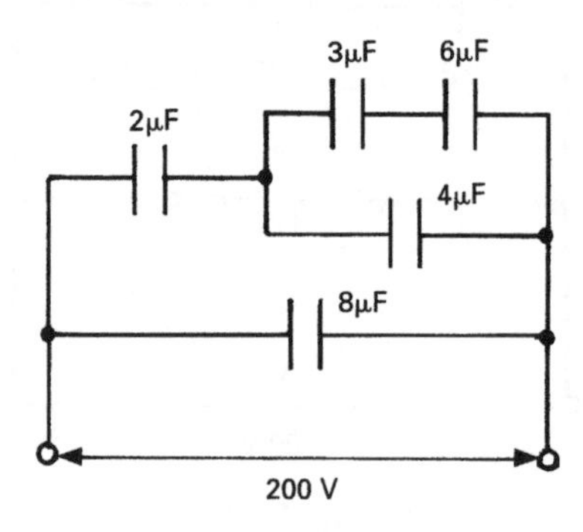

Fig. 3.14

Answer

The first task is to label the diagram as shown in Fig. 3.15.

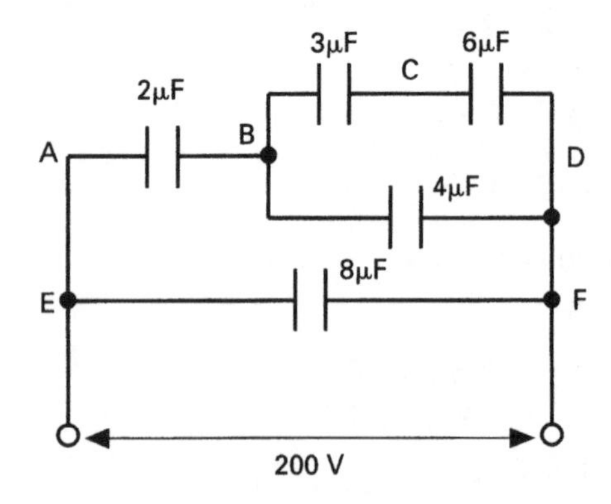

Fig. 3.15

(a) $C_{BCD} = \dfrac{3 \times 6}{3 + 6} = 2\,\mu\text{F}$ (see Fig. 3.16)

$C_{BD} = 2 + 4 = 6\,\mu\text{F}$ (see Fig. 3.17)

$$C_{AD} = \frac{6 \times 2}{6 + 2} = 1.5\,\mu\text{F} \quad \text{(see Fig. 3.18)}$$

$$C = C_{AD} + C_{EF} = 1.5 + 8$$

so $C = 9.5\,\mu\text{F}$

$$Q = VC \text{ coulomb} = 200 \times 9.5 \times 10^{-6}$$

hence, $Q = 1.9\,\text{mC}$ **Ans**

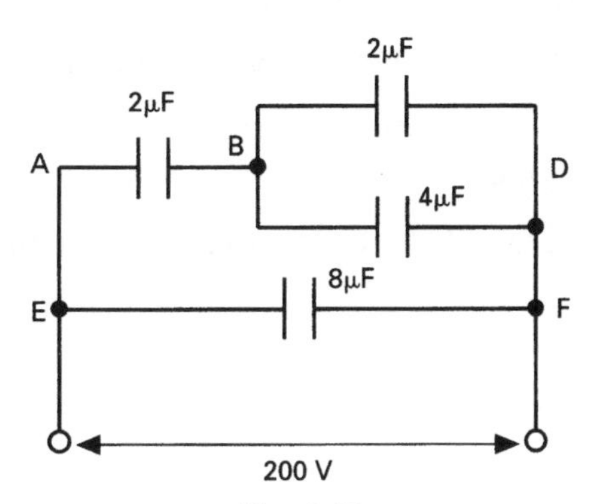

Fig. 3.16

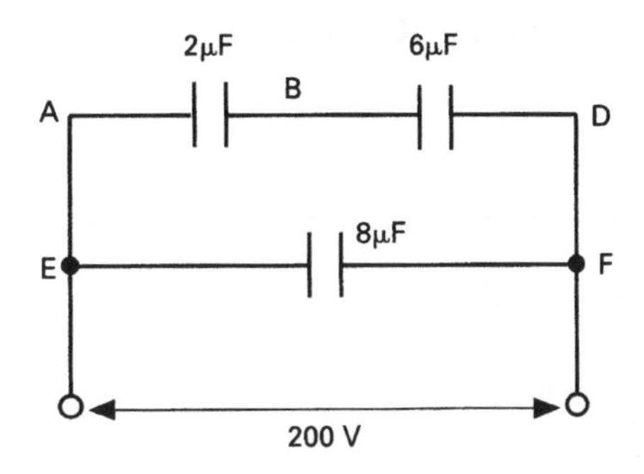

Fig. 3.17

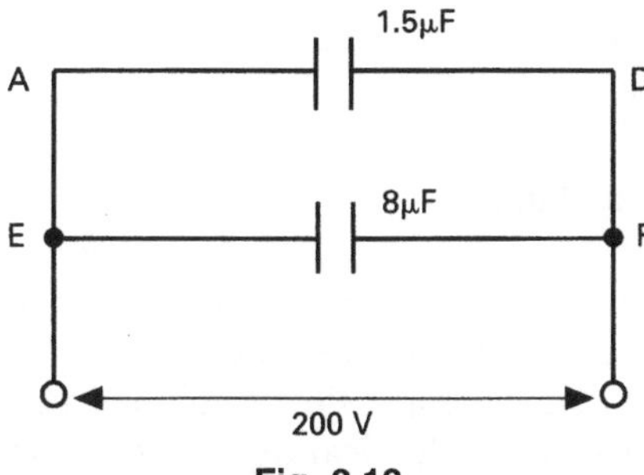

Fig. 3.18

Continued on p. 142

Worked Example 3.10 (*Continued*)

(b) $Q_{EF} = VC_{EF}$ coulomb $= 200 \times 8 \times 10^{-6}$

so $Q_{EF} = 1.6\,\text{mC}$ **Ans**

(c) Total charge $Q = 1.9\,\text{mC}$ and $Q_{EF} = 1.6\,\text{mC}$

so $Q_{AD} = 1.9 - 1.6 = 0.3\,\text{mC}$ (see Fig. 3.18).

and referring to Fig. 3.17, this will be the charge on both the capacitors shown, i.e.

$Q_{AB} = Q_{BD} = 0.3$ mC.

Thus, V_{BD} = p.d. across 4 μF capacitor (see Fig. 3.16)

$$V_{BD} = \frac{Q_{BD}}{C_{BD}}$$

$$= \frac{0.3 \times 10^{-3}}{6 \times 10^{-6}} \text{ volt}$$

so $V_{BD} = 50\,\text{V}$ **Ans**

(d) $Q_{BCD} = V_{BD}C_{BCD}$ (see Figs 3.16 and 3.15)

$$= 50 \times 2 \times 10^{-6}$$

$$= 100\,\mu\text{C}$$

and this will be the charge on both the 3 μF and 6 μF capacitors, i.e.

$$Q_{BC} = Q_{CD} = 100\,\mu\text{C}$$

Thus $V_{BC} = \dfrac{Q_{BC}}{C_{BC}}$ volt

$$= \frac{1 \times 10^{-4}}{3 \times 10^{-6}}$$

so $V_{BC} = 33.33\,\text{V}$ **Ans**

3.15 Multiplate Capacitors

Most practical capacitors consist of more than one pair of parallel plates, and in these cases they are referred to as multiplate capacitors. The sets of plates are often interleaved as shown in Fig. 3.19. The example illustrated has a total of five plates. It may be seen that this effectively forms four identical capacitors, in which the three inner plates are common to the two 'inner' capacitors. Since all the positive plates are joined together, and so too are the negative plates, then this arrangement is equivalent to four identical capacitors connected in parallel, as shown in Fig. 3.20. The total capacitance of

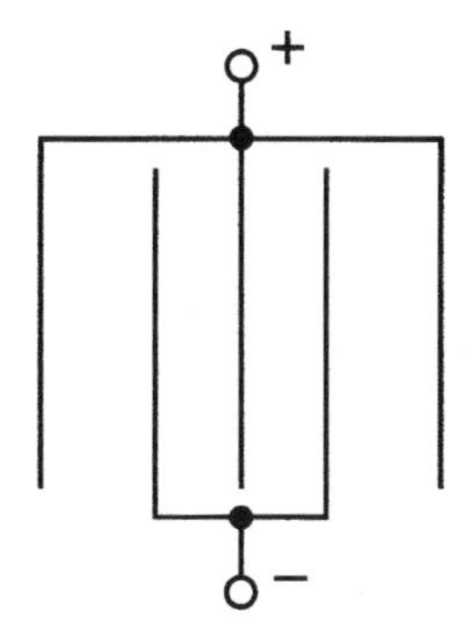

Fig. 3.19

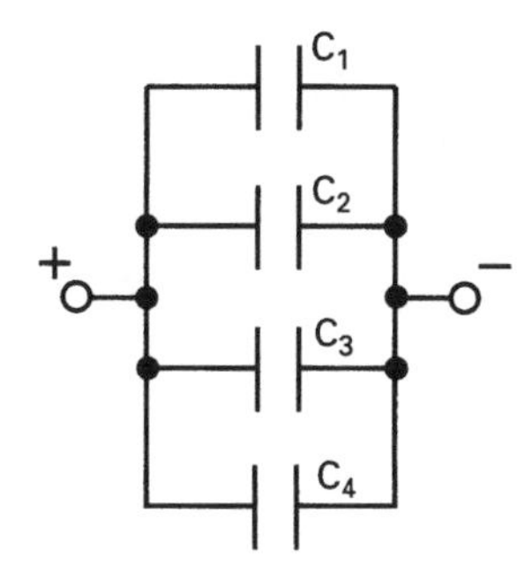

Fig. 3.20

four identical capacitors connected in parallel is simply four times the capacitance of one of them. Thus, this value will be the effective capacitance of the complete capacitor.

The capacitance between one adjacent pair of plates will be

$$C_1 = \frac{\epsilon_o \epsilon_r A}{d} \text{ farad}$$

so, the total for the complete arrangement $= C_1 \times 4$, but we can express 4 as (5–1) so the total capacitance is

$$C_1 = \frac{\epsilon_o \epsilon_r A(5-1)}{d} \text{ farad}$$

In general therefore, if a capacitor has N plates, the capacitance is given by the expression:

$$C_1 = \frac{\epsilon_o \epsilon_r A(N-1)}{d} \text{ farad} \tag{3.16}$$

Since the above equation applies generally, then it must also apply to capacitors having just one pair of plates as previously considered. This is correct, since if $N = 2$ then $(N - 1) = 1$, and the above equation becomes identical to equation (3.12) previously used.

Worked Example 3.11

Question

A capacitor is made from 20 interleaved plates each 80 mm by 80 mm separated by mica sheets 1.5 mm thick. If the relative permittivity for mica is 6.4, calculate the capacitance.

Answer

$N = 20$; $A = 80 \times 80 \times 10^{-6}\,\text{m}^2$; $d = 1.5 \times 10^{-3}\,\text{m}$; $\epsilon_r = 6.4$

$$C_1 = \frac{\epsilon_0 \epsilon_r A(N-1)}{d}\,\text{farad}$$

$$= \frac{8.854 \times 10^{-12} \times 6.4 \times 6400 \times 10^{-6} \times 19}{1.5 \times 10^{-3}}$$

so $C = 4.6\,\text{nF}$ **Ans**

Worked Example 3.12

Question

A 300 pF capacitor has nine parallel plates, each 40 mm by 30 mm, separated by mica of relative permittivity 5. Determine the thickness of the mica.

Answer

$N = 9$; $C = 3 \times 10^{-10}\,\text{F}$; $A = 40 \times 30 \times 10^{-6}\,\text{m}^2$; $\epsilon_r = 5$

$$C = \frac{\epsilon_0 \epsilon_r A(N-1)}{d}\,\text{farad}$$

$$d = \frac{\epsilon_0 \epsilon_r A(N-1)}{C}\,\text{metre}$$

$$= \frac{8.854 \times 10^{-12} \times 5 \times 1200 \times 10^{-6} \times 8}{3x10^{-10}}$$

so $d = 1.42\,\text{mm}$ **Ans**

3.16 Energy Stored

When a capacitor is connected to a voltage source of V volts we have seen that it will charge up until the p.d. between the plates is also V volts. If the capacitor is now disconnected from the supply, the charge and p.d. between its plates will be retained.

Consider such a charged capacitor, as shown in Fig. 3.21, which now has a resistor connected across its terminals. In this case the capacitor will behave as if it were a source of emf. It will therefore drive current through the resistor. In this way the stored charge will be dissipated as the excess electrons on its negative plate are returned to the positive plate. This discharge process will continue until the capacitor becomes completely discharged (both plates electrically neutral). Note that the discharge current marked on the diagram indicates *conventional* current flow.

However, if a discharge current flows then work must be done (energy is being dissipated). The only possible source of this energy in these circumstances must be the capacitor itself. Thus the charged capacitor must store energy.

If a graph is plotted of capacitor p.d. to the charge it receives, the area under the graph represents the energy stored. Assuming a constant charging current, the graph will be as shown in Fig. 3.22.

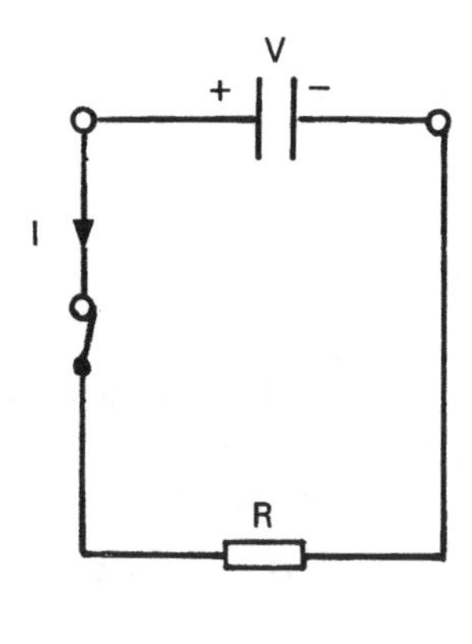

Fig. 3.21

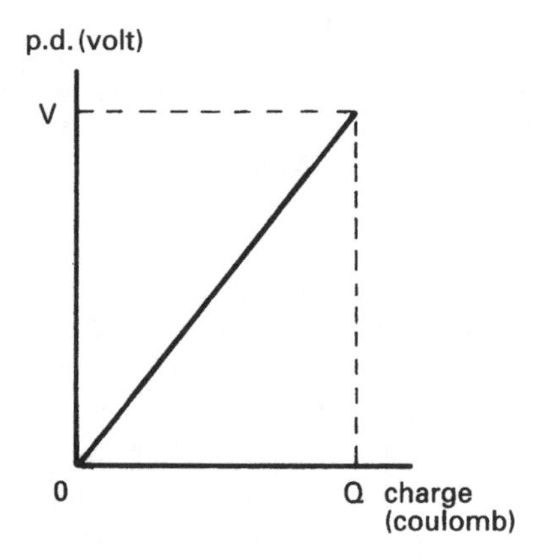

Fig. 3.22

$$\text{The area under the graph} = \frac{1}{2}QV$$

$$\text{but } Q = CV \text{ coulomb}$$

$$\text{so energy stored, } W = \frac{1}{2}CV^2 \text{ joule} \qquad (3.17)$$

Worked Example 3.13

Question

A 3 μF capacitor is charged from a 250 V d.c. supply. Calculate the charge and energy stored. The charged capacitor is now removed from the supply and connected across an uncharged 6 μF capacitor. Calculate the p.d. between the plates and the energy now stored by the combination.

Answer

$C_1 = 3\,\mu\text{F}$; $V_1 = 250\,\text{V}$; $C_2 = 6\,\mu\text{F}$

$$Q = V_1C_1 \text{ coulomb} = 250 \times 3 \times 10^{-6}$$

so $Q = 0.75\,\text{mC}$ **Ans**

$$W = \frac{1}{2}C_1V_1^2 \text{ joule} = \frac{1}{2} \times 3 \times 10^{-6} \times 250^2$$

so $W = 93.75\,\text{mJ}$ **Ans**

When the two capacitors are connected in parallel the 3 μF will share its charge with 6 μF capacitor. Thus the total charge in the system will remain unchanged, but the total capacitance will now be different:

$$\text{Total capacitance, } C = C_1 + C_2 \text{ farad} = 3 + 6$$

so $C = 9\,\mu\text{F}$

$$V = \frac{Q}{C} \text{ volt} = \frac{7.5 \times 10^{-4}}{9 \times 10^{-6}}$$

so $V = 83.33\,\text{V}$ **Ans**

$$\text{total energy stored, } W = \frac{1}{2}CV^2 \text{ joule} = \frac{1}{2} \times 9 \times 10^{-6} \times 83.33^2$$

so $W = 31.25\,\text{mJ}$ **Ans**

Note: The above example illustrates the law of conservation of charge, since the charge placed on the first capacitor is simply redistributed between the two capacitors when connected in parallel. The total charge therefore remains the same. However, the p.d. now existing between the plates has fallen, and so too has the total energy stored. But there is also a law of conservation of energy, so what has happened to the 'lost' energy? Well, in order for the 3 μF capacitor to share its charge with the 6 μF capacitor a charging current had to flow from one to the other. Thus this 'lost' energy was used in the charging process.

3.17 Dielectric Strength and Working Voltage

There is a maximum potential gradient that any insulating material can withstand before **dielectric breakdown** occurs.

There are of course some applications where dielectric breakdown is deliberately produced e.g. a sparking plug in a car engine, which produces an arc between its electrodes when subjected to a p.d. of several kilovolts. This then ignites the air/fuel mixture. However, it is obviously not a condition that is desirable in a capacitor, since it results in its destruction.

Capacitors normally have marked on them a maximum working voltage. When in use you must ensure that the voltage applied between its terminals does not exceed this value, otherwise dielectric breakdown will occur.

> Dielectric breakdown is the effect produced in an insulating material when the voltage applied across it is more than it can withstand. The result is that the material is forced to conduct. However when this happens, the sudden surge of current through it will cause it to burn, melt, vaporise or be permanently damaged in some other way

Another way of referring to this maximum working voltage is to quote the dielectric strength. This is the maximum voltage gradient that the dielectric can withstand, quoted in kV/m or in V/mm.

Worked Example 3.14

Question

A capacitor is designed to be operated from a 400 V supply, and uses a dielectric which (allowing for a factor of safety), has a dielectric strength of 0.5 MV/m. Calculate the minimum thickness of dielectric required.

Answer

$V = 400\,\text{V}$; $\mathbf{E} = 0.5 \times 10^6\,\text{V/m}$

$$\mathbf{E} = \frac{V}{d}\,\text{volt/metre}$$

$$\text{so } d = \frac{V}{\mathbf{E}}\,\text{metre} = \frac{400}{0.5 \times 10^6}$$

$d = 0.8\,\text{mm}$ **Ans**

3.18 Capacitor Types

The main difference between capacitor types is in the dielectric used. There are a number of factors that will influence the choice of capacitor type for a given application. Amongst these are the capacitance value, the working voltage, the **tolerance**, the stability, the leakage resistance, the size and the price.

Since $C = \epsilon A/d$, then any or all of these factors can be varied to suit particular requirements. Thus, if a large value of capacitance is required, a large csa and/or a small distance of separation will be necessary, together with a dielectric of high relative permittivity. However, if the area is to be large, then this can result in a device that is

> **Tolerance** is the deviation from the nominal value. This is normally expressed as a percentage. Thus a capacitor of nominal value 2 μF and a tolerance of $\pm$10%, should have an actual value of between 1.8 and 2.2 μF

unacceptably large. Additionally, the dielectric cannot be made too thin lest its dielectric strength is exceeded. The various capacitor types overcome these problems in a number of ways.

Paper This is the simplest form of capacitor. It utilises two strips of aluminium foil separated by sheets of waxed paper. The whole assembly is rolled up into the form of a cylinder (like a Swiss roll). Metal end caps make the electrical connections to the foils, and the whole assembly is then encapsulated in a case. By rolling up the foil and paper a comparatively large csa can be produced with reasonably compact dimensions. This type is illustrated in Fig. 3.23.

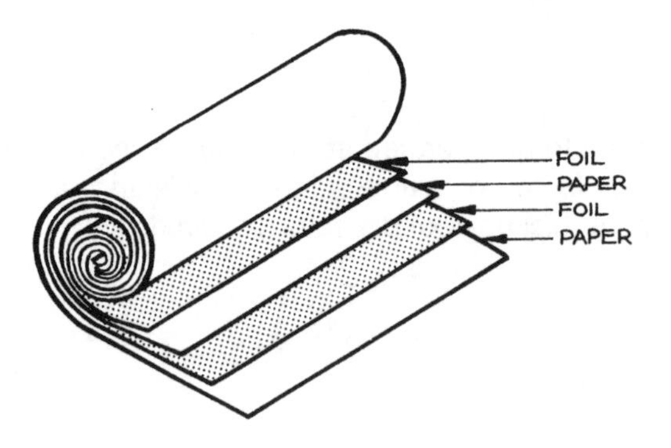

Fig. 3.23

Air Air dielectric capacitors are the most common form of variable capacitor. The construction is shown in Fig. 3.24. One set of plates is fixed, and the other set can be rotated to provide either more or less overlap between the two. This causes variation of the effective csa and hence variation of capacitance. This is the type of device connected to the station tuning control of a radio.

Fig. 3.24

'Plastic' With these capacitors the dielectric can be of polyester, polystyrene, polycarbonate or polypropylene. Each material has slightly different electrical characteristics which can be used to advantage, depending upon the proposed application. The construction takes much the same form as that for paper capacitors. Examples of these types are shown in Fig. 3.25, and their different characteristics are listed in Table 3.1.

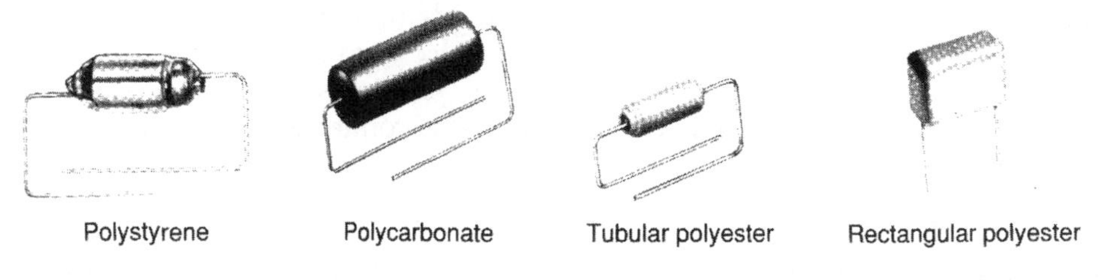

Fig. 3.25

Table 3.1 *Capacitor characteristics*

Type	*Capacitance*	*Tolerance (%)*	*Other characteristics*
Paper	1 nF–40 μF	±2	Cheap. Poor stability
Air	5 pF–1 nF	±1	Variable. Good stability
Polycarbonate	100 pF–10 μF	±10	Low loss. High temperature
Polyester	1 nF–2 μF	±20	Cheap. Low frequency
Polypropylene	100 pF–10 nF	±5	Low loss. High frequency
Polystyrene	10 pF–10 nF	±2	Low loss. High frequency
Mixed	1 nF–1 μF	±20	General purpose
Silvered mica	10 pF–10 nF	±1	High stability. Low loss
Electrolytic (aluminium)	1–100 000 μF	−20 to +80	High loss. High leakage. d.c. circuits only
Electrolytic (tantalum)	0.1–150 μF	±20	As for aluminium above
Ceramic	2 pF–100 nF	±10	Low temperature coefficient. High frequency

Silvered mica These are the most accurate and reliable of the capacitor types, having a low tolerance figure. These features are usually reflected in their cost. They consist of a disc or hollow cylinder of ceramic material which is coated with a silver compound. Electrical connections are affixed to the silver coatings and the whole assembly is placed into a casing or (more usually) the assembly is encased in a waxy substance.

Fig. 3.26

Mixed dielectric This dielectric consists of paper impregnated with polyester which separates two aluminium foil sheets as in the paper capacitor. This type makes a good general-purpose capacitor, and an example is shown in Fig. 3.26.

Electrolytic This is the form of construction used for the largest value capacitors. However, they also have the disadvantages of reduced working voltage, high leakage current, and the requirement to be **polarised**. Their terminals are marked + and −, and these polarities must be observed when the device is connected into a circuit. Capacitance values up to 100 000 μF are possible.

> A **polarised** capacitor is one in which the dielectric is formed by passing a d.c. current through it. The polarity of the d.c. supply used for this purpose must be subsequently observed in any circuit in which the capacitor is then used. Thus, they should be used only in d.c. circuits

The dielectric consists of either an aluminium oxide or tantalum oxide film that is just a few micrometres thick. It is this fact that allows such high capacitance values, but at the same time reduces the possible maximum working voltage. Tantalum capacitors are usually very much smaller than the aluminium types. They therefore cannot obtain the very high values of capacitance possible with the aluminium type. The latter consist of

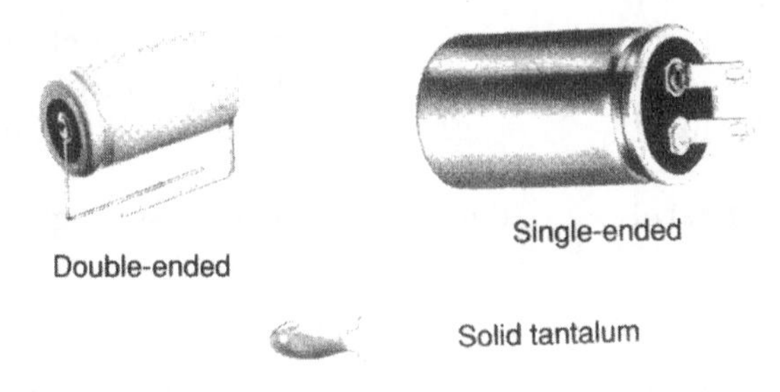

Fig. 3.27

two sheets of aluminium separated by paper impregnated with an electrolyte. These are then rolled up like a simple paper capacitor. This assembly is then placed in a hermetically sealed aluminium cannister. The oxide layer is formed by passing a charging current through the device, and it is the polarity of this charging process that determines the resulting terminal polarity that must be subsequently observed. If the opposite polarity is applied to the capacitor the oxide layer is destroyed. Examples of electrolytic capacitors are shown in Fig. 3.27.

Assignment Questions

1. Two parallel plates 25 cm by 35 cm receive a charge of 0.2 μC from a 250 V supply. Calculate (a) the electric flux and, (b) the electric flux density.
2. The flux density between two plates separated by a dielectric of relative permittivity 8 is 1.2 $\mu C/m^2$. Determine the potential gradient between them.
3. Calculate the electrical field strength between a pair of plates spaced 10 mm apart when a p.d. of 0.5 kV exists between them.
4. Two plates have a charge of 30 μC. If the effective area of the plates is 5 cm^2, calculate the flux density.
5. A capacitor has a dielectric 0.4 mm thick and operates at 50 V. Determine the electric field strength.
6. A 100 μF capacitor has a p.d. of 400 V across it. Calculate the charge that it has received.
7. A 47 μF capacitor stores a charge of 7.8 mC when connected to a d.c. supply. Calculate the supply voltage.
8. Determine the p.d. between the plates of a 470 nF capacitor if it stores a charge of 0.141 mC.
9. Calculate the capacitance of a pair of plates having a p.d. of 600 V when charged to 0.3 μC.
10. The capacitance of a pair of plates is 40 pF when the dielectric between them is air. If a sheet of glass is placed between the plates (so that it completely fills the space between them), calculate the capacitance of the new arrangement if the relative permittivity of the glass is 6.
11. A dielectric 2.5 mm thick has a p.d. of 440 V developed across it. If the resulting flux density is 4.7 $\mu C/m^2$ determine the relative permittivity of the dielectric.
12. State the factors that affect the capacitance of a parallel plate capacitor, and explain how the variation of each of these factors affects the capacitance.

    Calculate the value of a two plate capacitor with a mica dielectric of relative permittivity 5 and thickness 0.2 mm. The effective area of the plates is 250 cm^2.
13. A capacitor consists of two plates, each of effective area 500 cm^2, spaced 1 mm apart in air. If the capacitor is connected to a 400 V supply, determine (a) the capacitance, (b) the charge stored and (c) the potential gradient.
14. A paper dielectric capacitor has two plates, each of effective csa 0.2 m^2. If the capacitance is 50 nF calculate the thickness of the paper, given that its relative permittivity is 2.5.

15 A two plate capacitor has a value of 47 nF. If the plate area was doubled and the thickness of the dielectric was halved, what then would be the capacitance?

16 A parallel plate capacitor has 20 plates, each 50 mm by 35 mm, separated by a dielectric 0.4 mm thick. If the capacitance is 1000 pF determine the relative permittivity of the dielectric.

17 Calculate the number of plates used in a 0.5 nF capacitor if each plate is 40 mm square, separated by dielectric of relative permittivity 6 and thickness 0.102 mm.

18 A capacitor is to be designed to have a capacitance of 4.7 pF and to operate with a p.d. of 120 V across its terminals. The dielectric is to be teflon ($\epsilon_r = 2.1$) which, after allowing for a safety factor has a dielectric strength of 25 kV/m. Calculate (a) the thickness of teflon required and (b) the area of a plate.

19 Capacitors of 4 μF and 10 μF are connected (a) in parallel and (b) in series. Calculate the equivalent capacitance in each case.

20 Determine the equivalent capacitance when the following capacitors are connected (a) in series and (b) in parallel

(i) 3 μF, 4 μF and 10 μF

(ii) 0.02 μF, 0.05 μF and 0.22 μF

(iii) 20 pF and 470 pF

(iv) 0.01 μF and 220 pF.

21 Determine the value of capacitor which when connected in series with a 2 nF capacitor produces a total capacitance of 1.6 nF.

22 Three 15 μF capacitors are connected in series across a 600 V supply. Calculate (a) the total capacitance, (b) the p.d. across each and (c) the charge on each.

23 Three capacitors, of 6 μF, 8 μF and 10 μF respectively are connected in parallel across a 60 V supply. Calculate (a) the total capacitance, (b) the charge stored in the 8 μF capacitor and (c) the total charge taken from the supply.

24 For the circuit of Fig. 3.28 calculate (a) the p.d. across each capacitor and (b) the charge stored in the 3 nF.

25 Calculate the values of C_2 and C_3 shown in Fig. 3.29.

26 Calculate the p.d. across, and charge stored in, each of the capacitors shown in Fig. 3.30.

27 A capacitor circuit is shown in Fig. 3.31. With the switch in the open position, calculate the p.d.s across capacitors C_1 and C_2.
When the switch is closed, the p.d. across C_2 becomes 400 V. Calculate the value of C_3.

28 In the circuit of Fig. 3.32 the variable capacitor is set to 60 μF. Determine the p.d. across this capacitor if the supply voltage between terminals AB is 500 V.

29 A 50 pF capacitor is made up of two plates separated by a dielectric 2 mm thick and of relative permittivity 1.4. Calculate the effective plate area.

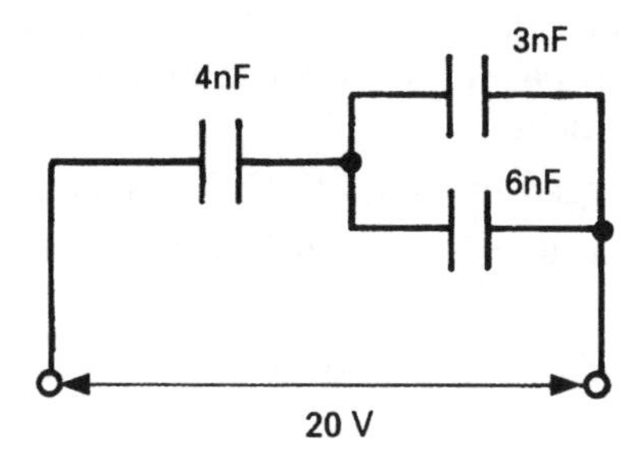

Fig. 3.28

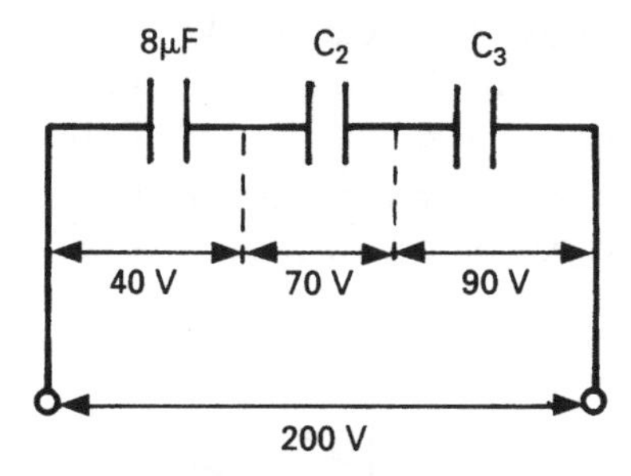

Fig. 3.29

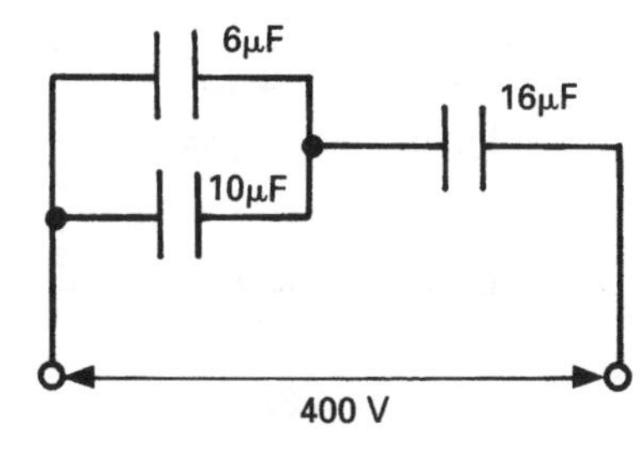

Fig. 3.30

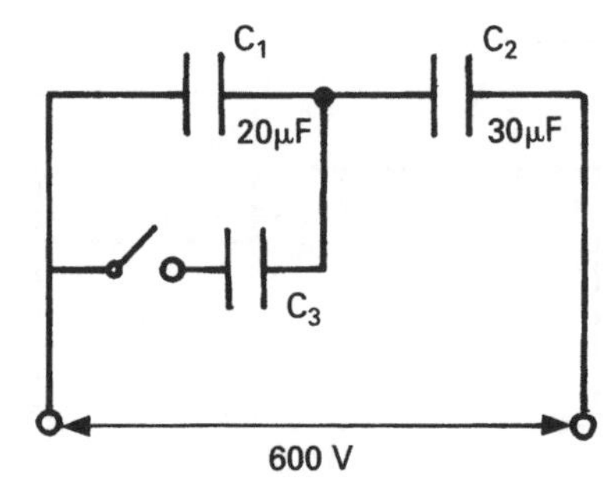

Fig. 3.31

30 For the circuit shown in Fig. 3.33 the total capacitance is 16 pF. Calculate (a) the value of the unmarked capacitor, (b) the charge on the 10 pF capacitor and (c) the p.d. across the 40 pF capacitor.

31 A 20 μF capacitor is charged to a p.d. of 250 V. Calculate the energy stored.

32 The energy stored by a 400 pF capacitor is 8 μJ. Calculate the p.d. between its plates.

33 Determine the capacitance of a capacitor that stores 4 mJ of energy when charged to a p.d. of 40 V.

34 When a capacitor is connected across a 200 V supply it takes a charge of 8 μC. Calculate (a) its capacitance, (b) the energy stored and (c) the electric field strength if the plates are 0.5 mm apart.

35 A 4 μF capacitor is charged to a p.d. of 400 V and then connected across an uncharged 2 μF capacitor. Calculate (a) the original charge and energy stored in the 4 μF, (b) the p.d. across, and energy stored in, the parallel combination.

36 Two capacitors, of 4 μF and 6 μF are connected in series across a 250 V supply. (a) Calculate the charge and p.d. across each. (b) The capacitors are now disconnected

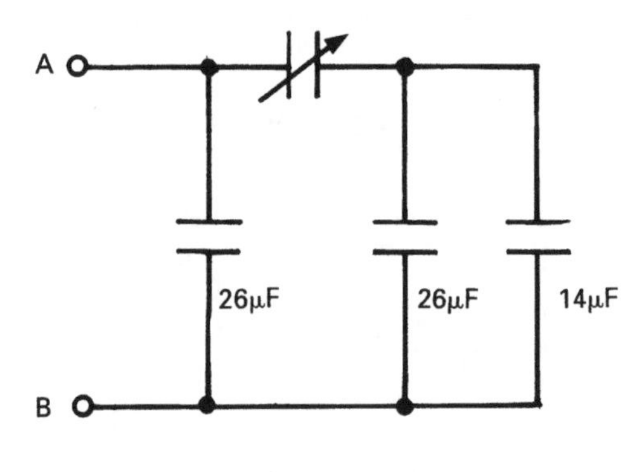

Fig. 3.32

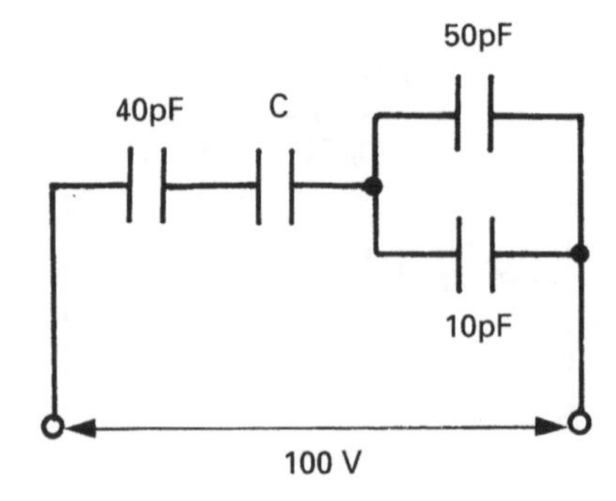

Fig. 3.33

from the supply and reconnected in parallel with each other, with terminals of similar polarity being joined together. Calculate the p.d. and charge for each.

37 A ceramic capacitor is to be made so that it has a capacitance of 100 pF and is to be operated from a 750 V supply. Allowing for a safety factor, the dielectric has a strength of 500 kV/m. Determine (a) the thickness of the ceramic, (b) the plate area if the relative permittivity of the ceramic is 3.2, (c) the charge and energy stored when the capacitor is connected to its rated supply voltage, and (d) the flux density under these conditions.

38 A large electrolytic capacitor of value 100 μF has an effective plate area of 0.942 m^2. If the aluminium oxide film dielectric has a relative permittivity of 6, calculate its thickness.

Suggested Practical Assignment

Assignment 1

To determine the total capacitance of capacitors, when connected in series, and in parallel.

Apparatus:

Various capacitors, of known values
1 × capacitance meter, or capacitance bridge

Method:

1. Using either the meter or bridge, measure the actual value of each capacitor.
2. Connect different combinations of capacitors in parallel, and measure the total capacitance of each combination.
3. Repeat the above procedure, for various series combinations.
4. Calculate the total capacitance for each combination, and compare these values to those previously measured.
5. Account for any difference between the actual and nominal values, for the individual capacitors.

Supplementary Worked Examples

Example 1: Two parallel metal plates, each having a csa of 400 mm^2, are charged from a constant current source of 50 μA for a time of 3 seconds. Calculate (a) the charge on the plates, and (b) the density of the electric field between them.

Solution: $A = 400 \times 10^{-6}\,m^2$; $I = 50 \times 10^{-6}\,A$; $t = 3\,s$

(a) $Q = It$ coulomb $= 50 \times 10^{-6} \times 3$

so, $Q = 150\,\mu\text{C}$ **Ans**

(b) $D = \frac{Q}{A}$ coulomb/metre^2

$= \frac{50 \times 10^{-6}}{400 \times 10^{-6}}$

so, $D = 0.125\,\text{C/m}^2$ **Ans**

Example 2: Two parallel plates, separated by an air space of 4 mm, receive a charge of 0.2 mC when connected to a 125 V source. Calculate (a) the electric field strength between the plates, (b) the csa of the field between the plates if the flux density is 15 C/m^2, and (d) the capacitance of the plates.

Solution: $d = 4 \times 10^{-3}$ m; $Q = 2 \times 10^{-4}$ C; $V = 125$ V; $D = 15\,\text{C/m}^2$

(a) $\mathbf{E} = \frac{V}{d}$ volt/metre $= \frac{125}{4 \times 10^{-3}}$

so, $\mathbf{E} = 31.25\,\text{kV/m}$ **Ans**

(b) $D = \frac{Q}{A}$ coulomb/metre^2

so, $A = \frac{Q}{D}$ metre^2 $= \frac{2 \times 10^{-4}}{15}$

thus, $A = 13.3 \times 10^{-6}\,\text{m}^2$ or $13.3\,\text{mm}^2$ **Ans**

(c) $Q = CV$ coulomb

so, $C = \frac{Q}{V}$ farad $= \frac{2 \times 10^{-4}}{125}$

thus, $C = 1.6\,\mu\text{F}$ **Ans**

Example 3: A parallel plate capacitor consists of 11 circular plates, each of radius 25 mm, with an air gap of 0.5 mm between each pair of plates. Calculate the value of the capacitor.

Solution: $n = 11$; $r = 25 \times 10^{-3}$ m; $d = 5 \times 10^{-4}$ m; $\varepsilon_r = 1$ (air)

$A = \pi r^2$ metre^2 $= \pi \times (25 \times 10^{-3})^2$

so, $A = 1.9635 \times 10^{-3}\,\text{m}^2$

$C = \frac{\varepsilon_0 \varepsilon_r A(n-1)}{d}$ farad

$= \frac{8.854 \times 10^{-12} \times 1.9635 \times 10^{-3} \times 10}{5 \times 10^{-4}}$

thus, $C = 3.48 \times 10^{-10}$ F or 348 pF **Ans**

Example 4: A 270pF capacitor is to be made from two metallic foil sheets, each of length 20 cm and width 3 cm, separated by a sheet of Teflon having a relative permittivity of 2.1. Determine (a) the thickness of Teflon sheet required, and (b)

the maximum possible working voltage for the capacitor if the Teflon has a dielectric strength of 350 kV/m.

Solution: $C = 270 \times 10^{-12}$ F; $A = 20 \times 3 \times 10^{-4}$ m^2; $\varepsilon_r = 2.1$; $\mathbf{E} = 350 \times 10^3$ V/m

(a)

$$C = \frac{\varepsilon_o \varepsilon_r A}{d} \text{ farad}$$

$$\text{so, } d = \frac{\varepsilon_o \varepsilon_r A}{C} \text{ metre}$$

$$= \frac{8.854 \times 10^{-12} \times 2.1 \times 60 \times 10^{-4}}{270 \times 10^{-12}}$$

thus, $d = 0.413$ mm **Ans**

(b) Dielectric strength is the same thing as electric field strength, expressed in volt/metre, so

$$\mathbf{E} = \frac{V}{d} \text{ volt/metre}$$

and $V = \mathbf{E}d$ volt $= 350 \times 10^3 \times 0.413 \times 10^{-3}$

so, $V = 144.6$ V **Ans**

Note: This figure is the voltage at which the dielectric will start to break down, so, for practical purposes, the maximum working voltage would be specified at a lower value. For example, if a factor of safety of 20% was required, then the maximum working voltage in this case would be specified as 115 V.

Example 5: Calculate the total (effective) capacitance for each of the three circuit arrangements shown in Figs. 3.34 (a), (b), and (c).

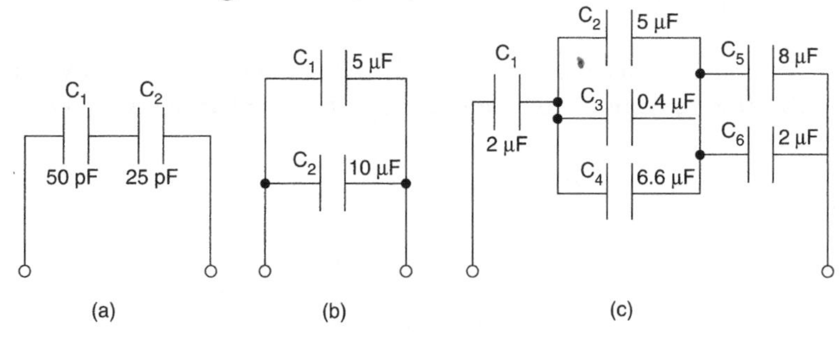

Fig. 3.34

Solution:

(a)

$$C = \frac{C_1 C_2}{C_1 + C_2} \text{ picofarad}$$

$$= \frac{50 \times 25}{50 + 25}$$

so, $C = 16.67$ pF **Ans**

(b) $C = C_1 + C_2$ microfarad
$= 5 + 10$
so, $C = 15\,\mu$F **Ans**

(c) A labelled circuit diagram is shown in Fig. 3.35.

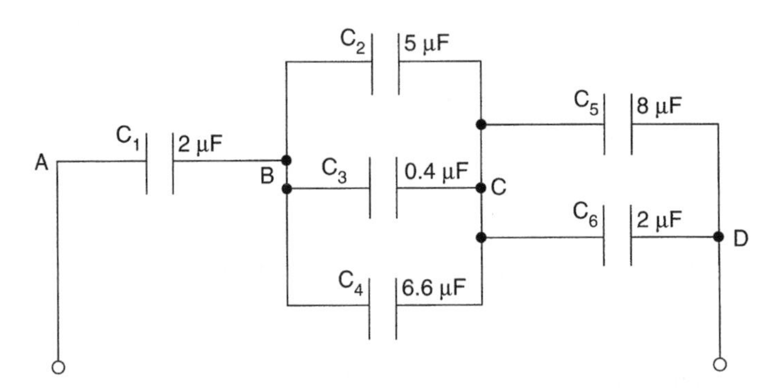

Fig. 3.35

$C_{CD} = C_5 + C_6$ microfarad $= 8 + 2 = 10\,\mu$F
$C_{BC} = C_2 + C_3 + C_4$ microfarad $= 5 + 0.4 + 6.6$
so, $C_{BC} = 12\,\mu$F

$$\frac{1}{C} = \frac{1}{C_{AB}} + \frac{1}{C_{BC}} + \frac{1}{C_{CD}} = \frac{1}{2} + \frac{1}{12} + \frac{1}{10}$$
$$= \frac{30 + 5 + 6}{60} = \frac{41}{60}$$

so, $C = 1.46\,\mu$F **Ans**

Example 6: A capacitor network connected to a d.c. supply is shown in Fig. 3.36. Calculate (a) the charge drawn from the supply, (b) the charge on C_3, and (c) the p.d. across C_1.

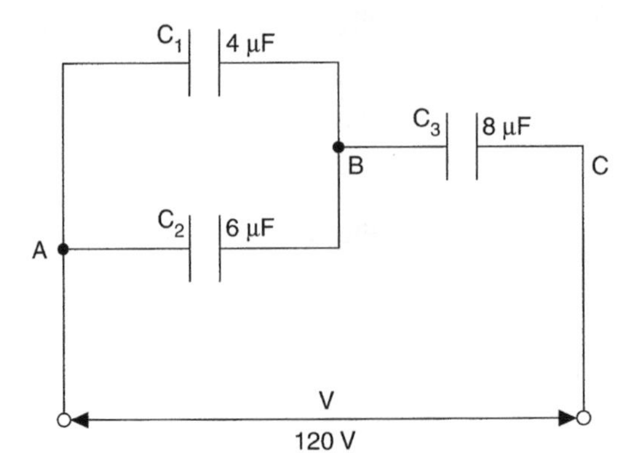

Fig. 3.36

Solution:

(a) $C_{AB} = C_1 + C_2$ microfarad $= 4 + 6$

so, $C_{AB} = 10\,\mu F$

$$C_{AC} = \frac{C_{AB}C_3}{C_{AB} \times C_3} \text{ microfarad} = \frac{10 \times 8}{10 + 8}$$

and $C_{AC} = 4.44\,\mu F$

$Q = VC_{AC}$ coulomb $= 120 \times 4.44 \times 10^{-6}$

so, $Q = 533\,\mu C$ **Ans**

(b) The equivalent circuit is shown in Fig. 3.37, in which each of the two effective capacitors will receive the same value of charge, Q coulomb, from the supply.

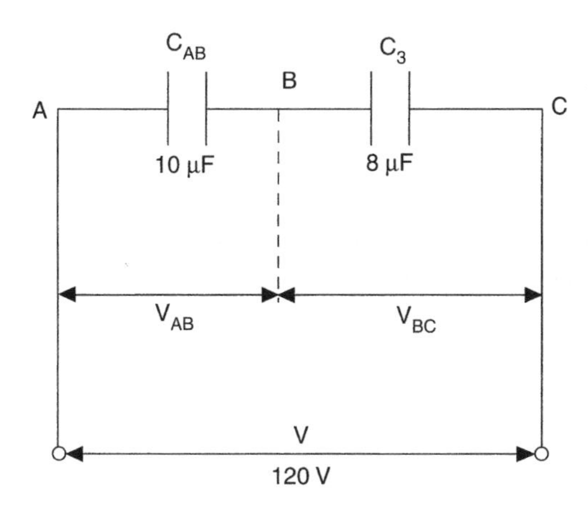

Fig. 3.37

Hence the charge on $C_3 = 533\,\mu C$ **Ans**

(c) From Fig. 3.37, $V_{AB} = \dfrac{Q}{C_{AB}}$ volt $= \dfrac{533 \times 10^{-6}}{10 \times 10^{-6}}$

so, p.d. across C_1 (and C_2) $= V_{AB} = 53.3\,V$ **Ans**

This result may be checked by calculating the p.d. across C_3, thus:

$$V_{BC} = \frac{533 \times 10^{-6}}{8 \times 10^{-6}} = 66.6\,V$$

and $V = V_{AB} + V_{BC} = 53.3 + 66.6 = 119.9\,V$

which, allowing for rounding error, is 120 V.

Example 7: For the circuit shown in Fig. 3.38 determine (a) the p.d. V_2, (b) the value of C_2, and (c) the value of C_3 and the energy stored in C_3.

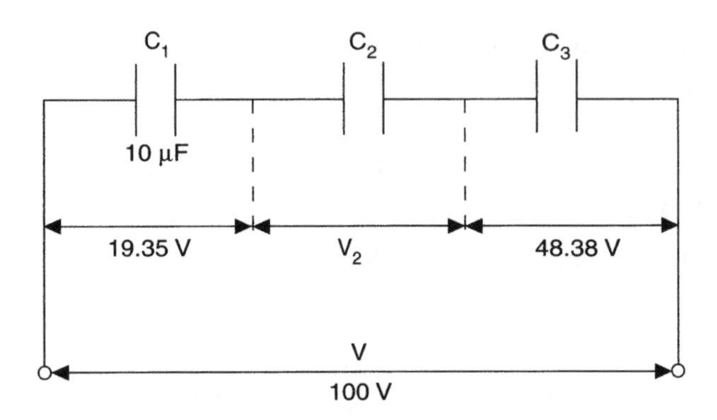

Fig. 3.38

Solution:

(a)

$$V_2 = 100 - (19.35 + 48.38)$$
$$= 100 - 67.73$$
$$\text{so, } V_2 = 32.27\,\text{V} \textbf{ Ans}$$

(b) Being a series circuit, all three capacitors will receive the same value of charge from the supply; i.e. the total circuit charge Q, where

$$Q = Q_3 = Q_2 = Q_1 = V_1C_1 \text{ coulomb}$$
$$= 19.35 \times 10 \times 10^{-6}$$
$$\text{Hence, } Q_2 = 193.5\,\mu\text{C}$$
$$C_2 = \frac{Q_2}{V_2} \text{ farad}$$
$$= \frac{193.5 \times 10^{-6}}{32.27}$$
$$\text{so, } C_2 = 6\,\mu\text{F} \textbf{ Ans}$$

(c)

$$C_3 = \frac{Q_3}{V_3} \text{ farad} = \frac{193.5 \times 10^{-6}}{48.38}$$
$$\text{so, } C_3 = 4\,\mu\text{F} \textbf{ Ans}$$
$$W_3 = \frac{1}{2}C_3V_3^2 \text{ joule} = 0.5 \times 4 \times 10^{-6} \times 48.38^2$$
$$\text{so, } W_3 = 4.68\,\text{mJ} \textbf{ Ans}$$

Example 8: Consider the circuit of Fig. 3.39, where initially all three capacitors are fully discharged, with the switch in position '1'.

(a) If the switch is now moved to position '2', calculate the charge and energy stored by C_1.

(b) Once C_1 is fully charged, the switch is returned to position '1'. Calculate the p.d. now existing across C_1 and the amount of energy used in charging C_2 and C_3 from C_1.

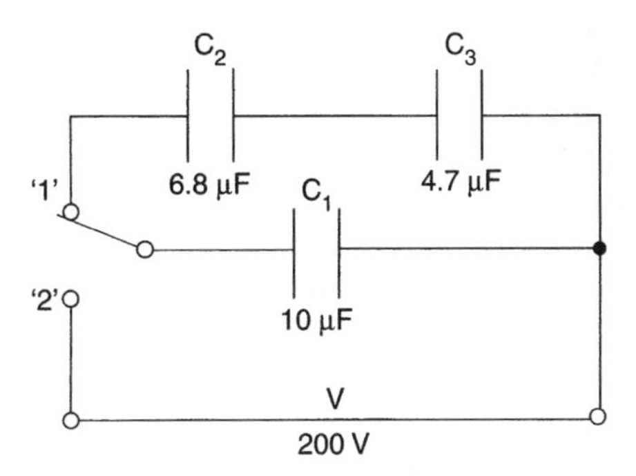

Fig. 3.39

Solution:

(a) $Q_1 = VC_1$ coulomb $= 200 \times 10 \times 10^{-6}$

so, $Q_1 = 2\,\text{mC}$ **Ans**

$$W_1 = \frac{1}{2}\,C_1V^2\,\text{joule} = 0.5 \times 10^{-5} \times 200^2$$

so, $W_1 = 0.2\,\text{J}$ **Ans**

(b) C_2 and C_3 in series is equivalent to $C_4 = \dfrac{C_2C_3}{C_2 + C_3}$ microfarad

$$\text{so, } C_4 = \frac{6.8 \times 4.7}{6.8 + 4.7} = 2.78\,\mu\text{F}$$

and total capacitance of the whole circuit,

$$C = C_1 + C_4 \text{ microfarad}$$
$$= 10 + 2.78$$

hence, $C = 12.78\,\mu\text{F}$

Now, the charge received by the circuit remains constant, although the total capacitance has increased.

$$\text{Thus, } V = \frac{Q}{C}\text{ volt} = \frac{2 \times 10^{-3}}{12.78 \times 10^{-6}}$$

and $V = 156.5\,\text{V}$ **Ans**

Total energy remaining in the circuit, $W = \dfrac{1}{2}\,CV^2$ joule

$$= 0.5 \times 12.78 \times 10^{-6} \times 156.5^2$$

and $W = 0.156\,\text{J}$

the energy used up must be the difference between the energy first stored by C_1 and the final energy stored in the system, hence

Energy used $= 0.2 - 0.156 = 44\,\text{mJ}$ **Ans**

Summary of Equations

Force between charges: $F = \dfrac{Q_1Q_2}{\varepsilon_o d^2}$ newton

Electric field strength: $\mathbf{E} = \dfrac{F}{q}$ newton/coulomb $= \dfrac{V}{d}$ volt/metre

Electric flux density: $D = \dfrac{Q}{A} = \varepsilon_o\varepsilon_r\,\mathbf{E}$ coulomb/metre^2

Permittivity: $\varepsilon = \varepsilon_o\varepsilon_r = \dfrac{D}{\mathbf{E}}$ farad/metre

Capacitance: $C = \dfrac{Q}{V}$ farad $= \dfrac{\varepsilon_o\varepsilon_r A\,(N-1)}{d}$ farad

Capacitors in parallel: $C = C_1 + C_2 + C_3 + \ldots$ farad

Capacitors in series: $\dfrac{1}{C} = \dfrac{1}{C_1} + \dfrac{1}{C_2} + \dfrac{1}{C_3} + \ldots$ farad^{-1}

For ONLY two in series, $C = \dfrac{C_1C_2}{C_1 + C_2}$ farad $\left(\dfrac{\text{product}}{\text{sum}}\right)$

Energy stored: $W = 0.5\,QV = 0.5\,CV^2$ joule

4 Magnetic Fields and Circuits

This chapter introduces the concepts and laws associated with magnetic fields and their application to magnetic circuits and materials.

On completion of the chapter you should be able to:

1 Describe the forces of attraction and repulsion between magnetised bodies.

2 Define the various magnetic properties and quantities, and use them to solve simple series magnetic circuit problems.

3 Appreciate the effect of magnetic hysteresis, and the properties of different types of magnetic material.

4.1 Magnetic Materials

All materials may be broadly classified as being in one of two groups. They may be magnetic or non-magnetic, depending upon the degree to which they exhibit magnetic effects. The vast majority of materials fall into the latter group, which may be further classified into diamagnetic and paramagnetic materials. The magnetic properties of these materials are very slight, and extremely difficult even to detect. Thus, for practical purposes, we can say that they are totally non-magnetic. The magnetic materials (based on iron, cobalt and ferrites) are the ferromagnetic materials, all of which exhibit very strong magnetic effects. It is with these materials that we will be principally concerned.

4.2 Magnetic Fields

Magnetic fields are produced by permanent magnets and by electric current flowing through a conductor. Like the electric field, a magnetic field may be considered as being the medium by which forces are transmitted. In this case, the forces between magnetised materials. A magnetic field is also represented by lines of force or magnetic flux. These are attributed with certain characteristics, listed below:

1 They always form complete closed loops.*

---

*Unlike lines of electric flux, which radiate from and terminate at the charged surfaces, lines of magnetic flux also exist all the way through the magnet.

2 They behave as if they are elastic. That is, when distorted they try to return to their natural shape and spacing.
3 In the space surrounding a magnet, the lines of force radiate from the north (N) pole to the south (S) pole.
4 They never intersect (cross).
5 Like poles repel and unlike poles attract each other.

Characteristics (1) and (3) are illustrated in Fig. 4.1 which shows the magnetic field pattern produced by a bar magnet.

Characteristics (2) and (4) are used to explain characteristic (5), as illustrated in Figs 4.2 and 4.3.

In the case of the arrangement of Fig. 4.2, since the lines behave as if they are elastic, then those lines linking the two magnets try to shorten themselves. This tends to bring the two magnets together.

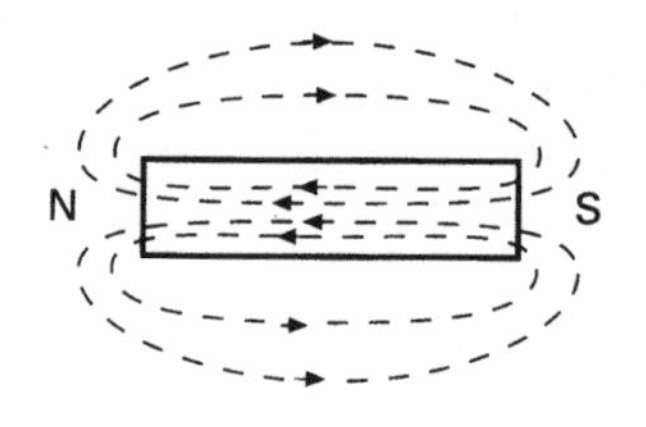

Fig. 4.1

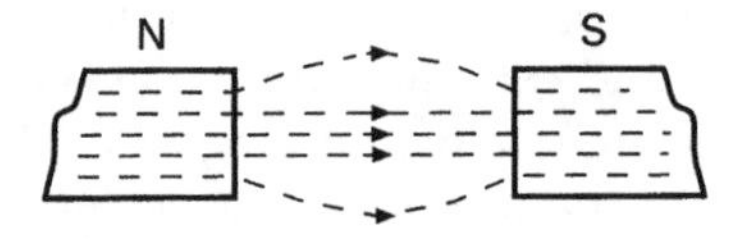

Fig. 4.2

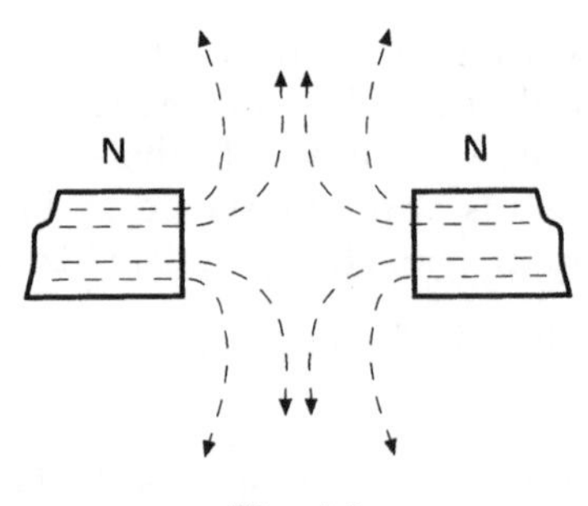

Fig. 4.3

The force of repulsion shown in Fig. 4.3 is a result of the unnatural compression of the lines between the two magnets. Once more, acting as if they are elastic, these lines will expand to their normal shape. This will tend to push the magnets apart.

Permanent magnets have the advantage that no electrical supply is required to produce the magnetic field. However, they also have several disadvantages. They are relatively bulky. The strength of the field cannot be varied. Over a period of time they tend to lose some of their magnetism (especially if subjected to physical shock or vibration). For many practical applications these disadvantages are unacceptable. Therefore a more convenient method of producing a magnetic field is required.

In addition to the heating effect, an electric current also produces a magnetic field. The strength of this field is directly proportional to the value of the current. Thus a magnetic field produced in this way may be turned on and off, reversed, and varied in strength very simply. A magnetic field is a vector quantity, as indicated by the arrows in the previous diagrams. The field pattern produced by a current flowing through a straight conductor is illustrated in Figs. 4.4(a) and (b). Note that *conventional* current flow is considered. The convention adopted to represent conventional current flowing away from the observer is a cross, and current towards the observer is marked by a dot. The direction of the arrows on the flux lines can easily be determined by considering the X as the head of a cross-head screw. In order to drive the screw away from you, the screw would be rotated *clockwise.* On the other hand, if you were to observe the point of the screw coming out towards you, it would be rotating *anticlockwise.* This convention is called the screw rule, and assumes a normal right-hand thread.

It should be noted that the magnetic flux actually extends the whole length of the conductor, in the same way that the insulation on a cable covers the whole length. In addition, the flux pattern extends outwards in concentric circles to infinity. However, as with electric and gravitational fields, the force associated with the field follows an inverse square law. It therefore diminishes very rapidly with distance.

The flux pattern produced by a straight conductor can be adapted to provide a field pattern like a bar magnet. This is achieved by winding the conductor in the form of a

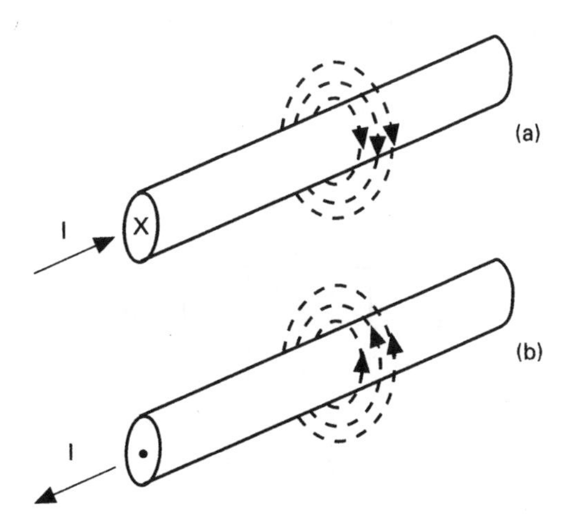

Fig. 4.4

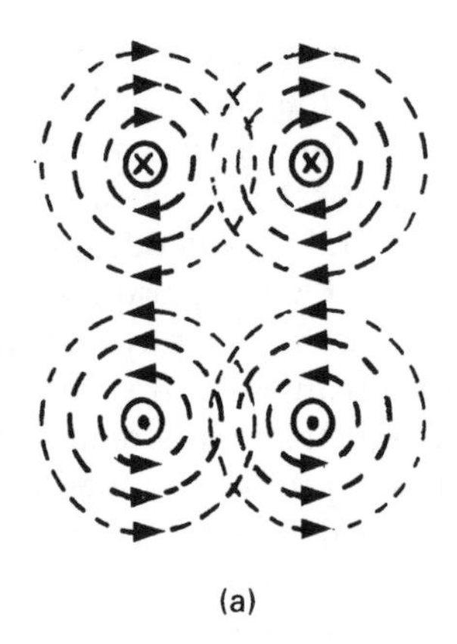

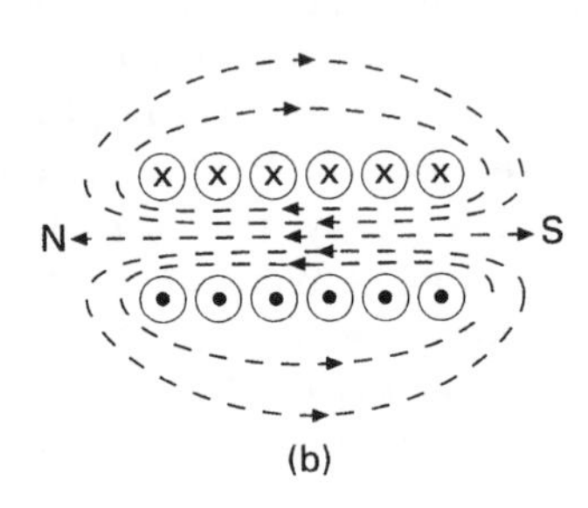

Fig. 4.5

coil. This arrangement is known as a solenoid. The principle is illustrated in Figs. 4.5(a) and (b), which show a cross-section of a solenoid. Figure 4.5(a) shows the flux patterns produced by two adjacent turns of the coil. However, since lines of flux will not intersect, the flux distorts to form complete loops around the whole coil as shown in Fig. 4.5(b).

4.3 The Magnetic Circuit

A magnetic circuit is all of the space occupied by the magnetic flux. Figure 4.6 shows an iron-cored solenoid, supplied with direct current, and the resulting flux pattern. This is what is known as a composite magnetic circuit, since the flux exists both in the iron core and in the surrounding air space. In addition, it can be seen that the spacing of the lines within the iron core is uniform, whereas it varies in the air space. Thus there is a uniform magnetic field in the core and a non-uniform field in the rest of the magnetic circuit. In order to make the design and analysis of a magnetic circuit easier, it is more

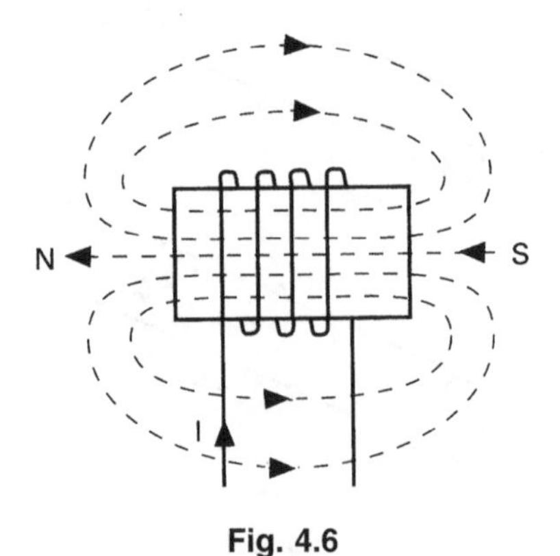

Fig. 4.6

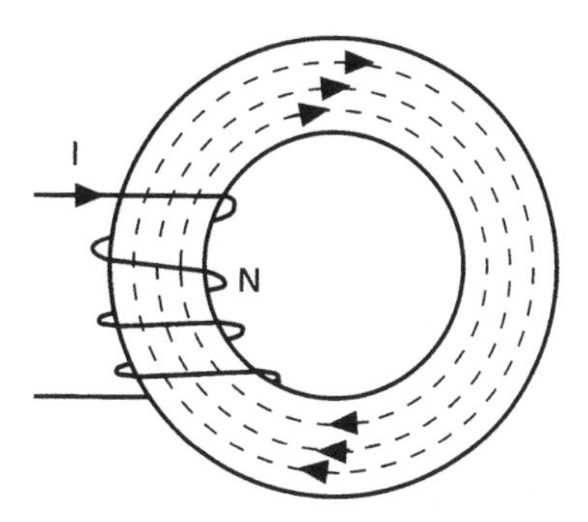

Fig. 4.7

convenient if a uniform field can be produced. This may be achieved by the use of a completely enclosed magnetic circuit. One form of such a circuit is an iron toroid, that has a current carrying coil wound round it. A toroid is a 'doughnut' shape having either a circular or a rectangular cross-section. Such an arrangement is shown in Fig. 4.7, and from this it can be seen that only the toroid itself forms the magnetic circuit. Provided that it has a uniform cross-section then the field contained within it will be uniform.

4.4 Magnetic Flux and Flux Density

The magnetic flux is what causes the observable magnetic effects such as attraction, repulsion etc. The unit of magnetic flux is the weber (Wb). This was the name of a German scientist so it is pronounced as 'vayber'.

The number of webers of flux per square metre of cross-section of the field is defined as the magnetic flux density (B), which is measured in tesla (T). This sometimes causes some confusion at first, since the logical unit would appear to be weber/metre^2. Indeed, this is the way in which it is calculated: the value of flux must be divided by the appropriate area. Tesla was the name of another scientist, whose name is thus commemorated. On reflection, it should not be particularly confusing, since the logical unit for electrical current would be coulomb/second; but it seems quite natural to use the term ampere.

The quantity symbols for magnetic flux and flux density are Φ and B respectively. Hence, flux density is given by the equation:

$$B = \frac{\Phi}{A} \text{ tesla} \tag{4.1}$$

Note: references have been made to iron as a core material and as the material used for toroids etc. This does not necessarily mean that pure iron is used. It could be mild steel, cast iron, silicon iron, or ferrite etc. The term 'iron circuit', when used in this context, is merely a simple way in which to refer to that part of the circuit that consists of a magnetic material. It is used when some parts of the circuit may be formed from non-magnetic materials.

Worked Example 4.1

Question

The pole face of a magnet is 3 cm by 2 cm and it produces a flux of 30 μWb. Calculate the flux density at the pole face.

Answer

$A = 3 \times 2 \times 10^{-4}\,\text{m}^2$; $\Phi = 30 \times 10^{-6}\,\text{Wb}$

$$B = \frac{\Phi}{A}\ \text{tesla}$$
$$= \frac{30 \times 10^{-6}}{6 \times 10^{-4}}$$

so $B = 50\,\text{mT}$ **Ans**

Worked Example 4.2

Question

A magnetic field of density 0.6 T has an effective csa of $45 \times 10^{-6}\,\text{m}^2$. Determine the flux.

Answer

$B = 0.6\,\text{T}$; $A = 45 \times 10^{-6}\,\text{m}^2$

Since $B = \frac{\Phi}{A}$ tesla, then $\Phi = BA$ weber

so $\Phi = 0.6 \times 45 \times 10^{-6} = 27\,\mu\text{Wb}$ **Ans**

4.5 Magnetomotive Force (mmf)

In an electric circuit, any current that flows is due to the existence of an emf. Similarly, in a magnetic circuit, the magnetic flux is due to the existence of an mmf. The concept of an mmf for permanent magnets is a difficult one. Fortunately it is simple when we consider the flux being produced by current flowing through a coil. This is the case for most practical magnetic circuits.

In section 4.2 we saw that each turn of the coil made a contribution to the total flux produced, so the flux must be directly proportional to the number of turns on the coil. The flux is also directly proportional to the value of current passed through the coil.

Putting these two facts together we can say that the mmf is the product of the current and the number of turns. The quantity symbol for mmf is F (the same as for mechanical force). The number of turns is just a number and therefore dimensionless. The SI unit for mmf is therefore simply ampere. However, this tends to cause considerable confusion to students new to the subject. For this reason, throughout *this book*, the unit will be quoted as ampere turns (At).

Thus mmf, $F = NI$ ampere turn (4.2)

Worked Example 4.3

Question

A 1500 turn coil is uniformly wound around an iron toroid of uniform csa 5 cm^2. Calculate the mmf and flux density produced if the resulting flux is 0.2 mWb when the coil current is 0.75 A.

Answer

$N = 1500$; $A = 5 \times 10^{-4}\,\text{m}^2$; $\Phi = 0.2 \times 10^{-3}\,\text{Wb}$; $I = 0.75\,\text{A}$

$$F = NI \text{ ampere turn} = 1500 \times 0.75$$

so $F = 1125$ At **Ans**

$$B = \frac{\Phi}{A} \text{ tesla} = \frac{0.2 \times 10^{-3}}{5 \times 10^{-4}}$$

so $B = 0.4$ T **Ans**

Worked Example 4.4

Question

Calculate the excitation current required in a 600 turn coil in order to produce an mmf of 1500 At.

Answer

$N = 600$; $F = 1500$ At

since $F = NI$ ampere turn, then $I = \dfrac{F}{N}$ ampere

so $I = \dfrac{1500}{600} = 2.5$ A **Ans**

4.6 Magnetic Field Strength

This is the magnetic equivalent to electric field strength in electrostatics. It was found that electric field strength is the same as potential gradient, and is measured in volt/metre. Now, the volt is the unit of emf, and we have just seen that mmf and emf are comparable quantities, i.e. mmf can be considered as the magnetic circuit equivalent of electric potential. Hence magnetic field strength is defined as the mmf per metre length of the magnetic circuit. The quantity symbol for magnetic field strength is H, the unit of measurement being ampere turn/metre.

$$H = \frac{F}{l} = \frac{NI}{l} \text{ ampere turn/metre} \tag{4.3}$$

where l is the *mean* or average length of the magnetic circuit. Thus, if the circuit consists of a circular toroid, then the mean length is the mean circumference. This point is illustrated in Figs. 4.8(a) and (b).

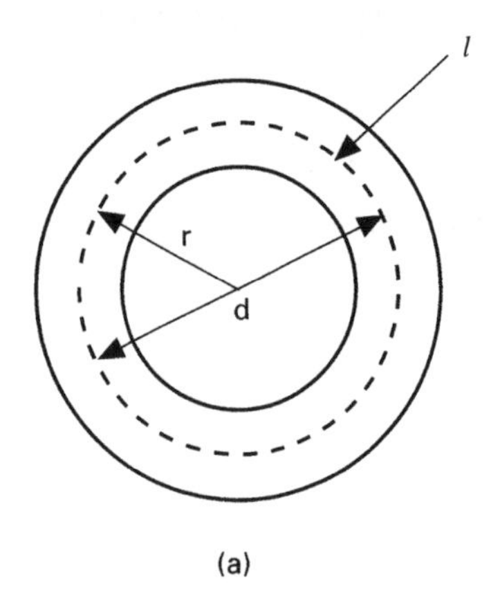

(a)

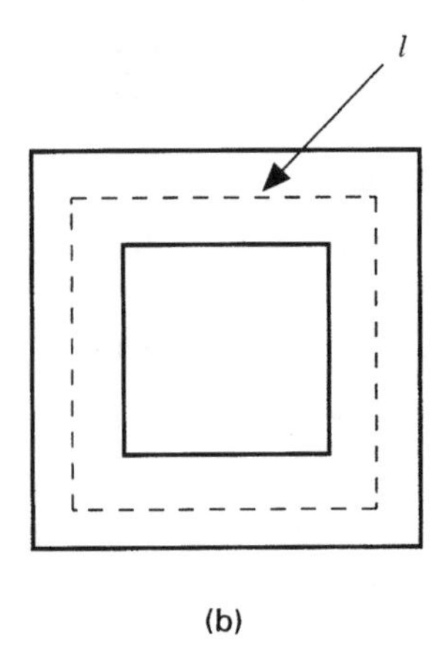

(b)

Fig. 4.8

Worked Example 4.5

Question

A current of 400 mA is passed through a 550 turn coil wound on a toroid of mean diameter 8 cm. Calculate the magnetic field strength.

Answer

$I = 0.4$ A; $N = 550$; $d = 8 \times 10^{-2}$ m

$$l = \pi d \text{ metre} = \pi \times 8 \times 10^{-2} = 0.251 \text{ m}$$

$$H = \frac{NI}{l} \text{ ampere turn/metre} = \frac{550 \times 0.4}{0.251}$$

so $H = 875.35$ At/m **Ans**

4.7 Permeability of Free Space (μ_0)

We have seen in electrostatics that the permittivity of the dielectric is a measure of the 'willingness' of the dielectric to allow an electric field to exist in it. In magnetic circuits the corresponding quantity is the permeability of the material.

If the magnetic field exists in a vacuum, then the ratio of the flux density to the magnetic field strength is a constant, called the permeability of free space.

$$\mu_0 = \frac{B}{H} \text{ henry/metre} \quad (4.4)$$

compare this to $\epsilon_0 = \dfrac{D}{E}$ farad/metre

The value for $\mu_0 = 4\pi \times 10^{-7}$ henry/metre

μ_0 is used as the reference or datum level from which the permeabilities of all other materials are measured.

4.8 Relative Permeability (μ_r)

Consider an air-cored solenoid with a fixed value of current flowing it. The mmf will produce a certain flux density in this air core. If an iron core was now inserted, it would be found that the flux density would be very much increased. To account for these different results for different core materials, a quantity known as the relative permeability is used. This is defined as the ratio of the flux density produced in the iron, to that produced in the air, for the same applied mmf.

$$\text{i.e. } \mu_r = \frac{B_2}{B_1} \quad (4.5)$$

where B_2 is the flux density produced in the iron and B_1 is the flux density produced in the air.

Compare this to the equation $\epsilon_r = \dfrac{C_2}{C_1}$ used in electrostatics.

As with ϵ_r, μ_r has no units, since it is simply a ratio.

Note: For air or any other NON-MAGNETIC material, $\mu_r = 1$. In other words, all non-magnetic materials have the same magnetic properties as a vacuum.

4.9 Absolute Permeability (μ)

The absolute permeability of a material is the ratio of the flux density to magnetic field strength, for a given mmf.

$$\text{Thus, } \mu = \frac{B}{H} \text{ henry/metre} \quad (4.6)$$

but since μ_0 is the reference value, then $\mu = \mu_0\mu_r$.

compare this to the equation $\epsilon = \epsilon_0 \epsilon_r$

$$\text{Therefore, } \mu_0 \mu_r = \frac{B}{H} \text{ so, } B = \mu_0 \mu_r H \text{ tesla} \quad (4.7)$$

This equation compares directly with $D = \epsilon_o \epsilon_r \mathbf{E}$ coulomb/m^2.

Worked Example 4.6

Question

A solenoid with an air core of csa of 15 cm^2 and relative permeability 65, produces a flux of 200 μWb. If the core material is changed to one of relative permeability 800 what will be the new flux and flux density?

Answer

$A = 15 \times 10^{-4}\,\text{m}^2$; $\mu_{r1} = 65$; $\Phi_1 = 2 \times 10^{-4}\,\text{Wb}$; $\mu_{r2} = 800$

$$B_1 = \frac{\Phi_1}{A}\,\text{tesla} = \frac{2 \times 10^{-4}}{15 \times 10^{-4}}$$

$$\text{so } B_1 = 0.133\,\text{T}$$

Now, the original core is 65 times more effective than air. The second core is 800 times more effective than air. Therefore, we can say that the second core will produce a greater flux density. The ratio of the two flux densities will be 800/65 = 12.31:1. Thus the second core will result in a flux density 12.31 times greater than produced by the first core

$$\text{Thus } B_2 = 12.31 B_1 = 12.31 \times 0.133$$

$$\text{so } B_2 = 1.641\,\text{T} \quad \textbf{Ans}$$

$$\text{and } \Phi_2 = B_2 A\,\text{weber} = 1.641 \times 15 \times 10^{-4}$$

$$\text{so } \Phi_2 = 2.462\,\text{mWb} \quad \textbf{Ans}$$

Worked Example 4.7

Question

A toroid of mean radius 40 mm, effective csa 3 cm^2, and relative permeability 150, is wound with a 900 turn coil that carries a current of 1.5 A. Calculate (a) the mmf, (b) the magnetic field strength and (c) the flux and flux density.

Answer

$r = 0.04\,\text{m}$; $A = 3 \times 10^{-4}\,\text{m}^2$; $\mu_r = 150$; $N = 900$; $I = 1.5\,\text{A}$

(a) $F = NI$ ampere turn $= 900 \times 1.5$

so $F = 1350$ At **Ans**

(b) $H = \frac{F}{\ell}$ ampere turn/metre, where $\ell = 2\pi r$ metre

$$= \frac{1350}{2\pi \times 0.04}$$

so, $H = 5371.5$ At/m **Ans**

(c) $B = \mu_0 \mu_r H$ tesla $= 4\pi \times 10^{-7} \times 150 \times 5371.5$

so $B = 1.0125$ T **Ans**

$\Phi = BA$ weber $= 1.0125 \times 3 \times 10^{-4}$

so $\Phi = 303.75\ \mu$Wb **Ans**

4.10 Magnetisation (B/H) Curve

A magnetisation curve is a graph of the flux density produced in a magnetic circuit as the magnetic field strength is varied. Since $H = NI/l$, then for a given magnetic circuit, the field strength may be varied by varying the current through the coil.

If the magnetic circuit consists entirely of air, or any other non-magnetic material then, the resulting graph will be a straight line passing through the origin. The reason for this is that since $\mu_r = 1$ for all non-magnetic materials, then the ratio B/H remains constant. Unfortunately, the relative permeability of magnetic materials does not remain constant for all values of applied field strength, which results in a curved graph.

This non-linearity is due to an effect known as magnetic saturation. The complete explanation of this effect is beyond the scope of this book, but a much simplified version of this afforded by Ewing's molecular theory. This states that each molecule in a magnetic material may be considered as a minute magnet in its own right. When the material is unmagnetised, these molecular magnets are orientated in a completely random fashion. Thus, the material has no overall magnetic polarisation. This is similar to a conductor in which the free electrons are drifting in a random manner. Thus, when no emf is applied, no current flows. This random orientation of the molecular magnets is illustrated in Fig. 4.9, where the arrows represent the north poles.

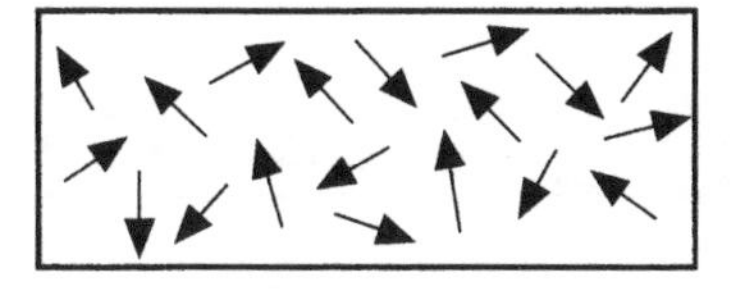

Fig. 4.9

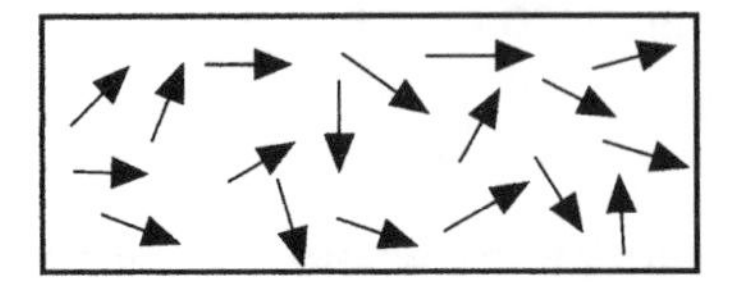

Fig. 4.10

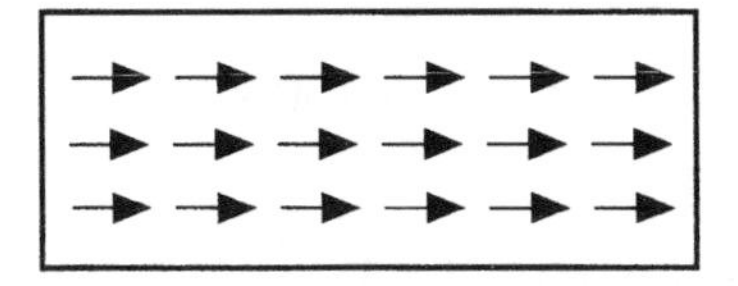

Fig. 4.11

However, as the coil magnetisation current is slowly increased, so the molecular magnets start to rotate towards a particular orientation. This results in a certain degree of polarisation of the material, as shown in Fig. 4.10. As the coil current continues to be increased, so the molecular magnets continue to become more aligned. Eventually, the coil current will be sufficient to produce complete alignment. This means that the flux will have reached its maximum possible value. Further increase of the current will produce no further increase of flux. The material is then said to have reached magnetic saturation, as illustrated in Fig. 4.11.

Typical magnetisation curves for air and a magnetic material are shown in Fig. 4.12. Note that the flux density produced for a given value of H is very much greater in the magnetic material. The slope of the graph is $B/H = \mu_0\mu_r$, and this slope varies. Since μ_0 is a constant, then the value of μ_r for the magnetic material must vary as the slope of the graph varies.

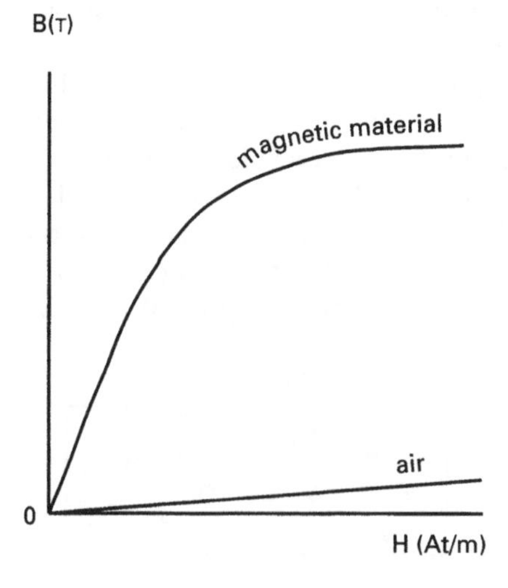

Fig. 4.12

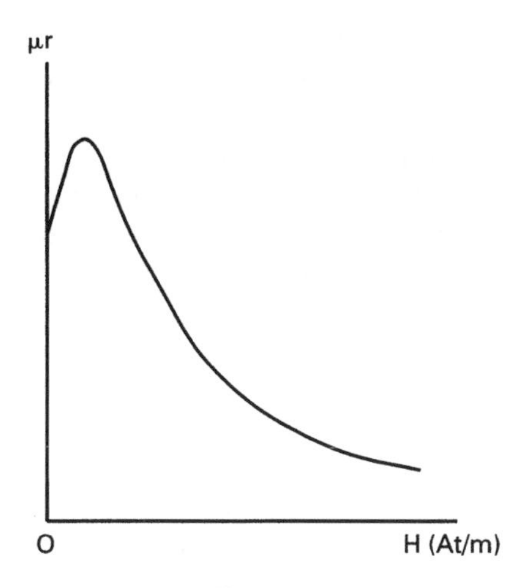

Fig. 4.13

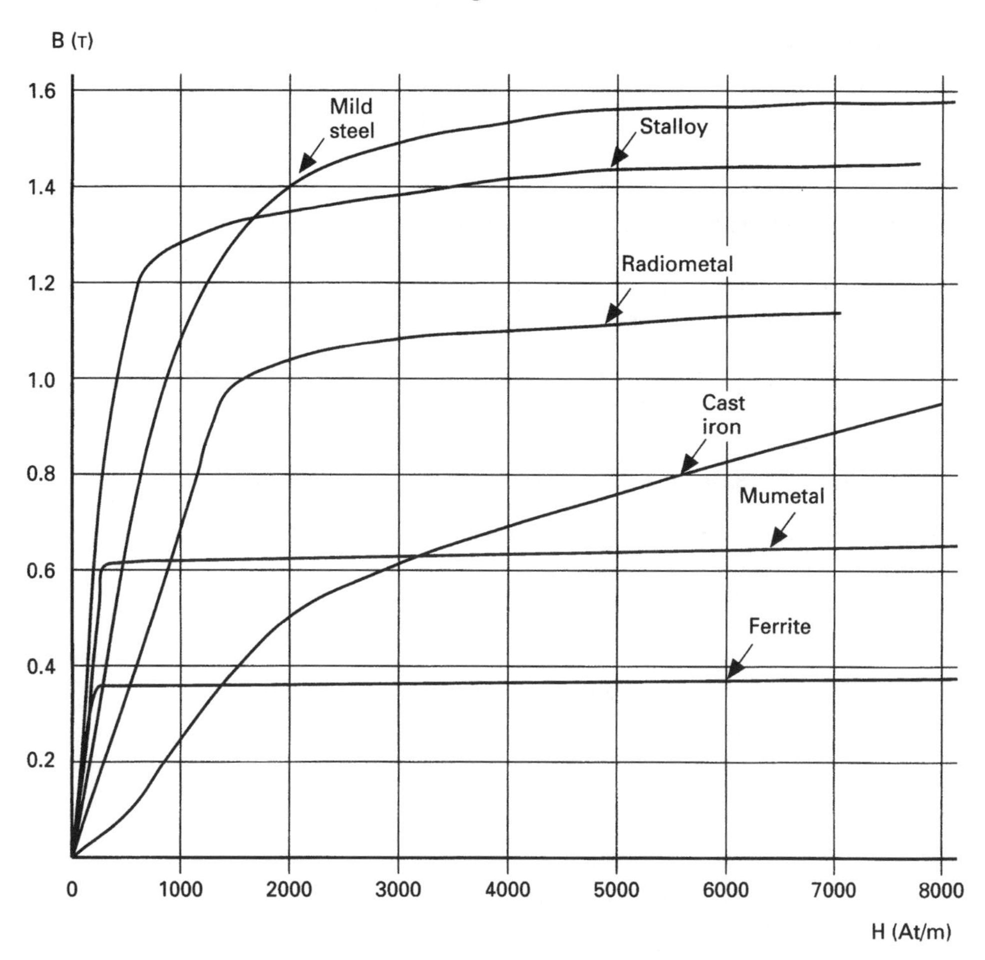

Fig. 4.14

The variation of μ_r with variation of H may be obtained from the B/H curve, and the resulting μ_r/H graph is shown in Fig. 4.13. The magnetisation curves for a range of magnetic materials is given in Fig. 4.14.

For a practical magnetic circuit, a single value for μ_r cannot be specified unless it is quoted for a specified value of B or H. Thus B/H data must be available. These may be presented either in the form of a graph as in Fig. 4.14, or in the form of tabulated data, from which the relevant section of the B/H curve may be plotted.

Worked Example 4.8

Question

An iron toroid having a mean radius of 0.1 m and csa of π cm^2 is wound with a 1000 turn coil. The coil current results in a flux of 0.1775 mWb in the toroid. Using the following data, determine (a) the coil current and (b) the relative permeability of the toroid under these conditions.

H(At/m)	80	85	90	95	100
B(T)	0.50	0.55	0.58	0.59	0.6

Answer

The first step in the solution of the problem is to plot the section of *B/H* graph from the given data.

Note: This *must* be plotted as accurately as possible on *graph* paper. The values used in this example have been obtained from such a graph.

$$B = \frac{\Phi}{A}\ \text{tesla} = \frac{0.1775 \times 10^{-3}}{\pi \times 10^{-4}}$$

so $B = 0.565\,\text{T}$

and from the plotted graph, when $B = 0.565$ T, $H = 88$ At/m.

(a) Now, the length of the toroid, $l = 2\pi r$ metre $= 0.2\pi$ m

$$\text{and } H = \frac{NI}{l}\ \text{ampere turn/metre}$$

$$\text{so } I = \frac{Hl}{N}\ \text{amp} = \frac{88 \times 0.2\pi}{1000}$$

hence $I = 55.3$ mA **Ans**

(b)

$$B = \mu_0\mu_r H\ \text{tesla}$$

$$\text{so } \mu_r = \frac{B}{\mu_0 H} = \frac{0.565}{4\pi \times 10^{-7} \times 88}$$

hence $\mu_r = 5109$ **Ans**

Worked Example 4.9

Question

A cast iron toroid of mean length 15 cm is wound with a 2500 turn coil, through which a magnetising current of 0.3 A is passed. Calculate the resulting flux density and relative permeability of the toroid under these conditions.

Answer

Since *B/H* data are necessary for the solution, but none have been quoted, then the *B/H* curve for cast iron shown in Fig. 4.14 will be used.

$l = 0.15\,\text{m}$; $N = 2500$; $I = 0.3\,\text{A}$

$$H = \frac{NI}{l}\ \text{ampere turn/metre} = \frac{2500 \times 0.3}{0.15}$$

so $H = 5000\,\text{At/m}$

and from the graph for cast iron in Fig. 4.14, the corresponding flux density is

$$B = 0.75\,\text{T}\ \textbf{Ans}$$

But, $B = \mu_0\mu_r H$ tesla

$$\text{so } \mu_r = \frac{B}{\mu_0 H} = \frac{0.75}{4\pi \times 10^{-7} \times 5000}$$

$$\mu_r = 119.4\ \textbf{Ans}$$

4.11 Composite Series Magnetic Circuits

Most practical magnetic circuits consist of more than one material in series. This may be deliberate, as in the case of an electric motor or generator, where there have to be air gaps between the stationary and rotating parts. Sometimes an air gap may not be required, but the method of construction results in small but unavoidable gaps.

In other circumstances it may be a requirement that two or more different magnetic materials form a single magnetic circuit.

Let us consider the case where an air gap is deliberately introduced into a magnetic circuit. For example, making a sawcut through a toroid, at right angles to the flux path.

Worked Example 4.10

Question

A mild steel toroid of mean length 18.75 cm and csa 0.8 cm^2 is wound with a 750 turn coil. (a) Calculate the coil current required to produce a flux of 112 μWb in the toroid. (b) If a 0.5 mm sawcut is now made across the toroid, calculate the coil current required to maintain the flux at its original value.

Answer

$l = 0.1875\,\text{m}$; $A = 8 \times 10^{-5}\,\text{m}^2$; $N = 750$; $\Phi = 112 \times 10^{-6}\,\text{Wb}$; $l_{gap} = 0.5 \times 10^{-3}\,\text{m}$

(a) $$B = \frac{\Phi}{A}\,\text{tesla} = \frac{112 \times 10^{-6}}{8 \times 10^{-5}}$$

so $B = 1.4$ T, and from the graph for mild steel in Fig. 4.14, the corresponding value for H is 2000 At/m

$$F_{Fe} = Hl\,\text{ampere turn} = 2000 \times 0.1875$$
$$\text{so } F_{Fe} = 375\,\text{At}$$

Fe is the chemical symbol for iron. In Worked Example 4.10 the mmf required to produce the flux in the 'iron' part of the circuit has been referred to as F_{Fe}. This will distinguish it from the mmf required for the air gap which is shown as F_{gap}

$$I = \frac{F_{Fe}}{N}\,\text{amp} = \frac{375}{750}$$
so $I = 0.5$ A **Ans**

(b) When the air gap is introduced into the steel the effective length of the steel circuit changes by only 0.27%. This is a negligible amount, so the values obtained in part (a) above for H and F_{Fe} remain unchanged. However, the introduction of the air gap will produce a considerable reduction of the circuit flux. Thus we need to calculate the extra mmf, and hence current, required to restore the flux to its original value.

Since the relative permeability for air is a constant (=1), then a *B/H* graph is not required. The csa of the gap is the same as that for the steel, and the same flux exists in it. Thus, the flux density in the gap must also be the same as that calculated in part (a) above. Hence the value of H required to maintain this flux density in the gap can be calculated from:

$$B = \mu_0 H_{gap}$$

$$\text{so } H_{gap} = \frac{B}{\mu_0} \text{ ampere turn/metre} = \frac{1.4}{4\pi \times 10^{-7}}$$

$$\text{and } H_{gap} = 1.11 \times 10^6 \text{ At/m}$$

$$\text{also, since } F_{gap} = H_{gap} l_{gap}$$

$$\text{then } F_{gap} = 1.11 \times 10^6 \times 0.5 \times 10^{-3} = 557 \text{ At}$$

$$\text{Total circuit mmf, } F = F_{Fe} + F_{gap} = 375 + 557$$

$$\text{so, } F = 932 \text{ At}$$

$$I = \frac{F}{N} = \frac{932}{750}$$

$$\text{so } I = 1.243 \text{ A } \textbf{Ans}$$

Worked Example 4.11

Question

A magnetic circuit consists of two stalloy sections A and B as shown in Fig. 4.15. The mean length and csa for A are 25 cm and 11.5 cm^2, whilst the corresponding values for B are 15 cm and 12 cm^2 respectively. A 1000 turn coil wound on section A produces a circuit flux of 1.5 mWb. Calculate the coil current required.

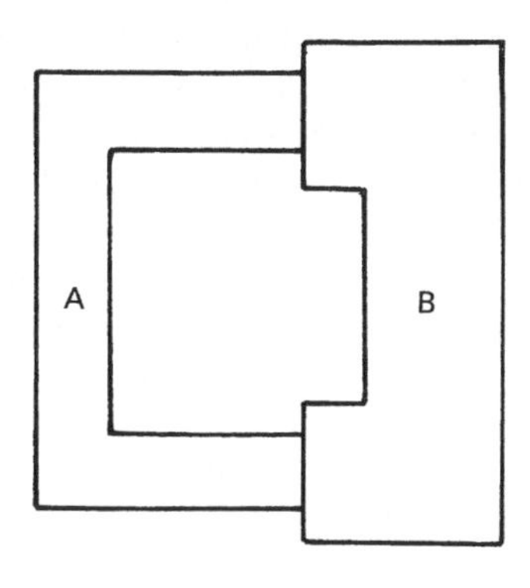

Fig. 4.15

Answer

$l_A = 0.25$ m; $A_A = 11.5 \times 10^{-4}$ m^2; $l_B = 0.15$ m

$A_B = 12 \times 10^{-4}$ m^2; $\Phi = 1.5 \times 10^{-3}$ Wb; $N = 1000$

Continued on p. 180

Worked Example 4.11 (*Continued*)

$$B_A = \frac{\Phi}{A_A} \text{ tesla} \quad \text{and} \quad B_B = \frac{\Phi}{A_B} \text{ tesla}$$

$$= \frac{1.5 \times 10^{-3}}{11.5 \times 10^{-4}} \quad\quad = \frac{1.5 \times 10^{-3}}{12 \times 10^{-4}}$$

so $B_A = 1.3\,\text{T}$ and $B_B = 1.25\,\text{T}$

From the *B/H* curve for stalloy in Fig. 4.14, the corresponding *H* values are:

$H_A = 1470$ At/m and $H_B = 845$ At/m

$F_A = H_A l_A$ ampere turn/metre, and $F_B = H_B l_B$ ampere turn/metre

$= 1470 \times 0.25$ $\quad = 845 \times 0.15$

$F_A = 367.5$ At $\quad F_B = 126.75$ At

total circuit mmf, $F = F_A + F_B$

so, $F = 494.25$ At

and $I = \frac{F}{N} = \frac{494.25}{1000}$

so $I = 0.494$ A **Ans**

From the last two examples it should now be apparent that in a series magnetic circuit the only quantity that is common to both (all) sections is the magnetic flux Φ. This common flux is produced by the current flowing through the coil, i.e. the total circuit mmf *F*. Also, if the lengths, csa and/or the materials are different for the sections, then their flux densities and *H* values must be different. For these reasons it is not legitimate to add together the individual *H* values. It *is* correct, however, to add together the individual mmfs to obtain the total circuit mmf *F*. This technique is equivalent to adding together the p.d.s across resistors connected in series in an electrical circuit. The sum of these p.d.s then gives the value of emf required to maintain a certain current through the circuit.

For example, if a current of 4 A is to be maintained through two resistors of 10 Ω and 20 Ω connected in series, then the p.d.s would be 40 V and 80 V respectively. Thus, the emf required would be 120 V.

4.12 Reluctance (S)

Comparisons have already been made between the electric circuit and the magnetic circuit. We have compared mmf to emf; **current** to **flux**; and potential gradient to magnetic field strength. A further comparison may be made, as follows.

The resistance of an electric circuit limits the current that can flow for a given applied emf. Similarly, in a magnetic circuit, the flux produced by a given mmf is limited by the reluctance of the circuit. Thus, the reluctance of a magnetic circuit is the opposition it offers to the existence of a magnetic flux within it.

> **Current** is a movement of electrons around an electric circuit. A magnetic **flux** merely *exists* in a magnetic circuit; it does not involve a flow of particles. However, both current and flux are the direct result of some form of applied force

$$\text{In an electric circuit, current} = \frac{\text{emf}}{\text{resistance}}$$

$$\text{so in a magnetic circuit, flux} = \frac{\text{mmf}}{\text{reluctance}}$$

$$\text{Thus, } \Phi = \frac{F}{S} = \frac{NI}{S} \text{ weber}$$

$$\text{So reluctance, } S = \frac{F}{\Phi} = \frac{NI}{\Phi} \text{ amp turn/weber} \qquad (4.8)$$

$$\text{but } NI = H\ell$$

$$\text{so } S = \frac{H\ell}{\Phi}$$

$$\text{also, } \Phi = BA$$

$$\text{so } S = \frac{H\ell}{BA}$$

$$\text{and since } \frac{H}{B} = \frac{1}{\mu} = \frac{1}{\mu_0 \mu_r}$$

$$\text{then } S = \frac{\ell}{\mu_0 \mu_r A} \text{ amp turn/metre} \qquad (4.9)$$

Let us continue the comparison between series electrical circuits and series magnetic circuits. We know that the total resistance in the electrical circuit is obtained simply by adding together the resistor values. The same technique may be used in magnetic circuits, such that the total reluctance of a series magnetic circuit, S is given by

$$S = S_1 + S_2 + S_3 + \ldots \qquad (4.10)$$

Assume that the physical dimensions of the sections, and the relative permeabilities (for the given operating conditions) of each section are known. In this case, equations (4.9), (4.10) and (4.8) enable an alternative form of solution.

Worked Example 4.12

Question

An iron ring of csa 8 cm^2 and mean diameter 24 cm, contains an air gap 3 mm wide. It is required to produce a flux of 1.2 mWb in the air gap. Calculate the mmf

Continued on p. 182

Worked Example 4.12 (*Continued*)

required, given that the relative permeability of the iron is 1200 under these operating conditions.

Answer

$A_{Fe} = A_{gap} = 8 \times 10^{-4}\,m^2$; $\Phi = 1.2 \times 10^{-3}$; $\mu_r = 1200$; $l_{gap} = 3 \times 10^{-3}\,m$;
$l_{Fe} = 0.24 \times \pi\,m$

For the iron circuit: $S_{Fe} = \frac{l_{Fe}}{\mu_0 \mu_r A}$ ampere turn/Wb

$$= \frac{0.24 \times \pi}{4\pi \times 10^{-7} \times 1200 \times 8 \times 10^{-4}}$$

so $S_{Fe} = 6.25 \times 10^5$ At/Wb

For the air gap: $S_{gap} = \frac{l_{gap}}{\mu_0 \mu_r A}$

$$= \frac{3 \times 10^{-3}}{4\pi \times 10^{-7} \times 1 \times 8 \times 10^{-4}}$$

so $S_{gap} = 2.984 \times 10^6$ At/Wb

Total circuit reluctance, $S = S_{Fe} + S_{gap}$

$$= 6.25 \times 10^5 + 2.984 \times 10^6$$

so $S = 3.61 \times 10^6$ At

Since $\Phi = \frac{F}{S}$ weber

then $F = \Phi S$ ampere turn (compare to $V = IR$ volt)

so $F = 4331$ At **Ans**

The above example illustrates the quite dramatic increase of circuit reluctance produced by even a very small air gap. In this example, the air gap length is only 0.4% of the total circuit length. Yet its reluctance is almost five times greater than that of the iron section. For this reason, the design of a magnetic circuit should be such as to try to minimise any unavoidable air gaps.

Worked Example 4.13

Question

A magnetic circuit consists of three sections, the data for which is given below. Calculate (a) the circuit reluctance and (b) the current required in a 500 turn coil, wound onto section 1, to produce a flux of 2 mWb.

Section	length (cm)	csa (cm^2)	μ_r
1	85	10	600
2	65	15	950
3	0.1	12.5	1

Answer

(a)

$$S_1 = \frac{l_1}{\mu_0 \mu_{r1} A_1} \text{ ampere turn/weber}$$

$$= \frac{0.85}{4 \times 10^{-7} \times \pi \times 600 \times 10^{-3}}$$

$$\text{so } S_1 = 1.127 \times 10^6 \text{ At/Wb}$$

$$S2 = \frac{l_2}{\mu_0 \mu_{r2} A_2}$$

$$= \frac{0.65}{4 \times 10^{-7} \times \pi \times 950 \times 15 \times 10^{-4}}$$

$$\text{so } S_2 = 3.63 \times 10^5 \text{ At/Wb}$$

$$S_3 = \frac{l_3}{\mu_0 \mu_{r3} A_3}$$

$$= \frac{10^{-3}}{4 \times 10^{-7} \times \pi \times 1 \times 12.5 \times 10^{-4}}$$

$$\text{so } S_3 = 6.37 \times 10^5 \text{ At/Wb}$$

$$\text{Total reluctance, } S = S_1 + S_2 + S_3$$

$$= 1.127 \times 10^6 + 3.63 \times 10^5 + 6.37 \times 10^5$$

so $S = 2.13 \times 10^6$ At/Wb **Ans**

(b)

$$F = \Phi S \text{ ampere turn} = 2 \times 10^{-3} \times 2.13 \times 10^6$$

$$\text{so } F = 4254 \text{ At}$$

$$\text{and } I = \frac{F}{N} \text{ amp} = \frac{4254}{500}$$

so $I = 8.51$ A **Ans**

Magnetic flux, like most other things in nature tends to take the easiest path available. For flux this means the lowest reluctance path. This is illustrated in Fig. 4.16. The reluctance of the soft iron bar is very much less than the surrounding air. For this reason, the flux will opt to distort from its normal pattern, and make use of this lower reluctance path.

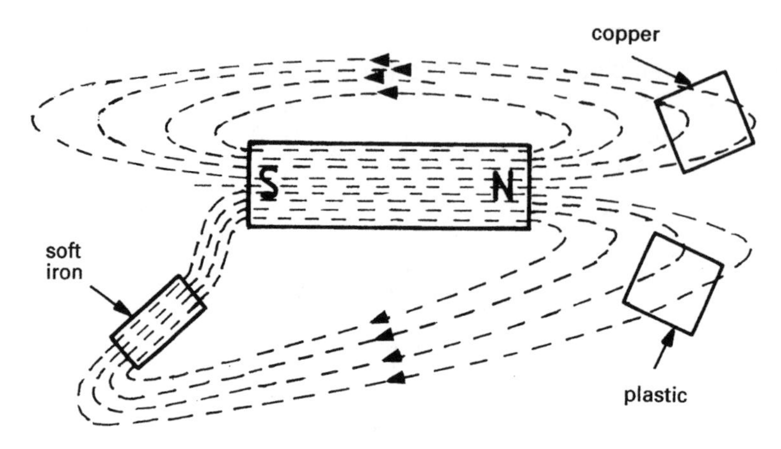

Fig. 4.16

4.13 Comparison of Electrical, Magnetic and Electrostatic Quantities

Although a number of comparisons have already been made in the text, it is useful to list these comparisons on one page. This helps to put the three systems into context, and Table 4.1 serves this purpose.

Table 4.1 *Comparison of Quantities*

Electrical			*Magnetic*			*Electrostatic*		
Quantity	*Symbol*	*Unit*	*Quantity*	*Symbol*	*Unit*	*Quantity*	*Symbol*	*Unit*
emf	E	V	mmf	F	At	emf	E	V
current	I	A	flux	Φ	Wb	flux	Q	C
resistance	R	Ω	reluctance	S	At/Wb	resistance	R	Ω
resistivity	ρ	Ωm	permeability	μ	H/m	permittivity	ϵ	F/m
potential gradient	—	V/m	field strength	H	At/m	field strength	$\mathbf{E}$	V/m
current density	J	A/m²	flux density	B	T	flux density	D	C/m²

Also listed below are some comparable equations

$E = IR$ volt; $F = \Phi S$ ampere turn

$\mathrm{J} = \dfrac{I}{A}$ amp/metre^2; $B = \dfrac{\Phi}{A}$ tesla; $D = \dfrac{Q}{A}$ coulomb/metre^2

$R = R_1 + R_2 + \ldots$ ohm; $S = S_1 + S_2 + \ldots$ ampere turn/weber

potential gradient $= \dfrac{E}{d}$ volt/metre; $H = \dfrac{F}{l}$ ampere turn/metre; $\mathbf{E} = \dfrac{V}{d}$ volt/metre

$B = \mu_0 \mu_r H$ tesla; $D = \epsilon_0 \epsilon_r \mathbf{E}$ coulomb/metre^2

$$\mu_r = \frac{B_2}{B_1};\ \epsilon_r = \frac{D_2}{D_1} \text{ or } \frac{C_2}{C_1}$$

$\mu_r = 1$ for all non-magnetic materials; $\epsilon_r = 1$ for air only

$\mu_0 = 4\pi \times 10^{-7}$ henry/metre; $\epsilon_0 = 8.854 \times 10^{-12}$ farad/metre

Note: Although the concept of current density has not been covered previously, it may be seen from Table 4.1 that it is simply the value of current flowing through a conductor divided by the csa of the conductor.

4.14 Magnetic Hysteresis

Hysteresis comes from a Greek word meaning 'to lag behind'. It is found that when the magnetic field strength in a magnetic material is varied, the resulting flux density displays a lagging effect.

Consider such a specimen of magnetic material that initially is completely unmagnetised. If no current flows through the magnetising coil then both H and B will initially be zero. The value of H is now increased by increasing the coil current in discrete steps. The corresponding flux density is then noted at each step. If these values are plotted on a graph until magnetic saturation is achieved, the dotted curve (the initial magnetisation curve) shown in Fig. 4.17 results.

Let the current now be reduced (in steps) to zero, and the corresponding values for B again noted and plotted. This would result in the section of graph from A to C. This shows that when the current is zero once more (so $H = 0$), the flux density has not reduced to zero. The flux density remaining is called the remanent flux density (OC). This property of a magnetic material, to retain some flux after the magnetising current is removed, is known as the remanance or retentivity of the material.

Let the current now be reversed, and increased in the opposite direction. This will have the effect of opposing the residual flux. Hence, the latter will be reduced until at

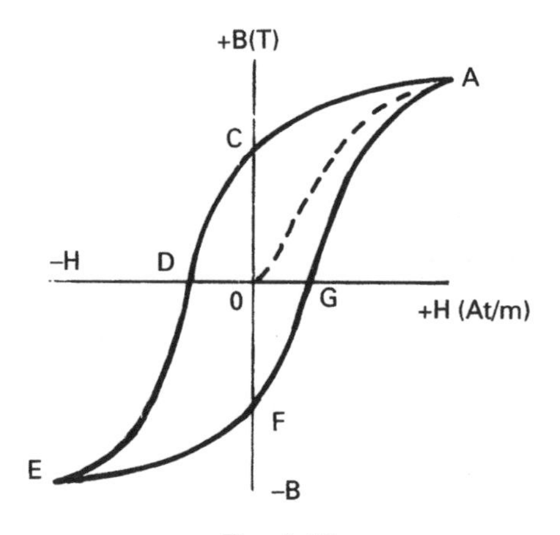

Fig. 4.17

some value of $-I$ it reaches zero (point D on the graph). The amount of reverse magnetic field strength required to reduce the residual flux to zero is known as the coercive force. This property of a material is called its coercivity.

If we now continue to increase the current in this reverse direction, the material will once more reach saturation (at point E). In this case it will be of the opposite polarity to that achieved at point A on the graph.

Once again, the current may be reduced to zero, reversed, and then increased in the original direction. This will take the graph from point E back to A, passing through points F and G on the way. Note that residual flux density shown as OC has the same value, but opposite polarity, to that shown as OF. Similarly, coercive force OD = OG.

In taking the specimen through the loop ACDEFGA we have taken it through one complete magnetisation cycle. The loop is referred to as the hysteresis loop. The degree to which a material is magnetised depends upon the extent to which the 'molecular magnets' have been aligned. Thus, in taking the specimen through a magnetisation cycle, energy must be expended. This energy is proportional to the area enclosed by the loop, and the rate (frequency) at which the cycle is repeated.

Magnetic materials may be subdivided into what are known as 'hard' and 'soft' magnetic materials. A hard magnetic material is one which possesses a large remanance and coercivity. It is therefore one which retains most of its magnetism, when the magnetising current is removed. It is also difficult to demagnetise. These are the materials used to form permanent magnets, and they will have a very 'fat' loop as illustrated in Fig. 4.18(a).

A soft magnetic material, such as soft iron and mild steel, retains very little of the induced magnetism. It will therefore have a relatively 'thin' hysteresis loop, as shown in Fig. 4.18(b). The soft magnetic materials are the ones used most often for engineering applications. Examples are the magnetic circuits for rotating electric machines (motors and generators), relays, and the cores for inductors and transformers.

When a magnetic circuit is subjected to continuous cycling through the loop a considerable amount of energy is dissipated. This energy appears as heat in the material. Since this is normally an undesirable effect, the energy thus dissipated is called the hysteresis loss. Thus, the thinner the loop, the less wasted energy. This is why 'soft' magnetic materials are used for the applications listed above.

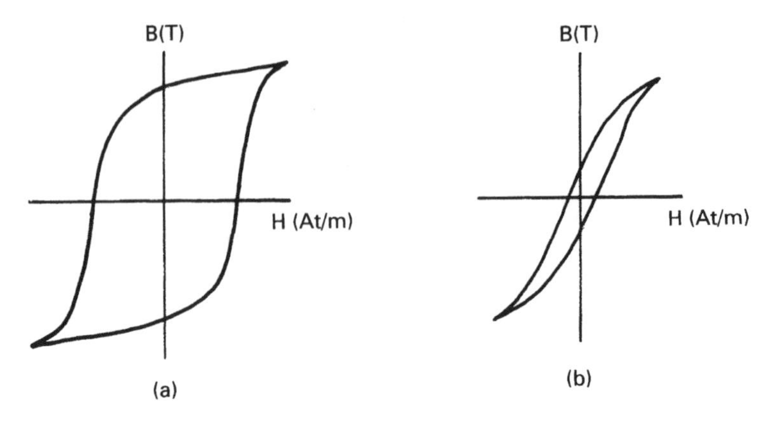

Fig. 4.18

4.15 Parallel Magnetic Circuits

The BTEC syllabus at this level does not require the treatment of parallel magnetic circuits, but since many practical circuits take this form, a brief coverage now follows.

We have seen that the magnetic circuit may be treated in much the same manner as its electrical circuit equivalent. The same is true for parallel circuits in the two systems. Two equivalent circuits are shown in Fig. 4.19, and from this the following points emerge:

1 In the electrical circuit, the current supplied by the source of emf splits between the two outer branches according to the resistances offered. In the magnetic circuit, the flux produced by the mmf splits between the outer limbs according to the reluctances offered.

2 If the two resistors in the outer branches are identical, the current splits equally. Similarly, if the reluctances of the outer limbs are the same then the flux splits equally between them.

However, a note of caution. In the electric circuit it has been assumed that the source of emf is ideal (no internal resistance) and that the connecting wires have no resistance. The latter assumption cannot be applied to the magnetic circuit. All three limbs will have a value of reluctance that *must* be taken into account when calculating the total circuit reluctance.

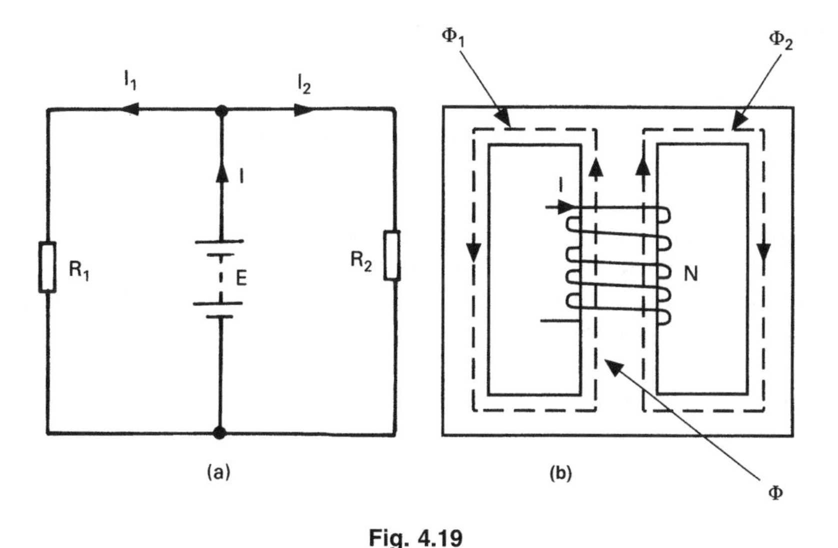

Fig. 4.19

Assignment Questions

1 The pole faces of a magnet are 4 cm × 3 cm and produce a flux of 0.5 mWb. Calculate the flux density.

2 A flux density of 1.8 T exists in an air gap of effective csa 11 cm^2. Calculate the value of the flux.

3 If a flux of 5 mWb has a density of 1.25 T, determine the csa of the field.

4 A magnetising coil of 850 turns carries a current of 25 mA. Determine the resulting mmf.

5 It is required to produce an mmf of 1200 At from a 1500 turn coil. What will be the required current?

6 A current of 2.5 A when flowing through a coil produces an mmf of 675 At. Calculate the number of turns on the coil.

7 A toroid has an mmf of 845 At applied to it. If the mean length of the toroid is 15 cm, determine the resulting magnetic field strength.

8 A magnetic field strength of 2500 At/m exists in a magnetic circuit of mean length 45 mm. Calculate the value of the applied mmf.

9 Calculate the current required in a 500 turn coil to produce an electric field strength of 4000 At/m in an iron circuit of mean length 25 cm.

10 A 400 turn coil is wound onto an iron toroid of mean length 18 cm and uniform csa 4.5 cm^2. If a coil current of 2.25 A results in a flux of 0.5 mWb, determine (a) the mmf, (b) the flux density, (c) the magnetic field strength.

11 An air-cored coil contains a flux density of 25 mT. When an iron core is inserted the flux density is increased to 1.6 T. Calculate the relative permeability of the iron under these conditions.

12 A magnetic circuit of mean diameter 12 cm has an applied mmf of 275 At. If the resulting flux density is 0.8 T, calculate the relative permeability of the circuit under these conditions.

13 A toroid of mean radius 35 mm, effective csa 4 cm^2 and relative permeability 200, is wound with a 1000 turn coil that carries a current of 1.2 A. Calculate (a) the mmf, (b) the magnetic field strength, (c) the flux density and (d) the flux in the toroid.

14 A magnetic circuit of square cross-section 1.5×1.5 cm and mean length 20 cm is wound with a 500 turn coil. Given the B/H data below, determine (a) the coil current required to produce a flux of 258.8 μWb and (b) the relative permeability of the circuit under these conditions.

B (T)	0.9	1.1	1.2	1.3
H (At/m)	250	450	600	825

15 For the circuit of Question 14 above, a 1.5 mm sawcut is made through it. Calculate the current now required to maintain the flux at its original value.

16 A cast steel toroid has the following B/H data. Complete the data table for the corresponding values of μ_r and hence plot the μ_r/H graph, and (a) from your graph determine the values of magnetic field strength at which the relative permeability of the steel is 520, and (b) the value of relative permeability when $H = 1200$ At/m.

B(T)	0.15	0.35	0.74	1.05	1.25	1.39
H(At/m)	250	500	1000	1500	2000	2500
μ_r						

17 A magnetic circuit made of radiometal is subjected to a magnetic field strength of 5000 At/m. Using the data given in Fig. 4.14, determine the relative permeability under this condition.

18 A magnetic circuit consists of two sections as shown in Fig. 4.20. Section 1 is made of mild steel and is wound with a 100 turn coil. Section 2 is made from cast iron. Calculate the coil current required to produce a flux of 0.72 mWb in the circuit. Use the B/H data given in Fig. 4.14.

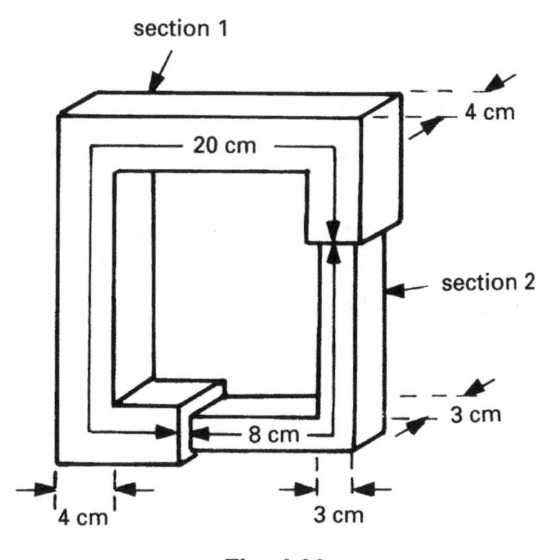

Fig. 4.20

19 A circular toroid of mean diameter 25 cm and csa 4 cm^2 has a 1.5 mm air gap in it. The toroid is wound with a 1200 turn coil and carries a flux of 0.48 mWb. If, under these conditions, the relative permeability of the toroid is 800, calculate the coil current required.

20 A closed magnetic circuit made from silicon steel consists of two sections, connected in series. One is of effective length 42 mm and csa 85 mm^2, and the other of length 17 mm and csa 65 mm^2. A 50 turn coil is wound on to the second section and carries a current of 0.4 A. Determine the flux density in the 17 mm length section if the relative permeability of the silicon iron under this condition is 3000.

21 A magnetic circuit of csa 0.45 cm^2 consists of one part 4 cm long and μ_r of 1200; and a second part 3 cm long and μ_r of 750. A 100 turn coil is wound onto the first part and a current of 1.5 A is passed through it. Calculate the flux produced in the circuit.

Suggested Practical Assignments

Note: These assignments are qualitative in nature.

Assignment 1

To compare the effectiveness of different magnetic core materials.

Apparatus:

1 × coil of wire of known number of turns
1 × d.c. psu
1 × ammeter
1 × set of laboratory weights
1 × set of different ferromagnetic cores, suitable for the coil used.

Method:

1. Connect the circuit as shown in Fig. 4.21.
2. Adjust the coil current carefully until the magnetic core *just* holds the smallest weight in place. Note the value of current and weight.
3. Using larger weights, in turn, increase the coil current until each weight is just held by the core. Record all values of weight and corresponding current.
4. Repeat the above procedure for the other core materials.
5. Tabulate all results. Calculate and tabulate the force of attraction and mmf in each case.
6. Write an assignment report, commenting on your findings, and comparing the relative effectiveness of the different core materials.

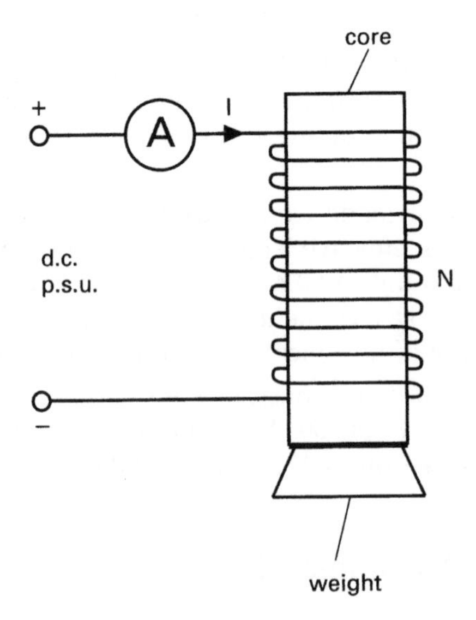

Fig. 4.21

Assignment 2

To plot a magnetisation curve, and initial section of a hysteresis loop, for a magnetic circuit.

Apparatus:

1 × magnetic circuit of known length, and containing a coil(s) of known number of turns
1 × variable d.c. psu
1 × Hall Effect probe
1 × ammeter
1 × DVM

Method:

1 Ensure that the core is completely demagnetised before starting.

2 Zero the Hall probe, monitoring its output with the DVM.

3 Connect the circuit as in Fig. 4.22.

4 Increase the coil current in 0.1 A steps, up to 2 A. Record the DVM reading at each step.
Note: If you 'overshoot' a desired current setting. DO NOT then reduce the current back to that setting. Record the value actually set, together with the corresponding DVM reading.

5 Reduce the current from 2 A to zero, in 0.1 A steps. Once more, if you overshoot a desired current setting, DO NOT attempt to correct it.

6 Reverse the connections to the psu, and increase the reversed current in small steps until the DVM indicates zero.
Note: The Hall effect probe output (as measured by the DVM) *represents* the flux density in the core. The magnetic field strength, H, may be calculated from NI/l.

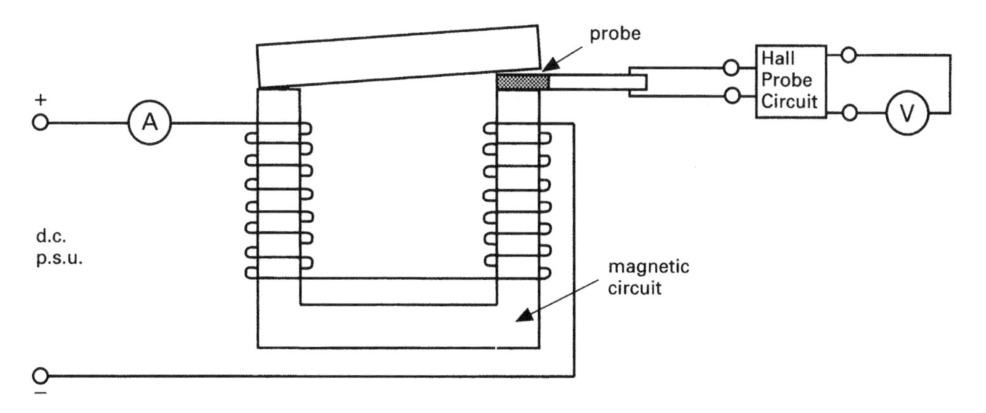

Fig. 4.22

7 Plot a graph of DVM readings (B) versus H.

8 Submit a full assignment report.

Supplementary worked examples

Example 1: A steel toroid of the dimensions shown in Fig. 4.23 is wound with a 500 turn coil of wire. What value of current needs to be passed through this coil in order to produce a flux of 250 μWb in the toroid, if under these conditions the relative permeability of the toroid is 300.

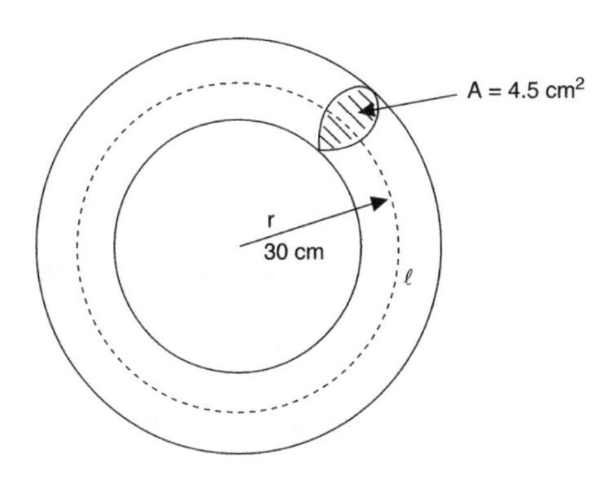

Fig. 4.23

Solution: $r = 3 \times 10^{-2}$ m; $A = 4.5 \times 10^{-4}$ m^2; $N = 500$; $\Phi = 250 \times 10^{-6}$ Wb; $\mu_r = 300$
Effective length of the toroid, $\ell = 2\pi r$ metre $= 2\pi \times 3 \times 10^{-2}$ m $= 0.188$ m

$$B = \frac{\Phi}{A}\text{ tesla} = \frac{250 \times 10^{-6}}{4.5 \times 10^{-4}}$$

so, $B = 0.556$ T

Now, $B = \mu_o \mu_r H$ tesla

$$\text{and, } H = \frac{B}{\mu_o \mu_r}\text{ ampere turns/weber}$$

$$= \frac{0.556}{4\pi \times 10^{-7} \times 300}$$

so, $H = 1474$ At/m

$F = H\ell$ ampere turns

$= 1474 \times 0.188$

thus, $F = 277$ At

$$\text{and, } I = \frac{F}{N}\text{ amp} = \frac{277}{500}$$

so, $I = 0.55$ A **Ans**

Example 2: A mild steel toroid of effective length 20 cm is wound with 164 turns of wire which carries a current of 1.5 A. Determine (a) the mmf produced, (b) the magnet field strength, (c) the flux density, and (c) the csa of the toroid if the flux produced is 154 μWb.

Solution: $\ell = 0.2\,\text{m}$; $N = 164$; $I = 1.5\,\text{A}$; $\Phi = 154 \times 10^{-6}\,\text{Wb}$

(a) $F = NI$ ampere turn $= 164 \times 1.5$

so, $F = 246$ At **Ans**

(b) $H = \dfrac{F}{\ell}$ ampere turn/metre $= \dfrac{246}{0.2}$

so, $H = 1230$ At/m **Ans**

(c) $B = \mu_o \mu_r H$ tesla; but since we do not know the value of μ_r we will need to use the B/H data for mild steel given in Fig. 4.14. From this diagram:

when $H = 1230$ At/m, $B = 1.2$ T **Ans**

(d) $B = \dfrac{\Phi}{A}$ tesla, so, $A = \dfrac{\Phi}{B}$ metre^2

$$= \frac{154 \times 10^{-6}}{1.2}$$

and, $A = 128.3\,\text{mm}^2$ **Ans**

Example 3: A coil is made by winding a single layer of 0.5 mm diameter wire onto a cylindrical wooden dowel, which is 5 cm long and of csa 7 cm^2. When a current of 0.2 A is passed through the coil, calculate (a) the mmf produced, (b) the flux density, and (c) the flux produced.

Solution: $I = 0.2\,\text{A}$; $\ell = 5 \times 10^{-2}\,\text{m}$; $A = 7 \times 10^{-4}\,\text{m}^2$; $d = 0.5 \times 10^{-3}$; $\mu_r = 1$ (wood)

Since $F = NI$ ampere turn, we first need to calculate the number of turns of wire on the coil. Consider Fig. 4.24 which represents the coil wound onto the dowel.

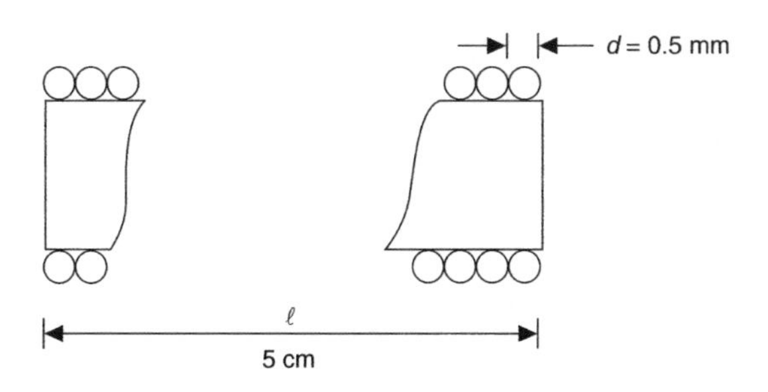

Fig. 4.24

From Fig. 4.24 it may be seen that the number of turns may be obtained by dividing the length of the dowel by the diameter (thickness) of the wire.

$$\text{Thus, } N = \frac{\ell}{d} = \frac{50}{0.5}$$

$$\text{so, } N = 100$$

$$\text{Thus, } F = 100 \times 0.2 = 20\,\text{At}\ \textbf{Ans}$$

(b) $B = \dfrac{\Phi}{A}$ tesla, or $B = \mu_o \mu_r H$ tesla

but since we do not yet know the value for the flux, but can calculate the value for H, then the second equation needs to be used.

$$H = \frac{F}{\ell} \text{ ampere turn/metre} = \frac{20}{5 \times 10^{-2}}$$

$$\text{so, } H = 400\,\text{At/m}$$

$$\text{and, } B = 4\pi \times 10^{-7} \times 1 \times 400 = 5.026 \times 10^{-4}$$

$$\text{so, } B = 503\,\mu\text{T}\ \textbf{Ans}$$

(c) $\Phi = BA$ weber $= 503 \times 10^{-6} \times 7 \times 10^{-4}$

$$\text{thus, } \Phi = 0.352\,\mu\text{Wb}\ \textbf{Ans}$$

Example 4: Considering the toroid and coil specified in Example 1, a saw cut is now made across the toroid. Under this condition it is found that the coil current has to be increased to 0.896 A in order to maintain the original flux density. Detemine the width of the saw cut.

Solution: We now have a composite magnetic circuit consisting of the original steel toroid and an air gap. However, the reduction of the length of the steel path will be negligible, so we may use figures for this part of the circuit which were calculated in Example 1.

Hence, $F_{Fe} = 277$ At/m; $B = 0.556$ T; and for the air gap, $\mu_r = 1$

$$\text{Total mmf, } F = NI \text{ ampere turn} = 500 \times 0.896 = 448\,\text{At}$$

$$\text{and, } F = F_{Fe} + F_{gap}, \text{ so } F_{gap} = F - F_{Fe}$$

$$\text{thus, } F_{gap} = 448 - 277 = 171\,\text{At}$$

$$H_{gap} = \frac{B}{\mu_o} \text{ ampere turn/metre } (\mu_r = 1)$$

$$= \frac{0.556}{4\pi \times 10^{-7}} = 442\,450\,\text{At/m}$$

$$\text{and, } H_{gap} = \frac{F_{gap}}{\ell_{gap}}, \text{ so, } \ell_{gap} = \frac{F_{gap}}{H_{gap}} \text{ metre}$$

$$= \frac{171}{442\,450}$$

$$\text{hence, } \ell_{gap} = 0.386\,\text{mm } \textbf{Ans}$$

Example 5: A magnetic circuit consists of two 'iron' sections in series, separated by an air gap of length 1.5 mm. Data for the two iron sections are given below.

Section	length (mm)	csa (mm²)	μ_r
1	200	800	402
2	150	1000	300

A coil wound onto section 1 carries a current of 5 A which produces a flux density of 1.5 T in the air gap. Determine the number of turns on the coil to achieve this.

Solution: $I = 5\,\text{A}$; $B_{gap} = 1.5\,\text{T}$; $\ell_{gap} = 1.5 \times 10^{-3}\,\text{m}$; $\ell_1 = 0.2\,\text{m}$; $A_1 = 8 \times 10^{-4}\,\text{m}^2$
$\ell_2 = 0.15\,\text{m}$; $A_2 = 10^{-3}\,\text{m}^2$

From the data given the reluctance of each section may be calculated thus.

$$S_1 = \frac{\ell_1}{\mu_o \mu_{r1} A_1} \text{ ampere turn/Wb} = \frac{0.2}{4\pi \times 10^{-7} \times 420 \times 8 \times 10^{-4}}$$

$$\text{and, } S_1 = 473\,675 = 0.474 \times 10^6\,\text{At/Wb}$$

$$S_2 = \frac{\ell_2}{\mu_o \mu_{r2} A_2} \text{ ampere turn/Wb} = \frac{0.15}{4\pi \times 10^{-7} \times 300 \times 10^{-3}}$$

$$\text{and, } S_2 = 397\,887 = 0.398 \times 10^6\,\text{At/Wb}$$

$$S_{gap} = \frac{\ell_{gap}}{\mu_o A_{gap}} = \frac{1.5 \times 10^{-3}}{4\pi \times 10^{-7} \times 9 \times 10^{-4}}$$

where $A_{gap} = 9 \times 10^{-4}\,\text{m}^2$,

i.e. the average of A_1 and A_2

$$\text{so, } S_{gap} = 1.326 \times 10^6\,\text{At/Wb}$$

$$\text{total circuit reluctance, } S = S_1 + S_2 + S_{gap} = (0.474 + 0.398 + 1.326) \times 10^6$$

$$\text{hence, } S = 2.198 \times 10^6\,\text{At/Wb}$$

$$\Phi = B_{gap} A_{gap} \text{ weber} = 1.5 \times 9 \times 10^{-4}$$

$$\text{so, } \Phi = 1.35\,\text{mWb}$$

$$\text{and, } F = \Phi S \text{ ampere turn} = 1.35 \times 10^{-3} \times 2.198 \times 10^6$$

$$\text{so, } F = 2963\,\text{At}$$

$$\text{and number of turns, } N = \frac{F}{I} = \frac{2963}{5}$$

$$\text{so, } N = 593 \textbf{ Ans}$$

Summary of Equations

Magnetic flux density: $B = \frac{\Phi}{A} = \mu_o \mu_r H$ tesla

Magnetomotive force (mmf): $F = NI = \Phi S$ ampere turn

Magnetic field strength: $H = \frac{F}{\ell} = \frac{NI}{\ell}$ ampere turn/metre

Permeability: $\mu = \mu_o \mu_r = \frac{B}{H}$ henry/metre

Reluctance: $S = \frac{\ell}{\mu_o \mu_r A} = \frac{NI}{\Phi} = \frac{F}{\Phi}$ ampere turn/weber

Series magnetic circuit: $S = S_1 + S_2 + S_3 + \ldots$ ampere turn/weber

5 Electromagnetism

This chapter concerns the principles and laws governing electromagnetic induction and the concepts of self and mutual inductance.

On completion of the chapter, you should be able to use these principles to:

1. Explain the basic operating principles of motors and generators.
2. Carry out simple calculations involving the generation of voltage, and the production of force and torque.
3. Appreciate the significance of eddy current loss.
4. Determine the value of inductors, and apply the concepts of self and mutual inductance to the operating principles of transformers.
5. Calculate the energy stored in a magnetic field.
6. Explain the principle of the moving coil meter, and carry out simple calculations for the instrument.
7. Describe the operation of a wattmeter and simple ohmmeter.

5.1 Faraday's Law of Electromagnetic Induction

It is mainly due to the pioneering work of Michael Faraday, in the nineteenth century, that the modern technological world exists as we know it. Without the development of the generation of electrical power, such advances would have been impossible. Thus, although the concepts involved with electromagnetic induction are very simple, they have far-reaching influence. Faraday's law is best considered in two interrelated parts:

1. The value of emf induced in a circuit or conductor is directly proportional to the rate of change of magnetic flux linking with it.
2. The polarity of such an emf, induced by an increasing flux, is opposite to that induced by a decreasing flux.

The key to electromagnetic induction is contained in part one of the law quoted above. Here, the words 'rate of change' are used. If there is no change in flux, or the way in which this flux links with a conductor, then no emf will be induced. The proof of the law can be very simply demonstrated. Consider a coil of wire, a permanent bar magnet and a galvanometer as illustrated in Figs. 5.1 and 5.2.

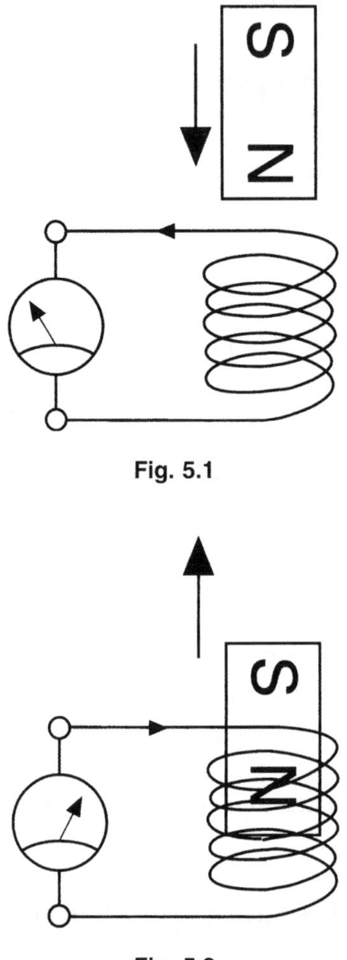

Fig. 5.1

Fig. 5.2

Consider the magnet being moved so that it enters the centre of the coil. When this is done it will be seen that the pointer of the galvo deflects in one direction. This deflection of the pointer is only momentary, since it only occurs whilst the magnet is moving. The galvo is of course a current measuring device. However, any current flowing through it must be due to a voltage between its terminals. Since there is no other source of emf in the circuit, then it must be concluded that an emf has been induced or created in the coil itself. The resulting current indicated by the galvo depends on the value of this emf. It will also be observed that when the magnet is stationary (either inside or outside the coil) the galvo does not deflect. Hence, emf is induced into the coil only when the magnet is in motion.

When the magnet is withdrawn from the coil, the galvo will again be seen to deflect momentarily. This time, the deflection will be in the opposite direction. Provided that the magnet is removed at the same rate as it was inserted, then the magnitudes of the deflections will be the same. The polarities of the induced emfs will be opposite to each other, since the current flow is reversed. Thus far, we have confirmation that an emf is induced in the coil when a magnetic flux is moving relative to it. We also have confirmation of part two of the law.

In order to deduce the relationship between the value of induced emf and the rate of change of flux, the magnet needs to be moved at different speeds into and out of the coil. When this is done, and the resulting magnitudes of the galvo deflection noted, it will be found that the faster the movement, the greater the induced emf.

This simple experiment can be further extended in three ways. If the magnet is replaced by a more powerful one, it will be found that for the same speed of movement, the corresponding emf will be greater. Similarly, if the coil is replaced with one having more turns, then for a given magnet and speed of movement, the value of the emf will again be found to be greater. Finally, if the magnet is held stationary within the coil, and the coil is then moved away, it will be found that an emf is once more induced in the coil. In this last case, it will also be found the emf has the same polarity as obtained when the magnet was first inserted into the stationary coil. This last effect illustrates the point that it is the *relative movement* between the coil and the flux that induces the emf.

The experimental procedure described above is purely qualitative. However, if it was refined and performed under controlled conditions, then it would yield the following results:

The magnitude of the induced emf is directly proportional to the value of magnetic flux, the rate at which this flux links with the coil, and the number of turns on the coil. Expressed as an equation we have:

$$e = \frac{-N\mathrm{d}\Phi}{\mathrm{d}t} \text{ volt} \tag{5.1}$$

Notes:

1. The symbol for the induced emf is shown as a lower-case letter e. This is because it is only present for the short interval of time during which there is relative movement taking place, and so has only a momentary value.
2. The term $\mathrm{d}\Phi/\mathrm{d}t$ is simply a mathematical means of stating 'the rate of change of flux with time'. The combination $N\Phi/\mathrm{d}t$ is often referred to as the 'rate of change of flux linkages'.
3. The minus sign is a reminder that Lenz's law applies. This law is described in the next section.
4. Equation (5.1) forms the basis for the definition of the unit of magnetic flux, the weber, thus:

The weber is that magnetic flux which, linking a circuit of one turn, induces in it an emf of one volt when the flux is reduced to zero at a uniform rate in one second.

In other words, 1 volt = 1 weber/second or 1 weber = 1 volt second.

5.2 Lenz's Law

This law states that the polarity of an induced emf is always such that it opposes the change which produced it. This is similar to the statement in mechanics, that for every force there is an opposite reaction.

5.3 Fleming's Righthand Rule

This is a convenient means of determining the polarity of an induced emf in a conductor. Also, provided that the conductor forms part of a complete circuit, it will indicate the direction of the resulting current flow.

The first finger, the second finger and the thumb of the *right* hand are held out mutually at right angles to each other (like the three edges of a cube as shown in Fig. 5.3). The ***F***irst finger indicates the direction of the ***F***lux, the thu***M***b the direction of ***M***otion of the conductor relative to the flux, and the s***EC***ond finger indicates the polarity of the induced ***E***mf, and direction of ***C***urrent flow. This process is illustrated in Fig. 5.4, which shows the cross-section of a conductor being moved vertically upwards at a constant velocity through the magnetic field.
Note: The thumb indicates the direction of motion of the *conductor relative to* the flux. Thus, the same result would be obtained from the arrangement of Fig. 5.4 if the conductor was kept stationary and the magnetic field was moved down

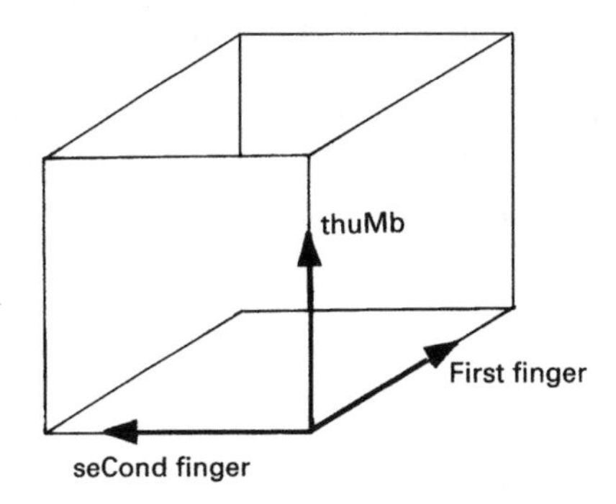

Fig. 5.3

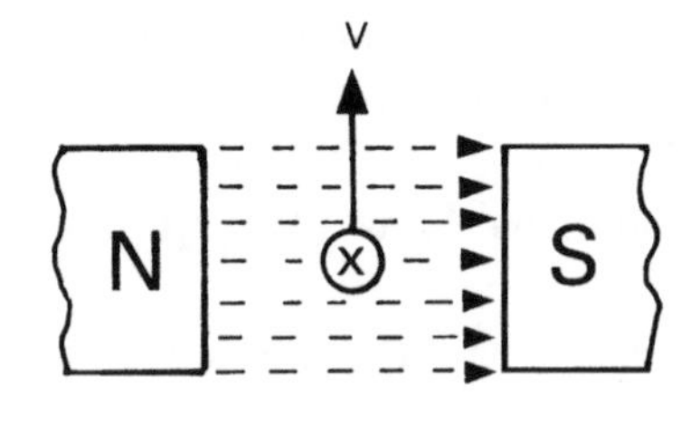

Fig. 5.4

Worked Example 5.1

Question

The flux linking a 100 turn coil changes from 5 mWb to 15 mWb in a time of 2 ms. Calculate the average emf induced in the coil; see Fig. 5.5.

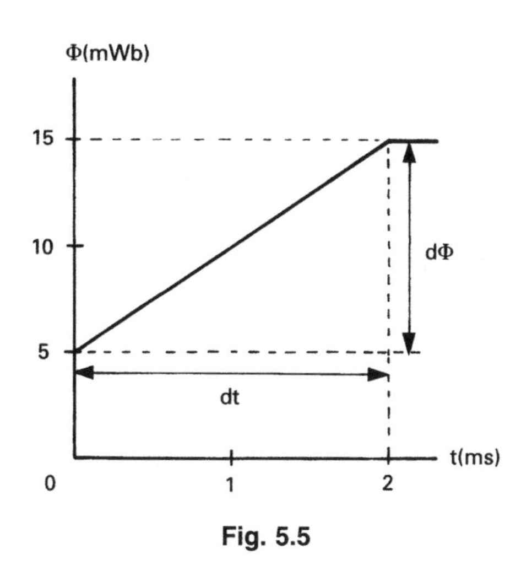

Fig. 5.5

Answer

$N = 100$; d $\Phi = (15 - 5) \times 10^{-3}$ Wb; d$t = 2 \times 10^{-3}$ s

$$e = \frac{-N\mathrm{d}\Phi}{\mathrm{d}t}\ \text{volt} = \frac{-100 \times (15 - 5) \times 10^{-3}}{2 \times 10^{-3}}$$

$$= \frac{-100 \times 10 \times 10^{-3}}{2 \times 10^{-3}}$$

so $e = -500$ V **Ans**

Note that if the flux was *reduced* from 15 mWb to 5 mWb, then the term shown in brackets above would be -10. The resulting emf would be $+500$ V. When quoting equation (5.1), the minus sign should always be included. However, since it is often the magnitude of the induced emf that is more important, it is normal practice to ignore the minus sign in the subsequent calculation. One of the major exceptions to this practice arises when considering the principles of operation of the transformer.

Worked Example 5.2

Question

A 250 turn coil is linked by a magnetic flux that varies as follows: an increase from zero to 20 mWb in a time of 0.05 s; constant at this value for 0.02 s; followed by a decrease to 4 mWb in a time of 0.01 s. Assuming that these changes are uniform, draw a sketch graph (i.e. not to an accurate scale) of the variation of the flux and the corresponding emf induced in the coil, showing all principal values.

Continued on p. 202

Worked Example 5.2 (*Continued*)

Answer

Firstly, the values of induced emf must be calculated for those periods when the flux changes.

$$d\Phi_1 = (20 - 0) \times 10^{-3}\ \text{Wb};\ dt_1 = 0.05\ \text{s}$$
$$d\Phi_2 = (4 - 20) \times 10^{-3}\ \text{Wb};\ dt_2 = 0.01\ \text{s}$$

$$e_1 = \frac{-Nd\Phi_1}{dt_1}\ \text{volt} \quad \text{and} \quad e_2 = \frac{-Nd\Phi_2}{dt_2}\ \text{volt}$$
$$= \frac{-250 \times 20 \times 10^{-3}}{0.05} \qquad = \frac{-250 \times (-16) \times 10^{-3}}{0.01}$$
$$\text{so } e_1 = -100\ \text{V} \quad \text{and} \quad e_2 = 400\ \text{V}$$

The resulting sketch graph is shown in Fig. 5.6.

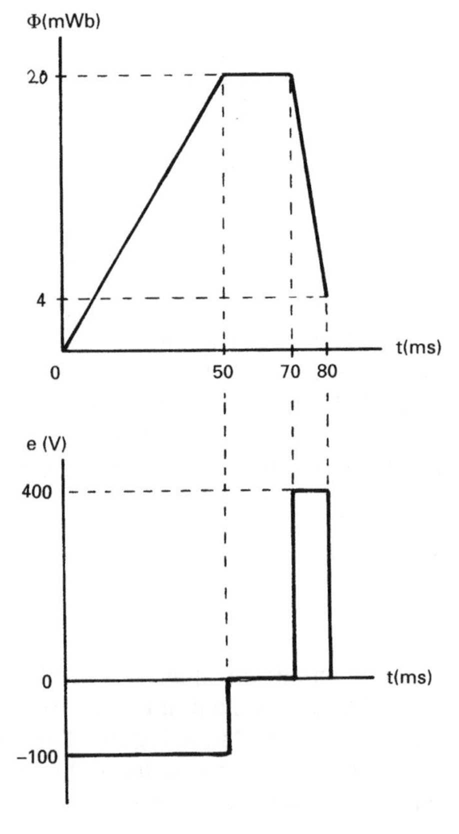

Fig. 5.6

Worked Example 5.3

Question

A coil when linked by a flux which changes at the rate of 0.1 Wb/s, has induced in it an emf of 100 V. Determine the number of turns on the coil.

Answer

$e = 100$ V; $d\Phi/dt = 0.1$ Wb/s

$$e = -N d\Phi/dt \text{ volt}$$

$$\text{so } N = \frac{e}{d\Phi/dt} \text{ turns} = \frac{100}{0.1}$$

$$N = 1000 \textbf{ Ans}$$

Note that the minus sign has been ignored in the calculation. A negative value for number of turns makes no sense.

5.4 EMF Induced in a Single Straight Conductor

Consider a conductor moving at a constant velocity v metre per second at right angles to a magnetic field having the dimensions shown in Fig. 5.7. The direction of the induced emf may be obtained using Fleming's righthand rule, and is shown in the diagram. Equation (5.1) is applicable, and in this case, the value for N is 1.

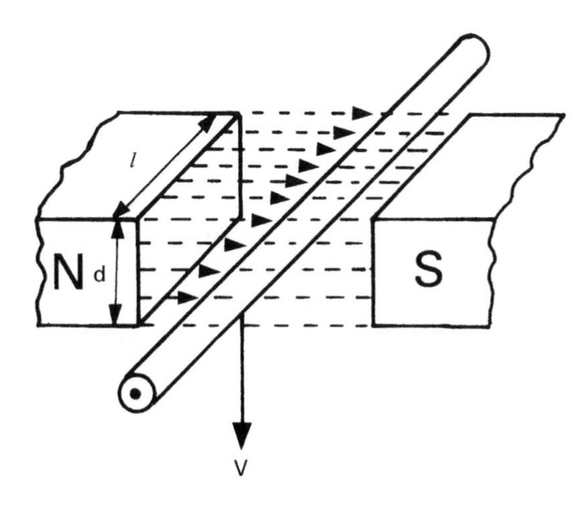

Fig. 5.7

Thus, $e = \frac{d\Phi}{dt}$ volt, and since Φ is constant

then $e = \frac{\Phi}{t}$ volt

but $\Phi = BA$ weber

so $e = \frac{BA}{t}$

also, the csa of the field, $A = \ell \times d$ metre^2

so $e = \frac{B\ell d}{t}$ and since $\frac{d}{t}$ = velocity v, then

$$e = B\ell v \text{ volt} \tag{5.2}$$

The above equation is only true for the case when the conductor is moving at right angles to the magnetic field. If the conductor moves through the field at some angle less than 90°, then the 'cutting' action between the conductor and the flux is reduced. This results in a consequent reduction in the induced emf. Thus, if the conductor is moved horizontally through the field, the 'cutting' action is zero, and so no emf is induced. To be more precise, we can say that *only the component of the velocity at 90° to the flux* is responsible for the induced emf. In general therefore, the induced emf is given by:

$$e = B\ell v \sin\theta \text{ volt} \tag{5.3}$$

where $v \sin\theta$ is the component of velocity at 90° to the field, as illustrated in Fig. 5.8.

This equation is simply confirmed by considering the previous two extremes; i.e. when conductor moves parallel to the flux, $\theta = 0°$; $\sin\theta = 0$; so $e = 0$. When it moves at right angles to the flux, $\theta = 90°$; $\sin\theta = 1$; so we are back to equation (5.2).

Note: ℓ is known as the *effective* length of the conductor, since it is that portion of the conductor that actually links with the flux. The total length of the conductor may be considerably greater than this, but those portions that may extend beyond the field at either end will not have any emf induced.

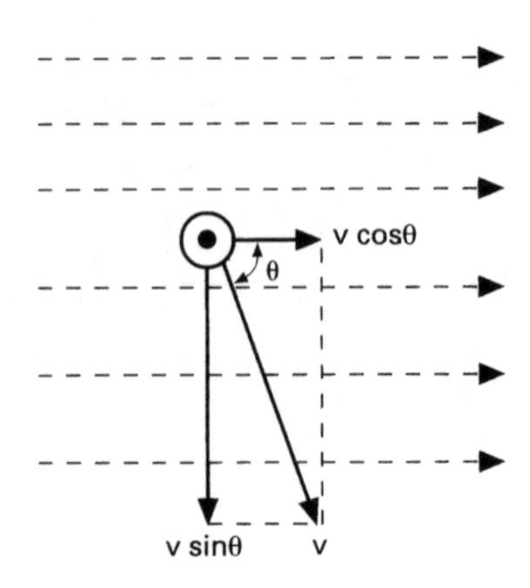

Fig. 5.8

Worked Example 5.4

Question

A conductor is moved at a velocity of 5 m/s at an angle of 60° to a uniform magnetic field of 1.6 mWb. The field is produced by a pair of pole pieces, the faces of which measure 10 cm by 4 cm. If the conductor length is parallel to the longer side of the field, calculate the emf induced; see Fig. 5.9.

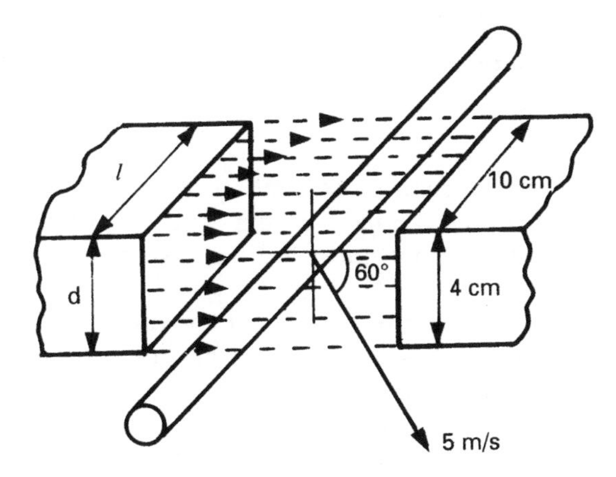

Fig. 5.9

Answer

$v = 5$ m/s; $\theta = 60°$; $\Phi = 1.6 \times 10^{-3}$ Wb; $l = 0.1$ m; $d = 0.04$ m (see Fig. 5.9)

$$B = \frac{\Phi}{A} \text{ tesla} = \frac{1.6 \times 10^{-3}}{0.1 \times 0.04} = 0.4 \text{ T}$$

$$e = B\ell v \sin\theta \text{ volt} = 0.4 \times 0.1 \times 5 \times \sin 60°$$

so $e = 0.173$ V **Ans**

Worked Example 5.5

Question

A conductor of effective length 15 cm, when moved at a velocity of 8 m/s at an angle of 55° to a uniform magnetic field, generates an emf of 2.5 V. Determine the density of the field.

Continued on p. 206

Worked Example 5.5 (*Continued*)

Answer

$\ell = 0.15$ m; $v = 8$ m/s; $\theta = 55°$; $e = 2.5$ V

$$e = B\ell v \sin\theta \text{ volt, so } B = \frac{e}{\ell v \sin\theta} \text{ tesla}$$

$$\text{hence, } B = \frac{2.5}{0.15 \times 8 \times \sin 55°}$$

$$B = 2.543 \text{ T } \textbf{Ans}$$

This section, covering the induction or generation of an emf in a conductor moving through a magnetic field, forms the basis of the generator principle. However, most electrical generators are rotating machines, and we have so far considered only linear motion of the conductor.

Consider the conductor now formed into the shape of a rectangular loop, mounted on to an axle. This arrangement is then rotated between the poles of a permanent magnet. We now have the basis of a simple generator as illustrated in Fig. 5.10.

The two sides of the loop that are parallel to the pole faces will each have an effective length ℓ metre. At any instant of time, these sides are passing through the field in opposite directions. Applying the righthand rule at the instant shown in Fig. 5.10, the directions of the induced emfs will be as marked, i.e. of opposite polarities. However, if we trace the path around the loop, it will be seen that both emfs are causing current to flow in the same direction around the loop. This is equivalent to two cells connected in series as shown in Fig. 5.11.

The situation shown in Fig. 5.10 applies only to one instant in one revolution of the loop (it is equivalent to a 'snapshot' at that instant). If we were to plot a graph of the total emf generated in the loop, for one complete revolution, it would be found to be

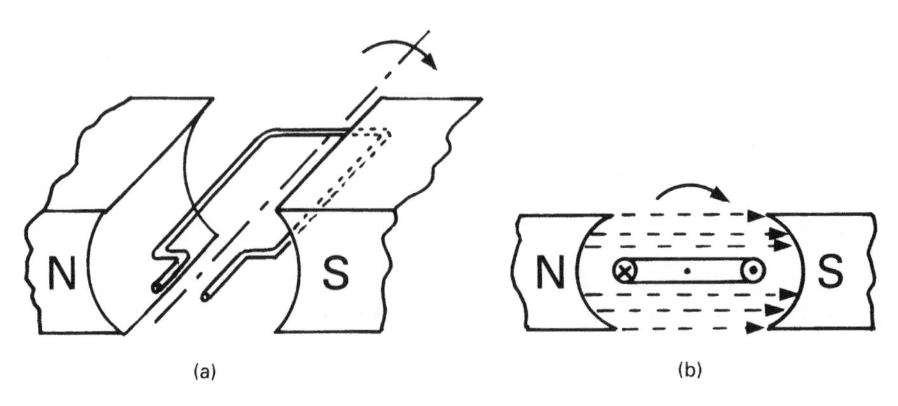

Fig. 5.10

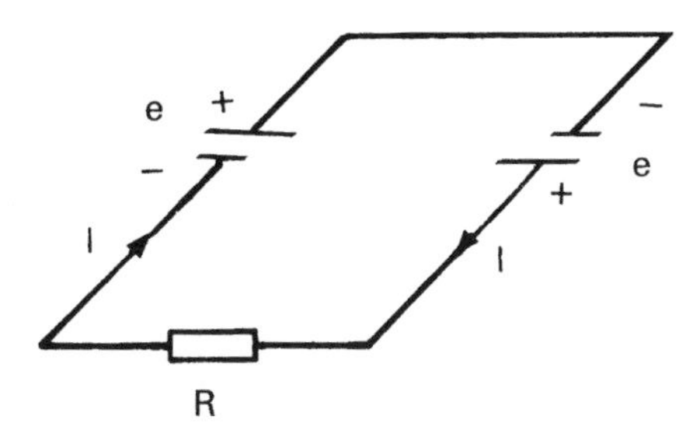

Fig. 5.11

one cycle of a sinewave, i.e. an alternating voltage. This result should not come as any surprise though, since the equation for the emf generated in each side of the loop is $e = Blv\ \mathbf{sin}\,\theta$ volt. This very simple arrangement therefore is the basis of a simple form of a.c. generator or alternator. Exactly the same principles apply to a d.c. generator, but the way in which the inherent a.c. voltage is converted into d.c. automatically by the machine is dealt with in detail in Volume 2 (*Further Electrical and Electronic Principles*).
Note: the 'ends' of the loop attached to the axle do not have emf induced in them, since they do not 'cut' the flux. Additionally, current can only flow around the loop provided that it forms part of a closed circuit.

5.5 Force on a Current-Carrying Conductor

Figure 5.12(a) shows the field patterns produced by two pole pieces, and the current flowing through the conductor. Since the lines of flux obey the rule that they will not intersect, then the flux pattern from the poles will be distorted as illustrated in Fig. 5.12(b). Also, since the lines of flux tend to act as if elastic, then they will try to straighten themselves. This results in a force being exerted on the conductor, in the direction shown.

The direction of this force may be more simply obtained by applying Fleming's *lefthand* rule. This rule is similar to the righthand rule. The major difference is of course that the fingers and thumb of the left hand are now used. In this case, the ***F***irst finger indicates the direction of the main ***F***lux (from the poles). The se***C***ond finger indicates

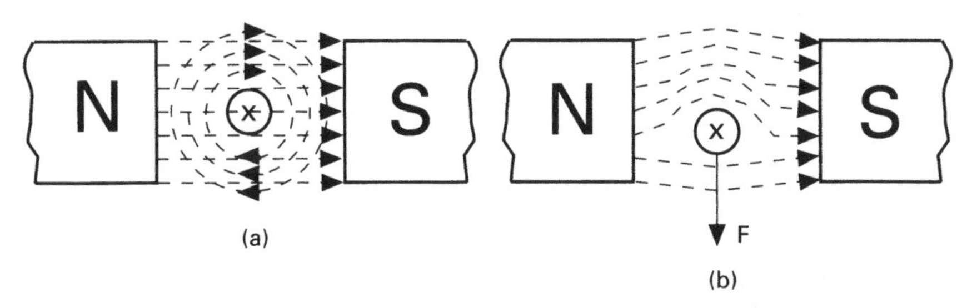

Fig. 5.12

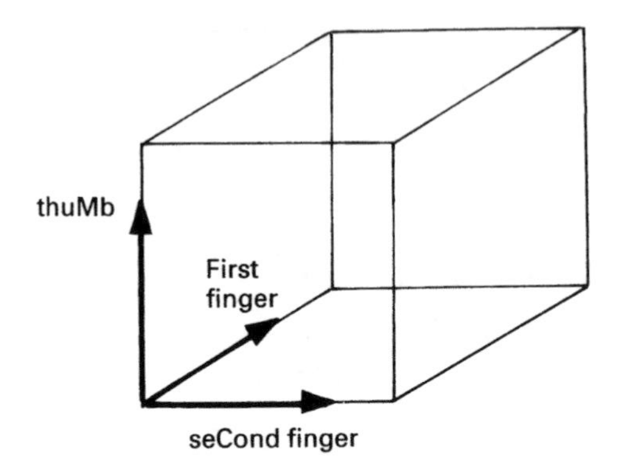

Fig. 5.13

the direction of ***C***urrent flow. The thu***M***b shows the direction of the resulting force and hence consequent ***M***otion. This is shown in Fig. 5.13.

Simple experiments can be used to confirm that the force exerted on the conductor is directly proportional to the flux density produced by the pole pieces, the value of current flowing through the conductor, and the length of conductor lying inside the field. This yields the following equation:

$$\text{Force, } F = BI\ell \text{ newton} \tag{5.4}$$

The determination of the effective length ℓ of the conductor is exactly the same as that for the generator principle previously considered. So any conductor extending beyond the main field does not contribute to the force exerted.

Equation (5.4) also only applies to the condition when the conductor is perpendicular to the main flux. If it lies at some angle less than 90°, then the force exerted on it will be reduced. Thus, in general, the force exerted is given by

$$F = BI\ell \sin\theta \text{ newton} \tag{5.5}$$

Worked Example 5.6

Question

A conductor of effective length 22 cm lies at right angles to a magnetic field of density 0.35 T. Calculate the force exerted on the conductor when carrying a current of 3 A.

Answer

$\ell = 0.22$ m; $B = 0.35$ T; $I = 3$ A; $\theta = 90°$

$$F = BI\ell \sin\theta \text{ newton} = 0.35 \times 3 \times 0.22 \times 1$$

so, $F = 0.231$ N **Ans**

Worked Example 5.7

Question

A pair of pole pieces 5 cm by 3 cm produce a flux of 2.5 mWb. A conductor is placed in this field with its length parallel to the longer dimension of the poles. When a current is passed through the conductor, a force of 1.25 N is exerted on it. Determine the value of the current.

If the conductor was placed at 45° to the field, what then would be the force exerted?

Answer

$\Phi = 2.5 \times 10^{-3}$ Wb; $\ell = 0.05$ m; $d = 0.03$ m; $F = 1.25$ N

$$\text{csa of the field, } A = 0.05 \times 0.03 = 1.5 \times 10^{-3}\ \text{m}^2$$

$$\text{flux density, } B = \frac{\Phi}{A}\ \text{tesla} = \frac{2.5 \times 10^{-3}}{1.5 \times 10^{-3}} = 1.667\ \text{T}$$

$$\text{and since } \theta = 90°, \text{ then } \sin\theta = 1$$

$$F = BI\ell \sin\theta \text{ newton, so } I = \frac{F}{B\ell} \text{ amp}$$

$$\text{therefore } I = \frac{1.25}{1.667 \times 0.05} = 15\ \text{A}\ \textbf{Ans}$$

$$F = BI\ell \sin\theta \text{ newton, where } \theta = 45°$$

$$\text{so } F = 1.667 \times 15 \times 0.05 \times 0.707$$

$$\text{hence, } F = 0.884\ \text{N}\ \textbf{Ans}$$

The principle of a force exerted on a current carrying conductor as described above forms the basis of operation of a linear motor. However, since most electric motors are rotating machines, the above system must be modified.

5.6 The Motor Principle

Once more, consider the conductor formed into the shape of a rectangular loop, placed between two poles, and current passed through it. A cross-sectional view of this arrangement, together with the flux patterns produced is shown in Fig. 5.14.

The flux patterns for the two sides of the loop will be in opposite directions because of the direction of current flow through it. The result is that the main flux from the poles is twisted as shown in Fig. 5.15. This produces forces on the two sides of the loop in opposite directions. Thus there will be a turning moment exerted on the loop, in a counterclockwise direction. The distance from the axle (the pivotal point) is r metre, so the torque exerted on each side of the loop is given by

$$T = Fr \text{ newton metre}$$

$$\text{but } F = BI\ell \sin\theta \text{ newton, and } \sin\theta = 1$$

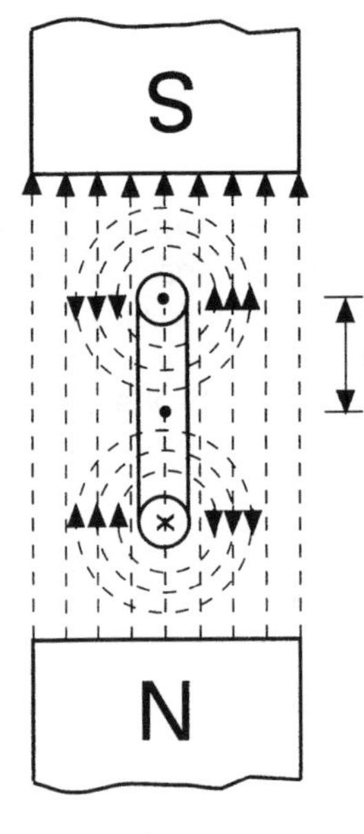

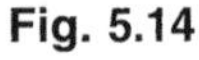
Fig. 5.14

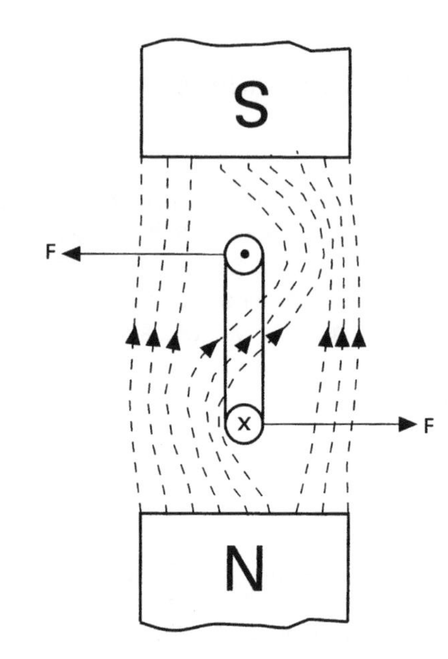

Fig. 5.15

so torque on each side $= BI\ell r$. Since the torque on each side is exerting a counter-clockwise turning effect then the total torque exerted on the loop will be

$$T = 2BI\ell r \text{ newton metre} \quad (5.6)$$

Worked Example 5.8

Question

A rectangular single-turn loop coil 1.5 cm by 0.6 cm is mounted between two poles, which produce a flux density of 1.2 T, such that the longer sides of the coil are parallel to the pole faces. Determine the torque exerted on the coil when a current of 10 mA is passed through it.

Answer

$\ell = 0.015$ m; $B = 1.2$ T; $I = 10^{-2}$ A

radius of rotation, $r = 0.006/2 = 0.003$ m

$$T = 2BI\ell r \text{ newton metre}$$

$$= 2 \times 1.2 \times 10^{-2} \times 0.015 \times 0.003$$

so $T = 1.08\ \mu$Nm **Ans**

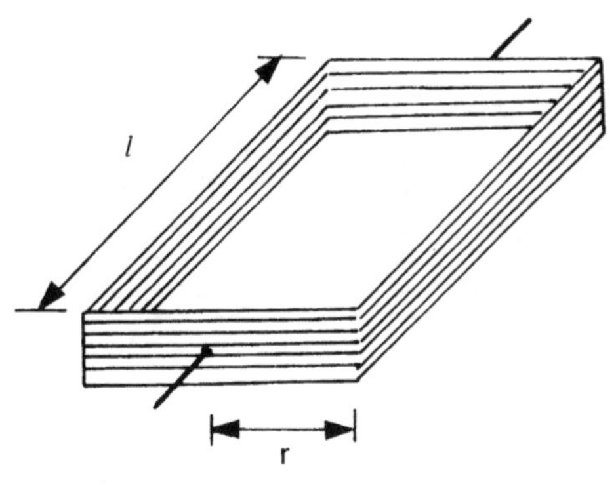

Fig. 5.16

From the above example it may be seen that a single-turn loop produces a very small amount of torque. It is acknowledged that the dimensions of the coil specified, and the current flowing through it, are also small. However, even if the coil dimensions were increased by a factor of ten times, and the current increased by a factor of a thousand times (to 10 A), the torque would still be only a very modest 0.108 Nm.

The practical solution to this problem is to use a multi-turn coil, as illustrated in Fig. 5.16. If the coil now has N turns, then each side has an effective length of $N \times l$. The resulting torque will be increased by the same factor. So, for a multi-turn coil the torque is given by

$$T = 2NBI\ell r \text{ newton metre}$$

The term $2\ell r$ in the above expression is equal to the area 'enclosed' by the coil dimensions, so this is the effective csa A, of the field affecting the coil.

Thus, $2\ell r = A$ metre^2, and the above equation may be written

$$T = BANI \text{ newton metre} \quad (5.7)$$

The principle of using a multi-turn current-carrying coil in a magnetic field is therefore used for rotary electric motors. However, the same principles apply to the operation of analogue instruments known as moving coil meters. The classic example of such an instrument is the AVO meter, mentioned earlier, in Chapter 1.

5.7 Force between Parallel Conductors

When two parallel conductors are both carrying current their magnetic fields will interact to produce a force of attraction or repulsion between them. This is illustrated in Fig. 5.17.

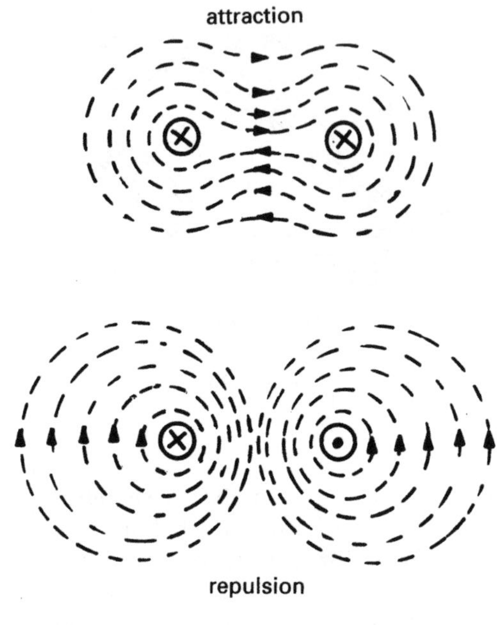

Fig. 5.17

In order to determine the value of such a force, consider first a single conductor carrying a current of I ampere. The magnetic field produced at some distance d from its centre is shown in Fig. 5.18.

In general, $H = \frac{NI}{\ell}$ ampere turn/metre

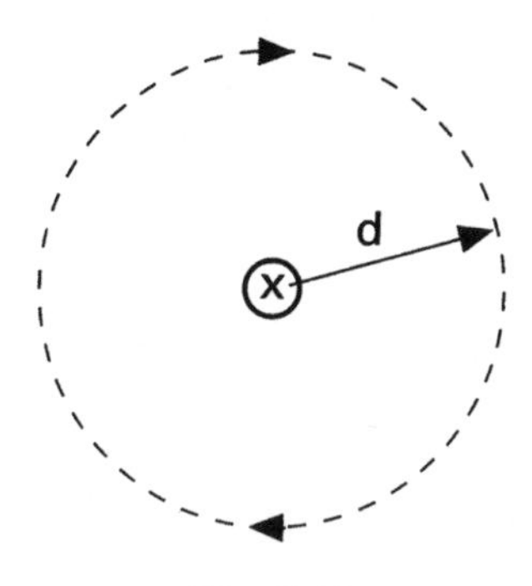

Fig. 5.18

but in this case, $N = 1$ (one conductor) and $\ell = 2\pi d$ metre (the circumference of the dotted circle), so

$$H = \frac{NI}{2\pi d}$$

Now, flux density $B = \mu_0\mu_r H$ tesla, and as the field exists in air, then $\mu_r = 1$. Thus, the flux density at distance d from the centre is given by

$$B = \frac{I\mu_0}{2\pi d} \text{ tesla} \ldots\ldots\ldots\ldots\ldots\ldots\ldots [1]$$

Consider now two conductors Y and Z carrying currents I_1 and I_2 respectively, at a distance of d metres between their centres as in Fig. 5.19.

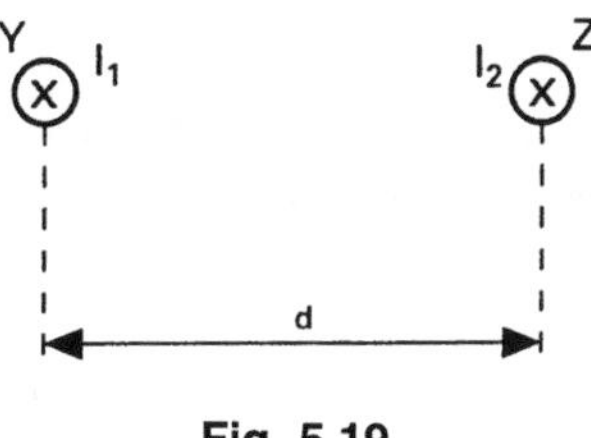

Fig. 5.19

Using equation [1] we can say that the flux density acting on Z due to current I_1 flowing in Y is:

$$B_1 = \frac{I_1\mu_0}{2\pi d} \text{ tesla}$$

and the force exerted on $Z = B_1 I_2 \ell$ newton, or $B_1 I_2$ newton per metre length of Z.

Hence, force/metre length acting on Z

$$= \frac{\mu_0 I_1 I_2}{2\pi d} \text{ newton}$$

$$= \frac{4\pi \times 10^{-7} \times I_1 I_2}{2\pi d}$$

so force/metre length acting on Z

$$= \frac{2 \times 10^{-7} I_1 I_2}{d} \text{ newton} \qquad (5.8)$$

Now, the current I_2 flowing in Z also produces a magnetic field which will exert a force on Y. Using the same reasoning as above, it can be shown that:

force/metre length acting on Y

$$= \frac{2 \times 10^{-7}}{d} I_1 I_2 \text{ newton}$$

so if $I_1 = I_2 = 1$ A, and $d = 1$m, then

force exerted on *each* conductor $= 2 \times 10^{-7}$ newton

This value of force forms the basis for the definition of the ampere, namely: that current, which when maintained in each of two infinitely long parallel conductors situated *in vacuo*, and separated one metre between centres, produces a force of 2×10^{-7} newton per metre length on each conductor.

Worked Example 5.9

Question

Two long parallel conductors are spaced 35 mm between centres. Calculate the force exerted between them when the currents carried are 50 A and 40 A respectively.

Answer

$d = 0.035$ m; $I_1 = 50$ A; $I_2 = 40$ A

$$F = \frac{2 \times 10^{-7} I_1 I_2}{d} \text{ newton} = \frac{2 \times 10^{-7} \times 50 \times 40}{0.035}$$

so $F = 11.4$ mN **Ans**

Worked Example 5.10

Question

Calculate the flux density at a distance of 2 m from the centre of a conductor carrying a current of 1000 A. If the centre of a second conductor, carrying 300 A, was placed at this same distance, what would be the force exerted?

Answer

$d = 2$ m; $I_1 = 1000$ A; $I_2 = 300$ A

$$B = \frac{\mu_0 I_1}{d} \text{ tesla} = \frac{4\pi \times 10^{-7} \times 1000}{2}$$

so, $B = 0.628$ mT **Ans**

$$F = \frac{2 \times 10^{-7} I_1 I_2}{d} \text{ newton} = \frac{2 \times 10^{-7} \times 300 \times 1000}{2}$$

so $F = 30$ mN **Ans**

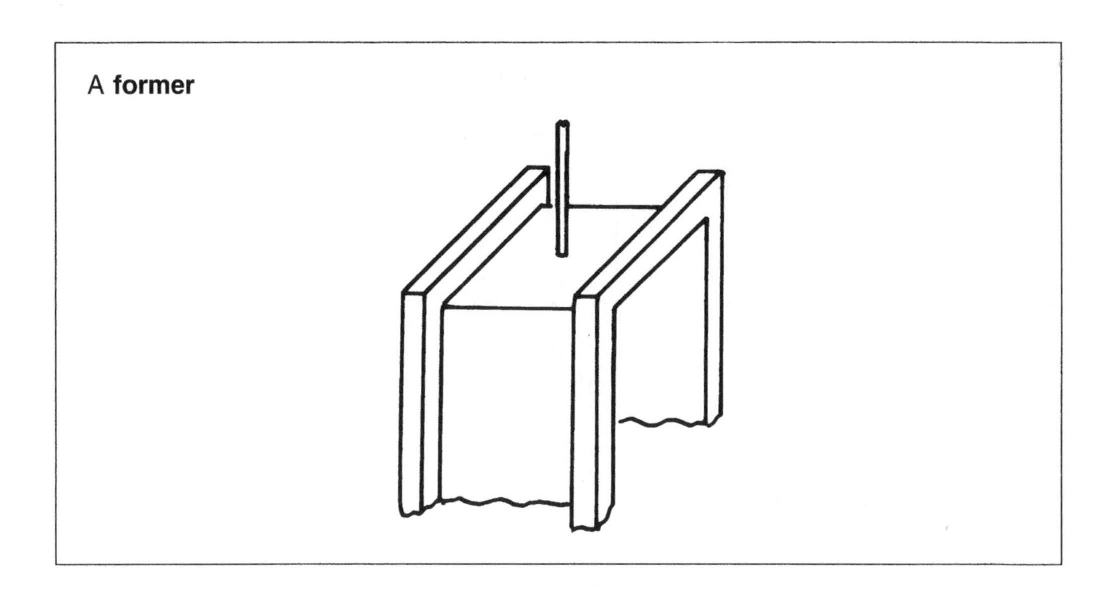

5.8 The Moving Coil Meter

Most analogue (pointer-on-scale) instruments rely on three factors for their operation: a *deflecting* torque; a *restoring* torque; and a *damping* torque.

Deflecting Torque Essentially, a moving coil meter is a current measuring device. The current to be measured is passed through a multiturn coil suspended between the poles of a permanent magnet. The coil is made from fine copper wire which is wound on to a light aluminium **former**. Thus the motor effect, described in section 5.6, is utilised. Since the wire is of small diameter, and the aluminium former is also very light, then this assembly has very little inertia. This is an essential requirement, to ensure that the instrument is sufficiently sensitive. This means that it can respond to very small deflecting torques. To illustrate this point, refer back to Worked Example 5.8, in which the torque exerted on a small coil with a current of 10 mA was found to be only 1.08 μNm. In order to improve the sensitivity further, the friction of the coil pivots must be minimised. This is achieved by the use of jewelled bearings as shown in Fig. 5.20.

When a current is passed through the coil it will rotate under the influence of the deflecting torque. However, if the 'starting' position of the coil is as shown in Fig. 5.21, then it will finally settle in the vertical position, regardless of the actual value of the current. The reason is that the effective perpendicular distance from the pivot point (the term $r \sin\theta$ in the expression $F = 2NBI\ell r \sin\theta$) decreases to zero at this position. Another way to explain this is to consider the forces acting on each side of the coil. These will always act at right angles to the main flux. Thus, when the coil reaches the vertical position, these forces are no longer producing a turning effect. This means that the instrument is capable of indicating only the presence of a current, but not the actual value. Hence, the deflecting torque (which is dependent upon the value of the coil current) needs to be counter-balanced by another torque.

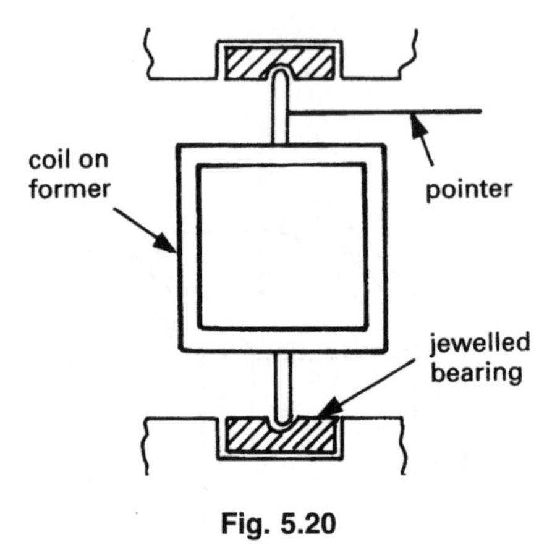

Fig. 5.20

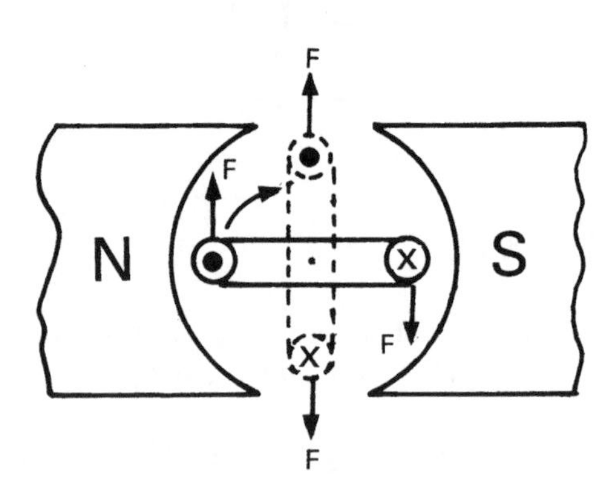

Fig. 5.21

Restoring Torque This is the counter-balancing torque mentioned above. It is provided by two contrawound* spiral springs, one end of each being connected to the top and bottom respectively, of the spindle that carries the coil former. The greater the current passed through the coil, the greater will be the deflecting torque. The restoring torque provided by the springs will increase in direct proportion to the deflection applied to them. The pointer, carried by the coil spindle will therefore move to a point at which the deflecting and restoring torques balance each other. The springs also serve as the means of passing the current to and from the coil. This avoids the problem of the coil having to drag around a pair of trailing leads.

We now have a system in which the deflection obtained depends upon the value of the coil current. There are still other problems to overcome though. One of these is that the deflecting torque, due to a given value of current, will vary with the coil position. This would have the effect of non-linear deflections for linear increments of current. If

* The word contrawound means 'wound in opposite directions'. The reason for winding them in this way is to prevent the pointer position from being affected by temperature changes. If the temperature increases, then both springs tend to expand, by the same amount. Since one spring is acting to push the pointer spindle in one direction, and the other one in the opposite direction, then the effects cancel out.

we could ensure that the coil always lies at right angles to the field, regardless of its rotary position, then this problem would be resolved. The way in which this is achieved is by the inclusion of a soft iron cylinder inside the coil former. This cylinder does not touch the former, but causes a radial flux pattern in the air gap in which the coil rotates. This pattern results from the fact that the lines of flux will take the path(s) of least reluctance, and so cross the gaps between the pole faces and iron cylinder by the shortest possible path, i.e. at 90° to the surfaces. This effect is illustrated in Fig. 5.22.

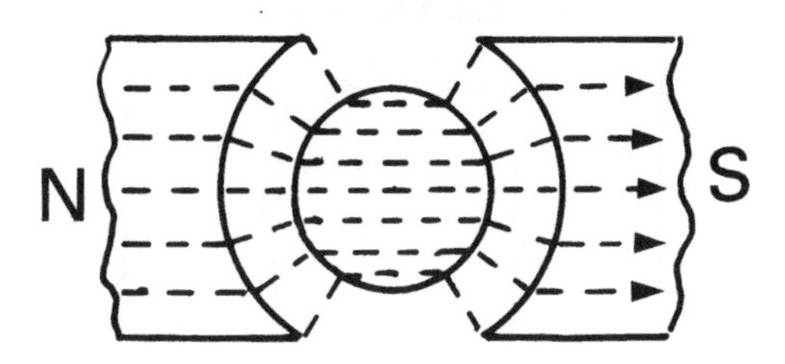

Fig. 5.22

Damping Torque When current is passed through the coil, the deflecting torque accelerates the pointer away from the zero position. Now, although the coil and pointer assembly is very light, it will still have sufficient inertia to 'overshoot' its final position on the graduated scale. It is also likely to under- and overshoot several times before settling. To prevent this from happening, the movement needs to be slowed down, or damped. This effect is achieved automatically by the generation of eddy currents in the aluminium coil former as it rotates in the magnetic field. The full description of eddy currents is dealt with later in this chapter. However, being induced currents means they are subject to Lenz's law. They will therefore flow in the coil former in such a direction as to oppose the change that produced them; that is the rapid deflection of the coil. If the dimensions of the former are correctly chosen, then the result will be either one very small overshoot, or the overshoots are *just* prevented from occurring. In the latter case the instrument is said to be critically damped, or 'dead beat'.

The main advantages of the moving coil instrument are:

1 Good sensitivity: this is due to the low inertia of the coil and pointer assembly. Typically, a current of 50 μA through the coil is sufficient to move the pointer to the extreme end of the scale (full-scale deflection or fsd).
2 Linear scale: from equation (5.7) we know that $T = BANI$. For a given instrument, B, A, and N are fixed values, so $T \propto I$. Thus the deflecting torque is *directly* proportional to the coil current.

The main disadvantage is the fact that the basic meter movement so far described can be used only for d.c. measurements. If a.c. was applied to the coil, the pointer would try to deflect in opposite directions alternately. Thus, if only a moderate frequency such as 50 Hz is applied, the pointer cannot respond quickly enough. In this case only a very small vibration of the pointer, about the zero position, might be observed.

The complete arrangement for a moving coil meter is illustrated in Fig. 5.23.

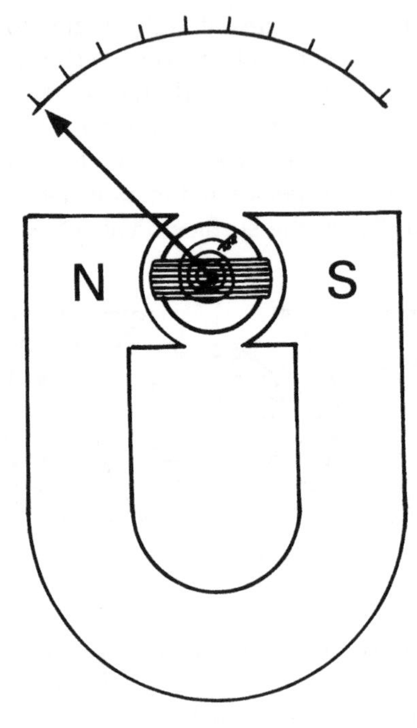

Fig. 5.23

5.9 Shunts and Multipliers

The basic instrument so far described is capable of measuring only small d.c. currents (up to 50 μA typically). This obviously severely limits its usefulness. The figure of 50 μA quoted is the value of coil full-scale deflection current (I_{fsd}) for an AVO meter. As you will now be aware, this meter is in fact capable of measuring up to 10 A, by using the current range switch. This is achieved by means of what are known as shunts. Similarly, this type of meter is also capable of measuring a range of voltages. This is achieved by means of multipliers. Both shunts and multipliers are incorporated into the instrument when it is assembled.

5.10 Shunts

A shunt is a device connected in parallel with the meter's moving coil, in order to extend the current reading range of the instrument. It consists of a very low resistance element, often made from a small strip of copper or aluminium. Being connected in parallel with the coil, it forms an alternative path for current flow. It can therefore divert excessive current from the coil itself. In this instance, 'excessive current' means current in excess of that required to produce full-scale deflection of the pointer. The latter is known as the full-scale deflection current, which is normally abbreviated to I_{fsd}.

The application for a shunt is best illustrated by a worked example.

Worked Example 5.11

Question

A moving coil meter has a coil resistance of 40 Ω, and requires a current of 0.5 mA to produce f.s.d. Determine the value of shunt required to extend the current reading range to 3 A.

Answer

A sketch of the appropriate circuit should be made, and such a diagram is shown in Fig. 5.24.

$R_C = 40\,\Omega$; $I_{fsd} = 5 \times 10^{-4}$ A; $I = 3$ A

$$V_C = I_{fsd} R_C \text{ volt} = 5 \times 10^{-4} \times 40$$
$$\text{so } V_C = 0.02\text{ V}$$
$$I_s = I - I_{fsd} \text{ amp} = 3 - (5 \times 10^{-4})$$
$$\text{so } I_s = 2.995\text{ A}$$
$$R_s = \frac{V_c}{I_s} \text{ ohm} = \frac{0.02}{2.9995}$$
$$\text{therefore, } R_s = 6.67\text{ m}\Omega \textbf{ Ans}$$

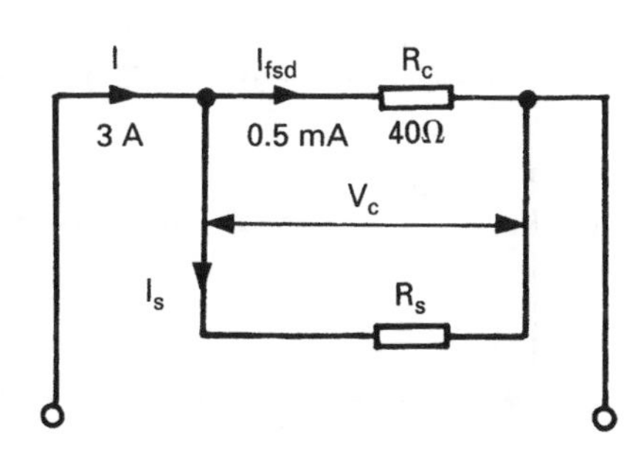

Fig. 5.24

5.11 Multipliers

From the previous worked example, it may be seen that the p.d. across the coil (V_c) when I_{fsd} flows through it is only 20 mV. This is therefore the maximum voltage that may be applied to the coil. It also represents the maximum voltage measurement that can be made.

In order to extend the voltage reading range, a multiplier must be employed. This is a high value resistance, connected in series with the meter coil. Thus, when a voltage in excess of V_c is applied to the meter terminals, the multiplier will limit the current to I_{fsd}.

Worked Example 5.12

Question

Considering the same meter movement specified in Worked Example 5.11, determine the multiplier required to extend the voltage reading range to 10 V.

Answer

$R_C = 40\,\Omega$; $I_{fsd} = 5 \times 10^{-4}$ A; $V = 10$ V

From worked Example 5.11, we know that the p.d. across the coil with I_{fsd} flowing is $V_C = 0.02$ V. The circuit diagram is shown in Fig. 5.25.

$$\text{Total resistance, } R = \frac{V}{I_{fsd}} \text{ volt} = \frac{10}{5 \times 10^{-4}}$$

$$\text{so } R = 20\,\text{k}\Omega$$

$$\text{but } R = R_m + R_c \text{ ohm}$$

$$\text{so } R_m = R - R_c = 20\,000 - 40$$

$$\text{therefore, } R_m = 19.96\,\text{k}\Omega \textbf{ Ans}$$

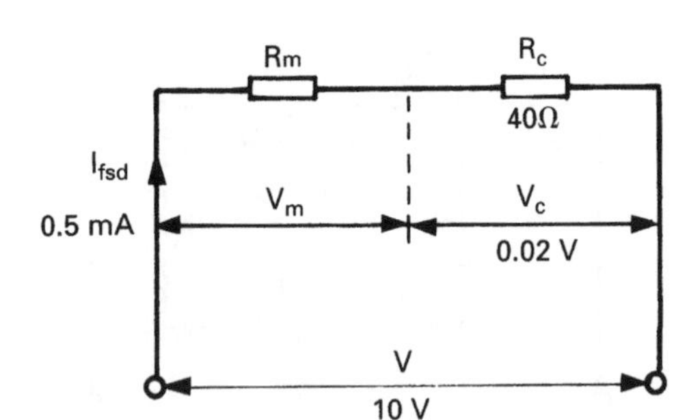

Fig. 5.25

This problem may also be solved by the following method:

$$V_m = V - V_c \text{ volt} = 10 - 0.02$$

$$\text{so } V_m = 9.98 \text{ V}$$

$$R_m = \frac{V_m}{I_{fsd}} \text{ ohm} = \frac{9.98}{5 \times 10^{-4}}$$

$$\text{therefore, } R_m = 19.96\,\text{k}\Omega \textbf{ Ans}$$

When the voltage range switch on an AVO is rotated, it simply connects the appropriate value of multiplier in series with the moving coil. Similarly, when the current range switch is operated, it connects the appropriate value shunt in parallel with the coil. In order to protect the instrument from an electrical overload, you must always start the measurement from the highest available range. Progressively lower range settings can then be selected, until a suitable deflection is obtained.
Note: The range setting marked on any instrument indicates the value of applied voltage or current that will cause full-scale deflection on that range.

5.12 Figure of Merit and Loading Effect

When a moving coil meter is used as a voltmeter, the total resistance between its terminals will depend upon the value of multiplier connected. This will of course vary with the range selected. This total resistance is called the input resistance of the instrument (R_{in}). An ideal voltmeter would draw zero current from the circuit to which it is connected. Thus, for a practical instrument, the higher the value of R_{in}, the closer it approaches the ideal.

Since R_{in} changes as the voltage ranges are changed, some other means of indicating the 'quality' of the instrument is required. This is the figure of merit, which is quoted in ohms/volt.

$$R_{in} = \frac{V}{I_{fsd}} \text{ ohm; where}$$

V is the voltage range selected

$$\text{So, } \frac{1}{I_{fsd}} = \frac{R_{in}}{V} \text{ ohms/volt;}$$

$$\text{and } \frac{1}{I_{fsd}} \text{ is the figure of merit}$$

Thus, the reciprocal of the full-scale deflection current gives the figure of merit for a moving coil instrument, when used as a voltmeter.

The lowest current range on an AVO is 50 μA, which happens to be its I_{fsd}. Hence, its figure of merit

$$= \frac{1}{50 \times 10^{-6}} = 20\,\text{k}\Omega/V.$$

Since R_{in} = figure of merit $\times V$, then on the 0.3 V range, $R_{in} = 20\,000 \times 0.3 = 6\,\text{k}\Omega$. The figures for other voltage ranges will therefore be:

1 V range, $R_{in} = 20\,\text{k}\Omega$; 100 V range, $R_{in} = 2\,\text{M}\Omega$; etc

It may therefore be seen that the higher the voltage range selected, the closer R_{in} approaches the ideal of infinity. In order to illustrate the practical significance of this, let us consider another example.

Worked Example 5.13

Question

A simple potential divider circuit is shown in Fig. 5.26. The p.d. V_2 is to be measured by an AVO, using the 10 V range. Calculate the p.d. indicated by this meter, and the percentage error in this reading.

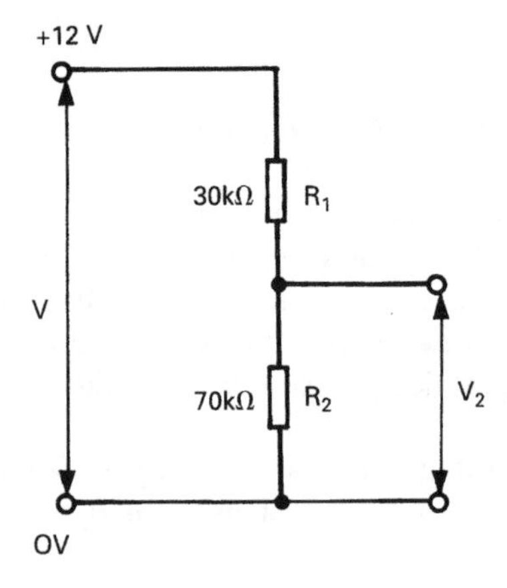

Fig. 5.26

Answer

Firstly, we can calculate the p.d. developed across R_2 by applying the potential divider theory, thus

$$V_2 = \frac{R_2}{R_1 + R_2}.V \text{ volt} = \frac{7}{10} \times 12$$

so, the true value of $V_2 = 8.4\,\text{V}$

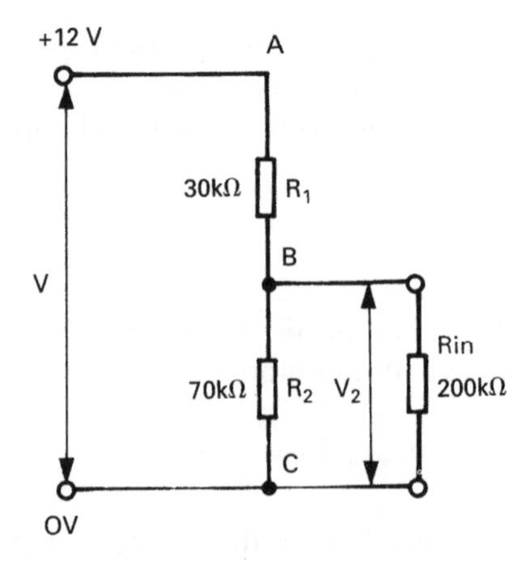

Fig. 5.27

However, when the voltmeter is connected across R_2, the circuit is effectively modified to that shown in Fig. 5.27. (Since it is an AVO, switched to the 10 V range, then the value for R_{in} is 200 kΩ.)

$$R_{BC} = \frac{R_2 R_{in}}{R_2 + R_{in}} \text{ ohm} = \frac{70 \times 200}{270} \text{ k}\Omega$$

so $R_{BC} = 51.85 \text{ k}\Omega$

and using the potential divider technique:

$$V_2 = \frac{51.85}{51.85 + 30} \times 12 = 7.6 \text{ V};$$

this is the p.d. that will be indicated by the voltmeter.

so, $V_2 = 7.6$ V **Ans**

The percentage error in the reading is defined as:

$$\text{error} = \frac{\text{indicated value} - \text{true value}}{\text{true value}} \times 100\%$$

$$\text{so, error} = \frac{7.6 - 8.4}{8.4} \times 100\% = -9.52\% \textbf{ Ans}$$

Hence, it can be seen that the meter does not indicate the true p.d. across R_2, since it indicates a lower value. This is known as the loading effect. The reason for this effect is that, in order to operate, the meter has to draw current from the circuit. The original circuit conditions have therefore been altered, as shown in Fig. 5.27.

The loading effect does not depend entirely on the value of R_{in}. The value of R_{in}, *relative* to the resistance of the component whose p.d. is being measured, is equally important. It is left to the reader to verify that, if resistors R_1 and R_2 in Fig. 5.26 were 300 Ω and 700 Ω respectively, the loading error would be only 0.004%.

Digital multimeters have an input resistance of 10 to 20 MΩ. This figure remains sensibly constant regardless of the voltage range selected. The loading effect is therefore much less than for an AVO, especially when using the lower voltage ranges. Also, as the indicated values are easier to read, they tend to be used more often than moving coil meters. However, there are other factors that need to be considered when selecting a meter for a given measurement. These are covered in Volume 2 *Further Electrical and Electronic Principles.*

5.13 The Ohmmeter

As the name implies, this instrument is used for the measurement of resistance. This feature is normally included in multimeters. In the case of the AVO, the moving coil movement is also utilised for this purpose. The theory is based simply on Ohm's law. If

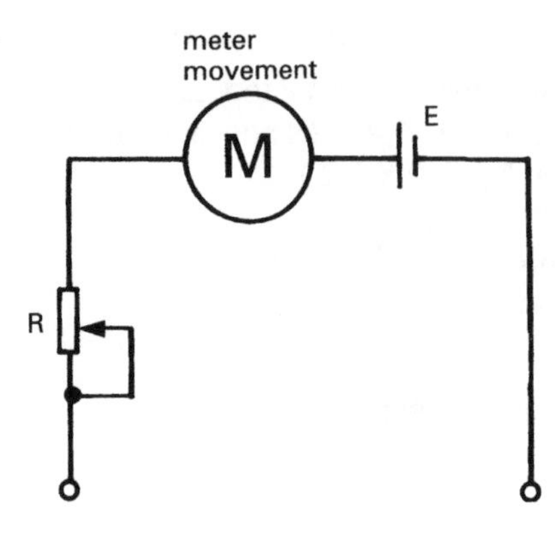

Fig. 5.28

a known emf is applied to a resistor, the resulting current flow is inversely proportional to the resistance value. The basic arrangement is shown in Fig. 5.28. The battery is incorporated into the instrument, as is the variable resistance, R.

To use the instrument in this mode, the terminals are first 'shorted' together, using the instrument leads. The resistor R is then adjusted so that the pointer indicates zero on the ohms scale. Note that the zero position on this scale is at the *righthand* extreme of the scale. Having 'zeroed' the meter, the 'short' is removed. Finally, the resistance to be measured is connected between the terminals. The resistance value will then be indicated by the position of the pointer on the ohms scale.

It is not recommended that resistance values be measured using this technique! The scale is extremely cramped over the higher resistance range, and is extremely non-linear. This makes accurate measurement of resistance virtually impossible. However, the use of this facility for continuity checking is useful.

If you wish to measure a resistance value fairly accurately, then use this facility on a digital multimeter. The same basic principle, of applying a known emf and measuring the resulting current, may still be employed. The internal electronic circuitry then converts this into a display of resistance value. By definition, the scale cannot be cramped, and is easy to read. If a resistance has to be measured with a high degree of accuracy, then a Wheatstone Bridge must be used.

5.14 Wattmeter

This instrument is used to measure electrical power. In its traditional form it is called a dynamometer wattmeter. It is also available in a purely electronic form. The dynamometer type utilises the motor principle. The meter consists of two sets of coils. One set is fixed, and is made in two identical parts. This is the current coil, and is made from heavy gauge copper wire. The resistance of this coil is therefore low. The voltage coil is wound from fine gauge wire, and therefore has a relatively high resistance. The voltage coil is mounted on a circular former, situated between the two parts of the current coil. The basic arrangement is illustrated in Fig. 5.29.

The current coil is connected to a circuit so as to allow the circuit current to flow through it. The voltage coil is connected in parallel with the load, in the same way as a voltmeter. Both coils produce magnetic fields, which interact to produce a deflecting torque on the voltage coil. Restoring torque is provided by contrawound spiral springs, as in the moving coil meter. The interacting fields are proportional to the circuit current and voltage respectively. The deflecting torque is therefore proportional to the product of these two quantities; i.e. VI which is the circuit power.

The voltage coil may be connected either on the supply side or the load side of the current coil. The choice of connection depends on other factors concerning the load. This aspect is dealt with in Volume 2. The manner of connecting the wattmeter into a circuit is shown in Fig. 5.30.

For d.c. circuits, a wattmeter is not strictly necessary. Since power $P = VI$ watt, then V and I may be measured separately by a multimeter. The power can then be calculated simply by multiplying together these two meter readings. However, in an a.c. circuit, this simple technique does not yield the correct value for the true power. Thus a wattmeter is required for the measurement of power in a.c. circuits.

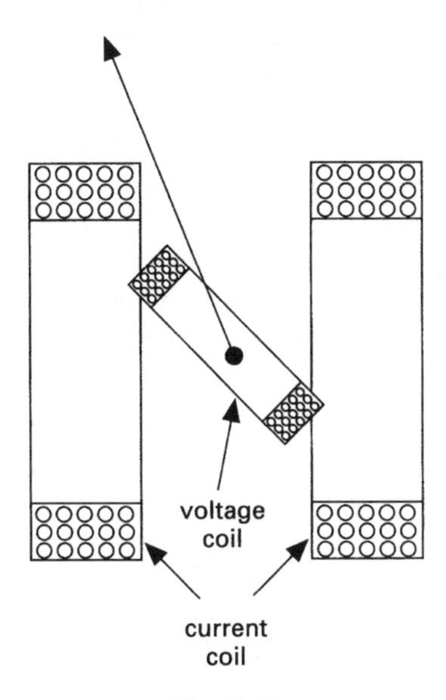

Fig. 5.29

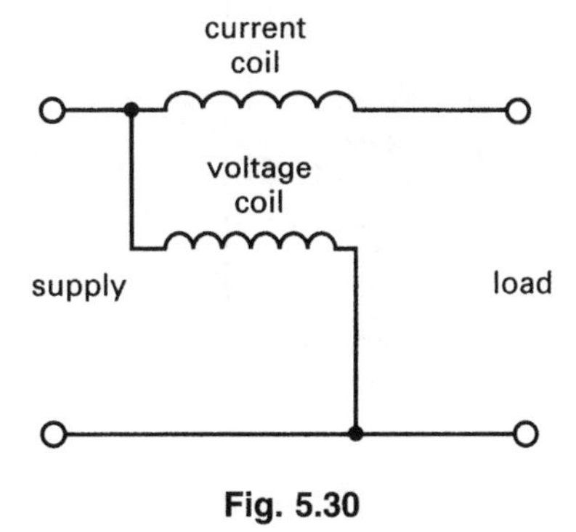

Fig. 5.30

5.15 Eddy Currents

Consider an iron-cored solenoid, as shown in cross-section in Fig. 5.31. Let the coil be connected to a source of emf via a switch. When the switch is closed, the coil current will increase rapidly to some steady value. This steady value will depend upon the resistance of the coil. The coil current will, in turn, produce a magnetic field. Thus, this flux pattern will increase from zero to some steady value. This changing flux therefore expands outwards from the centre of the iron core. This movement of the flux pattern is shown by the arrowed lines pointing outwards from the core.

Since there is a changing flux linking with the core, then an emf will be induced in the core. As the core is a conductor of electricity, then the induced emf will cause a current to be circulated around it. This is known as an eddy current, since it traces out a circular path similar to the pattern created by an eddy of water. The direction of the induced emf and eddy current will be as shown in Fig. 5.31. This has been determined by applying Fleming's righthand rule. Please note, that to apply this rule, we need to consider the movement of the *conductor relative to* the flux. Thus, the *effective* movements of the left and right halves of the core are *opposite* to the arrows showing the expansion of the flux pattern.

As the eddy current circulates in the core, it will produce a heating effect. This is normally an undesirable effect. The energy thus dissipated is therefore referred to as the eddy current loss. If the solenoid forms part of a d.c. circuit, this loss is negligible. This is because the eddy current will flow only momentarily—when the circuit is first connected, and again when it is disconnected. However, if an a.c. supply is connected to the coil, the eddy current will be flowing continuously, in alternate directions. Under these conditions, the core is also being taken through repeated magnetisation cycles. This will result in a hysteresis loss also.

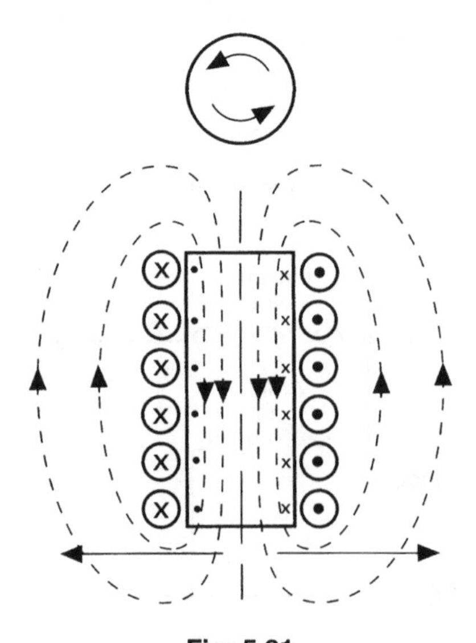

Fig. 5.31

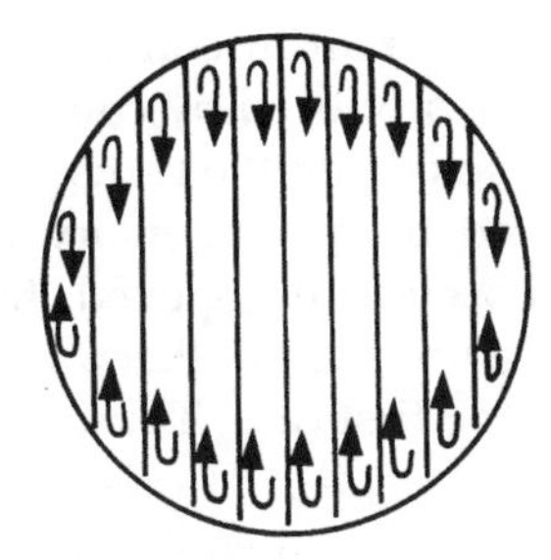

Fig. 5.32

In order to minimise the eddy current loss, the resistance of the core needs to be increased. On the other hand, the low reluctance needs to be retained. It would therefore be pointless to use an insulator for the core material, since we might just as well use an air core! The technique used for devices such as transformers, used at mains frequency, is to make the core from laminations of iron. This means that the core is made up of thin sheets (laminations) of steel, each lamination being insulated from the next. This is illustrated in Fig. 5.32. Each lamination, being thin, will have a relatively high resistance. Each lamination will have an eddy current, the circulation of which is confined to that lamination. If the values of these individual eddy currents are added together, it will be found to be less than that for the solid core.

The hysteresis loss is proportional to the frequency f of the a.c. supply. The eddy current loss is proportional to f^2. Thus, at higher frequencies (e.g. radio frequencies), the eddy current loss is predominant. Under these conditions, the use of laminations is not adequate, and the eddy current loss can be unacceptably high. For this type of application, iron dust cores or ferrite cores are used. With this type of material, the eddy currents are confined to individual 'grains', so the eddy current loss is considerably reduced.

5.16 Self and Mutual Inductance

The effects of self and mutual inductance can be demonstrated by another simple experiment. Consider two coils, as shown in Fig. 5.33. Coil 1 can be connected to a battery via a switch. Coil 2 is placed close to coil 1, but is not electrically connected to it. Coil 2 has a galvo connected to its terminals.

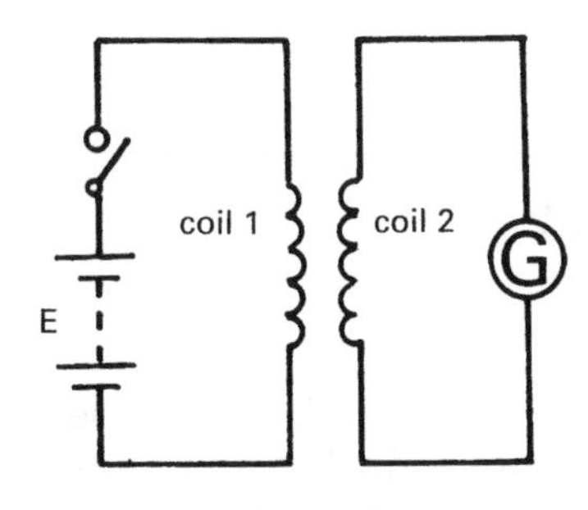

Fig. 5.33

When the switch is closed, the current in coil 1 will rapidly increase from zero to some steady value. Hence, the flux produced by coil 1 will also increase from zero to a steady value. This changing flux links with the turns of coil 2, and therefore induces an emf into it. This will be indicated by a momentary deflection of the galvo pointer. Similarly, when the switch is subsequently opened, the flux produced by coil 1 will collapse to zero. The galvo will again indicate that a momentary emf is induced in coil 2, but of the opposite polarity to the first case. Thus, an emf has been induced into coil 2, by a changing current (and flux) in coil 1. This is known as a mutually induced emf.

If the changing flux can link with coil 2, then it must also link with the turns of coil 1. Thus, there must also be a momentary emf induced in this coil. This is known as a self-induced emf. Any induced emf obeys Lenz's law. This self-induced emf must therefore be of the opposite polarity to the battery emf. For this reason, it is also referred to as a back emf. Unfortunately, it is extremely difficult to demonstrate the existence of this back emf. If a voltmeter was connected across coil 1, it would merely indicate the terminal voltage of the battery.

5.17 Self-Inductance

Self-inductance is that property of a circuit or component which causes a self-induced emf to be produced, when the current through it changes. The unit of self-inductance is the henry, which is defined as follows:

A circuit has a self-inductance of one henry (1 H) if an emf of one volt is induced in it, when the circuit current changes at the rate of one ampere per second (1 A/s).

The quantity symbol for self-inductance is L. From the above definition, we can state the following equation

$$L = \frac{-e}{\mathrm{d}i/\mathrm{d}t} \text{ henry}$$

$$\text{or, self-induced emf, } e = \frac{-L\mathrm{d}i}{\mathrm{d}t} \text{ volt} \qquad (5.9)$$

Notes:

1 The minus sign again indicates that Lenz's law applies.
2 The emf symbol is e, because it is only a momentary emf.
3 The current symbol is i, because it is the *change* of current that is important.
4 The term $\mathrm{d}i/\mathrm{d}t$ is the rate of change of current.

Worked Example 5.14

Question

A coil has a self-inductance of 0.25 H. Calculate the value of emf induced, if the current through it changes from 100 mA to 350 mA, in a time of 25 ms.

Answer

$L = 0.25$ H; $\mathrm{d}i = (350 - 100) \times 10^{-3}$ A; $\mathrm{d}t = 25 \times 10^{-3}$ s

$$e = \frac{-L\mathrm{d}i}{\mathrm{d}t}\ \text{volt} = \frac{0.25 \times 250 \times 10^{-3}}{25 \times 10^{-3}}$$

so $e = 2.5\,\text{V}$ **Ans**

Worked Example 5.15

Question

Calculate the inductance of a circuit in which an emf of 30 V is induced, when the circuit current changes at the rate of 200 A/s.

Answer

$e = 30$ V; $\mathrm{d}i/\mathrm{d}t = 200$ A/s

$$e = \frac{-L\mathrm{d}i}{\mathrm{d}t}\,\text{volt}$$

$$\text{so,}\quad L = \frac{e}{\mathrm{d}i/\mathrm{d}t}\,\text{henry} = \frac{30}{200}$$

therefore, $L = 0.15\,\text{H}$ **Ans**

Worked Example 5.16

Question

A circuit of self-inductance 50 mH has an emf of 8 V induced into it. Calculate the rate of change of the circuit current that induced this emf.

Answer

$L = 50 \times 10^{-3}$ H; $e = 8$ V

$$e = \frac{-L\mathrm{d}i}{\mathrm{d}t}\ \text{volt}$$

$$\text{so,}\ \mathrm{d}i/\mathrm{d}t = \frac{e}{L}\,\text{amp/s} = \frac{8}{50 \times 10^{-3}}$$

hence, $\mathrm{d}i/\mathrm{d}t = 160\,\text{A/s}$ **Ans**

5.18 Self-Inductance and Flux Linkages

Consider a coil of N turns, carrying a current of I amp. Let us assume that this current produces a flux of Φ weber. If the current now changes at a uniform rate of $\mathrm{d}i/\mathrm{d}t$ ampere per second, it will cause a corresponding change of flux of $\mathrm{d}\phi/\mathrm{d}t$ weber per second. Let us also assume that the coil has a self-inductance of L henry.

The self-induced emf may be determined from equation (5.9):

$$e = -L\mathrm{d}i/\mathrm{d}t \text{ volt} \ldots\ldots\ldots\ldots\ldots\ldots [1]$$

However, the induced emf is basically due to the rate of change of flux linkages. Thus, the emf may also be calculated by using equation (5.1), namely:

$$e = -N\mathrm{d}\phi/\mathrm{d}t \text{ volt} \ldots\ldots\ldots\ldots\ldots\ldots [2]$$

Since both equations [1] and [2] represent the same induced emf, then [1] must be equal to [2]. Thus

$$\frac{L\mathrm{d}i}{\mathrm{d}t} = \frac{N\mathrm{d}\phi}{\mathrm{d}t} \text{(the minus signs cancel out)}$$

$$\text{so, } L = \frac{N\mathrm{d}\phi}{\mathrm{d}i} \text{ henry} \qquad (5.10)$$

A coil which is designed to have a specific value of self-inductance is known as an inductor. An inductor is the third of the main **passive** electrical components. The other two are the resistor and the capacitor.

A **passive** component is one which (a) requires an external source of emf in order to serve a useful function, and (b) does not provide any amplification of current or voltage

Now, a resistor will have a specific value of resistance, regardless of whether it is in a circuit or not. Similarly, an inductor will have some value of self-inductance, even when the current through it is constant. In other words, an inductor does not have to have an emf induced in it, in order to possess the property of self-inductance. For this reason, equation (5.10) may be slightly modified as follows.

If the current I through an N turn coil produces a flux of Φ weber, then its self-inductance is given by the equation

$$L = \frac{N\Phi}{I} \text{ henry} \qquad (5.11)$$

In other words, although no *change* of current and flux is specified, the coil will still have some value of inductance. Strictly speaking, equation (5.11) applies only to an inductor with a non-magnetic core. The reason is that, in this case, the flux produced is directly proportional to the coil current. However, it is a very close approximation to the true value of inductance for an iron-cored inductor which contains an air gap in it.

Worked Example 5.17

Question

A coil of 150 turns carries a current of 10 A. This current produces a magnetic flux of 0.01 Wb. Calculate (a) the inductance of the coil, and (b) the emf induced when the current is uniformly reversed in a time of 0.1 s.

Answer

$N = 150$; $I = 10$ A; $\Phi = 0.01$ Wb; $dt = 0.1$ s

(a) $$L = \frac{N\Phi}{I} \text{ henry} = \frac{150 \times 0.10}{10}$$

so, $L = 1.5$ H **Ans**

(b) Since current is reversed then it will change from 10 A to −10 A, i.e. a *change* of 10 − (−10). So, $di = 20$ A.

$$e = \frac{-Ldi}{dt} \text{ volt} = \frac{1.5 \times 20}{0.1}$$

therefore, $e = 300$ V **Ans**

Worked Example 5.18

Question

A current of 8 A, when flowing through a 3000 turn coil, produces a flux of 4 mWb. If the current is reduced to 2 A in a time of 100 ms, calculate the emf thus induced in the coil. Assume that the flux is directly proportional to the current.

Answer

$I_1 = 8$ A; $N = 3000$; $\Phi_1 = 4 \times 10^{-3}$ Wb; $I_2 = 2$ A; $dt = 0.1$ s

This problem may be solved in either of two ways. Both methods will be demonstrated.

$$e = \frac{-Ldi}{dt} \text{ volt, where } L = \frac{N\Phi_1}{I_1} \text{ henry}$$

so, $L = \dfrac{3000 \times 4 \times 10^{-3}}{8} = 1.5$ H; $di = 8 - 2 = 6$ A

therefore, $e = \dfrac{1.5 \times 6}{0.1} = 90$ V **Ans**

Continued on p. 232

Worked Example 5.18 (*Continued*)

Alternatively, $\Phi \propto I$ *so* $\Phi_1 \propto I_1$ and $\Phi_2 \propto I_2$

$$\text{therefore, } \frac{\Phi_2}{\Phi_2} = \frac{I_2}{I_1} \text{ and } \Phi_2 = \frac{\Phi_1 I_2}{I_1}$$

$$\text{hence, } \Phi_2 = \frac{2 \times 4 \times 10^{-3}}{8} = 1 \times 10^{-3} \text{ Wb;}$$

$$\text{and } d\phi = (4 - 1) \times 10^{-3} \text{ Wb}$$

$$e = \frac{-Nd\phi}{dt} \text{ volt} = \frac{3000 \times 3 \times 10^{-3}}{0.1}$$

so, $e = 90\,\text{V}$ **Ans**

5.19 Factors Affecting Inductance

Consider a coil of N turns wound on to a non-magnetic core, of uniform csa A metre^2 and mean length l metre. The coil carries a current of I amp, which produces a flux of Φ weber. From equation (5.11), we know that the inductance will be

$$L = \frac{N\Phi}{I} \text{ henry, but } \Phi = BA \text{ weber}$$

$$\text{therefore, } L = \frac{NBA}{I} \text{ henry} \ldots\ldots\ldots\ldots\ldots\ldots [1]$$

Also, magnetic field strength, $H = \frac{NI}{l}$; so $I = \frac{Hl}{N}$

and substituting this expression for I into equation [1]

$$L = \frac{NBA}{Hl/N} = \frac{BAN^2}{Hl} \ldots\ldots\ldots\ldots\ldots\ldots [2]$$

Now, equation [2] contains the term $\frac{B}{H}$, which equals $\mu_0\mu_r$

$$\text{therefore, } L = \frac{\mu_0\mu_r N^2 A}{l} \text{ henry} \qquad (5.12)$$

We also know that $\frac{l}{\mu_0\mu_r A}$ = reluctance, S

$$\text{hence } L = \frac{N^2}{S} \text{ henry} \qquad (5.13)$$

Notes:

1 Equation (5.12) compares with $C = \frac{\epsilon_0 \epsilon_r A(N-1)}{d}$ farad for a capacitor.
2 If the number of turns is doubled, then the inductance is quadrupled, i.e. $L \propto N^2$.
3 The terms A and l in equation (5.12) refer to the dimensions of the core, and NOT the coil.

Worked Example 5.19

Question

A 600 turn coil is wound on to a non-magnetic core of effective length 45 mm and csa 4 cm^2. (a) Calculate the inductance. (b) The number of turns is increased to 900. Calculate the inductance value now produced. (c) The core of the 900 turn coil is now replaced by an iron core having a relative permeability of 75, and of the same dimensions as the original. Calculate the inductance in this case.

Answer

$N_1 = 600$; $l = 45 \times 10^{-3}$ m; $A = 4 \times 10^{-4}$ m^2; $\mu_{r1} = 1$

$N_2 = 900$; $\mu_{r2} = 1$; $N_3 = 900$; $\mu_{r3} = 75$

(a)

$$L_1 = \frac{\mu_0 \mu_{r1} N_1^2 A}{l} \text{ henry}$$

$$= \frac{4\pi \times 10^{-7} \times 1 \times 600^2 \times 4 \times 10^{-4}}{45 \times 10^{-3}}$$

so, $L_1 = 4.02$ mH **Ans**

(b) Since $L_1 \propto N_1^2$, and $L_2 \propto N_2^2$, then

$$\frac{L_2}{L_1} = \frac{N_2^2}{N_1^2}$$

$$\text{so, } L_2 = \frac{L_1 N_2^2}{N_1^2} = \frac{4.02 \times 10^{-3} \times 900^2}{600^2}$$

therefore, $L_2 = 9.045$ mH **Ans**

(c) Since $\mu_{r3} = 75 \times \mu_{r2}$, and there are no other changes, then $L_3 = 75 \times L_2$

therefore, $L_3 = 0.678$ H **Ans**

5.20 Mutual Inductance

When a changing current in one circuit induces an emf in another separate circuit, then the two circuits are said to possess mutual inductance. The unit of mutual inductance is the henry, and is defined as follows.

Two circuits have a mutual inductance of one henry, if the emf induced in one circuit is one volt, when the current in the other is changing at the rate of one ampere per second.

The quantity symbol for mutual inductance is M, and expressing the above definition as an equation we have

$$M = \frac{\text{induced emf in coil 2}}{\text{rate of change of current in coil 1}}$$

$$= -\frac{e_2}{\mathrm{d}i_1/\mathrm{d}t} \text{ henry}$$

and transposing this equation for emf e_2

$$e_2 = \frac{-M\mathrm{d}i_1}{\mathrm{d}t} \text{ volt} \qquad (5.14)$$

This emf may also be expressed in terms of the flux linking coil 2. If *all* of the flux from coil 1 links with coil 2, then we have what is called 100% flux linkage. In practice, it is more usual for only a proportion of the flux from coil 1 to link with coil 2. Thus the flux linkage is usually less than 100%. This is indicated by a factor, known as the coupling factor, k. 100% coupling is indicated by $k = 1$. If there is no flux linkage with coil 2, then k will have a value of zero. So if zero emf is induced in coil 2, the mutual inductance will also be zero. Thus, the possible values for the coupling factor k, lie between zero and 1. Expressed mathematically, this is written as

$$0 \leqslant k \leqslant 1$$

Consider two coils possessing mutual inductance, and with a coupling factor <1. Let a change of current $\mathrm{d}i_1/\mathrm{d}t$ amp/s in coil 1 produce a change of flux $\mathrm{d}\phi_1/\mathrm{d}t$ weber/s. The proportion of this flux change linking coil 2 will be $\mathrm{d}\phi_2/\mathrm{d}t$ weber/s. If the number of turns on coil 2 is N_2, then

$$e_2 = \frac{-N_2\mathrm{d}\phi_2}{\mathrm{d}t} \text{ volt} \qquad (5.15)$$

However, equations (5.14) and (5.15) both refer to the same induced emf. Therefore we can equate the two expressions

$$\frac{M\mathrm{d}i_1}{\mathrm{d}t} = \frac{N_2\mathrm{d}\phi_2}{\mathrm{d}t}$$

and transposing for M, we have

$$M = \frac{N_2\mathrm{d}\phi_2}{\mathrm{d}i_1} \text{ henry} \qquad (5.16)$$

As with self-inductance for a single coil, mutual inductance is a property of a pair of coils. They therefore retain this property, regardless of whether or not an emf is induced. Hence, equation (5.16) may be modified to

$$M = \frac{N_2\Phi_2}{I_1} \text{ henry} \tag{5.17}$$

Worked Example 5.20

Question

Two coils, A and B, have 2000 turns and 1500 turns respectively. A current of 0.5 A, flowing in A, produces a flux of 60 μWb. The flux linking with B is 83% of this value. Determine (a) the self-inductance of coil A, and (b) the mutual inductance of the two coils.

Answer

$N_A = 2000$; $N_B = 1500$; $I_A = 0.5$ A; $\Phi_A = 60 \times 10^{-6}$ Wb

(a) $$L_A = \frac{N_A\Phi_A}{I_A} \text{ henry} = \frac{2000 \times 60 \times 10^{-6}}{0.5}$$

so, $L_A = 0.24$ H **Ans**

(b) $$M = \frac{N_B\Phi_B}{I_A} \text{ henry} = \frac{1500 \times 0.83 \times 60 \times 10^{-6}}{0.5}$$

so, $M = 0.149$ H **Ans**

5.21 Relationship between Self- and Mutual-Inductance

Consider two coils of N_1 and N_2 turns respectively, wound on to a common non-magnetic core. If the reluctance of the core is S ampere turns/weber, and the coupling coefficient is unity, then

$$L_1 = \frac{N_1^2}{S} \text{ and } L_2 = \frac{N_2^2}{S}$$

therefore,

$$L_1L_2 = \frac{N_1^2N_2^2}{S^2} \quad \ldots\ldots\ldots\ldots [1]$$

$$M = \frac{N_2\Phi}{I_1} \text{ henry, and multiplying by } \frac{N_1}{N_1}$$

$$M = \frac{N_1N_2\Phi}{N_1I_1}$$

The above expression contains the term

$$\frac{\Phi}{N_1 I_1} = \frac{1}{S}$$

$$\text{so, } M = \frac{N_1 N_2}{S}$$

$$\text{therefore, } M^2 = \frac{N_1^2 N_2^2}{S^2} \quad\ldots\ldots\ldots\ldots\ldots\ldots\ldots\ldots [2]$$

and comparing equations [1] and [2]

$$M^2 = L_1 L_2$$

$$\text{therefore, } M = \sqrt{L_1 L_2} \text{ henry} \qquad (5.18)$$

The above equation is correct only provided that there is 100% coupling between the coils; i.e. $k = 1$. If $k < 1$, then the general form of the equation, shown below, applies.

$$M = k\sqrt{L_1 L_2} \text{ henry} \qquad (5.19)$$

5.22 Energy Stored

As with an electric field, a magnetic field also stores energy. When the current through an inductive circuit is interrupted, by opening a switch, this energy is released. This is the reason why a spark or arc occurs between the contacts of the switch, when it is opened.

Consider an inductor connected in a circuit, in which the current increases uniformly, to some steady value I amp. This current change is illustrated in Fig. 5.34. The magnitude of the emf induced by this change of current is given by

$$e = \frac{LI}{t} \text{ volt}$$

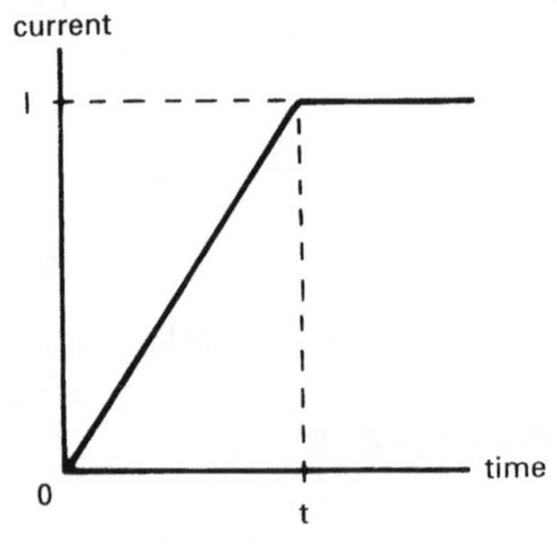

Fig. 5.34

The average power input to the coil during this time is:

average power $= e \times$ average current

From the graph, it may be seen that the average current over this time is $I/2$ amp. Therefore,

$$\text{average power} = \frac{1}{2}eI = \frac{LII}{2t} = \frac{LI^2}{2t} \text{ watt}$$

but, energy stored = average power × time

$$\text{power} = \frac{LI^2 t}{2t} \text{ joule}$$

$$\text{thus, energy stored, } W = \frac{1}{2}LI^2 \text{ joule} \qquad (5.20)$$

Worked Example 5.21

Question

Calculate the energy stored in a 50 mH inductor when it is carrying a current of 0.75 A.

Answer

$L = 50 \times 10^{-3}$ H; $I = 0.75$ A

$$W = \frac{1}{2}LI^2 \text{ joule} = \frac{50 \times 10^{-3} \times 0.75^2}{2}$$

therefore, $W = 14.1$ mJ **Ans**

Equation 5.20 applies to a single inductor. When two coils possess mutual inductance, and are connected in series, both will store energy. In this situation, the total energy stored is given by the equation

$$W = \frac{1}{2}L_1 I_1^2 + \frac{1}{2}L_2 I_2^2 \pm M I_1 I_2 \text{ joule} \qquad (5.21)$$

5.23 The Transformer Principle

A transformer is an **a.c.** machine, which utilises the mutual inductance between two coils, or windings. The two windings are wound on to a common iron core, but are not electrically connected to each other. The purpose of the iron core is to reduce the reluctance of the magnetic circuit. This ensures that the flux linkage between the coils is almost 100%.

> **a.c.** means alternating current, i.e. one which flows alternately, first in one direction, then in the opposite direction. It is normally a sinewave

Since it is an a.c. machine, an alternating flux is produced in the core. The core is therefore laminated, to minimise the eddy current loss. Indeed, the transformer is probably the most efficient of all machines. Efficiencies of 98% to 99% are typical. This high efficiency is due mainly to the fact that there are no moving parts.

The general arrangement is shown in Fig. 5.35. One winding, called the primary, is connected to an a.c. supply. The other winding, the secondary, is connected to a **load**. The primary will draw an alternating current I_1 from the supply. The flux, Φ, produced by this winding, will therefore also be alternating; i.e. it will be continuously changing. Assuming 100% flux linkage, then this flux is the only common factor

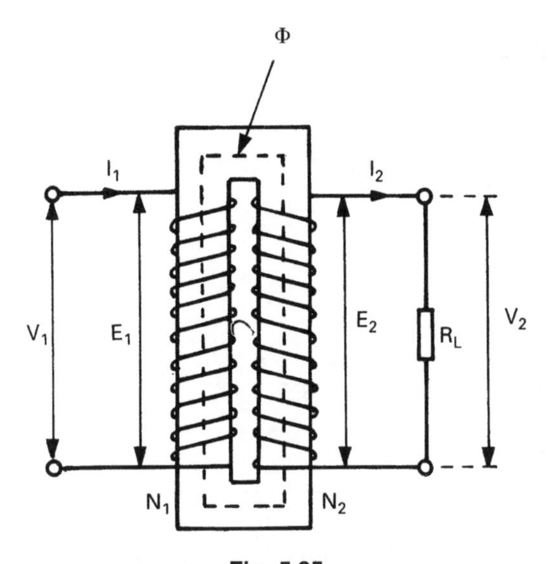

Fig. 5.35

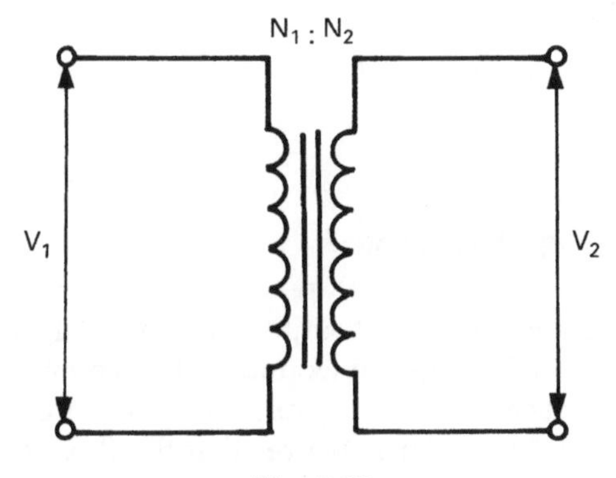

Fig. 5.36

linking the two windings. Thus, a mutually induced emf, E_2, will be developed across the secondary. Also, there will be a back emf, E_1, induced across the primary. If the secondary is connected to a load, then it will cause the secondary current I_2 to flow. This results in a secondary terminal voltage, V_2. Figure 5.36 shows the circuit symbol for a transformer.

> A **load** is any device or circuit connected to some source of emf. Thus, a load will draw current from the source. The term load is also loosely used to refer to the actual current drawn from a source

5.24 Transformer Voltage and Current Ratios

Let us consider an ideal transformer. This means that the resistance of the windings is negligible, and there are no core losses due to hysteresis and eddy currents. Also, let the secondary be connected to a purely resistive load, as shown in Fig. 5.37.

Under these conditions, the primary back emf, E_1, will be of the same magnitude as the primary applied voltage, V_1. The secondary terminal voltage, V_2, will be of the same magnitude as the secondary induced emf, E_2. Finally, the output power will be the same as the input power.

The two emfs are given by

$$E_1 = \frac{-N_1 \mathrm{d}\Phi}{\mathrm{d}t} \text{ volt, and } E_2 = \frac{-N_2 \mathrm{d}\Phi}{\mathrm{d}t} \text{ volt so,}$$

$$\frac{\mathrm{d}\Phi}{\mathrm{d}t} = \frac{E_1}{N_1} \ldots\ldots\ldots\ldots\ldots\ldots [1]$$

$$\text{and } \frac{\mathrm{d}\Phi}{\mathrm{d}t} = \frac{E_2}{N_2} \ldots\ldots\ldots\ldots\ldots\ldots [2]$$

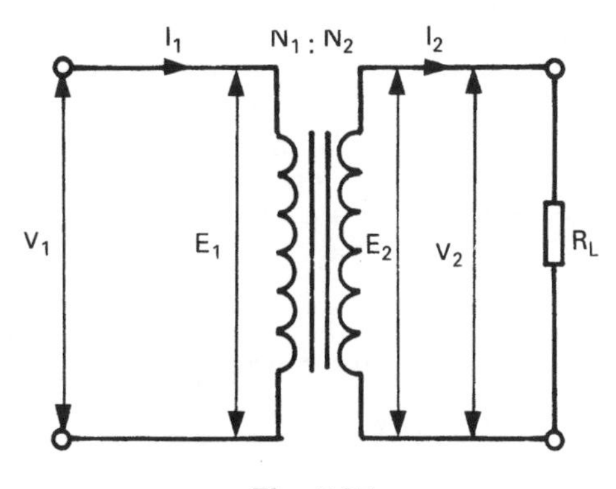

Fig. 5.37

Since both equations [1] and [2] refer to the same rate of change of flux in the core, then [1] = [2]:

$$\frac{E_1}{N_1} = \frac{E_2}{N_2}$$

$$\text{hence, } \frac{E_1}{E_2} = \frac{N_1}{N_2}$$

and since $E_1 = V_1$, and $E_2 = V_2$, then

$$\frac{V_1}{V_2} = \frac{N_1}{N_2} \qquad (5.22)$$

From this equation, it may be seen that the voltage ratio is the same as the turns ratio. This is perfectly logical, since the same flux links both windings, and each induced emf is directly proportional to its respective number of turns. This is the main purpose of the transformer. It can therefore be used to 'step up' or 'step down' a.c. voltages, depending upon the turns ratio selected.

Worked Example 5.22

Question

A transformer is to be used to provide a 60 V output from a 240 V a.c. supply. Calculate (a) the turns ratio required, and (b) the number of primary turns, if the secondary is wound with 500 turns.

Answer

$V_2 = 60\,\text{V}$; $V_1 = 240\,\text{V}$; $N_2 = 500$

(a)

$$\frac{V_1}{V_2} = \frac{N_1}{N_2} = \frac{240}{60}$$

$$\text{so, turns ratio, } \frac{N_1}{N_2} = \frac{4}{1} \text{ or } 4:1 \textbf{ Ans}$$

(b)

$$\frac{N_1}{500} = \frac{4}{1}$$

therefore, $N_1 = 2000$ **Ans**

Since the load is purely resistive, then the output power, P_2, is given by

$$P_2 = V_2 I_2 \text{ watt}$$

and the input power, $P_1 = V_1 I_1$ watt

Also since the transformer has been considered to be 100% efficient (no losses), then

$$P_2 = P_1$$

$$\text{therefore, } V_2 I_2 = V_1 I_2$$

$$\frac{I_1}{I_2} = \frac{V_2}{V_1} \quad \text{but} \quad \frac{V_2}{V_1} = \frac{N_2}{N_1}$$

$$\text{hence, } \frac{I_1}{I_2} = \frac{N_2}{N_1} \qquad (5.23)$$

i.e. The current ratio is the *inverse* of the turns ratio.

This result is also logical. For example, if the voltage was 'stepped up' by the ratio N_2/N_1, then the current must be 'stepped down' by the same ratio. If this was not the case, then we would get more power out than was put in! Although this result would be very welcome, it is a physical impossibility. It would require the machine to be *more* than 100% efficient.

Worked Example 5.23

Question

A 15 Ω resistive load is connected to the secondary of a transformer. The terminal p.d. at the secondary is 240 V. If the primary is connected to a 600 V a.c. supply, calculate (a) the transformer turns ratio, (b) the current and power drawn by the load, and (c) the current drawn from the supply. Assume an ideal transformer.

Answer

$R_L = 15\,\Omega$; $V_2 = 240\,\text{V}$; $V_1 = 600\,\text{V}$

The appropriate circuit diagram is shown in Fig. 5.37.

(a)

$$\frac{N_1}{N_2} = \frac{V_1}{V_2} = \frac{600}{240}$$

so, turns ratio, $N_1/N_2 = 2.5:1$ **Ans**

(b)

$$I_2 = \frac{V_2}{R_L} \text{ ohm} = \frac{240}{15}$$

so, $I_2 = 16\,\text{A}$ **Ans**

$$P_2 = V_2 I_2 \text{ watt} = 240 \times 16$$

therefore, $P_2 = 3.84\,\text{kW}$ **Ans**

Continued on p. 242

Worked Example 5.23 (*Continued*)

(c) $P_1 = P_2 = 3.84\,\text{kW}$

and, $P_1 = V_1 I_1$ watt

therefore, $I_1 = \frac{P_1}{V_1}\text{amp} = \frac{3840}{600}$

hence, $I_1 = 6.4$ A **Ans**

Alternatively, using the inverse of the turns ratio:

$$I_1 = I_2 \times \frac{N_2}{N_1} = \frac{16}{2.5}$$

so, $I_1 = 6.4$ A **Ans**

Assignment Questions

1 The flux linking a 600 turn coil changes uniformly from 100 mWb to 50 mWb in a time of 85 ms. Calculate the average emf induced in the coil.

2 An average emf of 350 V is induced in a 1000 turn coil when the flux linking it changes by 200 μWb. Determine the time taken for the flux to change.

3 A flux of 1.5 mWb, linking with a 250 turn coil, is uniformly reversed in a time of 0.02 s. Calculate the value of the emf so induced.

4 A coil of 2000 turns is linked by a magnetic flux of 400 μWb. Determine the emf induced in the coil when (a) this flux is reversed in 0.05 s, and (b) the flux is reduced to zero in 0.15 s.

5 When a magnetic flux linking a coil changes, an emf is induced in the coil. Explain the factors that determine (a) the magnitude of the emf, and (b) the direction of the emf.

6 State Lenz's law, and hence explain the term 'back emf'.

7 A coil of 15000 turns is required to produce an emf of 15 kV. Determine the rate of change of flux that must link with the coil in order to provide this emf.

8 A straight conductor, 8 cm long, is moved with a constant velocity at right angles to a magnetic field. If the emf induced in the conductor is 40 mV, and its velocity is 10 m/s, calculate the flux density of the field.

9 A conductor of effective length 0.25 m is moved at a constant velocity of 5 m/s, through a magnetic field of density 0.4 T. Calculate the emf induced when the direction of movement relative to the magnetic field is (a) 90°, (b) 60°, and (c) 45°.

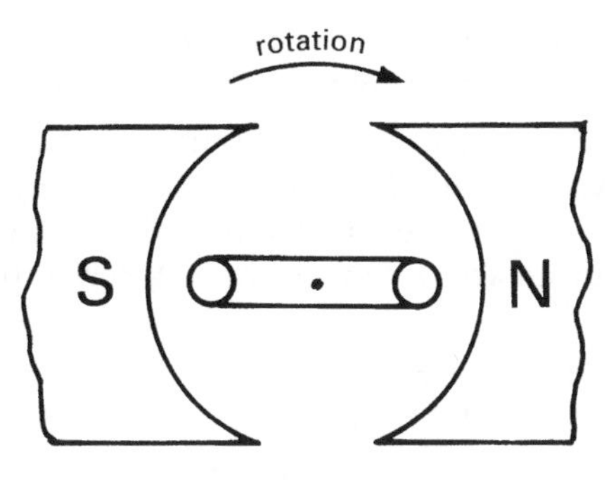

Fig. 5.38

10 Figure 5.38 represents two of the **armature** conductors of a d.c. generator, rotating in a clockwise direction. Copy this diagram and hence:

(a) Indicate the direction of the field pattern of the magnetic poles.
(b) Indicate the direction of induced emf in each side of the coil.
(c) If this arrangement was to be used as a motor, with the direction of rotation as shown, indicate the direction of current flow required through the coil.

> An **armature** is the rotating part of a d.c. machine. If the machine is used as a generator, it contains the coils into which the emf is induced. In the case of a motor, it contains the coils through which current must be passed, to produce the torque

11 A conductor of effective length 0.5 m is placed at right angles to a magnetic field of density 0.45 T. Calculate the force exerted on the conductor if it carries a current of 5 A.

12 A conductor of effective length 1.2 m is placed inside a magnetic field of density 250 mT. Determine the value of current flowing through the conductor, if a force of 0.75 N is exerted on the conductor.

13 A conductor, when placed at right angles to a magnetic field of density 700 mT, experiences a force of 20 mN, when carrying a current of 200 mA. Calculate the effective length of the conductor.

14 A conductor, 0.4 m long, lies between two pole pieces, with its length parallel to the pole faces. Determine the force exerted on the conductor, if it carries a current of 30 A, and the flux density is 0.25 T.

15 The coil of a moving coil meter is wound with 75 turns, on a former of effective length 2.5 cm, and diameter 2 cm. The former rotates at right angles to the field, which has a flux density of 0.5 T. Determine the deflecting torque when the coil current is 50 μA.

16 A moving coil meter has a coil of 60 turns wound on to a former of effective length 22.5 mm and diameter 15mm. If the flux density in the air gap is 0.2 T, and the coil current is 0.1 mA. Calculate (a) the force acting on each side of the coil, and (b) the restoring torque exerted by the springs for the resulting deflection of the coil.

17 Two long parallel conductors, are spaced 12 cm between centres. If they carry 100 A and 75 A respectively, calculate the force per metre length acting on them. If the currents are flowing in opposite directions, will this be a force of attraction or repulsion? Justify your answer by means of a sketch of the magnetic field pattern produced.

18 The magnetic flux density at a distance of 1.4 m from the centre of a current carrying conductor is 0.25 mT. Determine the value of the current.

19 A moving coil meter has a coil resistance of 25 Ω, and requires a current of 0.25 mA to produce full-scale deflection. Determine the values of shunts required to extend its current reading range to (a) 10 mA, and (b) 1 A. Sketch the relevant circuit diagram.

20 For the meter movement described in question 19 above, show how it may be adapted to serve as a voltmeter, with voltage ranges of 3 V and 10 V. Calculate the values for, and name, any additional components required to achieve this. Sketch the relevant circuit diagram.

21 Explain what is meant by the term 'loading effect'.

22 A voltmeter, having a figure of merit of 15 kΩ/volt, has voltage ranges of 0.1 V, 1 V, 3 V and 10 V. If the resistance of the moving coil is 30 Ω, determine the multiplier values required for each range. Sketch a circuit diagram, showing how the four ranges could be selected.

23 Figure 5.39 shows a circuit in which the p.d. across resistor R_2 is to be measured. The voltmeter available for this measurement has a figure of merit of 20 kΩ/V, and has voltage ranges of 1 V, 10 V and 100 V. Determine the percentage error in the voltmeter reading, when used to measure this p.d.

24 Calculate the self-inductance of a 700 turn coil, if a current of 5 A flowing through it produces a flux of 8 mWb.

25 A coil of 500 turns has an inductance of 2.5 H. What value of current must flow through it in order to produce a flux of 20 mWb?

26 When a current of 2.5 A flows through a 0.5 H inductor, the flux produced is 80 μWb. Determine the number of turns.

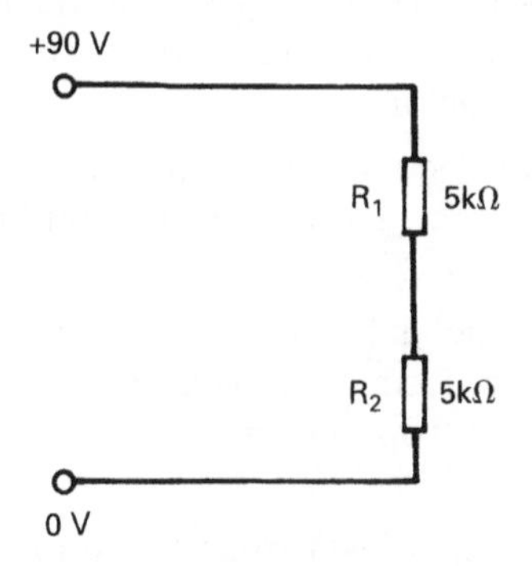

Fig. 5.39

27 A 1000 turn coil has a flux of 20 mWb linking it when carrying a current of 4 A. Calculate the coil inductance, and the emf induced when the current is reduced to zero in a time of 25 ms.

28 A coil has 300 turns and an inductance of 5 mH. How many turns would be required to produce an inductance of 0.8 mH, if the same core material were used?

29 If an emf of 4.5 V is induced in a coil having an inductance of 200 mH, calculate the rate of change of current.

30 An iron ring having a mean diameter of 300 mm and cross-sectional area of 500 mm^2 is wound with a 150 turn coil. Calculate the inductance, if the relative permeability of the ring is 50.

31 An iron ring of mean length 50 cm and csa 0.8 cm^2 is wound with a coil of 350 turns. A current of 0.5 A through the coil produces a flux density of 0.6 T in the ring. Calculate (a) the relative permeability of the ring, (b) the inductance of the coil, and (c) the value of the induced emf if the current decays to 20% of its original value in 0.01 s, when the current is switched off.

32 When the current in a coil changes from 2 A to 12 A in a time of 150 ms, the emf induced into an adjacent coil is 8 V. Calculate the mutual inductance between the two coils.

33 The mutual inductance between two coils is 0.15 H. Determine the emf induced in one coil when the current in the other decreases uniformly from 5 A to 3 A, in a time of 10 ms.

34 A coil of 5000 turns is wound on to a non-magnetic toroid of csa 100 cm^2, and mean circumference 0.5 m. A second coil of 1000 turns is wound over the first coil. If a current of 10 A flows through the first coil, determine (a) the self-inductance of the first coil, (b) the mutual inductance, assuming a coupling factor of 0.45, and (c) the average emf induced in the second coil if interruption of the current causes the flux to decay to zero in 0.05 s.

35 Two air-cored coils, A and B, are wound with 100 and 500 turns respectively. A current of 5 A in A produces a flux of 15 μWb. Calculate (a) the self-inductance of coil A, (b) the mutual inductance, if 75% of the flux links with B, and (c) the emf induced in each of the coils, when the current in A is reversed in a time of 10 ms.

36 Two coils, of self-inductance 50 mH and 85 mH respectively, are placed parallel to each other. If the coupling coefficient is 0.9, calculate their mutual inductance.

37 The mutual inductance between two coils is 250 mH. If the current in one coil changes from 14 A to 5 A in 15 ms, calculate (a) the emf induced in the other coil, and (b) the change of flux linked with this coil if it is wound with 400 turns.

38 The mutual inductance between the two windings of a car ignition coil is 5 H. Calculate the average emf induced in the high tension winding, when a current of 2.5 A, in the low tension winding, is reduced to zero in 1 ms. You may assume 100% flux linkage between the two windings.

39 Sketch the circuit symbol for a transformer, and explain its principle of operation. Why is the core made from laminations? Is the core material a 'hard' or a 'soft' magnetic material? Give the reason for this.

40 A transformer with a turns ratio of 20 : 1 has 240 V applied to its primary. Calculate the secondary voltage.

41 A 4 : 1 voltage 'step-down' transformer is connected to a 110 V a.c. supply. If the current drawn from this supply is 100 mA, calculate the secondary voltage, current and power.

42 A transformer has 450 primary turns and 80 secondary turns. It is connected to a 240 V a.c. supply. Calculate (a) the secondary voltage, and (b) the primary current when the transformer is supplying a 20 A load.

43 A coil of self-inductance 0.04 H has a resistance of 15 Ω. Calculate the energy stored when it is connected to a 24 V d.c. supply.

44 The energy stored in the magnetic field of an inductor is 68 mJ, when it carries a current of 1.5 A. Calculate the value of self-inductance.

45 What value of current must flow through a 20 H inductor, if the energy stored in its magnetic field, under this condition, is 60 J?

Suggested Practical Assignments

Note: The majority of these assignments are only qualitative in nature.

Assignment 1

To investigate Faraday's laws of electromagnetic induction.

Apparatus:

Several coils, having different numbers of turns
2 × permanent bar magnets
1 × galvanometer

Method:

1 Carry out the procedures outlined in section 5.1 at the beginning of this chapter.
2 Write an assignment report, explaining the procedures carried out, and stating the conclusions that you could draw from the observed results.

Assignment 2

Force on a current carrying conductor.

Apparatus:

1 × current balance

1 × variable d.c. psu
1 × ammeter

Method:

1 Assemble the current balance apparatus.
2 Adjust the balance weight to obtain the balanced condition, prior to connecting the psu.
3 With maximum length of conductor, and all the magnets in place, vary the conductor current in steps. For each current setting, re-balance the apparatus, and note the setting of the balance weight.
4 Repeat the balancing procedure with a constant current, and maximum magnets, but varying the effective length of the conductor.
5 Repeat once more, this time varying the number of magnets. The current must be maintained constant, as must the conductor length.
6 Tabulate all results obtained, and plot the three resulting graphs.
7 Write an assignment report. This should include a description of the procedures carried out, and conclusions drawn, regarding the relationships between the force produced and I, l, and B.

Assignment 3

To investigate the loading effect of a moving coil meter.

Apparatus:

1 × 10 kΩ rotary potentiometer, complete with circular scale
1 × d.c. psu
1 × Heavy Duty AVO
1 × DVM (digital voltmeter)

Method:

1 Connect the circuit as shown in Fig. 5.40, with the psu set to 10 V d.c.
2 Start with the potentiometer moving contact at the 'zero' end. Measure the p.d. indicated, in turn, by both the DVM and the AVO, for every 30° rotation of the moving contact.

Note: Do NOT connect both meters at the same time; connect them IN TURN.

3 Tabulate both voltmeter readings. Plot graphs (on the same axes) for the voltage readings versus angular displacement.
4 Determine the percentage loading error of the AVO, for displacements of 0°, 180°, and 270°. Write an assignment report, and include comment regarding the variation of loading error found.

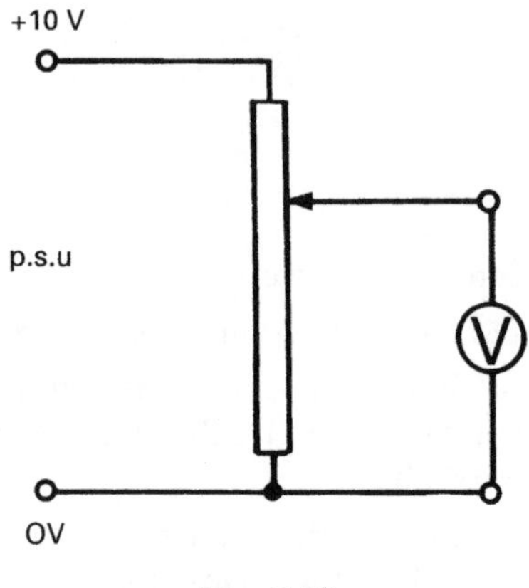

Fig. 5.40

Assignment 4

To demonstrate mutual inductance and coupling coefficient.

Apparatus:

Several coils, having different numbers of turns.
Ferromagnet core
1 × galvo
1 × d.c. psu

Method:

1. Place the two coils as close together as possible. Connect the galvo to one coil, and connect the other coil to the psu via a switch.
2. Close the switch, and note the deflection obtained on the galvo.
3. Repeat this procedure for increasing distances of separation, and for different coils.
4. Mount two of the coils on a common magnetic core, and repeat the procedure.
5. Write an assignment report, explaining the results observed.

Assignment 5

To determine the relationship between turns ratio and voltage ratio for a simple transformer.

Apparatus:

Either 1 × single-phase transformer with tappings on both windings;
or Several different coils with a ferromagnetic core.

Either a low voltage a.c. supply;
or $1 \times$ a.c. signal generator.
$1 \times$ DVM (a.c. voltage ranges)

Method:

1 Connect the primary to the a.c. source.
2 Measure both primary and secondary voltages, and note the corresponding number of turns on each winding.
3 Vary the number of turns on each winding, and note the corresponding values of the primary and secondary voltages.
4 Tabulate all results. Write a brief report, explaining your findings.

Supplementary Worked Examples

Example 1: A flux of 0.5 mWb is reversed in a time of 5 ms. If this flux links with a 200 turn coil, determine the emf thus induced.

Solution: Since flux is *reversed*, then, $d\Phi = 2 \times 5 \times 10^{-4}$ Wb; $dt = 5 \times 10^{-3}$ S; $N = 200$

$$e = -N\frac{d\Phi}{dt} \text{ volt} = \frac{200 \times 2 \times 5 \times 10^{-4}}{5 \times 10^{-3}}$$

so, $e = 40$ V **Ans**

Example 2: A 10 000 turn coil is required to have an emf of 8 kV induced in it when a change of flux of 1 mWb links with it. Calculate the time interval required for the flux change.

Solution: $N = 10^4$; $e = 8 \times 10^3$ V; $d\Phi = 10^{-3}$ Wb

$$e = -N\frac{d\Phi}{dt} \text{ volt}$$

$$\text{so, } dt = N\frac{d\Phi}{e} \text{ second} = \frac{10^4 \times 10^{-3}}{8 \times 10^3}$$

and, $dt = 1.25$ ms **Ans**

Example 3: The axle of a lorry is 2.2 m long, and the vertical component of the Earth's magnetic field density, through which the lorry is travelling, is 38 μT. If the speed of the lorry is 80 km/h, then calculate the emf induced in the axle.

Solution: $\ell = 2.2$ m; $v = \frac{80 \times 10^3}{60 \times 60}$ m/s; $B = 38 \times 10^{-6}$ T; $\sin\theta = 1$, since $\theta = 90°$

$$e = B\ell v \sin\theta \text{ volt}$$

$$= \frac{38 \times 10^{-6} \times 2.2 \times 80 \times 10^3}{3600}$$

so, $e = 1.86$ mV **Ans**

Example 4: A conductor 400 mm long is moved at right angles to a uniform magnetic field of density 0.25 T. If the emf generated in the conductor is 1.8 V, and the conductor forms part of a closed circuit of total resistance 0.6 Ω, calculate (a) the velocity of the conductor, and (b) the force acting on the conductor.

Solution: $\ell = 0.4$ m; $\sin\theta = 1$; $B = 0.25$ T; $e = 1.8$ V; $R = 0.6\,\Omega$

(a)

$$e = B\ell v \sin\theta \text{ volt}$$

$$\text{so, } v = \frac{e}{B\ell \sin\theta} \text{ metre/second}$$

$$= \frac{1.8}{0.25 \times 0.4 \times 1}$$

and, $v = 18$ m/s **Ans**

(b)

$$F = BI\ell \sin\theta \text{ newton; where } I = \frac{e}{R} \text{ amp} = \frac{1.8}{0.6}$$

so, $I = 3$A

and $F = 0.25 \times 3 \times 0.4 \times 1$

thus, $F = 0.3$ N **Ans**

Example 5: The coil of a moving coil meter consists of 80 turns of wire wound on a former of length 2 cm and radius 1.2 cm. When a current of 45 μA is passed through the coil the movement comes to rest when the springs exert a restoring torque of 1.4 μNm. Calculate the flux density produced by the pole pieces.

Solution: $N = 80$; $\ell = 0.02$ m; $r = 0.012$ m; $I = 45 \times 10^{-6}$; $T = 1.4 \times 10^{-6}$ Nm

The meter movement comes to rest when the deflecting torque exerted on the coil is balanced by the restoring torque of the springs.

$$T = BANI \text{ newton metre}$$

$$\text{so, } B = \frac{T}{ANI} \text{ tesla}$$

$$= \frac{1.4 \times 10^{-6}}{0.02 \times 2 \times 0.012 \times 80 \times 45 \times 10^{-6}}$$

so, $B = 0.81$ T **Ans**

Example 6: The coil of a moving coil multimeter has a resistance of 1.5 kΩ, and it requires a current of 75 μA through it in order to produce full scale deflection. Calculate (a) the value of shunt required to enable the meter to indicate current up to the value of 5 A, and (b) the value of multiplier required to enable it to indicate voltage up to 10 V.

Solution: Coil resistance, $R_C = 1500\,\Omega$; $I_{fsd} = 75 \times 10^{-6}$ A; (a) $I = 5$ A; (b) $V = 10$ V

(a)

$$V_C = I_{fsd} R_C \text{ volt} = 75 \times 10^{-6} \times 1500$$

so, $V_C = 0.1125$ V

$$I_S = I - I_{fsd} \text{ amp} = 5 - (75 \times 10^{-6})$$

and, $I_S = 4.999\,925$ A

$$R_S = \frac{V_C}{I_S} \text{ ohm}$$

$$= \frac{0.1125}{4.999\,925}$$

and, $R_S = 22.5$ mΩ **Ans**

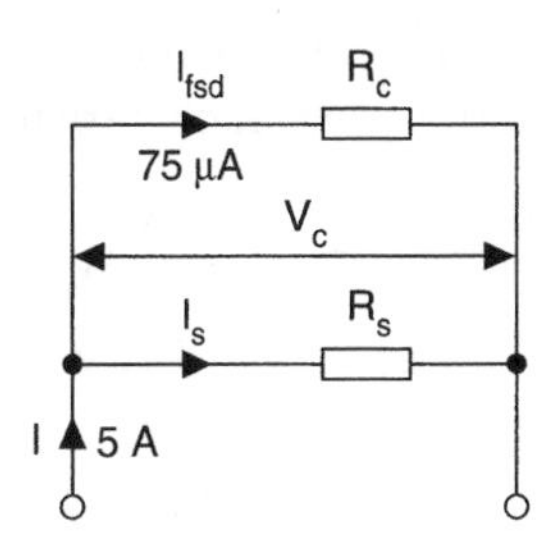

Fig. 5.41

Note: Since the value of R_S will be very small, all the digits obtained for I_S are retained in the calculation in order to minimise any rounding error.

(b) From part (a), $V_C = 0.1125$ V, and the circuit is now equivalent to that shown in Fig. 5.42.

Thus, $V_m = V - V_C = 10 - 0.1125$

so, $V_m = 9.8875$ V

$$R_m = \frac{V_m}{I_{fsd}} \text{ ohm} = \frac{9.8875}{75 \times 10^{-6}}$$

so, $R_m = 131.83$ kΩ **Ans**

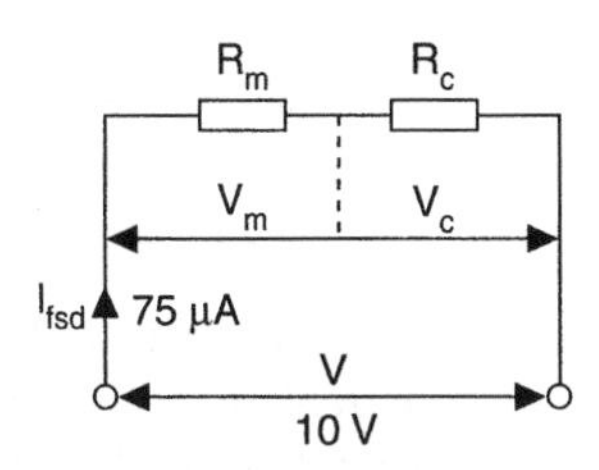

Fig. 5.42

Alternatively, total resistance, $R = R_m + R_C$ ohm

and also, $R = \frac{V}{I_{fsd}}$ ohm $= \frac{10}{75 \times 10^{-6}}$

so, $R = 133.33$ kΩ

and, $R_m = R - R_C = (133.33 - 1.5) \times 10^3$

thus, $R_m = 131.83$ kΩ **Ans**

Example 7: Two resistors, of 10 kΩ and 47 kΩ respectively, are connected in series across a 10 V d.c. supply. The p.d. developed across each resistor is measured with a moving coil multimeter having a figure of merit of 20 kΩ/V, and switched to its 10 V range. For each measurement calculate (a) the p.d. indicated by the meter, and (b) the loading error involved in each case.

Solution: With a figure of merit of 20 kΩ/V the meter internal resistance, R_{in} will be $R_{in} = 20\,\text{k}\Omega \times 10 = 200\,\text{k}\Omega$

Without the meter connected, the circuit will be as shown in Fig. 5.43, and the actual p.d. across each resistor will be V_1 and V_2 respectively.

$$V_1 = \frac{R_1}{R_1 + R_2} \times V \text{ volt} = \frac{10 \times 10}{10 + 47}$$

$$\text{so, } V_1 = 1.754 \text{ V}$$

$$V_2 = \frac{R_2}{R_1 + R_2} \times V \text{ volt} = \frac{47 \times 10}{10 + 47}$$

$$\text{and, } V_2 = 8.246 \text{ V}$$

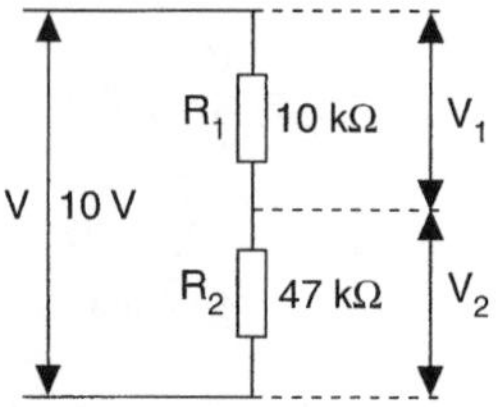

Fig. 5.43

(a) With the meter connected across R_1, the meter will indicate V_{AB} volt as shown in Fig. 5.44.

$$R_{AB} = \frac{R_{in}R_1}{R_{in} + R_1} \text{ ohm} = \frac{200 \times 10}{210} \text{ k}\Omega$$

$$\text{so, } R_{AB} = 9.524 \text{ k}\Omega$$

$$V_{AB} = \frac{R_{AB}}{R_{AB} + R_2} \times V \text{ volt} = \frac{9.524 \times 10}{9.524 + 47}$$

$$\text{and, } V_{AB} = 1.69 \text{ V } \textbf{Ans}$$

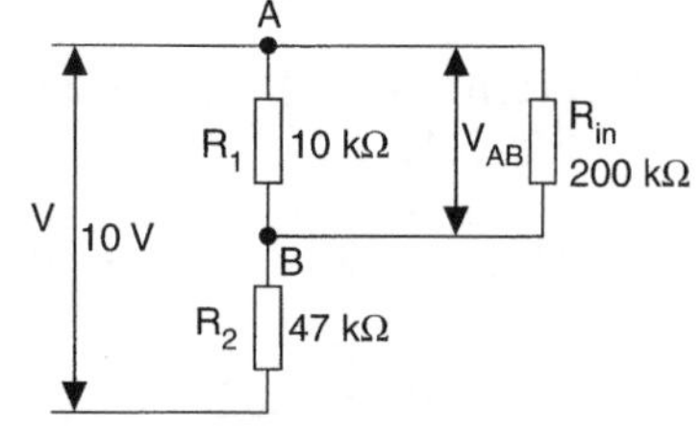

Fig. 5.44

In this case the voltmeter reading has been rounded up because it would not be practical to read the scale to two decimal places. Indeed, the indication would *probably* be read as 1.7 V.

With the meter now connected across R_2, the circuit is now modified to that as shown in Fig. 5.45.

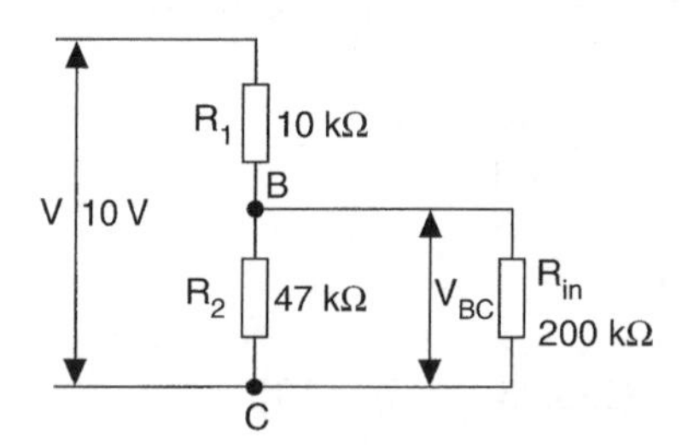

Fig. 5.45

$$R_{BC} = \frac{R_2 R_{in}}{R_2 + R_{in}} \text{ ohm} = \frac{47 \times 200}{247} \text{ k}\Omega$$

$$\text{so, } R_{BC} = 38.057 \text{ k}\Omega$$

$$V_{BC} = \frac{R_{BC}}{R_{BC} + R_1} \times V \text{ volt} = \frac{38.057 \times 10}{48.057}$$

$$\text{and, } V_{BC} = 7.92 \text{ V } \textbf{Ans}$$

(b) $\text{error} = \dfrac{\text{indicated value} - \text{true value}}{\text{true value}} \times 100\%$

so, error for V_1 measurement $= \dfrac{1.69 - 1.754}{1.754} \times 100\%$

hence, error $= -3.65\%$ **Ans**

error for V_2 measurement $= \dfrac{7.92 - 8.246}{8.246} \times 100\%$

and, error $= -3.95\%$ **Ans**

Note: Although the two error figures are fairly close to each other in value, the percentage error in the second measurement is about 10% greater than that for the first. This illustrates the fact that the higher the meter internal resistance, compared with the resistance across which it is placed, the smaller the loading effect, and hence higher accuracy is obtained. Since the internal resistance of DVMs is usually in the order of megohms, in many cases they are used in preference to the traditional moving coil instrument.

Example 8: A coil of self-inductance 0.5 H has 300 turns. Calculate (a) the current flowing through the coil if the flux produced is 2 mWb, and (b) the self-induced emf if this current is reduced to zero in a time of 8 ms.

Solution: $L = 0.5$ H; $N = 300$; $\Phi = 2 \times 10^{-3}$ Wb; $t = 8 \times 10^{-3}$ s

(a) $L = \dfrac{N\Phi}{I}$ henry

so, $I = \dfrac{N\Phi}{L}$ amp $= \dfrac{300 \times 2 \times 10^{-3}}{0.5}$

and, $I = 1.2$ A **Ans**

(b) $e = -\dfrac{N\Phi}{t}$ volt $= \dfrac{300 \times 2 \times 10^{-3}}{8 \times 10^{-3}}$

thus, $e = 75$ V **Ans**

Alternatively, $e = -\dfrac{LI}{t}$ volt $= \dfrac{0.5 \times 1.2}{8 \times 10^{-3}}$

so, $e = 75$ V **Ans**

Example 9: A flux of 0.2 mWb links with a coil when the current through the coil is 1 A. When this current is reduced to 0.1 A in a time of 5 ms, an emf of 45 V is induced in the coil. Calculate (a) the number of turns on the coil, and (b) the self-inductance of the coil. You may assume that the flux is directly proportional to the current flowing.

Solution: $\Phi_1 = 2 \times 10^{-4}$ Wb; $dt = 5 \times 10^{-3}$ s; $I_1 = 1$ A; $I_2 = 0.1$ A

(a) Since $\Phi \propto I$, then $\Phi_1 \propto I_1$ and $\Phi_2 \propto I_2$

$$\text{so, } \frac{\Phi_2}{\Phi_1} = \frac{I_2}{I_1}$$

$$\text{and, } \Phi_2 = \frac{\Phi I_2}{I_1} \text{ weber} = \frac{2 \times 10^{-4} \times 0.1}{1}$$

thus, $\Phi_2 = 2 \times 10^{-5}$ Wb or 0.2×10^{-4} Wb

and $\mathrm{d}\Phi = \Phi_1 - \Phi_2 = (2 - 0.2) \times 10^{-4} = 1.8 \times 10^{-4}$ Wb

$$\mathrm{d}i = I_1 - I_2 \text{ amp} = 1 - 0.1 = 0.9\,\text{A}$$

$$e = -N\frac{\mathrm{d}\Phi}{\mathrm{d}t} \text{ volt, so } N = e\,\frac{\mathrm{d}t}{\mathrm{d}\Phi}$$

$$\text{and, } N = \frac{45 \times 5 \times 10^{-3}}{1.8 \times 10^{-4}}$$

so, $N = 1250$ **Ans**

(b) $L = N\dfrac{\mathrm{d}\Phi}{\mathrm{d}i}$ henry $= \dfrac{1250 \times 1.8 \times 10^{-4}}{0.9}$

hence $L = 0.25$ H **Ans**

or, from $e = -L\dfrac{\mathrm{d}i}{\mathrm{d}t}$ volt

so, $L = e\dfrac{\mathrm{d}t}{\mathrm{d}i}$ henry $= \dfrac{45 \times 5 \times 10^{-3}}{0.9}$

and, $L = 0.25$ H **Ans**

Example 10: A 10 mH inductor is formed by winding a coil of wire onto a wooden former 4 cm long and of csa 5 cm^2. Calculate (a) the number of turns required, and (b) the inductance value if the wooden former is replaced with a ferromagnetic core of relative permeability 250.

Solution: $L = 10^{-2}$ H; $\ell = 0.04$ m; $A = 5 \times 10^{-4}$ cm^2; $\mu_r = 1$ (wood)

(a)
$$L = \frac{\mu_o \mu_r N^2 A}{\ell} \text{ henry}$$

$$\text{so, } N^2 = \frac{L\ell}{\mu_o \mu_r A} = \frac{10^{-2} \times 0.04}{4\pi \times 10^{-7} \times 5 \times 10^{-4}}$$

$$N^2 = 636\,619$$

so, $N = 798$ **Ans**

(b) Since the permeability of the iron is 250 times that of wood, then the self-inductance will be increased by the same factor, hence

$$L = 250 \times 10^{-2} = 2.5\,\text{H}\ \textbf{Ans}$$

Example 11: A 400 turn coil is wound onto a cast steel toroid having an effective length of 25 cm and csa 4.5 cm². If the steel has a relative permeability of 180 under the operating conditions, calculate the self-inductance of the coil.

Solution: $N = 400$; $\ell = 0.25$ m; $A = 4.5 \times 10^{-4}$ m²; $\mu_r = 180$

$$L = \frac{\mu_o \mu_r N^2 A}{\ell}\,\text{henry} = \frac{4\pi \times 10^{-7} \times 180 \times 400^2 \times 4.5 \times 10^{-4}}{0.25}$$

so, $L = 65\,\text{mH}$ **Ans**

Example 12: Considering Example 11, a second coil of 650 turns is wound over the first, and the current through coil 1 is changed from 2 A to 0.5 A in a time of 3 ms. If 95% of the flux thus produced links with coil 2, then calculate (a) the self-inductance of coil 2, (b) the value of mutual inductance, (c) the self-induced emf in coil 1, and (d) the mutually induced emf in coil 2.

Solution: $L_1 = 65 \times 10^{-3}$ H; $di_1 = 2 - 0.5 = 1.5$ A; $dt = 3 \times 10^{-3}$ s; $k = 0.95$

(a) From the equation for inductance we know that $L \propto N^2$, and since all other factors for the two coils are the same, then

$$\frac{L_2}{L_1} = \frac{N_2^2}{N_1^2}$$

$$\text{so, } L_2 = \frac{N_2^2 L_1}{N_1^2}\,\text{henry}$$

$$= \frac{650^2 \times 65}{400^2}\,\text{mH}$$

and, $L_2 = 172\,\text{mH}$ **Ans**

It is left to the student to confirm this answer by using the equation $L = \dfrac{\mu_o \mu_r N^2 A}{\ell}$

(b) $$M = k\sqrt{L_1 L_2}\,\text{henry} = 0.95\sqrt{65 \times 172}\,\text{mH}$$

thus, $M = 100\,\text{mH}$ **Ans**

(c) $$e_1 = -L_1 \frac{di}{dt}\,\text{volt} = \frac{65 \times 10^{-3} \times 1.5}{3 \times 10^{-3}}$$

so, $e_1 = 32.5\,\text{V}$ **Ans**

(d) $$e_2 = -M\,\frac{di_1}{dt}\,\text{volt} = \frac{100 \times 1.5}{3 \times 10^{-3}}$$

and, $e_2 = 50\,\text{V}$ **Ans**

Example 13: Two inductors, of inductance 25 mH and 40 mH respectively, are wound on a common ferromagnetic core, and are connected in series with each other. The coupling coefficient, *k*, between them is 0.8. When the current flowing through the two coils is 0.25 A, calculate (a) the energy stored in each, (b) the total energy stored when the coils are connected (i) in series aiding, and (ii) in series opposition.

Solution: $L_1 = 25 \times 10^{-3}$ H; $L_2 = 40 \times 10^{-3}$ H; $I_1 = I_2 = 0.25$ A; $k = 0.8$

(a) $$W_1 = \frac{1}{2}L_1I_1^2 \text{ joule} = 0.5 \times 25 \times 10^{-3} \times 0.25^2$$

so, $W_1 = 0.78$ mJ **Ans**

$$W_2 = \frac{1}{2}L_2I_2^2 \text{ joule} = 0.5 \times 40 \times 10^{-3} \times 0.25^2$$

so, $W_2 = 1.25$ mJ **Ans**

(b) The general equation for the energy stored by two inductors with flux linkage between them is:

$$W = \frac{1}{2}L_1I_1^2 + \frac{1}{2}L_2I_2^2 \pm MI_1I_2 \text{ joule}$$

When the coils are connected in series such that the two fluxes produced act in the same direction, the total flux is increased and the coils are said to be connected in series aiding. In this case the total energy stored in the system will be increased, so the last term in the above equation is added, i.e. the + sign applies. If, however, the connections to one of the coils are reversed, then the two fluxes will oppose each other, the total flux will be reduced, and the coils are said to be in series opposition. In this case the − sign is used. These two connections are shown in Fig. 5.46.

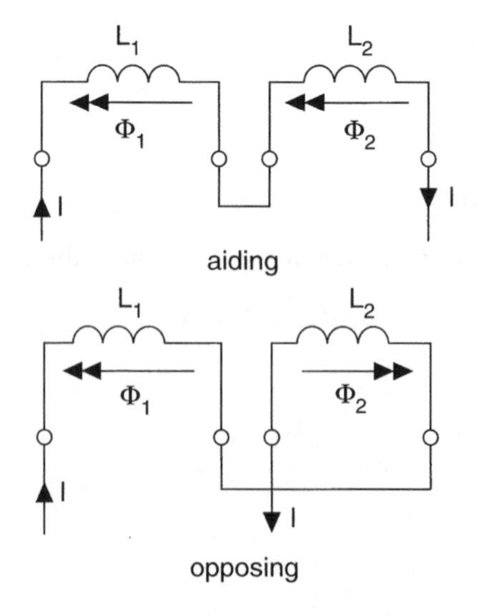

Fig. 5.46

$$M = k\sqrt{L_1L_2} \text{ henry} = 0.8\sqrt{25 \times 40} \text{ mH}$$

so, $M = 25.3$ mH

The values for $\frac{1}{2}L_1I_1^2$ and $\frac{1}{2}L_2I_2^2$ have been calculated in part (a) and

$$MI_1I_2 = 25.3 \times 10^{-3} \times 0.25^2 = 1.58 \text{ mJ}$$

(i) For series aiding:

$$W = 0.78 + 1.25 + 1.58 \text{ mJ}$$

so, $W = 3.6$ mJ **Ans**

(ii) For series opposition:

$$W = 0.78 + 1.25 - 1.58 \text{ mJ}$$

so, $W = 0.45$ mJ **Ans**

Summary of Equations

Self-induced emf: $e = -N\dfrac{d\phi}{dt}$ volt

Emf in a straight conductor: $e = B\ell v \sin\phi$ volt

Force on a current carrying conductor: $F = BI\ell \sin\phi$ volt

Motor principle: $T = BANI$ newton metre

Force between current carrying conductors: $F = \dfrac{2 \times 10^{-7} I_1I_2}{d}$ newton

Voltmeter figure of merit: $\dfrac{1}{I_{fsd}}$ ohm/volt

Self-inductance: Self-induced emf, $e = -L\dfrac{di}{dt}$ volt

$$L = N\frac{d\phi}{di} = \frac{N\Phi}{I} \text{ henry}$$

$$L = \frac{N^2}{S} = \frac{\mu_o\mu_r N^2 A}{\ell} \text{ henry}$$

Energy stored: $W = 0.5LI^2$ joule

Mutual inductance: Mutually induced emf, $e_2 = -M\dfrac{di_1}{dt}$ volt

$$M = N_2\frac{d\phi_2}{di_1} = \frac{N_2\Phi_2}{I_1} \text{ henry}$$

$$M = k\sqrt{L_1L_2} \text{ henry}$$

Energy stored: $W = 0.5L_1I_1^2 + 0.5L_2I_2^2 \pm MI_1I_2$ joule

Transformer: Voltage ratio, $\dfrac{V_2}{V_1} = \dfrac{N_2}{N_1}$

Current ratio, $\dfrac{I_2}{I_1} = \dfrac{N_1}{N_2}$

6 Alternating Quantities

This chapter deals with the concepts, terms and definitions associated with alternating quantities. The term alternating quantities refers to any quantity (current, voltage, flux, etc.), whose polarity is reversed alternately with time. For convenience, they are commonly referred to as a.c. quantities. Although an a.c. can have any waveshape, the most common waveform is a sinewave. For this reason, unless specified otherwise, you may assume that sinusoidal waveforms are implied.

On completion of this chapter, you should be able to:

1 Explain the method of producing an a.c. waveform.
2 Define all of the terms relevant to a.c. waveforms.
3 Obtain values for an a.c., both from graphical information and when expressed in mathematical form.
4 Understand and use the concept of phase angle.
5 Use both graphical and phasor techniques to determine the sum of alternating quantities.

6.1 Production of an Alternating Waveform

From electromagnetic induction theory, we know that the average emf induced in a conductor, moving through a magnetic field, is given by

$e = B\ell v \sin\theta$ volt .... [1]

Where B is the flux density of the field (in tesla)
ℓ is the effective length of conductor (in metre)
v is the velocity of the conductor (in metre/s)
θ is the angle at which the conductor 'cuts' the lines of magnetic flux (in degrees or radians)

i.e. $v \sin\theta$ is the component of velocity at right angles to the flux.

Consider a single-turn coil, rotated between a pair of poles, as illustrated in Figs. 6.1(a) and (b). Figure (a) shows the general arrangement. Figure (b) shows a cross-section at one instant in time, such that the coil is moving at an angle of $\theta°$ to the flux.

Considering Fig. 6.1(b), each side of the coil will have the same value of emf induced, as given by equation [1] above. The polarities of these emfs will be as shown (Fleming's righthand rule). Although these emfs are of opposite polarities, they both tend to cause

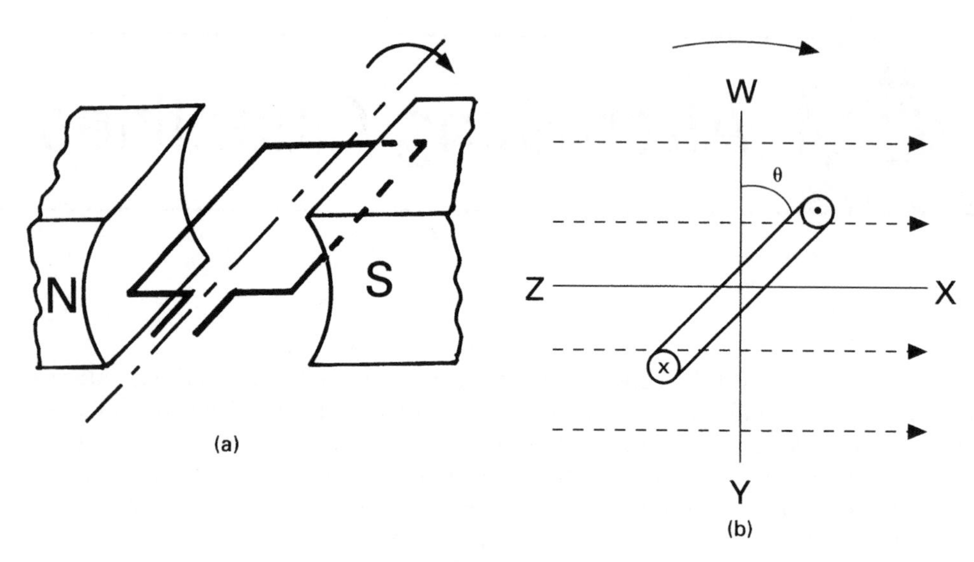

Fig. 6.1

current to flow in the same direction around the coil. Thus, the total emf generated is given by:

$e = 2 \times B\ell v \sin\theta$ volt.................. [2]

Still considering Fig. 6.1(b), at the instant the coil is in the plane W–Y, angle $\theta = 0°$. Thus the emf induced is zero. At the instant that it is in the plane X–Z, $\theta = 90°$. Thus, the emf is at its maximum possible value, given by:

$e = 2 \times B\ell v$ volt....................... [3]

Let us consider just one side of the coil, starting at position W. After 90° rotation (to position X), the emf will have increased from zero to its maximum value. During the next 90° of rotation (to position Y), the emf falls back to zero. During the next 180° rotation (from Y to Z to W), the emf will again increase to its maximum, and reduce once more to zero. However, during this half revolution, the polarity of the emf is reversed.

If the instantaneous emf induced in the coil is plotted, for one complete revolution, the sinewave shown in Fig. 6.2 will be produced. For convenience, it has been assumed that the maximum value of the coil emf is 1 V, and that the plot starts with the coil in position W.

When the coil passes through one complete revolution, the waveform returns to its original starting point. The waveform is then said to have completed one cycle. Note that one cycle is the interval between any two *corresponding* points on the waveform. The number of cycles generated per second is called the *frequency, f,* of the waveform. The unit for frequency is the hertz (Hz). Thus, one cycle per second is equal to 1 Hz.

For the simple two-pole arrangement considered, one cycle of emf is generated in one revolution. The frequency of the waveform is therefore the same as the speed of rotation, measured in revolution per second (rev/s). This yields the following equation

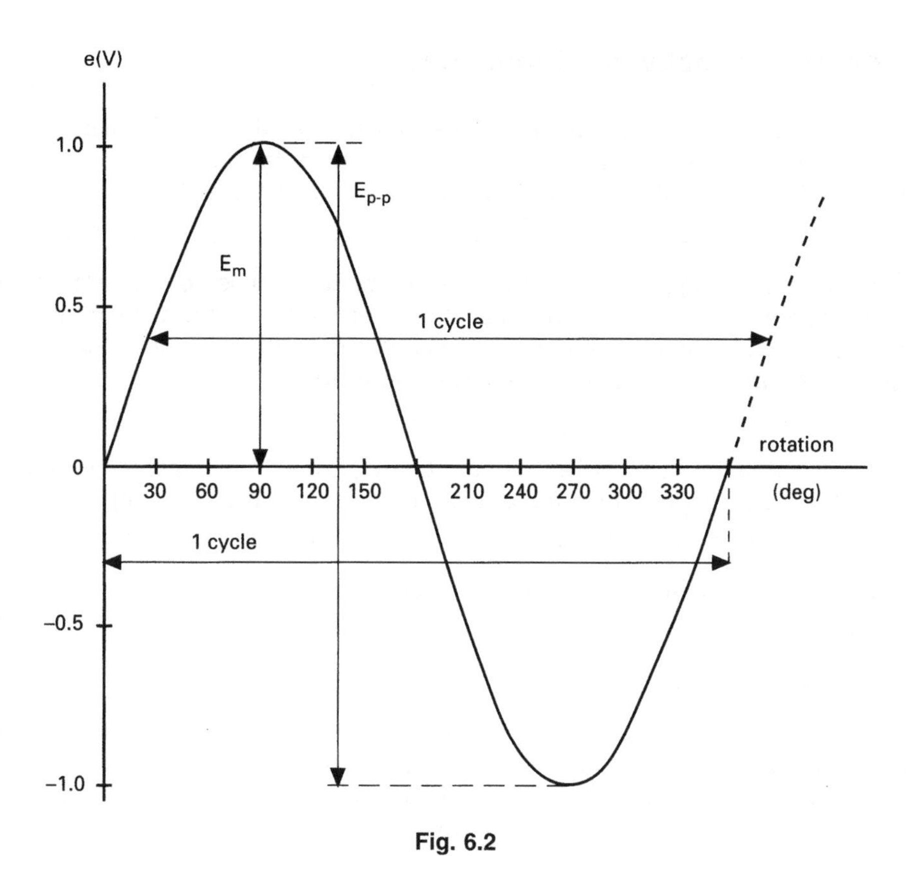

Fig. 6.2

$$f = np \text{ hertz} \tag{6.1}$$

where p = the number of pole *pairs*

Therefore, if the coil is rotated at 50 rev/s, the frequency will be

$$f = 50 \times 1\,\text{Hz } (\textit{one pair} \text{ of poles}) = 50\,\text{Hz}.$$

The time taken for the waveform to complete one cycle is called the periodic time, T. Thus, if 50 cycles are generated in one second, then one cycle must be generated in 1/50 of a second. The relationship between frequency and period is therefore

$$T = \frac{1}{f} \text{ second, or } f = \frac{1}{T} \text{ hertz} \tag{6.2}$$

The maximum value of the emf in one cycle is shown by the peaks of the waveform. This value is called, either the maximum or peak value, or the *amplitude* of the waveform. The quantity symbol used may be either $\hat{E}$, or E_m.

The voltage measured between the positive and negative peaks is called the peak-to-peak value. This has the quantity symbol E_{pk-pk}, or E_{p-p}.

6.2 Angular Velocity and Frequency

In SI units, angles are measured in **radians**, rather than degrees. Similarly, angular velocity is measured in radian per second, rather than revolutions per second. The quantity symbol for angular velocity is ω (lower case Greek omega).

> A **radian** is the angle subtended at the centre of a circle, by an arc on the circumference, which has length equal to the radius of the circle. Since the circumference $= 2\pi r$, then there must be 2π such arcs in the circumference. Hence there are 2π radians in one complete circle; i.e. $2\pi\,\text{rad} = 360°$

If the coil is rotating at n rev/s, then it is rotating at $360 \times n$ degrees/second. Since there are 2π radians in 360°, then the coil must be rotating at $2\pi n$ radian per second. Thus, angular velocity, $\omega = 2\pi n\,\text{rad/s}$

but for a 2-pole system, n = frequency, f hertz,

$$\text{therefore, } \omega = 2\pi f\,\text{rad/s} \qquad (6.3)$$

$$\text{and, } f = \frac{\omega}{2\pi}\,\text{hertz} \qquad (6.4)$$

If the coil is rotating at ω rad/s, then in a time of t seconds, it will rotate through an angle of ωt radian. Hence the waveform diagram may be plotted to a base of degrees, radians, or time. In the latter case, the time interval for one cycle is, of course, the periodic time, T, These are shown in Fig. 6.3.

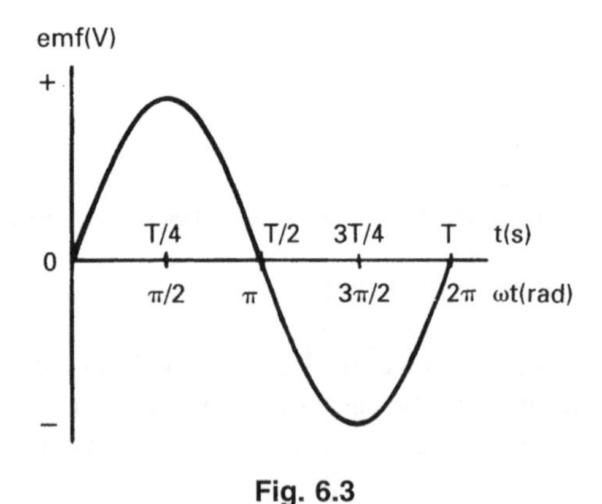

Fig. 6.3

6.3 Standard Expression for an Alternating Quantity

All the information regarding an a.c. can be presented in the form of a graph. The information referred to here is the amplitude, frequency, period, and value at any instant. The last is normally called the instantaneous value. However, presenting the information in this way is not always very convenient. Firstly, the graph has to be plotted accurately, on graph paper. This in itself is a time consuming procedure. In addition, obtaining precise information from the graph is difficult. The degree of accuracy depends on the suitability of the scales chosen, and the individual's interpretation. For example, if several people are asked to obtain a particular value from the graph, their answers are likely to differ slightly from one another. To overcome these difficulties, the a.c. needs to be expressed in a more convenient form. This results in an equation, sometimes referred to as the algebraic form of the a.c. More correctly, it should be called the trigonometric form. Since many students are put off by these terms, we shall refer to it simply as the standard expression for a waveform.

The emf for an N-turn coil is:

$$e = 2 \times NB\ell v \sin\theta \text{ volt, where } \theta \text{ is in degrees}$$

$$\text{or, } e = 2 \times NB\ell v \sin(\omega t) \text{ volt, where } \omega t \text{ is in radians}$$

and the emf is at its maximum value when $\sin(\omega t)$, or $\sin\theta$ is equal to 1.

Therefore, $E_m = 2 \times NB\ell v$ volt, so the expression becomes:

$$e = E_m \sin\theta \text{ volt} \qquad (6.5)$$

$$\text{or, } e = E_m \sin(\omega t) \text{ volt} \qquad (6.6)$$

$$\text{or, } e = E_m \sin(2\pi f t) \text{ volt} \qquad (6.7)$$

All three of the above equations are the so-called standard expressions for this a.c. voltage. Equations (6.6), and (6.7) in particular, are those most commonly used. Using these, all the relevant information concerning the waveform is contained in a neat mathematical expression. There is also no chance of ambiguity.

Worked Example 6.1

Question

An alternating voltage is represented by the expression $v = 35\sin(314.2t)$ volt. Determine, (a) the maximum value, (b) the frequency, (c) the period of the waveform, and (d) the value 3.5 ms after it passes through zero, going positive.

Answer

(a) v=35 sin (314.2t) volt
and comparing this to the standard,
$v = V_m \sin(2\pi f t)$ volt we can see that :

$V_m = 35$ V **Ans**

(b) Again, comparing the two expressions:

$2\pi f = 314.2$

so $f = \dfrac{314.2}{2\pi} = 50$ Hz **Ans**

Continued on p. 264

Worked Example 6.1 (*Continued*)

(c) $T = \frac{1}{f} = \frac{1}{50}$ second

so, $T = 20$ ms **Ans**

(d) When $t = 3.5$ ms; then:

$$v = 35 \sin(2\pi \times 50 \times 3.5 \times 10^{-3}) \text{ volt}$$
$$= 35 \sin(1.099)^*$$
$$= 35 \times 0.891$$

therefore, $v = 31.19$ V **Ans**

*The term inside the brackets is an angle in RADIAN. You must therefore remember to switch your calculator into the RADIAN MODE.

So far, we have dealt only with an alternating voltage. However, all of the terms and definitions covered are equally applicable to any alternating quantity. Thus, exactly the same techniques apply to a.c. currents, fluxes, etc. The same applies to mechanical alternating quantities involving oscillations, vibrations, etc.

Worked Example 6.2

Question

For a current, $i = 75 \sin(200\pi t)$ milliamp, determine (a) the frequency, and (b) the time taken for it to reach 35 mA, for the first time, after passing through zero.

Answer

(a) $i = 75 \sin(200\pi t)$ milliamp

$i = I_m \sin(2\pi ft)$ amp

so, $2\pi f = 200\pi$

and $f = \frac{200\pi}{2\pi} = 100$ Hz **Ans**

(b) $35 = 75 \sin(200\pi t)$ milliamp

$\frac{35}{75} = \sin(200\pi t) = 0.4667$

therefore, $200\pi t = \sin^{-1} 0.4667^*$

$$= 0.4855 \text{ rad}$$

$$\text{so, } t = \frac{0.4855}{200\pi} = 0.773 \text{ ms } \textbf{Ans}$$

**Remember*, use RADIAN mode on your calculator.

6.4 Average Value

Figure 6.4 shows one cycle of a sinusoidal current.

From this it is apparent that the area under the curve in the positive half is exactly the same as that for the negative half. Thus, the average value over one complete cycle must be zero. For this reason, the average value is taken to be the average over one half cycle. This average may be obtained in a number of ways. These include, the mid-ordinate rule, the trapezoidal rule, Simpson's rule, and integral calculus. The simplest of these is the mid-ordinate rule, and this will be used here to illustrate average value; see Fig. 6.5.

A number of equally spaced intervals are selected, along the time axis of the graph. At each of these intervals, the instantaneous value is determined. This results in values for a number of ordinates, $i_1, i_2, \ldots, i_n$, where n is the number of ordinates chosen. The larger the number of ordinates chosen, the more accurate will be the final average value obtained. The average is simply found by adding together all the ordinate values, and then dividing this figure by the number of ordinates chosen, thus

$$I_{av} = \frac{i_1 + i_2 + i_3 + \ldots + i_n}{n}$$

The average value will of course depend upon the shape of the waveform, and *for a sinewave only* it is

$$I_{av} = \frac{2}{\pi} I_m = 0.637\, I_m$$

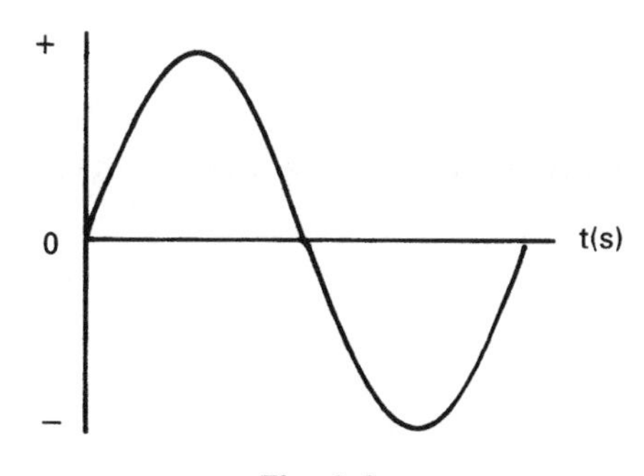

Fig. 6.4

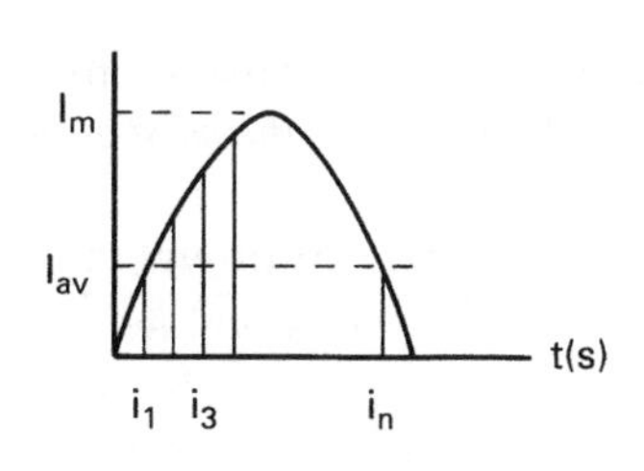

Fig. 6.5

6.5 r.m.s. Value

The r.m.s. value of an alternating current is equivalent to that value of direct current, which when passed through an identical circuit, will dissipate exactly the same amount of power. The r.m.s. value of an a.c. thus provides a means of making a comparison between a.c. and d.c. systems.

The term r.m.s. is an abbreviation of the square Root of the Means Squared. The technique for finding the r.m.s. value may be based on the same ways as were used to find the average value. However, the r.m.s. value applies to the complete cycle of the waveform. For simplicity, we will again consider the use of the mid-ordinate rule technique.

Considering Fig. 6.5, the ordinates would be selected and measured in the same way as before. The value of each ordinate is then squared. The resulting values are then summed, and the average found. Finally, the square root of this average (or mean) value is determined. This is illustrated below:

$$I_{rms} = \sqrt{\frac{i_1^2 + i_2^2 + i_3^2 + \ldots + i_n^2}{n}}$$

and, *for a sinewave only*, $I_{rms} = \frac{1}{\sqrt{2}} I_m = 0.707 I_m$

Other waveforms will have a different ratio between r.m.s. and peak values.
Note: The r.m.s. value of an a.c. is the value normally used and quoted. For example, if reference is made to a 240 V a.c. supply, then 240 V is the r.m.s. value. In general therefore, if an unqualified value for an a.c. is given, then the assumption is made that this is the r.m.s. value. Since r.m.s. values are those commonly used, the subscript letters r.m.s. are not normally included. I_{rms} has been used above, simply for emphasis. The following convention is used:

i, v, e, represent instantaneous values

I_{av}, V_{av}, E_{av}, represent average values

I_m, V_m, E_m, represent maximum or peak values, or *amplitude*

I, V, E, represent r.m.s. values

6.6 Peak Factor

This is defined as the ratio of the peak or maximum value, to the r.m.s. value, of a waveform. Thus, *for a sinewave only*

$$\text{peak factor} = \frac{\text{maximum value}}{\text{r.m.s. value}}$$

$$= \frac{V_m}{0.707\, V_m} = \sqrt{2} \text{ or } 1.414$$

Worked Example 6.3

Question

Calculate the amplitude of the household 240 V supply.

Answer

Since this supply is sinusoidal, then the peak factor will be $\sqrt{2}$, so

$$V_m = \sqrt{2} \times V \text{ volt} = \sqrt{2} \times 240$$

So $V_m = 339.4$ V **Ans**

Worked Example 6.4

Question

A non-sinusoidal waveform has a peak factor of 2.5, and an r.m.s. value of 240 V. It is proposed to use a capacitor in a circuit connected to this supply. Determine the minimum safe working voltage rating required for the capacitor.

Answer

peak factor = 2.5; $V = 240$ V

$$V_m = 2.5 \times V \text{ volt} = 2.5 \times 240$$

so, $V_m = 600$ V

Thus the absolute minimum working voltage must be 600 V **Ans**

In practice, a capacitor having a higher working voltage would be selected. This would then allow a factor of safety.

6.7 Form Factor

As the name implies, this factor gives an indication of the form or shape of the waveform. It is defined as the ratio of the r.m.s. value to the average value.

Thus, *for a sinewave,*

$$\text{form factor} = \frac{\text{r.m.s. value}}{\text{average value}} = \frac{0.707}{0.637}$$

so, form factor = 1.11

For a rectangular waveform (a squarewave), form factor = 1, since the r.m.s. value, the peak value, and the average value are all the same.

Worked Example 6.5

Question

A rectangular coil, measuring 25 cm by 20 cm, has 80 turns. The coil is rotated, about an axis parallel with its longer sides, in a magnetic field of density 75 mT. If the speed of rotation is 3000 rev/min, calculate, from first principles, (a) the amplitude, r.m.s. and average values of the emf, (b) the frequency and period of the generated waveform, (c) the instantaneous value, 2 ms after it is zero.

Answer

$\ell = 0.25\,\text{m};\ d = 0.2\,\text{m};\ N = 80;\ B = 0.075\,\text{T}$

$$n = \frac{3000}{60}\ \text{rev/s};\ t = 2 \times 10^{-3}\,\text{s}$$

(a) $e = 2 \times NB\ell v \sin(2\pi ft)$ volt

Now, we know the rotational speed n, but the above equation requires the tangential velocity, v, in metre per second. This may be found as follows. Consider Fig. 6.6,

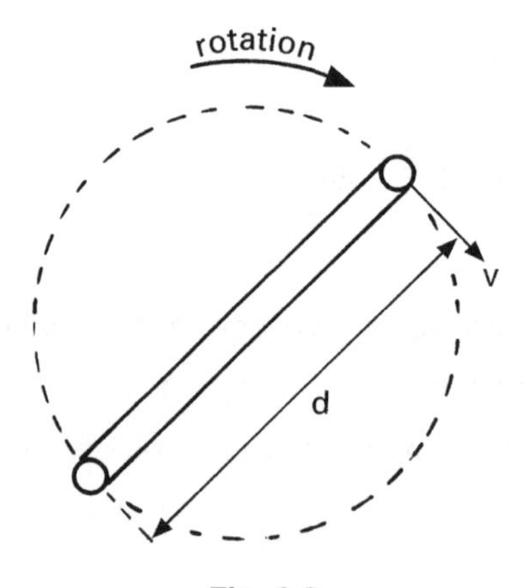

Fig. 6.6

which shows the path travelled by the coil sides.
The circumference of rotation = πd metre = 0.2π metre.
The coil sides travel this distance in one revolution.
The rotational speed $n = 3000/60 = 50$ rev/s.
Hence the coil sides have a velocity, $v = 50 \times 0.2\pi$ m/s.
Therefore, $e = 2 \times 80 \times 0.075 \times 0.25 \times 50 \times 0.2\pi \sin(2\pi ft)$ volt
and emf is a maximum value when $\sin(2\pi ft) = 1$

so, $E_m = 2 \times 80 \times 0.075 \times 0.25 \times 50 \times 0.2\pi$
$E_m = 94.25\,\text{V}$ **Ans**

Assuming a sinusoidal waveform:

$E = 0.707\ E_m = 0.707 \times 94.25$
so, $E = 66.64\,\text{V}$ **Ans**
$E_{av} = 0.637 E_m = 0.637 \times 94.25$
so, $E_{av} = 60.04\,\text{V}$ **Ans**

(b) Assuming a 2-pole field system, then $f = n$ therefore $f = 50\,\text{Hz}$ **Ans**

$$T = \frac{1}{f} = \frac{1}{50}\,\text{s}$$
so $T = 20\,\text{ms}$ **Ans**

(c) $e = E_m \sin(2\pi ft)$ volt
$= 94.25 \sin(2\pi \times 50 \times 2 \times 10^{-3})$
$= 94.25 \times 0.5878$
so, $e = 55.4\,\text{V}$ **Ans**

6.8 Rectifiers

A rectifier is a circuit which converts a.c. to d.c. The essential component of any rectifier circuit is a diode. This is a semiconductor device, which allows current to flow through it in one direction only. It is the electronic equivalent of a mechanical valve, for example the valve in a car tyre. This device allows air to be pumped into the tyre, but prevents the air from escaping.

The circuit symbol for a diode is shown in Fig. 6.7. The 'arrow head' part of the symbol is known as the anode. This indicates the direction in which *conventional* current can flow through it. The 'plate' part of the symbol is the cathode, and indicates that conventional current is prevented from entering at this terminal. Thus, provided that the anode is more positive than the cathode, the diode will conduct. This is known as the forward bias condition. If the cathode is more positive than the anode, the diode is in its blocking mode, and does not conduct. This is known as reverse bias.

Note: The potentials at anode and cathode do not have to be positive and negative. Provided that the anode is *more* positive than the cathode, the diode will conduct. So

Fig. 6.7

if the anode potential is (say) + 10 V, and the cathode potential is + 8 V, then the diode will conduct. Similarly, if these potentials are reversed, the diode will not conduct.

6.9 Half-wave Rectifier

This is the simplest form of rectifier circuit. It consists of a single diode, placed between an a.c. supply and the load, for which d.c. is required. The arrangement is shown in Fig. 6.8, where the resistor *R* represents the load.

Let us assume that, in the first half cycle of the applied voltage, the instantaneous polarities at the input terminals are as shown in Fig. 6.9. Under this condition, the diode is forward biased. A half sinewave of current will therefore flow through the load resistor, in the direction shown.

In the next half cycle of the input waveform, the instantaneous polarities will be reversed. The diode is therefore reverse biased, and no current will flow. This is illustrated in Fig. 6.10.

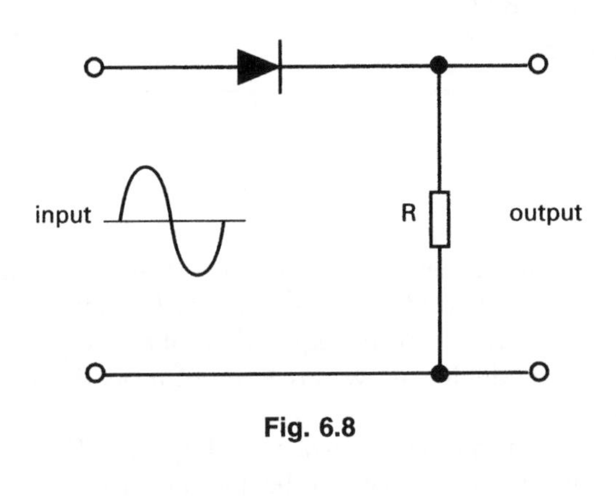

Fig. 6.8

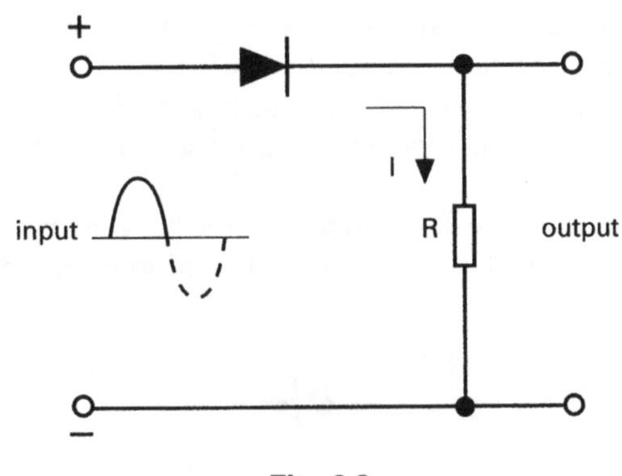

Fig. 6.9

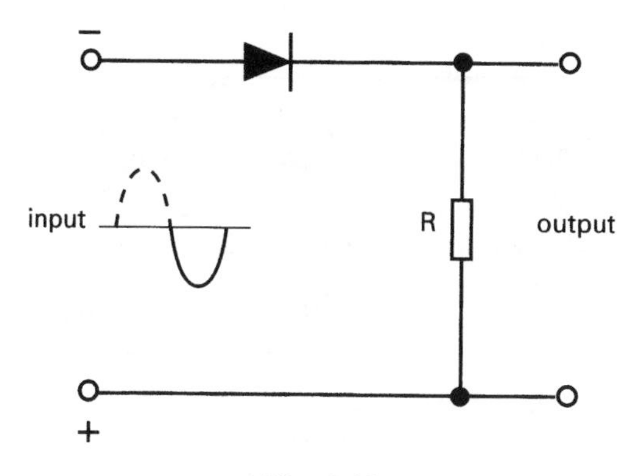

Fig. 6.10

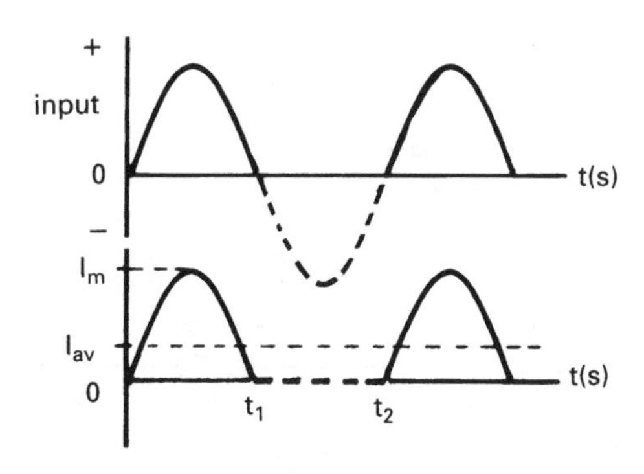

Fig. 6.11

The graphs of the applied a.c. voltage, and the corresponding load current, are shown in Fig. 6.11. The load p.d. will be of exactly the same waveshape as the load current. Both of these quantities are uni-directional, and so by definition, are d.c. quantities. The 'quality' of the d.c. so produced is very poor, since it exists only in pulses of current. The average value of this current is determined over the time period, 0 to t_2. The average value from 0 to time t_1 will be $0.637I_m$. From t_1 to t_2 it will be zero. The average value of the d.c. will therefore be, $I_{av} = 0.318I_m$.

6.10 Full-wave Bridge Rectifier

Both the 'quality' and average value of the d.c. need to be improved. This may be achieved by utilising the other half cycle of the a.c. supply. The circuit consists of four diodes, connected in a 'bridge' configuration, as shown in Fig. 6.12.

We will again assume the instantaneous polarities for the first half cycle as shown. In this case, diodes D_1 and D_4 will be forward biased. Diodes D_2 and D_3 will be reverse biased. Thus, D_1 and D_4 allow current to flow, as shown. In the next half cycle, the polarities are reversed. Hence, D_2 and D_3 will conduct, whilst D_1 and D_4 are reverse

+
I
input
D1
D2
R
D3
D4
−

Fig. 6.12

−
I
input
D1
D2
R
D3
D4
+

Fig. 6.13

+
input
0
−
t(s)
I_m
I_{av}
0
t(s)

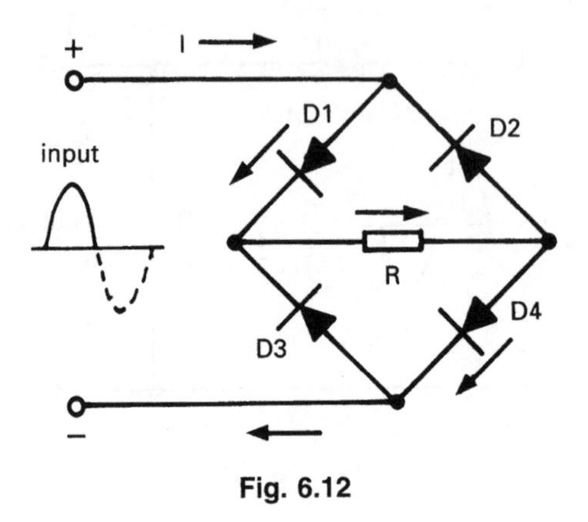

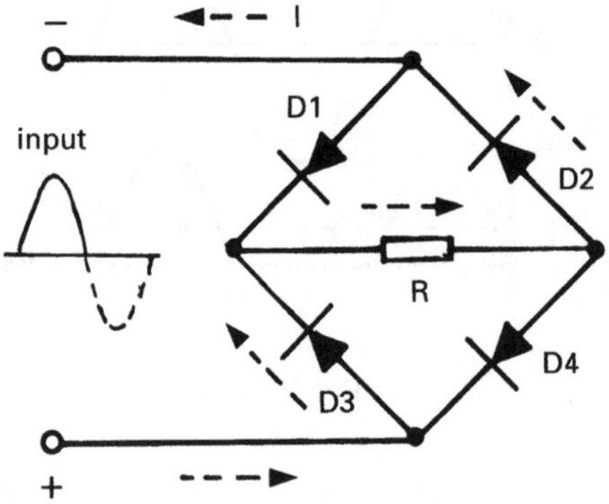

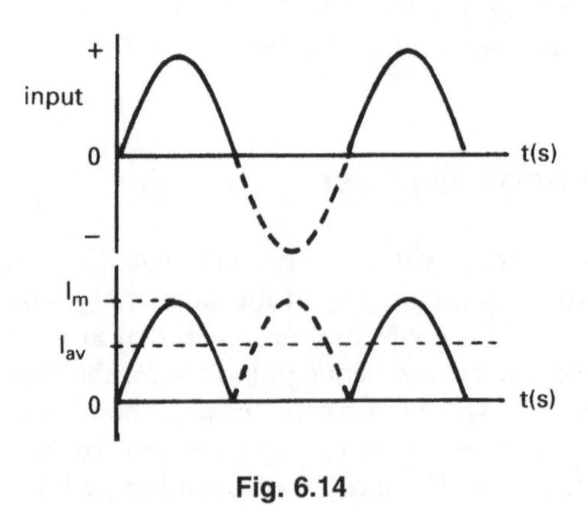

Fig. 6.14

biased. Current will therefore flow as shown in Fig. 6.13. Notice that the current through the load resistor is in the same direction, for the whole cycle of the a.c. supply.

The relevant waveforms are shown in Fig. 6.14. It should be apparent that the average value of the a.c. will now be twice that in the previous circuit. That is $I_{av} = 0.637I_m$.

It is this type of rectifier circuit that is incorporated in multimeters. This enables the measurement of a.c. voltages and currents.

6.11 Rectifier Moving Coil Meter

The circuit arrangement for such a meter is illustrated in Fig. 6.15. The symbol, with the letter M in it, represents the moving coil movement. The current through the coil will therefore be a series of half-sinewave pulses, as in Fig. 6.14.

Due to the inertia of the meter movement, the coil will respond to the average value of this current. Thus, the pointer would indicate the *average* value of the a.c. waveform being measured. However, the normal requirement is for the meter to indicate the *r.m.s.* value. Thus, some form of 'correction factor' is required.

The majority of a.c. quantities to be measured are sinewaves. For this reason, the meter is calibrated to indicate the r.m.s. values of sinewaves. Now, the ratio between r.m.s. and average values is the form factor. For the sinewave, the form factor has a value of 1.11. The values chosen for shunts and multipliers, used on the a.c. ranges, are therefore modified by this factor. Thus, although the pointer position corresponds to the average value, the scale indication will be the r.m.s. value.

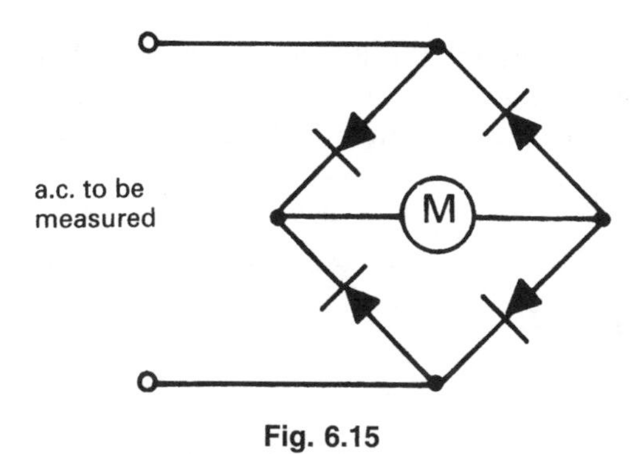

Fig. 6.15

Worked Example 6.6

Question

A moving coil meter has a figure of merit of 10 kΩ/*V*. The coil has a resistance of 50 Ω. Calculate the value of multiplier required for (a) the 10 V d.c. range, (b) the 10 V a.c. range.

Continued on p. 274

Worked Example 6.6 (*Continued*)

Answer

$R_c = 50\,\Omega$; figure of merit $= 10\,\text{k}\Omega/\text{V}$; $V = 10\,\text{V}$

(a) $I_{fsd} = 1/10\,000 = 100\,\mu\text{A}$

total meter resistance, $R = \dfrac{V}{I_{fsd}}$ ohm $= \dfrac{10}{10^{-4}}$

so, $R = 100\,\text{k}\Omega$; and since $R = R_m + R_c$, then:

$R_m = 100\,000 - 50 = 99.95\,\text{k}\Omega$ **Ans**

(b) $I_{fsd} = 100\,\mu\text{A} = I_{av}$

therefore, r.m.s. value,

$$I = 1.11 \times 100 = 111\,\mu\text{A}$$

$$\text{so, } R = \frac{10}{111 \times 10^{-6}} = 90.09\,\text{k}\Omega$$

therefore, $R_m = 90\,090 - 50$

$= 90.04\,\text{k}\Omega$ **Ans**

The meter is therefore calibrated for the measurement of sinewaves. If any other waveform is measured, the meter reading will be in error, because the waveform will have a different form factor. Provided that the form factor is known, then the true r.m.s. value can be calculated.

Worked Example 6.7

Question

A moving coil meter is used to measure both a squarewave and a triangular wave voltage. The meter reading is 5 V in each case. Calculate the true r.m.s. values.

Answer

(a) The form factor for a squarewave is 1. The meter, however, has been calibrated for a form factor of 1.11. Thus the indicated reading will be too high by a factor of 1.11:1

$$\text{Therefore the true value} = 5 \times \frac{1}{1.11}$$

$$= 4.504\,\text{V} \textbf{ Ans}$$

(b) If the form factor for the triangular wave is 1.15, then the meter reading will be too low, by a factor of 1.15:1.11.

$$\text{Therefore the true value} = 5 \times \frac{1.15}{1.11}$$
$$= 5.18\,\text{V } \textbf{Ans}$$

A further complication arises when the meter is used to measure small a.c. voltages. When conducting, each diode will have a small p.d. developed across it. This forward voltage drop will be in the order of 0.6 to 0.7 V. At any instant, two diodes are conducting. Thus, there will be a total forward voltage drop of 1.2 to 1.4 V across the diodes. This voltage drop is 'lost' as far as the meter coil is concerned. This effect also must be taken into account when determining the values of multipliers, for the lower a.c. voltage ranges.

6.12 Phase and Phase Angle

Consider two a.c. voltages, of the same frequency, as shown in Fig. 6.16. Both voltages pass through the zero on the horizontal axis at the same time. They also reach their positive and negative peaks at the same time. Thus, the two voltages are exactly synchronised, and are said to be *in phase* with each other.

Figure 6.17 shows the same two voltages, but in this case let v_2 reach its maximum value $\pi/2$ radian (90°) after v_1. It is necessary to consider one of the waveforms as the *reference* waveform. It is normal practice to consider the waveform that passes through zero, going positive, at the beginning of the cycle, as the reference waveform. So, for the two waveforms shown, v_1 is taken as the reference. In this case, v_2 is said to *lag* v_1, by $\pi/2$ radian or 90°. The standard expressions for the two voltages would therefore be written as follows:

$$v_1 = V_{m1} \sin(2\pi ft), \text{ or } V_{m1} \sin\theta \text{ volt}$$
$$v_2 = V_{m2} \sin(2\pi ft - \pi/2), \text{ or } V_{m2}(\sin\theta - 90°) \text{ volt}$$

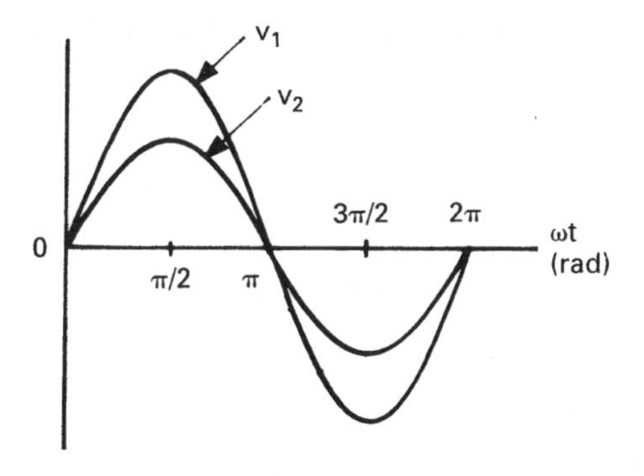

Fig. 6.16

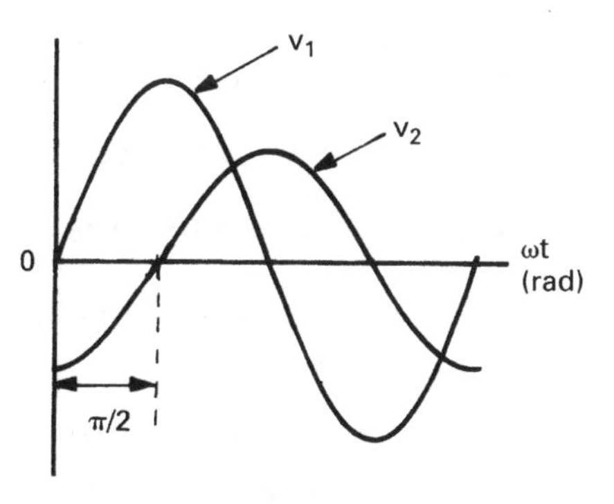

Fig. 6.17

The minus signs, in the brackets of the above expressions, indicate that v_2 lags the reference by the angle quoted. This angle is known as the *phase angle*, or *phase difference*, between the two waveforms.

In general, the standard expression for an a.c. voltage is:

$$v = V_m \sin(2\pi ft \pm \phi) \text{ volt,}$$
$$\text{or } v = V_m \sin(\omega t \pm \phi) \text{ volt} \quad (6.8)$$

Although it would be usual to take v_1 as the reference in the above example, it is not mandatory. Thus, if for some good reason v_2 was chosen as the reference, the expressions would be written as:

$$v_2 = V_{m2} \sin(\omega t), \text{ or } V_{m2} \sin\theta \text{ volt}$$
$$v_1 = V_{m1} \sin(\omega t + \pi/2), \text{ or } V_{m1} \sin(\theta + 90°) \text{ volt}$$

Note: when the relevant phase angle, ϕ, is quoted in the standard expression, do *not* mix degrees with radians. Thus, if the initial angular data is in radian (ωt or $2\pi ft$), then ϕ must also be expressed in radian. Similarly, if the angular data is initially in degrees (θ), the ϕ must also be quoted in degrees.

Worked Example 6.8

Question

Three alternating currents are specified below. Determine the frequency, and for each current, determine its phase angle, and amplitude.

$$i_1 = 5 \sin(80\pi t + \pi/6) \text{ amp}$$
$$i_2 = 3 \sin 80\pi t \text{ amp}$$
$$i_3 = 6 \sin(80\pi t - \pi/4) \text{ amp}$$

Answer

All three waveforms have the same value of ω, namely 80π rad/s. Thus all three have the same frequency:

$$\omega = 2\pi f = 80\pi \text{ rad/s}$$
$$\text{therefore, } f = \frac{80\pi}{2\pi} = 40 \text{ Hz } \textbf{Ans}$$

Since zero phase angle is quoted for i_2, then this is the reference waveform, of amplitude 3 A **Ans**

$I_{m1} = 5$ A, and leads i_2 by $\pi/6$ rad (30°) **Ans**

$I_{m3} = 6$ A, and lags i_2 by $\pi/4$ rad (45°) **Ans**

The majority of people can appreciate the relative magnitudes of angles, when they are expressed in degrees. Angles expressed in radians are more difficult to appreciate. Some of the principal angles encountered are listed below. This should help you to gain a better 'feel' for radian measure.

degrees	*radians*	*radians*	*degrees*
360	$2\pi \approx 6.28$	0.1	5.73
270	$3\pi/2 \approx 4.71$	0.2	11.46
180	$\pi \approx 3.14$	0.3	17.19
120	$2\pi/3 \approx 2.09$	0.4	22.92
90	$\pi/2 \approx 1.57$	0.5	28.65
60	$\pi/3 \approx 1.05$	1.0	57.30
45	$\pi/4 \approx 0.79$	1.5	85.94
30	$\pi/6 \approx 0.52$	2.0	114.60

6.13 Phasor Representation

A phasor is a rotating vector. Apart from the fact that a phasor rotates at a constant velocity, it has exactly the same properties as any other vector. Thus its length corresponds to the magnitude of a quantity. It has one end arrowed, to show the direction of action of the quantity.

Consider two such rotating vectors, v_1 and v_2, rotating at the same angular velocity, ω rad/s. Let them rotate in a counterclockwise direction, with v_2 lagging behind v_1 by $\pi/6$ radian (30°). This situation is illustrated in Fig. 6.18.

The instantaneous vertical height of each vector is then plotted for one complete revolution. The result will be the two sinewaves shown.

Notice that the angular difference between v_1 and v_2 is also maintained throughout the waveform diagram. Also note that the peaks of the two waveforms correspond to the magnitudes, or amplitudes, of the two vectors. In this case, these two waveforms could equally well represent either two a.c. voltages, or currents. If this were the case,

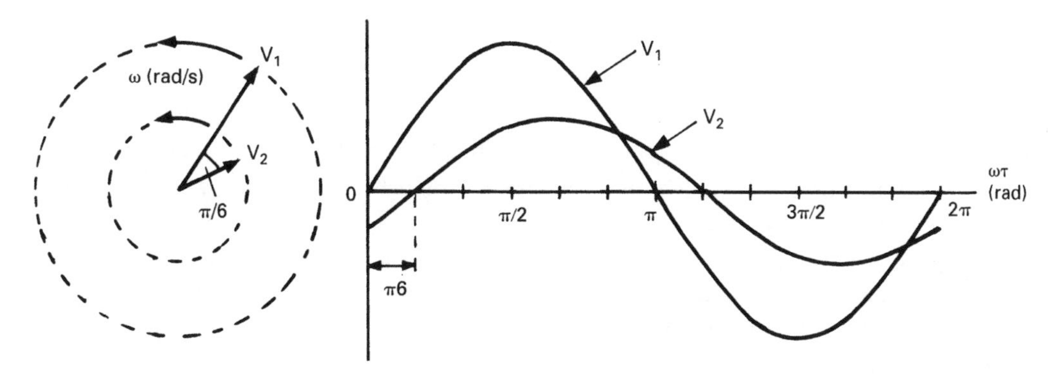

Fig. 6.18

then the two a.c. quantities would be of the *same frequency*. This is because the value of ω is the same for both. The angular difference, of $\pi/6$ radian, would then be described as the phase difference between them.

We can therefore, represent an alternating quantity by means of a phasor. The length of the phasor represents the amplitude. Its angle, with respect to some reference axis, will represent its phase angle. Considering the two waveforms in Fig. 6.18, the plot has been started with V_1 in the horizontal position (vertical component of $V_1 = 0$). This horizontal axis is therefore taken as being the reference axis. Thus, if these waveforms represent two voltages, v_1 and v_2, then the standard expressions would be:

$$v_1 = V_{m1} \sin(\omega t) \text{ volt}$$
$$\text{and } v_2 = V_{m2} \sin(\omega t - \pi/6) \text{ volt}$$

The inconvenience of representing a.c. quantities in graphical form was pointed out earlier, in section 6.3. This section introduced the concept of using a standard mathematical expression for an a.c. However, a visual representation is also desirable. We now have a much simpler means of providing a visual representation. It is called a phasor diagram. Thus the two voltages we have been considering above may be represented as in Fig. 6.19.

Notice that v_1 has been chosen as the reference phasor. This is because the standard expression for this voltage has a phase angle of zero (there is no $\pm\phi$ term in the bracket). Also, since the phasors are rotating counterclockwise, and v_2 is lagging v_1 by $\pi/6$ radian, then v_2 is shown at this angle *below* the reference axis.

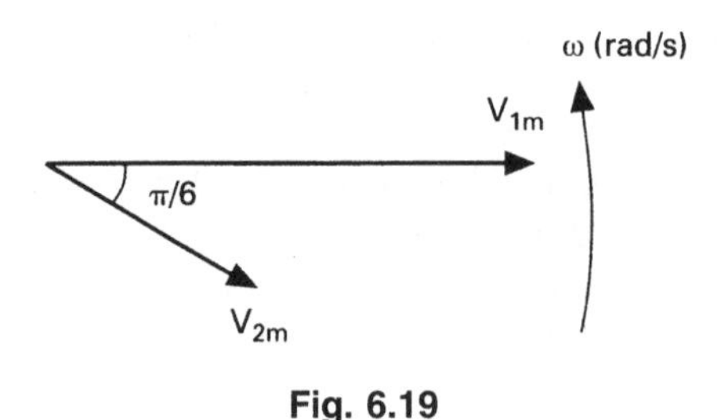

Fig. 6.19

Notes

1 Any a.c. quantity can be represented by a phasor, *provided that it is a sinewave.*

2 Any number of a.c. voltages and/or currents may be shown on the same phasor diagram, *provided that they are all of the same frequency.*

3 Figure 6.19 shows a *counterclockwise* arrow, with ω rad/s. This has been shown here to emphasise the point that phasors *must* rotate in this direction only. It is normal practice to omit this from the diagram.

4 When dealing with a.c. circuits (Chapter 7), r.m.s. values are used almost exclusively. In this case, it is normal to draw the phasors to lengths that correspond to r.m.s. values.

Worked Example 6.9

Question

Four currents are as shown below. Draw to scale the corresponding phasor diagram.

$i_1 = 2.5\sin(\omega t + \pi/4)$ amp; $i_2 = 4\sin(\omega t - \pi/3)$ amp;
$i_3 = 6\sin\omega t$ amp; $i_4 = 3\cos\omega t$ amp

Answer

Before the diagram is drawn, we need to select a reference waveform (if one exists). The currents i_1 and i_2 do not meet this criterion, since they both have an associated phase angle.

This leaves the other two currents. Neither of these has a phase angle shown. However, i_3 is a sinewave, whilst i_4 is a *cosine* waveform. Now, a cosine wave *leads* a sinewave by 90°, or $\pi/2$ radian.

Therefore, i_4 may also be expressed as $i_4 = 3\sin(\omega t + \pi/2)$ amp. Thus i_3 is chosen as the reference waveform, and will therefore be drawn along the horizontal axis.

The resulting phasor diagram is shown in Fig. 6.20.

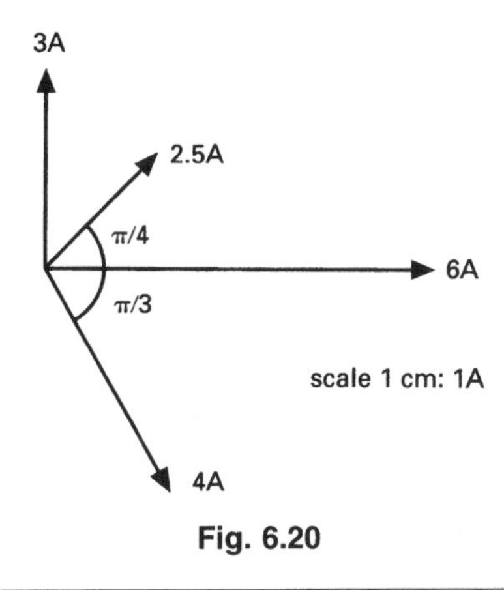

Fig. 6.20

6.14 Addition of Alternating Quantities

Consider two alternating currents, $i_1 = I_{m1}\sin\omega t$ amp and $i_2 = I_{m2}\sin(\omega t - \pi/4)$ amp, that are to be added together. There are three methods of doing this, as listed below.

(a) Plotting them on graph paper. Their ordinates are then added together, and the resultant waveform plotted. This is illustrated in Fig. 6.21. The amplitude, I_m, and the phase angle, ϕ, of the resultant current are then measured from the two axes.

Thus, $i = I_m \sin(\omega t - \phi)$ amp

Note: Although $i = i_1 + i_2$, the AMPLITUDE of the resultant is NOT $I_{m1} + I_{m2}$ amp. This would only be the case if i_1 and i_2 were in phase with each other.

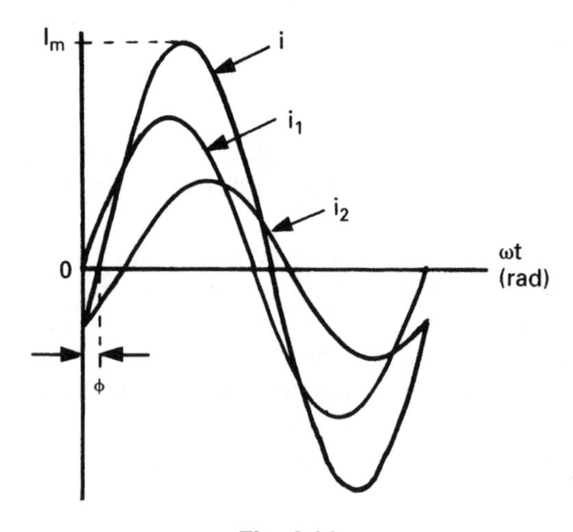

Fig. 6.21

(b) Drawing a scaled phasor diagram, as illustrated in Fig. 6.22. The resultant is found by completing the parallelogram of vectors. The amplitude and phase angle are then measured on the diagram.

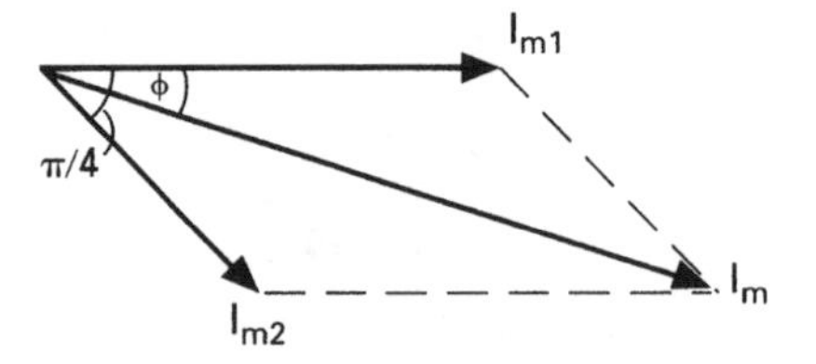

Scale: 1cm ≡ x amp.

Fig. 6.22

(c) Resolving the two currents, into horizontal and vertical components, and applying Pythagoras' theorem. This method involves using a *sketch* of the phasor diagram, followed by a purely mathematical process. This phasor diagram, including the identification of the horizontal and vertical components, is shown in Fig. 6.23.

Horizontal Component (H.C.):

$$\text{H.C.} = I_{m1}\cos 0 + I_{m2}\cos \pi/4$$
$$\text{so, H.C.} = I_{m1} + (0.707 \times I_{m2})\,\text{amp}$$

Vertical Component (V.C.):

$$\text{V.C.} = I_{m1}\sin 0 + I_{m2}\sin \pi/4$$
$$\text{so, V.C.} = 0 + (0.707 \times I_{m2})\,\text{amp}$$

The triangle of H.C., V.C., and the resultant current, is shown in Fig. 6.24. From this, we can apply Pythagoras' theorem to determine the amplitude and phase angle, thus:

$$I_m = \sqrt{\text{H.C.}^2 + \text{V.C.}^2}\,\text{amp}$$
$$\text{and } \tan\phi = \frac{\text{V.C.}}{\text{H.C.}},\ \text{so } \phi = \tan^{-1}\frac{\text{V.C.}}{\text{H.C.}}$$

The final answer, regardless of the method used, would then be expressed in the form $i = I_m \sin(\omega t \pm \phi)$ amp.

Let us now compare the three methods, for speed, convenience, and accuracy.

The graphical technique is very time-consuming (even for the addition of only two quantities). The accuracy also leaves much to be desired; in particular, determining the exact point for the maximum value of the resultant. The determination of the precise phase angle is also very difficult. This method is therefore not recommended.

A phasor diagram, drawn to scale, can be the quickest method of solution. However, it does require considerable care, in order to ensure a reasonable degree of accuracy. Even so, the precision with which the length—and (even more so) the angle—can be measured, leaves a lot to be desired. This is particularly true when three or more

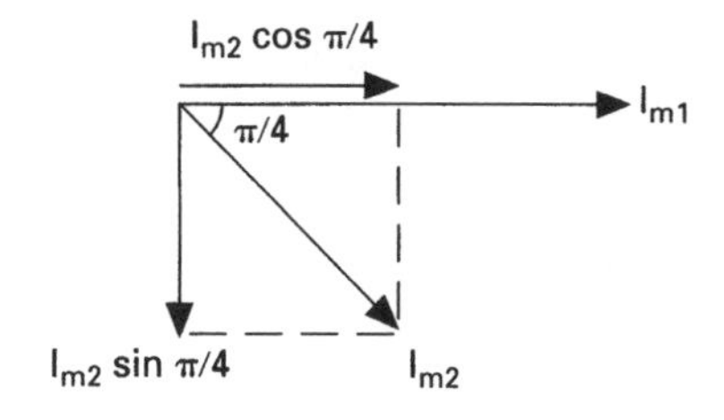

Fig. 6.23

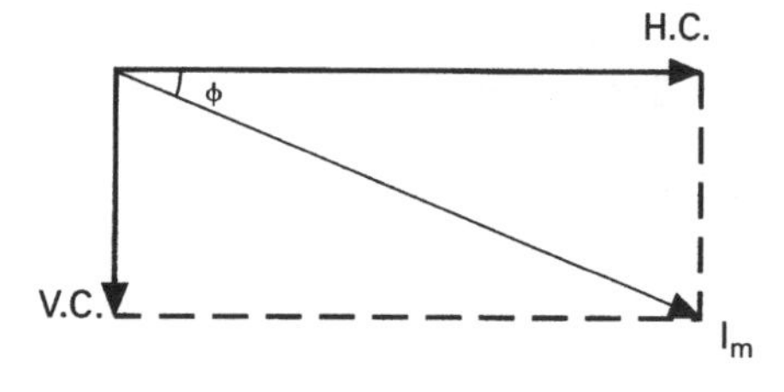

Fig. 6.24

phasors are involved. This method is therefore recommended only for a rapid estimate of the answer.

The use of the resolution of phasors is, with practice, a rapid technique, and yields a high degree of accuracy. Unless specified otherwise it is the technique you should use. Although, at first acquaintance, it may seem to be rather a complicated method, this is not the case. With a little practice, the technique will be found to be relatively simple and quick. Two worked examples now follow.

Worked Example 6.10

Question

Determine the phasor sum of the two voltages specified below.

$v_1 = 25 \sin(314t + \pi/3)$, and $v_2 = 15 \sin(314t - \pi/6)$ volt

Answer

Figure 6.25 shows the sketch of the phasor diagram.

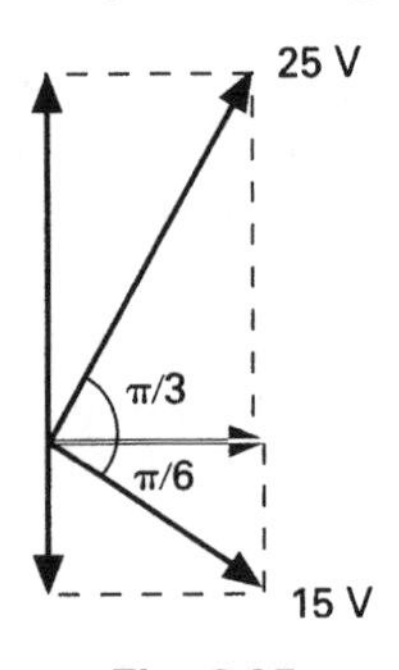

Fig. 6.25

Note: Always sketch a phasor diagram.

$$\text{H.C.} = 25 \cos \pi/3 + 15 \cos[-\pi/6]$$

$$= (25 \times 0.5) + (15 \times 0.866)$$

$$= 12.5 + 12.99$$

$$\text{so, H.C.} = 25.49\,\text{V}$$

$$\text{V.C.} = 25 \sin \pi/3 + 15 \sin[-\pi/6]$$

$$= (25 \times 0.866) + (15 \times [-0.5])$$

$$= 21.65 - 7.5$$

$$\text{so V.C.} = 14.15\,\text{V}$$

Figure 6.26 shows the phasor diagram for H.C., V.C. and V_m.

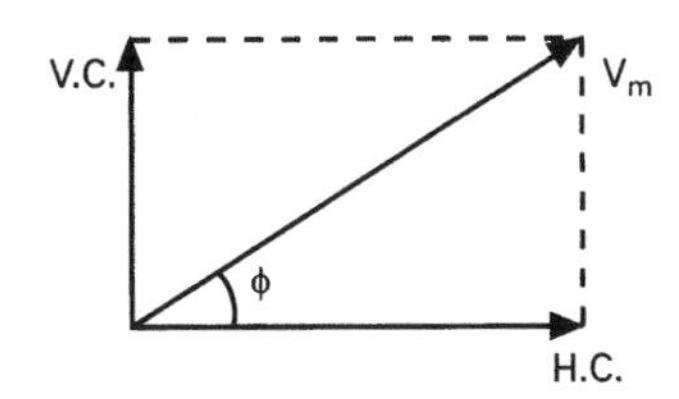

Fig. 6.26

$$V_m = \sqrt{\text{H.C.}^2 + \text{V.C.}^2} = \sqrt{25.49^2 + 14.15^2}$$
$$\text{so, } V_m = 29.15\,\text{V}$$
$$\tan\phi = \frac{\text{V.C.}}{\text{H.C.}} = \frac{14.15}{25.49} = 0.555^*$$
$$\text{so, } \phi = \tan^{-1} 0.555 = 0.507\,\text{rad}$$
$$\text{therefore, } v = 29.15 \sin(314t + 0.507)\,\text{volt } \textbf{Ans}$$

*radian mode required

Worked Example 6.11

Question

Calculate the phasor sum of the three currents listed below.

$i_1 = 6 \sin \omega t$ amp
$i_2 = 8 \sin(\omega t - \pi/2)$ amp
$i_3 = 4 \sin(\omega t + \pi/6)$ amp

Answer

The relevant phasor diagrams are shown in Figs. 6.27 and 6.28.

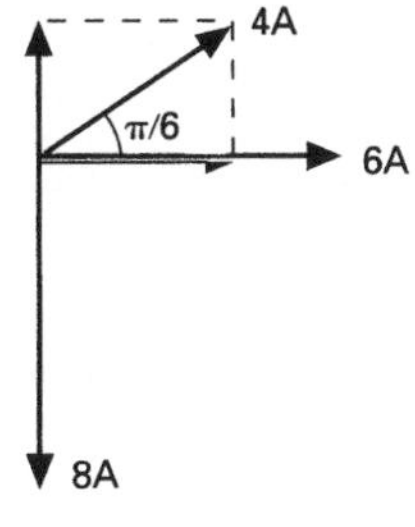

Fig. 6.27

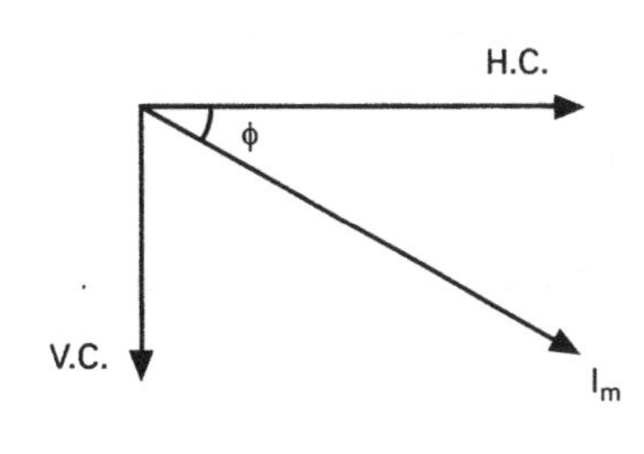

Fig. 6.28

Continued on p. 284

Worked Example 6.11 (*Continued*)

$$\text{H.C.} = 6\cos 0 + 8\cos[-\pi/2] + 4\cos \pi/6$$
$$= (6 \times 1) + (8 \times 0) + (4 \times 0.866)$$
$$= 6 + 3.46$$
$$\text{so, H.C.} = 9.46\ \text{A}$$
$$\text{V.C.} = 6\sin 0 + 8\sin[-\pi/2] + 4\sin \pi/6$$
$$= (6 \times 0) + (8 \times [-1]) + (4 \times 0.5)$$
$$= -8 + 2$$
$$\text{so, V.C.} = -6\ \text{A}$$
$$I_m = \sqrt{\text{H.C.}^2 + \text{V.C.}^2} = \sqrt{9.46^2 + (-6)^2}$$
$$\text{so, } I_m = 11.2\ \text{A}$$
$$\phi = \tan^{-1}\frac{\text{V.C.}}{H.C.} = \tan^{-1}\frac{-6}{9.46} = \tan^{-1} -0.6342$$
$$\text{so, } \phi = -0.565\ \text{rad}$$
therefore, $i = 11.2\sin(\omega t - 0.565)$ amp **Ans**

6.15 The Cathode Ray Oscilloscope

The name of this instrument is more often abbreviated to the oscilloscope, the 'scope, or CRO. It is a very versatile instrument, that may be used to measure both a.c. and d.c. voltages. For d.c. measurements, a voltmeter is usually more convenient to use. The principal advantages of the oscilloscope when used to measure a.c. quantities are:

1 A visual indication of the waveform is produced.
2 The frequency, period and phase angle of the waveform(s) can be determined.
3 It can be used to measure very high frequency waveforms.
4 Any waveshape can be displayed, and measured with equal accuracy.
5 The input resistance (impedance) is of the same order as a digital voltmeter. It therefore applies minimal loading effect to a circuit to which it is connected.
6 Some oscilloscopes can display two or more waveforms simultaneously.

Cathode Ray Tube The basic arrangement of a crt is shown in Fig. 6.29. The main components are contained within an evacuated glass tube. These components are: the electron gun; a focusing system; a beam deflection system; and a screen. Each of these will be very briefly described.

Electron gun assembly This component produces a beam of electrons. This beam can then be accelerated, down the axis of the tube, by a series of high potential anodes.

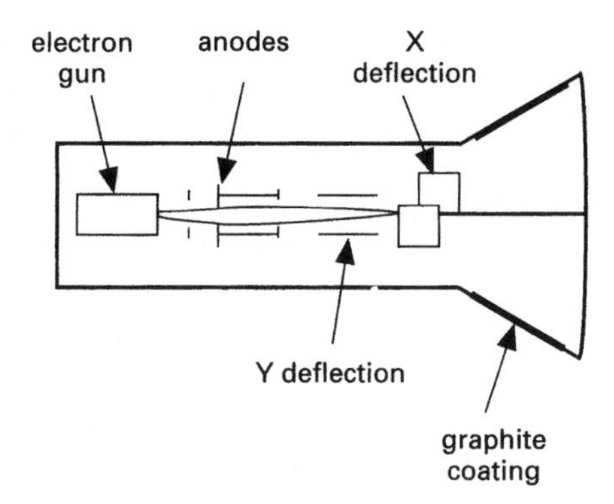

Fig. 6.29

Focusing system The beam consists entirely of electrons. Since they are all negatively charged, then they will tend to repel each other. The beam will therefore tend to spread out, and this would result in a very fuzzy display. The focusing system consists of a series of high potential anodes. These also provide the acceleration for the electron beam. Each successive anode along the tube, towards the screen, is at a higher potential than the previous one. The electric fields, between these anodes, will be of the same shape as a double convex optical lens. This is referred to as an electron lens, and causes the beam to converge to a small spot by the time it reaches the screen.

Deflection system Two sets of parallel plates are situated after the last anode. One set is mounted in the horizontal, and the other set in the vertical plane. These are the X-plates and the Y-plates. When a p.d. is developed between a pair of plates, an electric field will exist between them. This electric field will cause the electrons in the beam to be deflected, towards the more positive of the two plates. Thus, the beam can be made to deflect in both planes. This effect is illustrated in Fig. 6.30.

The screen The inner surface of the screen is coated with a phosphor. Wherever the electron beam strikes the phosphor, it will glow very briefly. This is because the kinetic energy of the bombarding electrons is converted into 'light' energy. On the inside of the 'bell' shape of the tube is a graphite coating. This provides a conducting path to return the electrons to the internal power supply, and hence complete the circuit back to the electron gun.

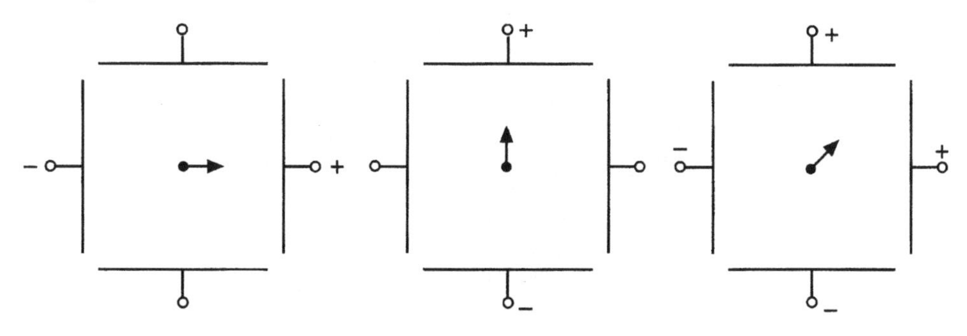

Fig. 6.30

In order for a waveform to be displayed on the screen, the beam must be swept at a constant speed across the screen. At the same time, the beam has to be deflected up and down. You can demonstrate this for yourself, as follows. Take a pencil and a sheet of paper. At the left-hand edge of the paper, move the pencil up and down, at as constant a rate as possible. Maintaining this up-and-down rhythm, now move the pencil across the page, again at as constant a rate as possible. The pattern traced on to the paper should now resemble a sinewave.

The electron beam in the crt is subjected to similar forces, exerted by the deflecting plates, when displaying a sinusoidal waveform. The X-plates cause the beam to be swept across the screen at a constant rate. The Y-plates cause the beam to deflect up and down, in sympathy with the voltage being displayed.

6.16 Operation of the Oscilloscope

In addition to the crt, the other main components in the oscilloscope enable the user to adjust the display, by means of controls on the front panel. A simplified block diagram of the oscilloscope is shown in Fig. 6.31.

The Power Supply This provides the high potentials required for the anodes. It also provides the d.c. supplies for the amplifiers, the electron gun, the timebase and trigger pulse generators.

The Focus Control This control allows the potentials applied to the anodes to be varied. This allows the shape of the electron lens to be altered, and hence achieve a sharp clear trace on the screen.

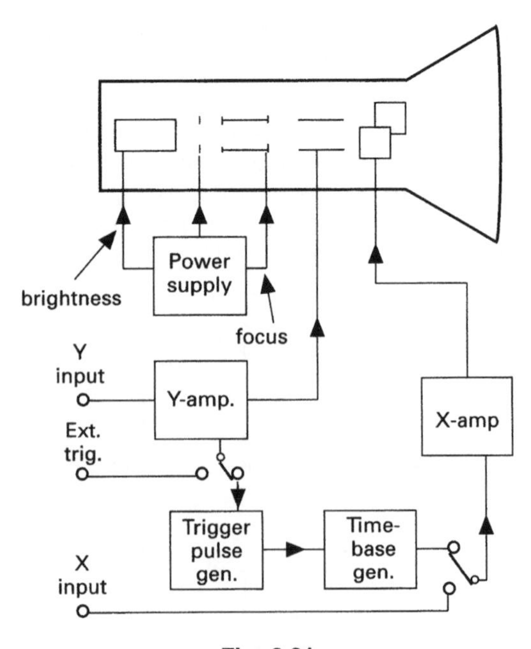

Fig. 6.31

The Brightness Control This varies the potentials applied to the electron gun. The number of electrons forming the beam are thus controlled, which determines the brightness of the display.

The Timebase Control This controls the speed at which the beam is swept across the screen, from left to right. It does this by setting the 'sweep' time of the Timebase generator.

This generator produces a sawtooth voltage, as illustrated in Fig. 6.32. This waveform is applied to the X-plates. During the sweep time, the beam is steadily deflected across the screen. During the 'flyback' time, the beam is rapidly returned to the left-hand side of the screen, ready for the next sweep. The steepness of the sweep section of this waveform determines the speed of the sweep.

At the front of the screen is a graticule, marked out in centimetre squares. The timebase control is marked in units of time/cm. Thus, if this control is set to (say) 10 ms/cm, then each centimetre graduation across the graticule represents a time interval of 10 ms. This facility enables the measurement of the periodic time (and hence frequency), of the displayed waveform.

Trigger Control This enables the user to obtain a single stationary image of the trace on the screen. If this control is incorrectly set, then the trace will scroll continuously across the screen. Alternatively, multiple overlapping traces are displayed, which may also be scrolling. In either case, measurement of the periodic time is impossible. This control determines the point in time at which the sweep cycle of the sawtooth waveform commences. If required an external trigger input may be used.

X-Amplifier This amplifies the sawtooth waveform. This ensures that the voltage applied to the X-plates is sufficiently large to deflect the beam across the full width of the screen. There is also provision for the application of an external timebase signal.

Y-Amplifier The waveform to be displayed is applied to this amplifier. Thus, small amplitude signals can be amplified to give a convenient height of the trace. The gain or amplification of this amplifier is determined by the user. The control on the front panel is marked in units of volt/cm. Thus, if this control was set to (say) 100 mV/cm, then each vertical graduation on the screen graticule represents a voltage of 100 mV. This enables the measurement of the amplitude of the displayed waveform.

X and Y Shift Controls These controls enable the trace position on the screen to be adjusted. This makes the measurement of period and amplitude easier.

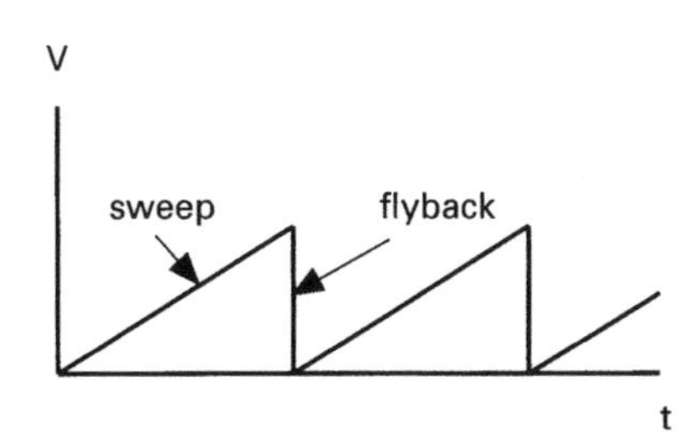

Fig. 6.32

6.17 Dual Beam Oscilloscopes

These instruments are widely used, and are more versatile than the single beam type described. They have the advantage that two waveforms can be displayed simultaneously. This enables waveforms to be compared, in terms of their amplitudes, shape, phase angle or frequency.

The principles of operation are exactly the same as for the single beam instrument. They contain two electron gun assemblies, which have common brightness and focusing controls. The timebase generator is also common to both channels. There will be two separate Y-amplifiers, each controlling its own set of vertical deflection plates. The inputs to these two amplifiers are usually marked as channel 1 and channel 2, or as channels A and B.

Worked Example 6.12

Question

The traces obtained on a double beam oscilloscope are shown in Fig. 6.33. The graticule is marked in 1 cm squares. The channel 1 input is displayed by the upper trace. If settings of the controls for the two channels are as follows, determine the amplitude, r.m.s. value, and frequency of each input.
Channel 1: timebase of 0.1 ms/cm; Y-amp setting of 5 V/cm
Channel 2: timebase of 10 μs/cm; Y-amp setting of 0.5 V/cm

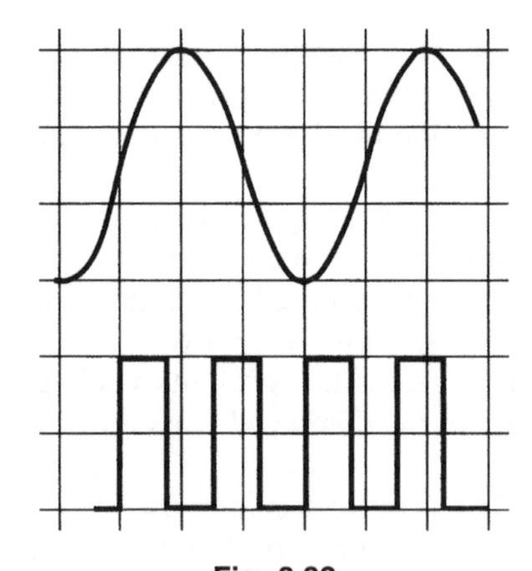

Fig. 6.33

Answer

Channel 1: peak to peak occupies 3 cm, so

$$V_{p-p} = 3 \times 5 = 15\,\text{V}$$

$$\text{the amplitude} = \frac{V_{p-p}}{2} - \frac{15}{2}$$

therefore, $V_m = 7.5\,\text{V}$ **Ans**

As the waveform is a sinewave, then r.m.s. value $V = V_m/\sqrt{2}$

therefore, $V = \dfrac{7.5}{\sqrt{2}} = 5.3\,\text{V}$ **Ans**

1 cycle occupies 4 cm, so $T = 4 \times 0.1 = 0.4\,\text{ms}$

$$f = \frac{1}{T}\,\text{Hz} = \frac{1}{0.4 \times 10^{-3}}$$

so, $f = 2.5\,\text{kHz}$ **Ans**

Channel 2: peak to peak occupies 2 cm, so

$V_{p-p} = 2 \times 0.5 = 1\,\text{V}$, and $V_m = 0.5\,\text{V}$ **Ans**

Since it is a squarewave, then r.m.s. value = amplitude,

hence $V = 0.5\,\text{V}$ **Ans**

2 cycles occur in 3 cm, so 1 cycle occurs in 2/3 cm

therefore, $T = 0.6667 \times 10 = 6.667\,\mu\text{s}$

$$f = \frac{1}{T}\,\text{Hz} = \frac{1}{6.667 \times 10^{-6}}$$

so, $f = 150\,\text{kHz}$ **Ans**

Assignment Questions

1 A coil is rotated between a pair of poles. Calculate the frequency of the generated emf if the rotational speed is (a) 150 rev/s, (b) 900 rev/minute, (c) 200 rad/s.

2 An alternator has 8 poles. If the motor winding is rotated at 1500 rev/minute, determine (a) the frequency of the generated emf, and (b) the speed of rotation required to produce frequency of 50 Hz.

3 A frequency of 240 Hz is to be generated by a coil, rotating at 1200 rev/min. Calculate the number of poles required.

4 A sinewave is shown in Fig. 6.34. Determine its amplitude, periodic time and frequency.

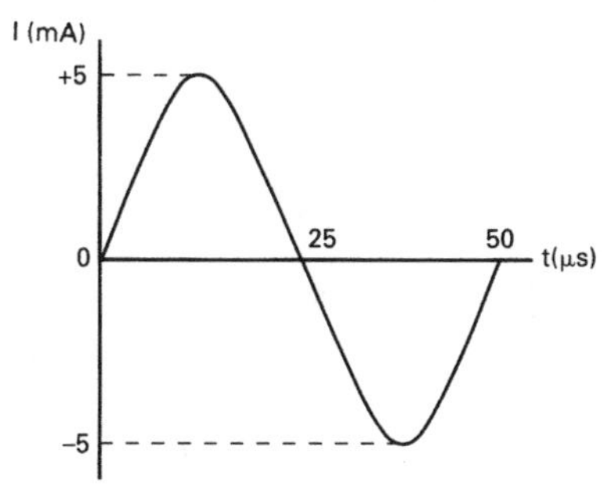

Fig. 6.34

5 A sinusoidal current has a peak-to-peak value of 15 mA and a frequency of 100 Hz. (a) Plot this waveform, to a base of time, and (b) write down the standard expression for the waveform.

6 A sinusoidal voltage is generated by an 85 turn coil, of dimensions 20 cm by 16 cm. The coil is rotated at 3000 rev/min, with its longer sides parallel to the faces of a pair of poles. If the flux density produced by the poles is 0.5 T, calculate (a) the amplitude of the generated emf, (b) the frequency, (c) the r.m.s. and average values.

7 Write down the standard expression for a voltage, of r.m.s. value 45 V, and frequency 1.5 kHz. Hence, calculate the instantaneous value, 38 μs after the waveform passes through its zero value.

8 For each of the following alternating quantities, determine (a) the amplitude and r.m.s. value, and (b) the frequency and period.

(i) $e = 250 \sin 50\pi t$ volt

(ii) $i = 75 \sin 628.3t$ milliamp

(iii) $\phi = 20 \sin 100\pi t$ milliweber

(iv) $v = 6.8 \sin (9424.8t + \phi)$ volt.

9 For a current of r.m.s. value 5 A, and frequency 2 kHz, write down the standard expression. Hence, calculate (a) the instantaneous value 150 μs after it passes through zero, and (b) the time taken for it to reach 4 A, after passing through zero for the first time.

10 Calculate the peak and average values for a 250 V sinusoidal supply.

11 A sinusoidal current has an average value of 3.8 mA. Calculate its r.m.s. and peak values.

12 An alternating voltage has an amplitude of 500 V, and an r.m.s. value of 350 V. Calculate the peak factor.

13 A waveform has a form factor of 1.6, and an average value of 10 V. Calculate its r.m.s. value.

14 A moving coil voltmeter, calibrated for sinewaves, is used to measure a voltage waveform having a form factor of 1.25. Determine the true r.m.s. value of this voltage, if the meter indicates 25 V. Explain why the meter does not indicate the true value.

15 Explain why only sinusoidal waveforms can be represented by phasors.

16 Sketch the phasor diagram for the two waveforms shown in Fig. 6.35.

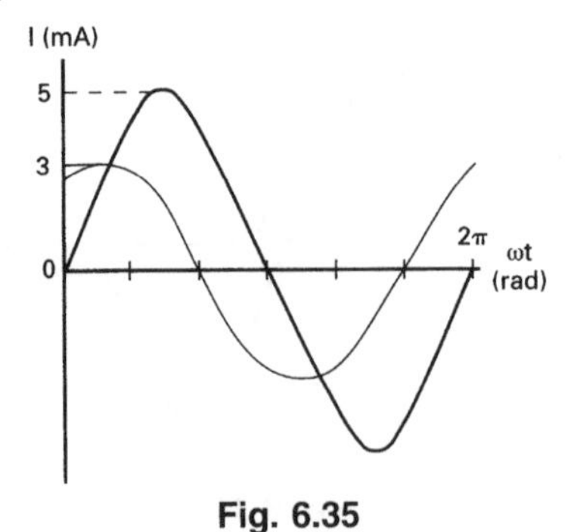

Fig. 6.35

17 Sketch the phasor diagram for the two voltages represented by the following expressions:

$$v_1 = 12 \sin 314t \text{ volt},$$
$$v_2 = 8 \sin (314t + \pi/3) \text{ volt}.$$

18 Determine the phasor sum of the two voltages specified in Question 17 above.

19 Three currents, in an a.c. circuit, meet at a junction. Calculate the phasor sum, if the currents are:

$$i_i = 10 \sin \omega t \text{ amp}$$
$$i_2 = 5 \sin (\omega t + \pi/4) \text{ amp}$$
$$i_3 = 14 \sin (\omega t - \pi/3) \text{ amp}.$$

20 Determine the phasor sum of the following voltages, all of which are sinewaves of the same frequency:

v_1 has an amplitude of 25 volt, and a phase angle of zero.

v_2 has an amplitude of 13.5 volt, and lags v_1 by 25°.

v_3 has an amplitude of 7.6 volt, and leads v_2 by 40°.

21 By means of a phasor diagram, drawn to scale, check your answer to Question 19 above.

22 Plot, on the same axes, the graphs of the following two voltages. By adding ordinates, determine the sum of these voltages. Express the result in the form $v = V_m \sin (\omega t \pm \phi)$

$$v_1 = 12 \sin t, \text{ and } v_2 = 8 \sin (\omega t - \pi/6) \text{ volt}.$$

23 The waveform displayed on an oscilloscope is as shown in Fig. 6.36. The timebase is set to 100 μs/cm, and the Y-amp is set to 2 V/cm. Determine the amplitude, r.m.s. value, periodic time and frequency of this waveform.

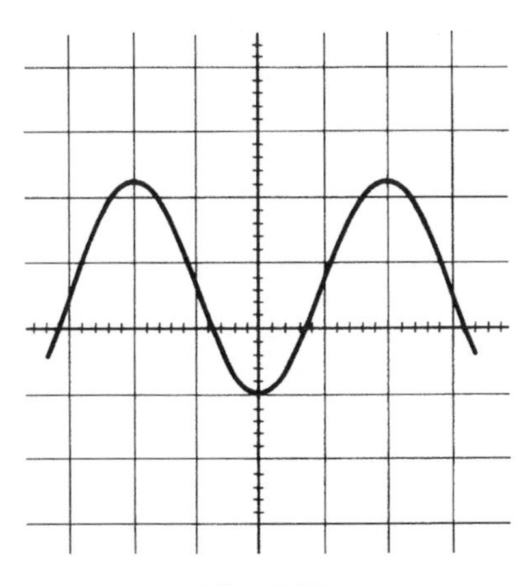

Fig. 6.36

Suggested Practical Assignments

The principal practical exercise relating to this chapter is the usage of the oscilloscope. The actual exercises carried out are left to the discretion of your teacher. Using an oscilloscope is not difficult, but does require some practice; particularly in obtaining a clear, stationary trace, from which measurements can be made.

Supplementary Worked Examples

Example 1: A sinusoidal alternating voltage has an average value of 3.5 V and a period of 6.67 ms. Write down the standard (trigonometrical) expression for this voltage.

Solution: $V_{av} = 3.5\,\text{V};\ T = 6.67 \times 10^{-3}\,\text{s}$

The standard expression is of the form $v = V_m \sin(2\pi ft)$ volt

$$V_{av} = 0.637 V_m \text{ volt}$$

$$\text{so, } V_m = \frac{V_{av}}{0.637} \text{ volt} = \frac{3.5}{0.637}$$

$$V_m = 5.5\,\text{V}$$

$$f = \frac{1}{T} \text{ hertz} = \frac{1}{6.67 \times 10^{-3}} \text{ Hz}$$

$$\text{and, } f = 150\,\text{Hz}$$

$$v = 5.5 \sin(2\pi \times 150 \times t) \text{ volt}$$

so, $v = 5.5 \sin(300\pi t)$ volt **Ans**

Example 2: For the waveform specified in Example 1 above, after the waveform passes through zero, going positive, determine its instantaneous value (a) 0.5 ms later, (b) 4.5 ms later, and (c) the time taken for the voltage to reach 3 V for the first time.

Solution: (a) $t = 0.5 \times 10^{-3}$ s; (b) $t = 4.5 \times 10^{-3}$ s; (c) $v = 3$ V

(a)

$$v = V_m \sin(300\pi \times 0.5 \times 10^{-3}) \text{ volt}$$
$$= 5.5 \sin 0.4712$$
$$= 5.5 \times 0.454$$

thus, $v = 2.5$ V **Ans**

(b)

$$v = 5.5 \sin(300\pi \times 4.5 \times 10^{-3}) \text{ volt}$$
$$= 5.5 \sin 4.241$$
$$= 5.5 \times (-0.891)$$

and, $v = -4.9$ V **Ans**

Note: Remember that the expression inside the brackets is an angle in RADIAN.

(c) $3 = 5.5 \sin(300\pi t)$ volt

$$\text{so, } \sin(300\pi t) = \frac{3}{5.5} = 0.5455$$

$$300\pi t = \sin^{-1} 0.5455 = 0.5769 \text{ rad}$$

$$t = \frac{0.5769}{300\pi} = 6.12 \times 10^{-4}$$

and, $t = 0.612$ ms **Ans**

A sketch graph illustrating these answers is shown in Fig. 6.37.

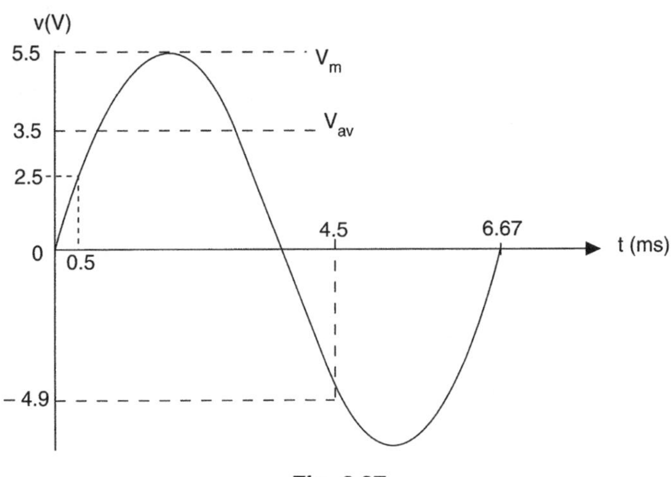

Fig. 6.37

Example 3: Three alternating voltages and one current are as specified in the expressions below.

$\mathbf{v_1 = 10 \sin(628t - \pi/6)}$ **volt**

$\mathbf{v_2 = 8 \sin(628t + \pi/3)}$ **volt**

$\mathbf{v_3 = 12 \sin(628t + \pi/4)}$ **volt**

$\mathbf{i = 6 \sin(628t)}$ **amp**

(a) For each waveform determine the frequency, phase angle and amplitude.
(b) Determine the phasor sum of the three voltages.

Solution:

(a) All four waveforms have the same value of $\omega = 628$ rad/s, so they are all of the same frequency, hence

$$\omega = 2\pi f = 628$$

$$\text{so, } f = \frac{628}{2\pi} \text{ Hz}$$

and, $f = 100$ Hz **Ans**

for v_1, $\phi_1 = -\pi/6\,\text{rad}$ or $-30°$; and $V_m = 10\,\text{V}$ **Ans**

for v_2, $\phi_2 = +\pi/3\,\text{rad}$ or $+60°$; and $V_m = 8\,\text{V}$ **Ans**

for v_3, $\phi_3 = +\pi/4\,\text{rad}$ or $+45°$; and $V_m = 12\,\text{V}$ **Ans**

(b) Firstly the phasor diagram (Fig. 6.38) is sketched, very roughly to scale. In order to do this a reference waveform needs to be selected, and since the current has a zero phase angle, this is chosen as the reference. However, if the current waveform had not been specified, the horizontal axis would still be taken as the reference from which all phase angles are measured. Since v_2 and v_3 have positive phase angles, and phasors rotate anticlockwise, then these two phasors will appear above the reference axis. The voltage v_1, having a negative phase angle will appear below the reference axis. Also shown on the phasor diagram are the horizontal and vertical components of each voltage.

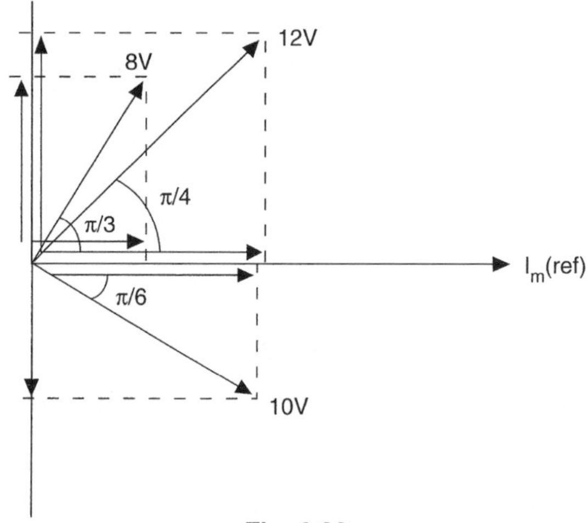

Fig. 6.38

$$\text{H.C.} = 12\cos\pi/4 + 8\cos\pi/3 + 10\cos\pi/6$$

$$= (12 \times 0.707) + (8 \times 0.5) + (10 \times 0.866)$$

$$= 8.48 + 4 + 8.66$$

$$\text{so, H.C.} = 21.44\,\text{V}$$

$$\text{V.C.} = 12\sin\pi/4 + 8\sin\pi/3 - 10\sin\pi/6$$

$$= (12 \times 0.707) + (8 \times 0.866) - (10 \times 0.5)$$

$$= 8.48 + 6.928 - 5$$

$$\text{and V.C.} = 10.412\,\text{V}$$

The phasor diagram for H.C. and V.C., and the resultant phasor sum is Fig 6.39.

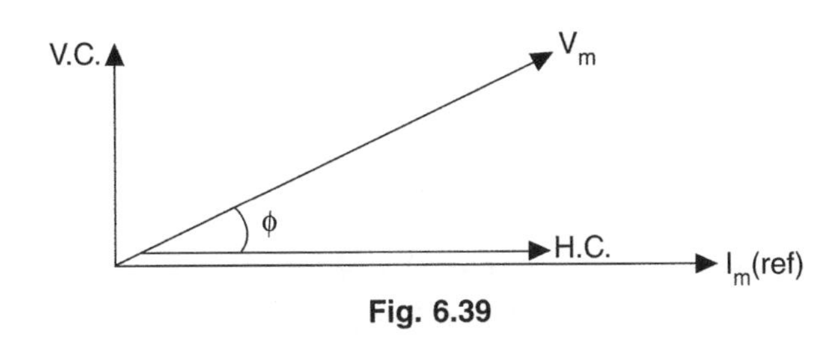

Fig. 6.39

$$V_m = \sqrt{\text{H.C.}^2 + \text{V.C.}^2} = \sqrt{21.44^2 + 10.412^2}$$

so, $V_m = 23.83\,\text{V}$

$$\phi = \tan^{-1}\frac{\text{V.C.}}{\text{H.C.}} = \tan^{-1}\frac{10.412}{21.44} = \tan^{-1} 0.4856$$

and, $\phi = 0.452\,\text{rad}$

Hence, the phasor sum, $v = 23.83 \sin(628t + 0.452)$ volt **Ans**

Example 4: The phasor diagram representing four alternating currents is shown in Fig. 6.40, where the length of each phasor represents the amplitude of that waveform. Write down the standard expression for (a) each waveform, and (b) the phasor sum of the four currents.

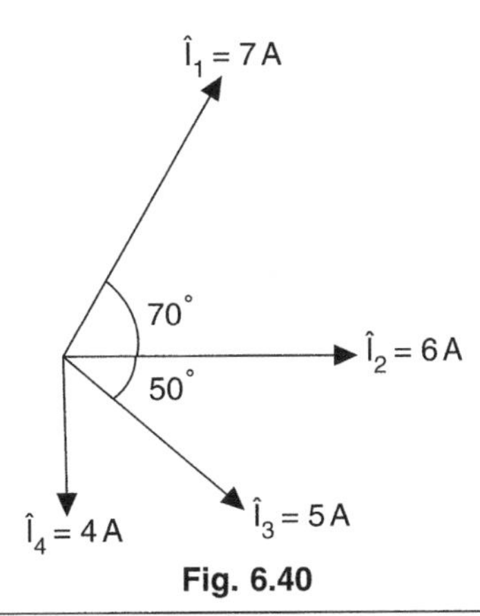

Fig. 6.40

Solution:

(a) $I_{m1} = 7\,\text{A};\ \phi_1 = +70° = \dfrac{70\pi}{180} = 1.22\,\text{rad}$

$I_{m2} = 6\,\text{A};\ \phi_2 = 0° = 0\,\text{rad}$

$I_{m3} = 5\,\text{A};\ \phi_3 = -50° = \dfrac{-50\pi}{180} = -0.873\,\text{rad}$

$I_{m4} = 4\,\text{A};\ \phi_4 = -90° = \dfrac{-90\pi}{180} = -1.571\,\text{rad}$

hence, $i_1 = 7\sin(\omega t + 1.22)$ amp **Ans**

$i_2 = 6\sin\omega t$ amp **Ans**

$i_3 = 5\sin(\omega t - 0.873)$ amp **Ans**

$i_4 = 4\sin(\omega t - 1.571)$ amp or $-4\cos\omega t$ amp **Ans**

(b)
$$\text{H.C.} = 6\cos 0° + 7\cos 70° + 5\cos 50° + 4\cos 90°$$
$$= (6\times 1) + (7\times 0.342) + (5\times 0.643) + (4\times 0)$$
$$= 6 + 2.394 + 3.214 + 0$$

so, H.C. = 11.608 A

$$\text{V.C.} = 6\sin 0° + 7\sin 70° - 5\sin 50° - 4\sin 90°$$
$$= (6\times 0) + (7\times 0.940) - (5\times 0.766) - (4\times 1)$$
$$= 0 + 6.578 - 3.83 - 4$$

and V.C. = −1.252 A

From Fig. 6.41

$$I_m = \sqrt{\text{H.C.}^2 + \text{V.C.}^2}\ \text{amp} = \sqrt{11.608^2 + 1.252^2}$$

so, $I_m = 11.68$ A

$$\phi = \tan^{-1}\frac{\text{V.C.}}{\text{H.C.}} = \tan^{-1}\frac{-1.252}{11.608} = -0.108\ \text{rad}$$

thus, $i = 11.68\sin(\omega t - 0.108)$ amp **Ans**

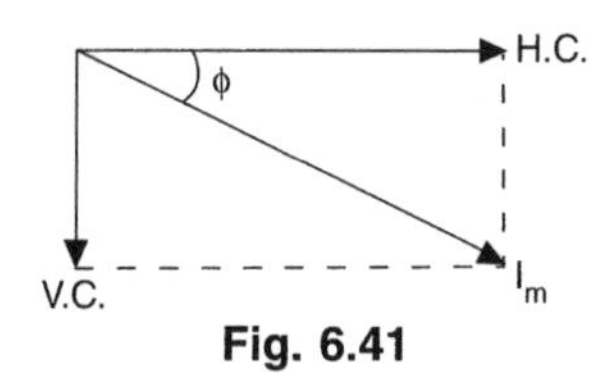

Fig. 6.41

Summary of Equations

Frequency generated: $f = np$ hertz

Periodic time: $T = \frac{1}{f}$ second

Angular velocity: $\omega = 2\pi f$ rad/second

Standard expression for a sinewave: $e = E_m\sin(\theta \pm \phi) = E_m\sin(\omega t \pm \phi)$
$= E_m\sin(2\pi ft \pm \phi)$ volt

Average value for a sinewave: $I_{ave} = \frac{2I_m}{\pi} = 0.637I_m$

R.m.s. value for a sinewave: $I = \frac{I_m}{\sqrt{2}} = 0.707I_m$

Peak factor for a sinewave: $\frac{\text{max. value}}{\text{r.m.s. value}} = 1.414$

Form factor for a sinewave: $\frac{\text{r.m.s.value}}{\text{ave value}} = 1.11$

7 Single Phase a.c. Circuits

This chapter concerns the effect of resistors, inductors and capacitors when connected to an a.c. supply. It also deals with the methods used to analyse simple series a.c. circuits. At the end of the chapter, the concept of series resonance is introduced.

On completion of this chapter you should be able to:

1. Draw the relevant phasor diagrams and waveform diagrams of voltage and current, for pure resistance, inductance and capacitance.
2. Understand and use the concepts of reactance and impedance to analyse simple a.c. series circuits.
3. Derive and use impedance and power triangles.
4. Calculate the power dissipation of an a.c. circuit, and understand the concept of power factor.
5. Explain the effect of series resonance, and its implications for practical circuits.

7.1 Pure Resistance

A pure resistor is one which exhibits only electrical resistance. This means that it has no inductance or capacitance. In practice, a carbon or metal film resistor is virtually perfect in these respects. Large wire-wound resistors can have a certain inductive and capacitive effect.

Consider a perfect (pure) resistor, connected to an a.c. supply, as shown in Fig. 7.1. The current flowing at any instant is directly proportional to the instantaneous applied

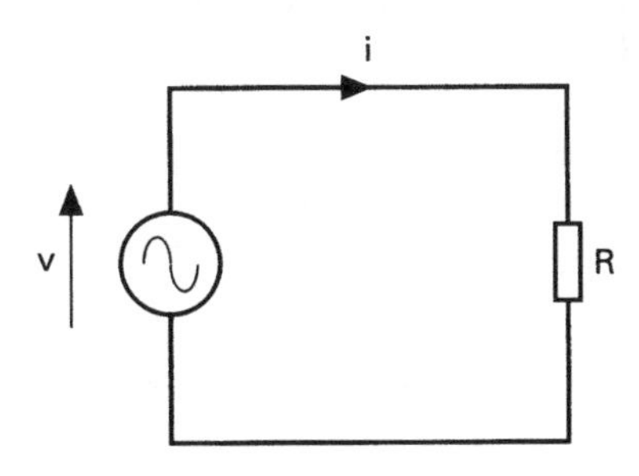

Fig. 7.1

voltage, and inversely proportional to the resistance value. The voltage is varying sinusoidally, and the resistance is a constant value. Thus the current flow will also be sinusoidal, and will be in phase with the applied voltage. This can be written as follows

$$i = \frac{v}{R} \text{ amp}$$

$$\text{but, } v = V_m \sin \omega t \text{ volt}$$

$$\text{therefore, } i = \frac{V_m}{R} \sin \omega t \text{ amp}$$

$$\text{but, } i = I_m \sin \omega t \text{ amp}$$

Thus, the current is a sinewave, of maximum value V_m/R, is of the same frequency as the voltage, and is in phase with it.

$$\text{Hence, } I_m = \frac{V_m}{R} \text{ amp, or } I = \frac{V}{R} \text{ amp} \qquad (7.1)$$

The relevant waveform and phasor diagrams are shown in Figs. 7.2 and 7.3 respectively.

The instantaneous power (p) is given by the product of the instantaneous values of voltage and current. Thus $p = vi$. The waveform diagram is shown in Fig. 7.4. From this diagram, it is obvious that the power reaches its maximum and minimum values at the same time as both voltage and current. Therefore

$$P_m = V_m I_m$$

$$\text{hence, } P = VI = I^2R = \frac{V^2}{R} \text{ watt} \qquad (7.2)$$

Note: When calculating the power, the *r.m.s.* values must be used.

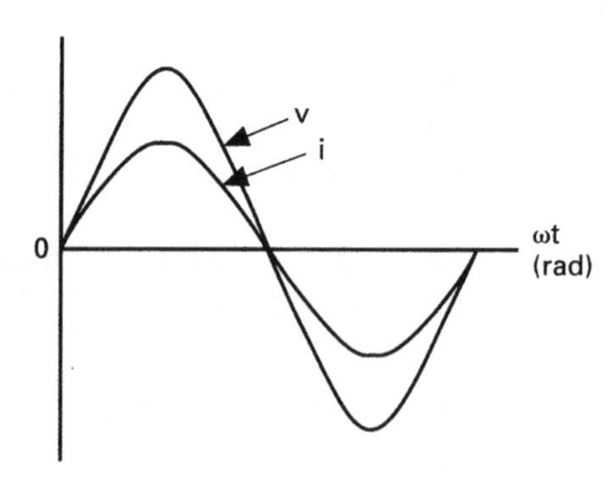

Fig. 7.2

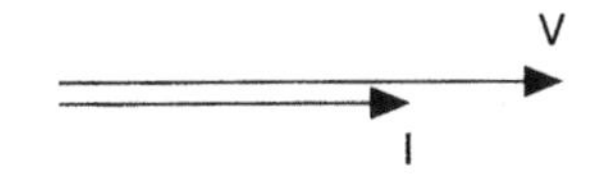

Fig. 7.3

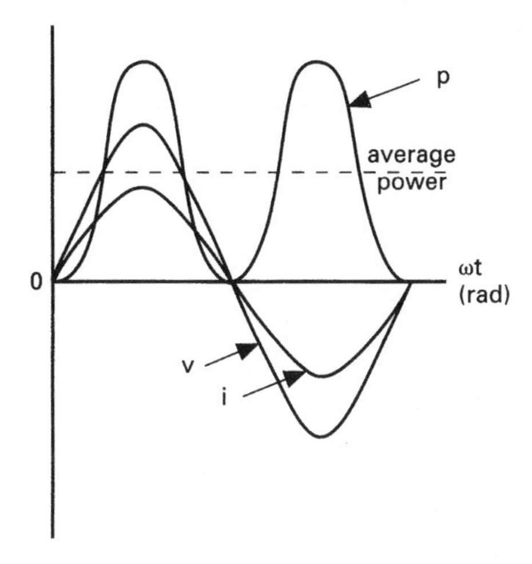

Fig. 7.4

From these results, we can conclude that a pure resistor, in an a.c. circuit, behaves in exactly the same way as in the equivalent d.c. circuit.

Worked Example 7.1

Question

Calculate the power dissipated by a 560 Ω resistor, when connected to a $v = 35 \sin 314t$ volt supply.

Answer

$R = 560\,\Omega$; $V_m = 35$ V

The r.m.s. value for the voltage, $V = 0.707\,V_m$ volt

so, $V = 0.707 \times 35 = 24.75$ V

$$P = \frac{V^2}{R} \text{ watt} = \frac{24.75^2}{560}$$

therefore, $P = 1.09$ W **Ans**

7.2 Pure Inductance

A pure inductor is one which possesses only inductance. It therefore has no electrical resistance or capacitance. Such a device is not practically possible. Since the inductor consists of a coil of wire, then it must possess a finite value of resistance in addition to a very small amount of capacitance. However, let us assume for the moment that an inductor having zero resistance is possible. Consider such a perfect inductor, connected to an a.c. supply, as shown in Fig. 7.5.

An alternating current will now flow through the circuit. Since the current is continuously changing, then a back emf, e will be induced across the inductor. In this case, e will be exactly equal and opposite to the applied voltage, v. The equation for this back emf is

$$e = -L\mathrm{d}i/\mathrm{d}t \text{ volt}$$

e will have its maximum values when the rate of change of current, $\mathrm{d}i/\mathrm{d}t$, is at its maximum values. These maximum rates of change occur as the current waveform passes through the zero position. Similarly, e will be zero when the rate of change is zero. This occurs when the current waveform is at its positive and negative peaks.

Thus e will reach its maximum *negative* value when the current waveform has its maximum *positive* slope. Similarly, e will be at its maximum positive value when i reaches its maximum negative slope. Also, since $v = -e$, then the applied voltage waveform will be the 'mirror image' of the waveform for the back emf. These waveforms are shown in Fig. 7.6. From this waveform diagram, it may be seen that the applied voltage, V *leads* the circuit current, I, by $\pi/2$ rad, or 90°. The corresponding phasor diagram is shown in Fig. 7.7.

Once more, the instantaneous power is given by the product of instantaneous values of voltage and current. In the first quarter cycle, both v and i are positive quantities. The power is therefore in the positive half of the diagram. In the next quarter cycle, i is still positive, but v is negative. The power waveform is therefore negative. This sequence is repeated every half cycle of the waveform. The average power is therefore zero. Thus, *a perfect inductor dissipates zero power*. These waveforms are shown in Fig. 7.8.

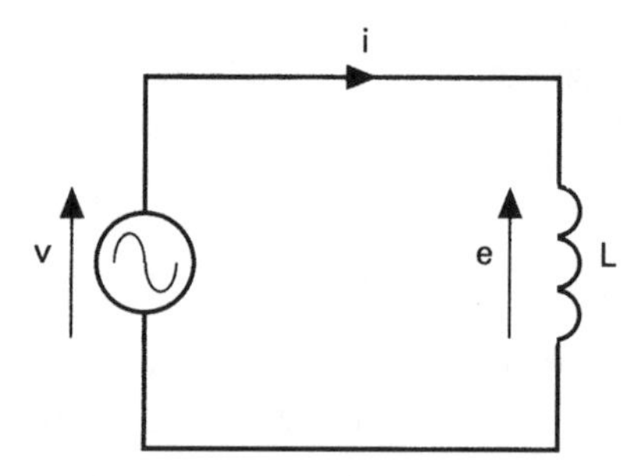

Fig. 7.5

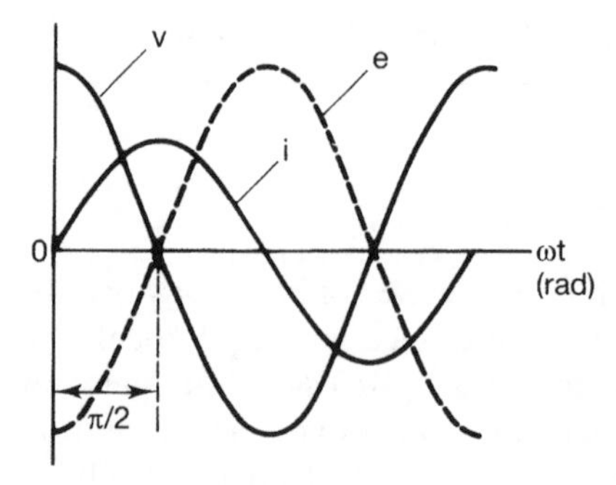

Fig. 7.6

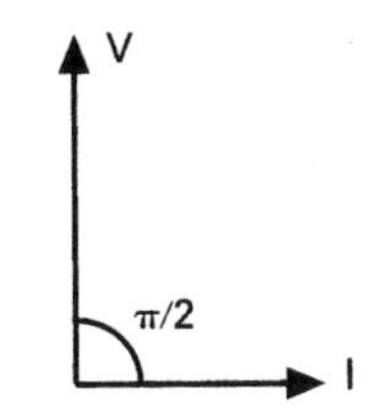

Fig. 7.7

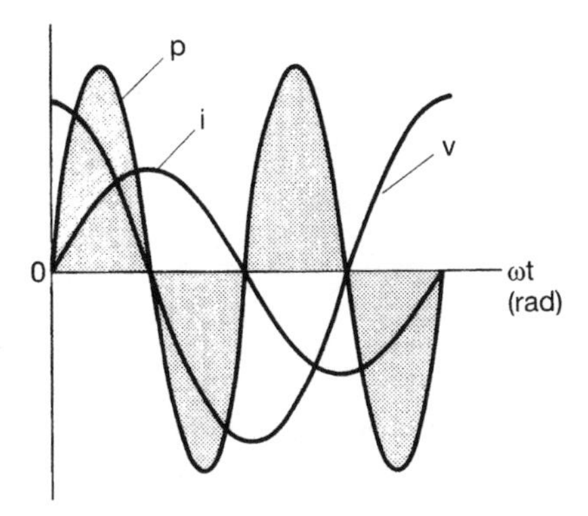

Fig. 7.8

In practical terms, the sequence is as follows. In the first quarter cycle, the magnetic field produced by the coil current, stores energy. In the next quarter cycle, the collapsing field returns all this energy back to the circuit. This sequence is repeated every half cycle. The net result is that the inductor returns as much energy as it receives. Thus no net energy is dissipated, so the power consumption is zero.

7.3 Inductive Reactance

A perfect inductor has no electrical resistance. However, there is some opposition to the flow of a.c. current through it. This opposition is of course due to the back emf induced in the coil. It would be most inconvenient to have to always express this opposition in terms of this emf. It is much better to be able to express the opposition in a quantity that is measured in ohms. This quantity is called the inductive reactance.

Inductive reactance is defined as the opposition offered to the flow of a.c., by a perfect inductor. It is measured in ohms, and the quantity symbol is X_L. The reactance value for an inductor, at a particular frequency, can be determined from a simple equation. This equation may be derived either mathematically or graphically. Both methods will be shown here.

$$e = -L\mathrm{d}i/\mathrm{d}t \text{ volt, and } e = -v$$

therefore, $v = L\mathrm{d}i/\mathrm{d}t$ volt

Now, $i = I_m \sin \omega t$ amp, so, $v = L\dfrac{\mathrm{d}}{\mathrm{d}t}(I_m \sin \omega t)$

therefore, $v = \omega L I_m \cos \omega t$
at time $t = 0$, $v = V_m$; and $\cos \omega t = 1$
hence, $V_m = \omega L I_m$; and dividing by I_m

$$\frac{V_m}{I_m} = \frac{V}{I} = \omega L \text{ ohm}$$

so, inductive reactance is:

$$X_L = \omega L = 2\pi f L \text{ ohm} \quad (7.3)$$

Alternatively, consider the first quarter cycle of the waveform diagram of Fig. 7.9.
The average emf induced in the coil,

$$e_{av} = \frac{-Ldi}{dt} \text{ volt}$$

$$\text{therefore, } e_{av} = \frac{-L(I_m - 0)}{T/4}; \text{ but } T = \frac{1}{f}$$

$$\text{so, } e_{av} = -4LI_m f \text{ volt; therefore } v_{av} = 4LI_m f \text{ volt}$$

$$\text{also, } v_{av} = 0.637V_m = \frac{2V_m}{\pi} \text{ so, } V_m = v_{av} \times \frac{\pi}{2}$$

$$\text{therefore, } V_m = 4LI_m f \times \frac{\pi}{2} = 2\pi f L I_m$$

$$\text{hence, } \frac{V_m}{I_m} = \frac{V}{I} = 2\pi f L \text{ ohm}$$

i.e. inductive reactance, $X_L = 2\pi f L = \omega L$ ohm
From equation (7.3), it can be seen that the inductive reactance is directly proportional to both the inductance value and the frequency of the supply. This is logical, since the greater the frequency, the greater the rate of change of current, and the greater the back emf. Figure 7.10 shows the relationship between X_L and frequency f. Notice that the graph goes through the origin (0, 0), confirming that a *perfect* inductor has zero resistance. That is, a frequency of 0 Hz is *d.c.*, so no opposition would be offered to the flow of d.c.

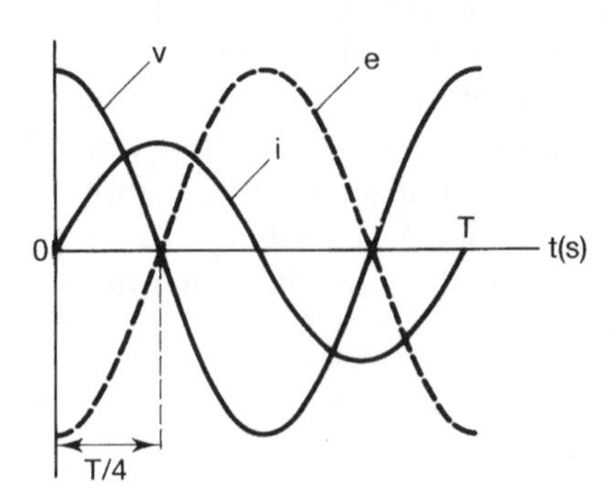

Fig. 7.9

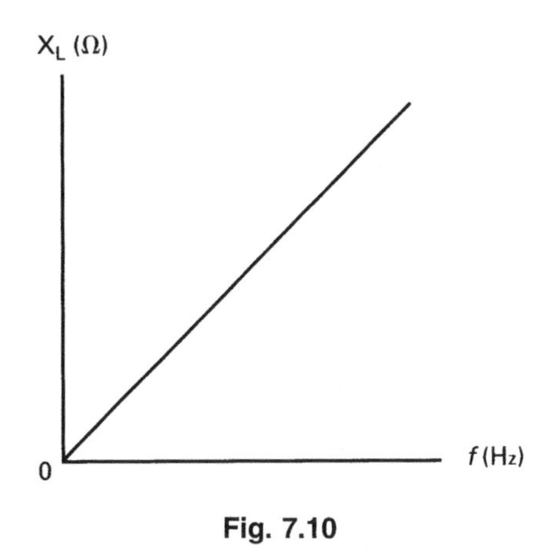

Fig. 7.10

Worked Example 7.2

Question

A pure 20 mH inductor is connected to a 30 V, 50 Hz supply. Calculate (a) the reactance at this frequency, and (b) the resulting current flow.

Answer

$L = 20 \times 10^{-3}$ H; $V = 30$ V; $f = 50$ Hz

(a) $X_L = 2\pi f L$ ohm $= 2 \times \pi \times 50 \times 20 \times 10^{-3}$
so $X_L = 6.283\,\Omega$ **Ans**

(b) $I = \dfrac{V}{X_L}$ amp $= \dfrac{30}{6.283}$
so $I = 4.77$ A **Ans**

Worked Example 7.3

Question

A current of 250 mA flows through a perfect inductor, when it is connected to a 5 V, 1 kHz supply. Determine the inductance value.

Answer

$I = 0.25$ A; $V = 5$ V; $f = 1000$ Hz

Continued on p. 304

Worked Example 7.3 (*Continued*)

Firstly, the inductive reactance must be calculated:

$$X_L = \frac{V}{I} \text{ ohm} = \frac{5}{0.25}$$

therefore, $X_L = 20\,\Omega$

Since $X_L = 2\pi fL$ ohm, then $L = \dfrac{X_L}{2\pi f}$ henry

$$L = \frac{20}{2 \times \pi \times 1000}$$

therefore, $L = 3.18$ mH **Ans**

Worked Example 7.4

Question

A coil of inductance 400 μH, and of negligible resistance, is connected to a 5 kHz supply. If the current flow is 15 mA, determine the supply voltage.

Answer

$L = 400 \times 10^{-6}$ H; $f = 5 \times 10^3$ Hz; $I = 15 \times 10^{-3}$ A

$$X_L = 2\pi fL \text{ ohm} = 2 \times \pi \times 5 \times 10^3 \times 400 \times 10^{-6}$$

so, $X_L = 12.57\,\Omega$

$$V = IX_L \text{ volt} = 15 \times 10^{-3} \times 12.57$$

so, $V = 188.5$ mV **Ans**

7.4 Pure Capacitance

Consider a perfect capacitor, connected to an a.c. supply, as shown in Fig. 7.11. The charge on the capacitor is directly proportional to the p.d. across it. Thus, when the voltage is at its maximum, so too will be the charge, and so on. The waveform for the capacitor charge will therefore be in phase with the voltage. Current is the rate of change of charge. This means that when the rate of change of charge is a maximum, then the current will be at a maximum, and so on. Since the rate of change of charge is maximum as it passes through the zero axis, the current will be at its maximum values at these points. The resulting waveforms are shown in Fig. 7.12. It may therefore be seen that the current now leads the voltage by $\pi/2$ rad, or 90°. To maintain consistency with the inductor previously considered, the circuit current will again be considered as the reference phasor. Thus, we can say that the voltage *lags* the current by $\pi/2$ rad. This is illustrated in Fig. 7.13.

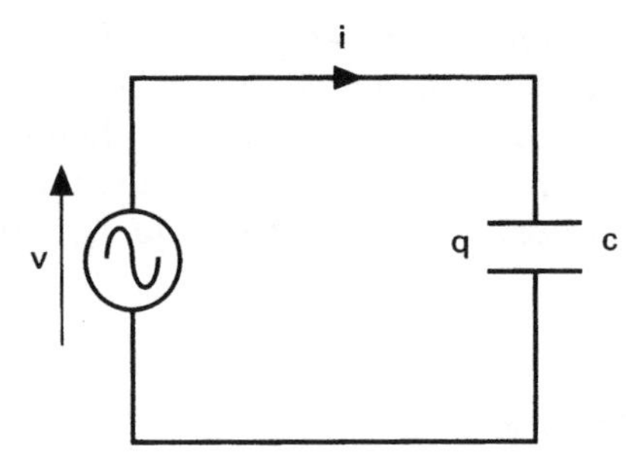

Fig. 7.11

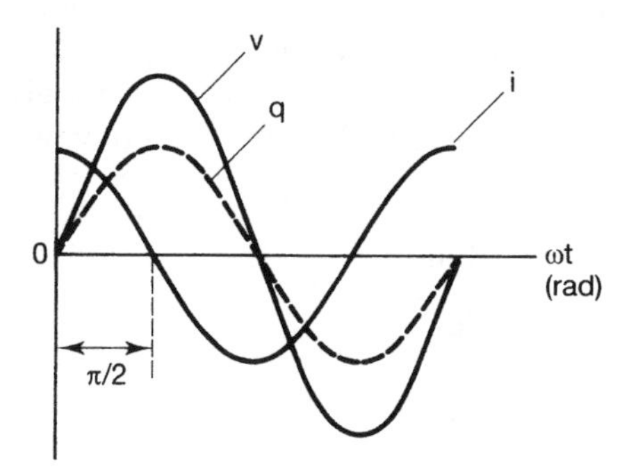

Fig. 7.12

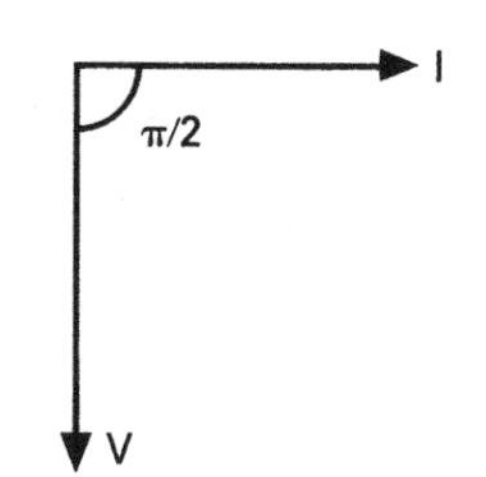

Fig. 7.13

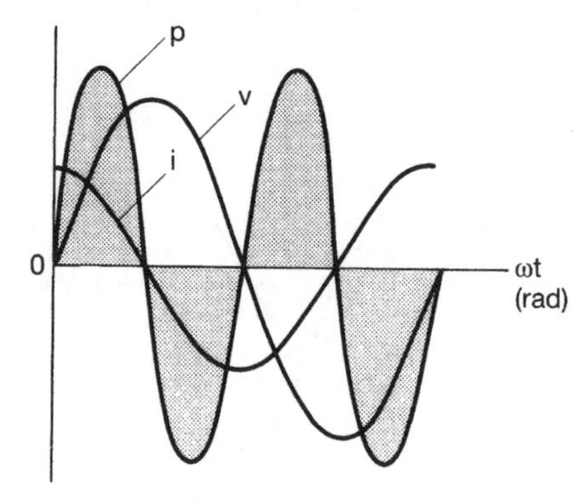

Fig. 7.14

The power at any instant is once more the product of the voltage and current at that instant. When the power waveform is plotted, it will be as shown in Fig. 7.14. Again, it can be seen that the average power is zero. We can therefore conclude that *a perfect capacitor dissipates zero power*. In a similar manner to the inductor, the energy stored in the electric field of the capacitor, in one quarter cycle, is returned to the supply in the next quarter cycle.

7.5 Capacitive Reactance

The opposition offered to the a.c. current is due to the p.d. developed between the capacitor plates. Again, it is more convenient to refer to this opposition in terms of a quantity measured in ohms. This is known as the capacitative reactance, X_C, which is defined as the opposition offered to the flow of a.c. through a perfect capacitor. The equation for this reactance can be derived in one of two ways. Once more, both will be demonstrated here.

$q = vC$ coulomb; and $i = \mathrm{d}q/\mathrm{d}t$ amp

therefore, $i = C\mathrm{d}V/\mathrm{d}t$

and since $v = V_m \sin \omega t$ volt, then

$$i = C\frac{\mathrm{d}}{\mathrm{d}t}(V_m \sin \omega t)$$

$$= \omega C V_m \cos \omega t$$

when time $t = 0$, $I = I_m$; and cos $\omega t = 1$ (see Fig. 7.12)

therefore, $I_m = \omega C V_m$

$$\text{and } \frac{V_m}{I_m} = \frac{V}{I} = \frac{1}{\omega C} \text{ ohm}$$

$$\text{capacitive reactance, } X_c = \frac{1}{\omega C} = \frac{1}{2\pi f C} \text{ ohm} \qquad (7.4)$$

Alternatively, consider the waveforms shown in Fig. 7.15.

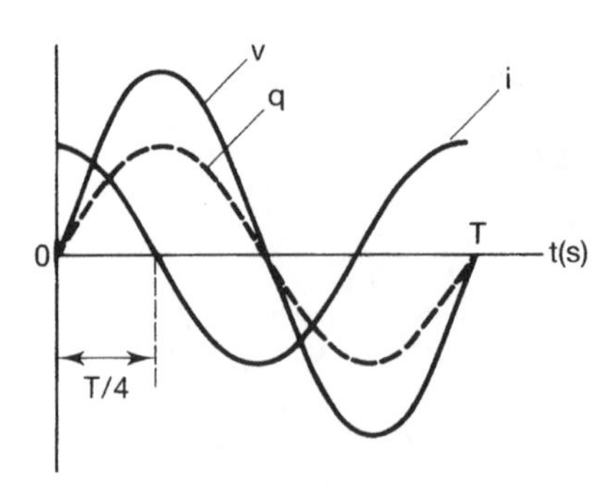

Fig. 7.15

In general:
average charge = average current × time
therefore, in first quarter cycle:

$$q = i_{av} \times T/4 \text{ coulomb}$$

$$\text{so } q = \frac{i_{av}}{4f}$$

$$\text{but } i_{av} = \frac{2I_m}{\pi} \text{ amp}$$

$$\text{therefore, } q = I_m 2\pi f; \text{ and since } q = CV$$

$$\text{then, } CV_m = \frac{I_m}{2\pi f}$$

$$\text{and } \frac{V_m}{I_m} = \frac{V}{I} = \frac{1}{2\pi fC} = \frac{1}{\omega C} \text{ ohm}$$

i.e. capacitive reactance, $$X_C = \frac{1}{2\pi fC} = \frac{1}{\omega C} \text{ ohm} \qquad (7.4)$$

Examination of equation (7.4) shows that the capacitive reactance is *inversely* proportional to both the capacitance and the frequency. Consider the two extremes of frequency. If $f = 0$ Hz (d.c.), then X_c = infinity. This must be correct, since a capacitor does not allow d.c. current to flow through it. At the other extreme, if f = infinity, then $X_c = 0$. The graph of capacitive reactance versus frequency is therefore in the shape of a rectangular hyperbole, as shown in Fig. 7.16.

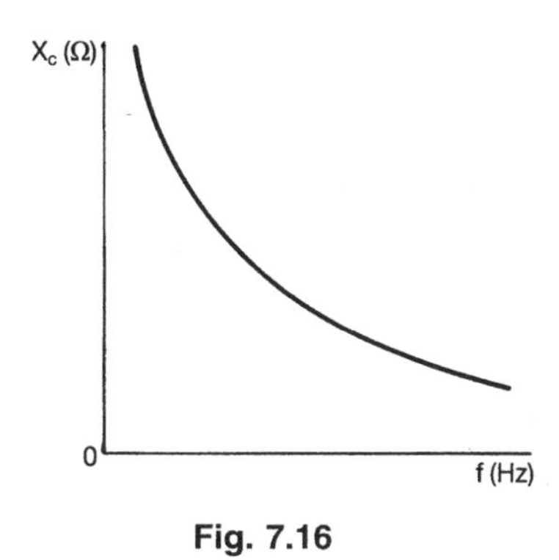

Fig. 7.16

Worked Example 7.5

Question

A 0.47 μF capacitor is connected to a variable frequency signal generator, which provides an output voltage of 25 V. Calculate the current flowing when the frequency is set to (a) 200 Hz, and (b) 4 kHz.

Continued on p. 308

Worked Example 7.5 (*Continued*)

Answer

$C = 0.47 \times 10^{-6}$ F; $V = 25$ V; $f_1 = 200$ Hz; $f_2 = 4000$ HZ

(a)

$$X_{C_1} = \frac{1}{2\pi f_1 C} \text{ ohm} = \frac{1}{2 \times \pi \times 200 \times 0.47 \times 10^{-6}}$$

$$X_{C_1} = 1.693 \text{ k}\Omega$$

$$I_1 = \frac{V}{X_C} \text{ amp} = \frac{25}{1693}$$

therefore, $I_1 = 14.77$ mA **Ans**

(b)

$$X_{C_2} = \frac{1}{2\pi f_2 C} \text{ ohm} = \frac{1}{2 \times \pi \times 4000 \times 0.47^{-6}}$$

$$X_{C_2} = 84.66\,\Omega$$

$$I_2 = \frac{V}{X_{C_2}} \text{ amp} = \frac{25}{84.66}$$

therefore, $I_2 = 295.3$ mA **Ans**

Alternatively, we can say that since

$f_2 = 5 \times f_1$, then $X_{C_2} = X_{C_1}/5 = 84.66\,\Omega$.

Hence, $I_2 = 5 \times I_1 = 295.3$ mA **Ans**

Worked Example 7.6

Question

At what frequency will the reactance of a 22 pF capacitor be 500 Ω?

Answer

$C = 22 \times 10^{-12}$ F; $X_C = 500\,\Omega$

$$X_C = \frac{1}{2\pi fC} \text{ ohm; so } f = \frac{1}{2\pi C X_C} \text{ Hz}$$

$$\text{therefore, } f = \frac{1}{2 \times \pi \times 500 \times 22 \times 10^{-12}}$$

hence, $f = 14.47$ MHz **Ans**

Summary

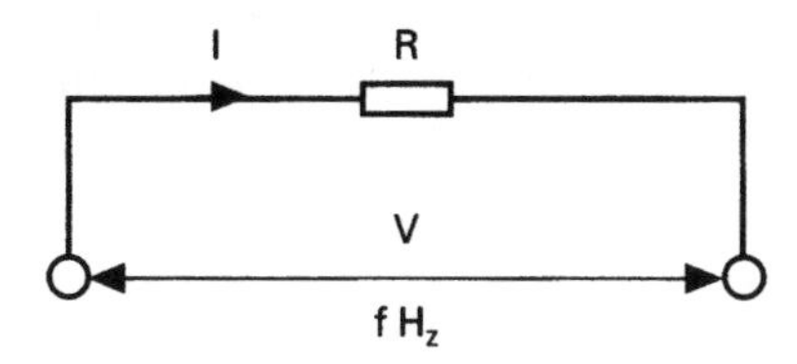

Fig. 7.17

$I = \dfrac{V}{R}$ amp; V in phase with I; $\phi = 0$; $P = I^2R$ watt

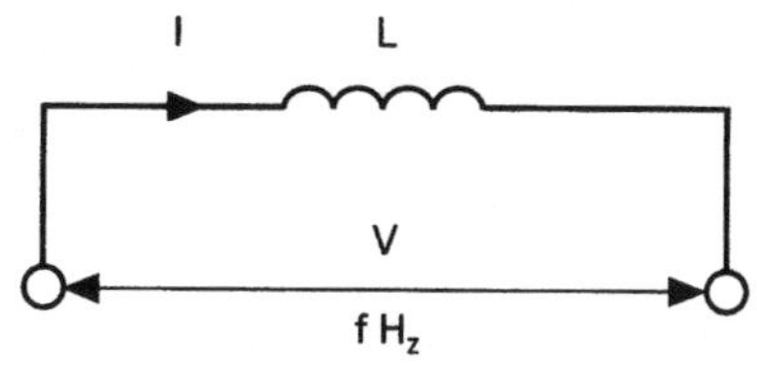

Fig. 7.18

$I = \dfrac{V}{X_L}$ amp; V leads I by 90°; $\phi = +\pi/2$ rad $(+90°)$

$X_L = 2\pi fL$ ohm; $P =$ ZERO watt

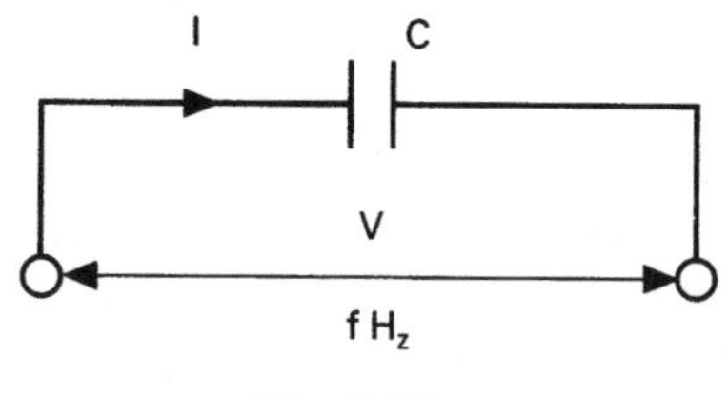

Fig. 7.19

$I = \dfrac{V}{X_C}$ amp; V lags I by 90°; $\phi = -\pi/2$ rad $(-90°)$

$X_C = \dfrac{1}{2\pi fc}$ ohm; $P =$ ZERO watt

C I V I L

In a capacitive (**C**) circuit	**I** before **V** (*I* leads *V*)	**V** before **I** in inductive (*V* leads *I*)(**L**) circuit

The 'keyword' CIVIL is a convenient means of remembering whether voltage either leads or lags the circuit current, depending upon whether the circuit is capacitive or inductive. It is particularly useful when making the transition from series to parallel a.c. circuits. In a series circuit, the current is used as the reference phasor. In parallel circuits, the voltage is used as the reference phasor.

7.6 Impedance

This is the total opposition, offered to the flow of a.c. current, by a circuit that contains both resistance and reactance. It is measured in ohms, and has the quantity symbol Z.

$$\text{Thus, } Z = \frac{V}{I} \text{ ohm} \tag{7.3}$$

where V is the circuit applied voltage, and I is the resulting circuit current.

In order to describe the concept of impedance, let us consider a circuit comprising a pure resistor connected in series with a pure inductor. Please note that this arrangement could equally well represent a practical inductor. That is, one which possesses both resistance and inductance. In this latter case, it makes the circuit analysis simpler if we consider the component to consist of the combination of these two separate, pure elements.

7.7 Inductance and Resistance in Series

A pure resistor and a pure inductor are shown connected in series in Fig. 7.20. The circuit current, I,will produce the p.d. V_R across the resistor, due to its resistance, R. Similarly, the p.d. V_L results from the inductor's opposition, the inductive reactance, X_L. Thus, the only circuit quantity that is common, to both the resistor and the inductor, is the circuit current, I. For this reason, the current is chosen as the reference phasor.

The p.d. across the resistor will be in phase with the current through it ($\phi = 0$). The p.d. across the inductor will lead the current by 90° ($\phi = +90°$). The total applied voltage, V, will be the *phasor sum* of V_R and V_L. This last statement may be considered as the 'a.c. version' of Kirchhoff's voltage law. In other words, the term 'phasor sum' has replaced the term 'algebraic sum', as used in d.c. circuits. The resulting phasor diagram is shown in Fig. 7.21. The angle ϕ, shown on this diagram, is the angle between the circuit applied voltage, V, and the circuit current, I. It is therefore known as the circuit phase angle.

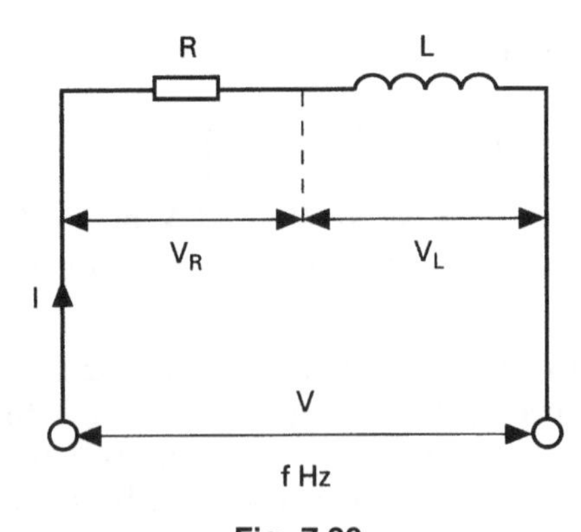

Fig. 7.20

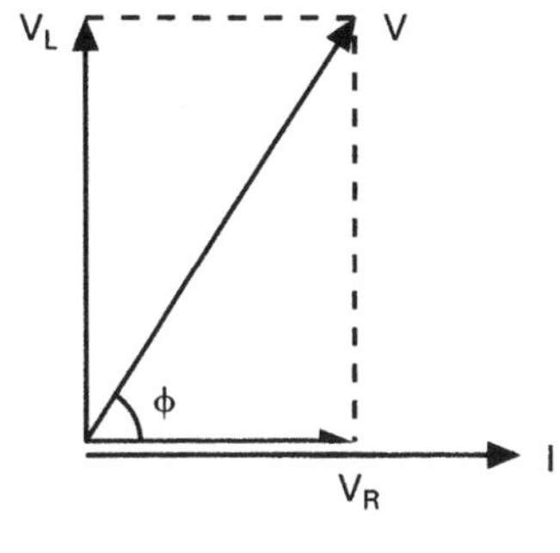

Fig. 7.21

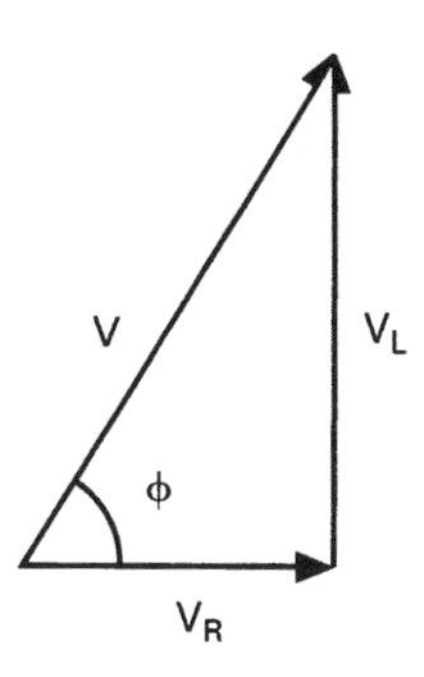

Fig. 7.22

Phasors are vector quantities. They may therefore be drawn anywhere on the page, provided that they are of appropriate length, and are drawn in the appropriate direction. Thus, the voltage triangle, involving V, V_R and V_L, as in Fig. 7.22, may be 'extracted' from the phasor diagram.

Considering Fig. 7.22, Pythagoras' Theorem may be applied, as follows:

$$V^2 = V_R^2 + V_L^2 \dots\dots\dots\dots [1]$$

Now, $V_R = IR$; $V_L = IX_L$; and $V = IZ$ volt

and substituting these into equation [1] we have:

$$(IZ)^2 = (IR)^2 + (IX_L)^2$$

and dividing through by I^2 we have:

$$Z^2 = R^2 + X_L^2$$

$$\text{therefore, } Z = \sqrt{R^2 + X_L^2} \text{ ohm} \qquad (7.6)$$

From the last equation, it may be seen that Z, R and X_L also form a right-angled triangle. This is known as the impedance triangle, and is shown in Fig. 7.23.

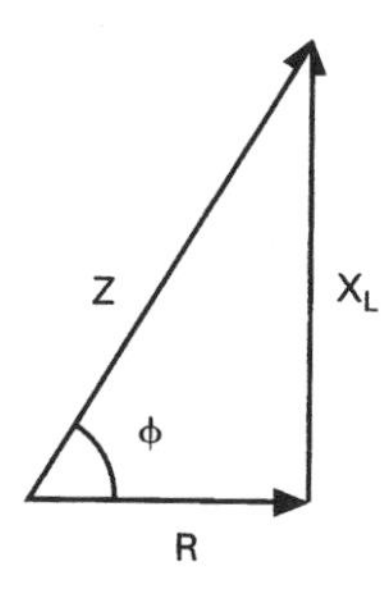

Fig. 7.23

From both the voltage and impedance triangles, the following expressions for the circuit phase angle, ϕ, are obtained:

$$\cos\phi = \frac{R}{Z} = \frac{V_R}{V} \tag{7.7}$$

$$\text{or,}\ \sin\phi = \frac{X_L}{Z} = \frac{V_L}{V}$$

$$\text{or,}\ \tan\phi = \frac{X_L}{R} = \frac{V_L}{V_R}$$

Any one of the above three expressions may be used to obtain the value of ϕ. However, as will be seen later, equation (7.7) is most frequently used.

Worked Example 7.7

Question

A resistor of 250 Ω, is connected in series with a 1.5 H inductor, across a 100 V, 50 Hz supply. Calculate (a) the current flowing, (b) the circuit phase angle, (c) the p.d. developed across each component and (d) the power dissipated.

Answer

$R = 250\,\Omega$; $L = 1.5\,\text{H}$; $V = 100\,\text{V}$; $f = 50\,\text{Hz}$

The relevant circuit and phasor diagrams are shown in Figs. 7.24 and 7.25.

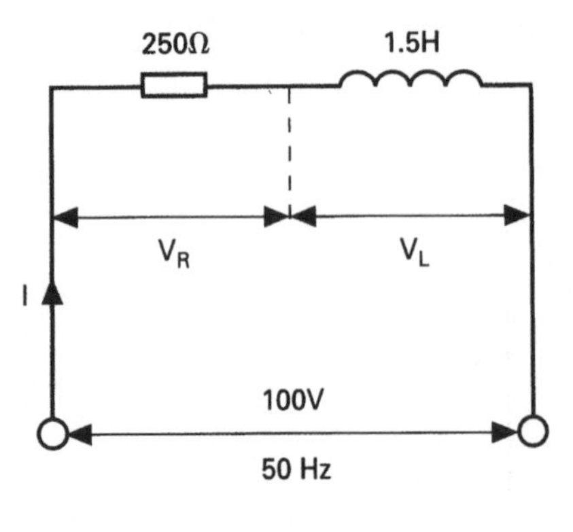

Fig. 7.24

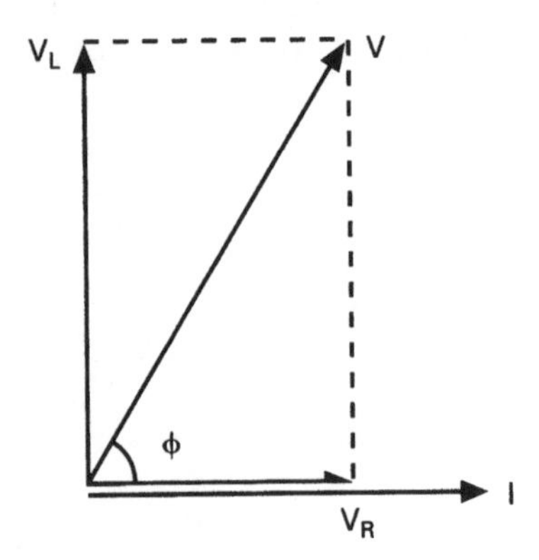

Fig. 7.25

Note: A sketch of these diagrams should normally accompany the written answer.

(a) In order to calculate the current, the impedance must be found. In order to calculate the impedance, the inductive reactance must be known. This then is the starting point in the solution.

$$X_L = 2\pi fL \text{ ohm} = 2 \times \pi \times 50 \times 1.5$$

$$\text{so, } X_L = 471.2\,\Omega$$

$$Z = \sqrt{R^2 + X_L^2} \text{ ohm} = \sqrt{250^2 + 471.2}$$

$$\text{so, } Z = 533.45\,\Omega$$

$$I = \frac{V}{Z} \text{ amp} = \frac{100}{533.45}$$

therefore, $I = 187.5\,\text{mA}$ **Ans**

(b)

$$\phi = \cos^{-1}\frac{R}{Z} = \frac{250}{533.45}$$

therefore, $\phi = +62.05^\circ$ or $+1.083$ rad **Ans**

(c)

$$V_R = IR \text{ volt} = 0.1875 \times 250$$

therefore, $V_R = 46.88\,\text{V}$ **Ans**

$$V_L = IX_L \text{ volt} = 0.1875 \times 471.2$$

therefore, $V_L = 88.35\,\text{V}$ **Ans**

(d)

$$P = I^2R \text{ watt} = 0.1875^2 \times 250$$

therefore $P = 8.789\,\text{W}$ **Ans**

Alternatively, the power could be calculated using the V^2/R form, BUT in this case, the voltage must be the actual p.d. across the *resistor*, i.e. V_R. To avoid the possibility of, mistakenly, using V, or V_L, it is better to use the form of equation used above; so use $P = I^2R$ watt.

Worked Example 7.8

Question

A coil, of resistance 35 Ω and inductance 0.02 H, carries a current of 326.5 mA, when connected to a 400 Hz a.c. supply. Determine the supply voltage.

Answer

$R = 35\,\Omega$; $L = 0.02\,\text{H}$; $I = 0.3265\,\text{A}$; $f = 400\,\text{Hz}$

$$X_L = 2\pi fL \text{ ohm} = 2 \times \pi \times 400 \times 0.02$$

therefore, $X_L = 50.27\,\Omega$

$$Z = \sqrt{R^2 + X_L^2} \text{ ohm} = \sqrt{35^2 + 50.27^2}$$

therefore, $Z = 61.25\,\Omega$

$$V = IZ \text{ volt} = 0.3265 \times 61.25$$

hence, $V = 20\,\text{V}$ **Ans**

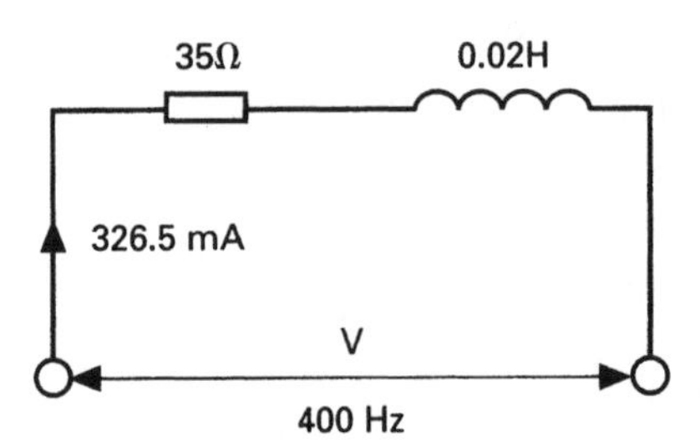

Fig. 7.26

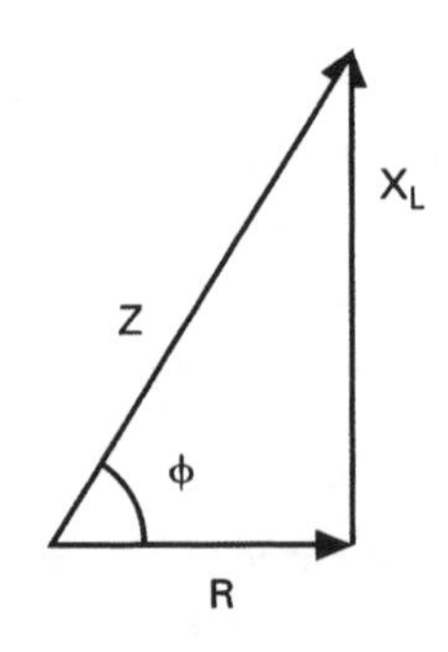

Fig. 7.27

Worked Example 7.9

Question

Two separate tests were carried out on a coil. In the first test, a current of 2 A flowed through it, when connected to a 24 V d.c. supply. In the second test, it was connected to a 24 V, 200 Hz supply. In the latter test, the current flow was 1.074 A. Calculate the resistance and inductance of the coil.

Answer

When the coil is connected to a d.c. supply, the only opposition to current flow is the coil resistance. Thus the result of this test can be used to determine the resistance.

$$R = \frac{V}{I} \text{ ohm} = \frac{24}{2}$$

therefore, $R = 12\,\Omega$ **Ans**

When connected to the a.c. supply, the opposition to current flow is the *impedance* of the coil. Thus:

$$Z = \frac{V}{I} \text{ ohm} = \frac{24}{1.074}$$

therefore, $Z = 22.35\,\Omega$

Since $Z^2 = R^2 + X_L^2$, then $X_L = \sqrt{Z^2 - R^2}$ ohm

therefore, $X_L = \sqrt{22.35^2 - 12^2}$ ohm $= 18.85\,\Omega$

$$L = \frac{X_L}{2\pi f} \text{ henry} = \frac{18.85}{2 \times \pi \times 200}$$

hence, $L = 15$ mH **Ans**

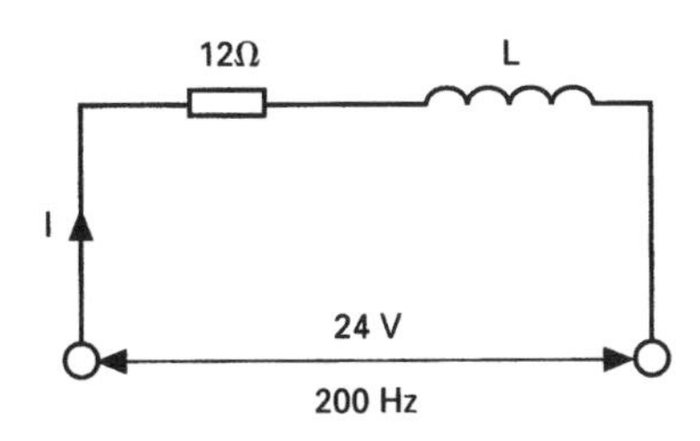

Fig. 7.28

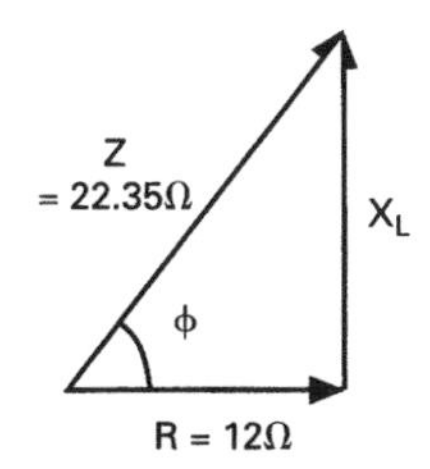

Fig. 7.29

7.8 Resistance and Capacitance in Series

Figure 7.30 shows a pure capacitor and resistor connected in series, across an a.c. supply. Again, being a series circuit, the circuit current is common to both components. Each will have a p.d. developed. In this case however, the p.d. across the capacitor will *lag* the current by 90°.

The resulting phasor diagram is shown in Fig. 7.31.

Exactly the same techniques used previously, for the inductor–resistor circuit, may be used here. Thus, both voltage and impedance triangles can be derived. These are shown in Figs. 7.32 and 7.33.

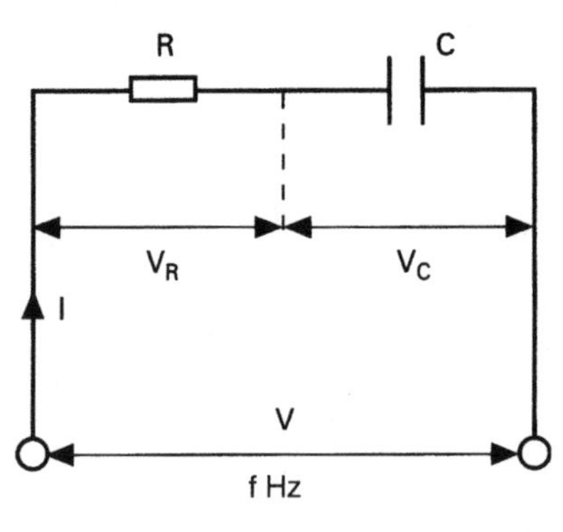

Fig. 7.30

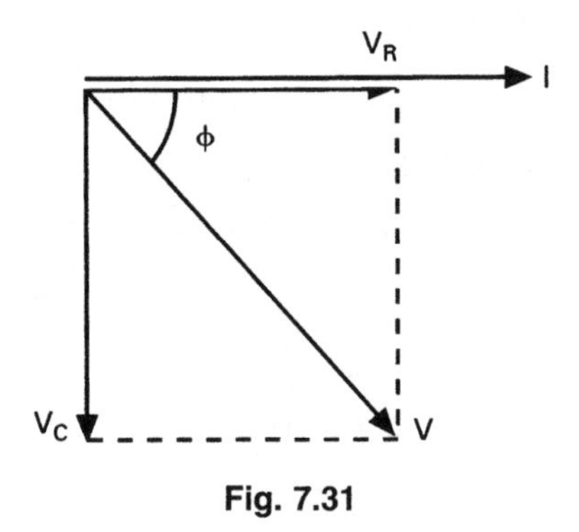

Fig. 7.31

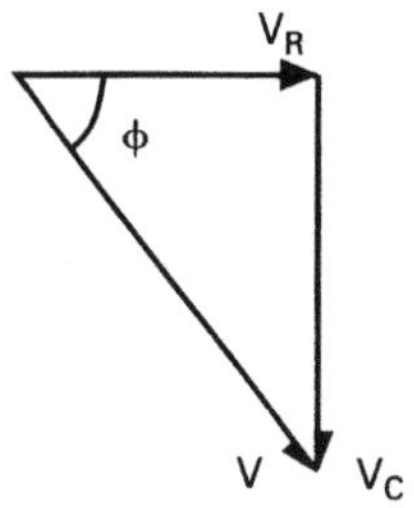

Fig. 7.32

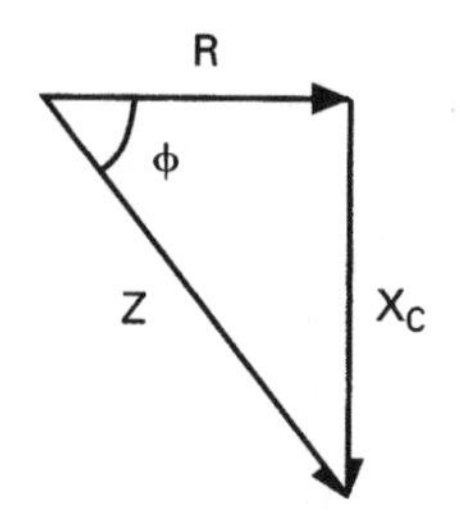

Fig. 7.33

The following equations result:

$$Z = \sqrt{R^2 + X_C^2} \text{ ohm} \tag{7.8}$$

$$\phi = \cos^{-1}\frac{R}{Z} \tag{7.9}$$

$$\sin\phi = \frac{X_C}{Z} = \frac{V_C}{V}$$

$$\text{and } \tan\phi = \frac{X_C}{R} = \frac{V_C}{V_R}$$

Worked Example 7.10

Question

A 22 nF capacitor, and a 3.9 kΩ resistor, are connected in series across a 40 V, 1 kHz supply. Determine, (a) the circuit current, (b) the circuit phase angle and (c) the power dissipated.

Answer

$C = 22 \times 10^{-9}$ F; $R = 3900\,\Omega$; $V = 40$ V; $f = 10^3$ Hz

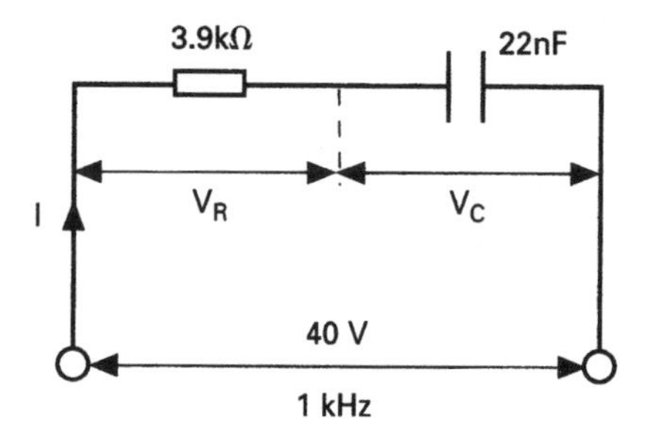

Fig. 7.34

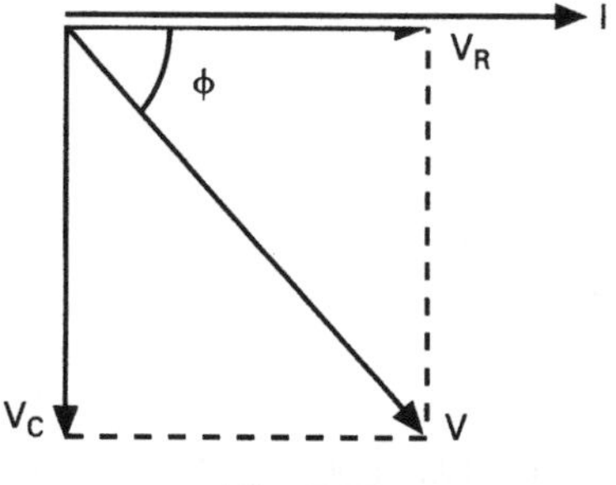

Fig. 7.35

Continued on p. 318

Worked Example 7.10 (*Continued*)

(a)

$$X_C = \frac{1}{2\pi fC} \text{ ohm} = \frac{1}{2 \times \pi \times 10^3 \times 22 \times 10^{-9}}$$

so, $X_C = 7.234\,\text{k}\Omega$

$$Z\sqrt{R^2 + X_C^2} \text{ ohm} = \sqrt{(3.9 \times 10^3)^2 + (7.234 \times 10^3)^2}$$

hence, $Z = 8.218\,\text{k}\Omega$

$$I = \frac{V}{Z} \text{ amp} = \frac{40}{8.218 \times 10^3}$$

therefore, $I = 4.87$ mA **Ans**

(b)

$$\phi = \cos^{-1}\frac{R}{Z} = \frac{3.9 \times 10^3}{8.218 \times 10^3}$$

therefore, $\phi = -61.67°$ or -1.076 rad **Ans**

(c)

$$P = I^2R \text{ watt} = (4.87 \times 10^{-3})^2 \times 3.9 \times 10^3$$

therefore, $P = 92.5$ mW **Ans**

Worked Example 7.11

Question

A 470 Ω resistor is connected in series with a 0.02 μF capacitor, across a 15 V a.c. supply. If the current is 28 mA, calculate (a) the frequency of the supply, and (b) the p.d. across the capacitor.

Answer

$C = 0.02 \times 10^{-6}$ F; $R = 470\,\Omega$; $V = 15$ V; $I = 0.028$ A

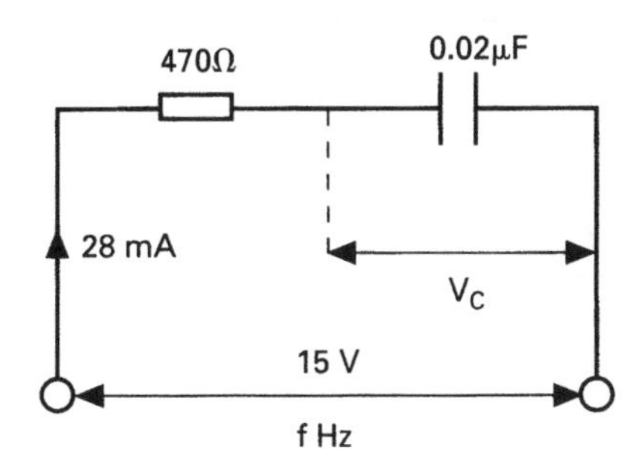

Fig. 7.36

(a) $$Z = \frac{V}{I} \text{ ohm} = \frac{15}{0.028}$$

therefore, $Z = 535.7\,\Omega$

$$\text{Now, } X_C = \sqrt{Z^2 - R^2} \text{ ohm} = \sqrt{535.7^2 - 470^2}$$

hence, $X_C = 257.08\,\Omega$

$$\text{Since } X_C = \frac{1}{2\pi fC} \text{ ohm, then } f = \frac{1}{2\pi CX_c} \text{ Hz}$$

$$\text{therefore, } f = \frac{1}{2 \times \pi \times 257.08 \times 0.02 \times 10^{-6}}$$

hence, $f = 30.95$ kHz **Ans**

(b) $$V_C = IX_C \text{ volt} = 0.028 \times 257.08$$

therefore, $V_C = 7.2$ V **Ans**

7.9 Resistance, Inductance and Capacitance in Series

These three elements, connected in series, are shown in Fig. 7.37. Of the three p.d.s, V_R will be in phase with the current I, V_L will lead I by 90°, and V_C will lag I by 90°. The associated phasor diagram is shown in Fig. 7.38.

The applied voltage V is the phasor sum of the circuit p.d.s. These p.d.s form horizontal and vertical components. It may be seen that the only horizontal voltage component is V_R. The vertical components, V_L and V_C, are in direct opposition to each other. The resulting vertical component is therefore $(V_L - V_C)$. Applying Pythagoras' theorem, we can say that:

$$V^2 = V_R^2 + (V_L - V_C)^2$$

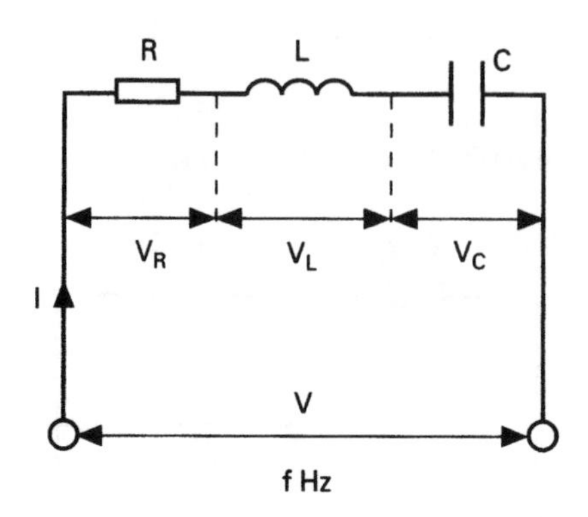

Fig. 7.37

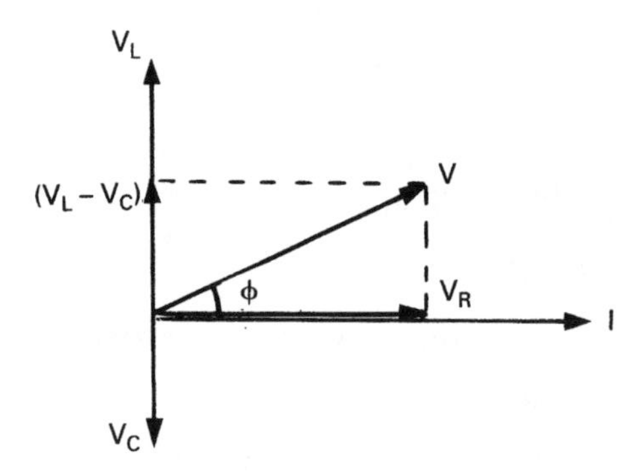

Fig. 7.38

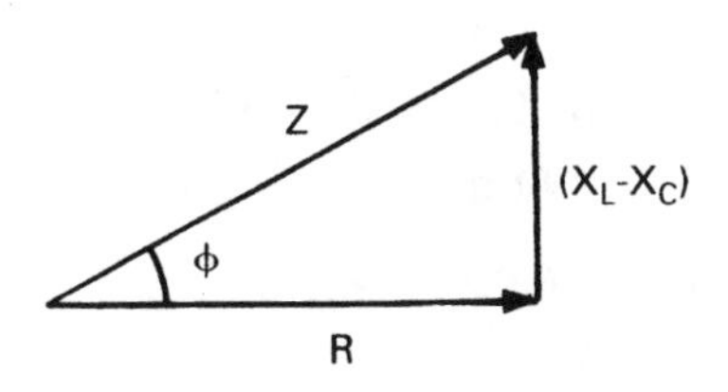

Fig. 7.39

but, $V = IZ$, $V_R = IR$, $V_L = IX_L$ and $V_C = IX_C$ volt

therefore, $(IZ)^2 = (IR)^2 + (IX_L - IX_C)^2$

hence, $Z^2 = R^2 + (X_L - X_C)^2$

$$\text{and, } Z = \sqrt{R^2 + (X_L - X_C)^2} \text{ ohm} \qquad (7.10)$$

The associated impedance triangle is shown in Fig. 7.39. Note that if $X_C > X_L$, then the circuit phase angle ϕ will be lagging, instead of leading as shown.

$$\cos\phi = \frac{R}{Z}; \ \sin\phi = \frac{X_L - X_C}{Z};$$

$$\text{and } \tan\phi = \frac{X_L - X_C}{R}$$

Worked Example 7.12

Question

A coil of resistance 8 Ω and inductance 150 mH, is connected in series with a 100 μF capacitor, across a 240 V, 50 Hz a.c. supply. Calculate (a) the circuit current, (b) the circuit phase angle, (c) the p.d. across the coil, (d) the p.d. across the capacitor, and (e) the power dissipated.

Answer

$R = 8\,\Omega$; $L = 0.15\,\text{H}$; $C = 10^{-4}\,\text{F}$; $V = 240\,\text{V}$; $f = 50\,\text{Hz}$

The circuit diagram is shown in Fig. 7.40. Note that the coil has been considered as the combination of a pure resistor and a pure inductor.

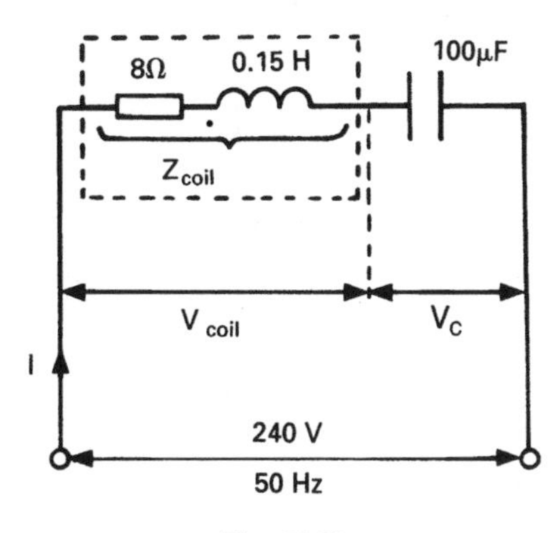

Fig. 7.40

(a)

$$X_L = 2\pi fL \text{ ohm} = 2 \times \pi \times 50 \times 0.15$$

$$\text{so, } X_L = 47.12\,\Omega$$

$$X_C = \frac{1}{2\pi fC} \text{ ohm} = \frac{1}{2 \times \pi \times 50 \times 10^{-4}}$$

$$\text{so } X_C = 31.83\,\Omega$$

$$Z = \sqrt{R^2 + (X_L - X_C)^2} \text{ ohm}$$

$$= \sqrt{18^2 + (47.12 - 31.83)^2}$$

$$\text{therefore } Z = 17.26\,\Omega$$

$$I = \frac{V}{Z} \text{ amp} = \frac{240}{17.26}$$

therefore $I = 13.91$ A **Ans**

Continued on p. 322

Worked Example 7.12 (*Continued*)

(b) $$\phi = \cos^{-1}\frac{R}{Z} = \cos^{-1}\frac{8}{17.26}$$

therefore, $\phi = +62.38°$ or $+1.089$ rad **Ans**

(c) The coil has both inductance and reactance. The coil itself therefore possesses impedance, Z_{coil} ohm. The p.d. across the coil is therefore due to its impedance, and NOT only R or X_L.

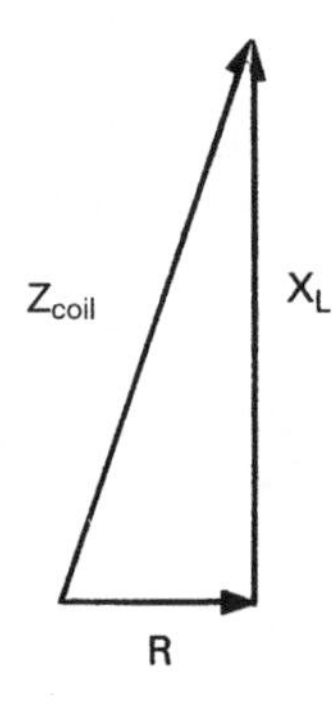

Fig. 7.41

$$Z_{coil} = \sqrt{R^2 + X_L^2} \text{ ohm} = \sqrt{8^2 + 47.12^2}$$

hence, $Z_{coil} = 47.8\,\Omega$

$$V_{coil} = IZ_{coil} \text{ volt} = 13.91 \times 47.8$$

therefore, $V_{coil} = 664.9$ V **Ans**

(d) $$V_C = IXc \text{ volt} = 13.91 \times 31.83$$

hence, $V_C = 442.8$ V **Ans**

(e) $$P = I^2R \text{ watt} = 13.91^2 \times 8$$

therefore, $P = 1.547$ kW **Ans**

Note: From the above calculations, we have the result that the p.d.s, across both the coil and the capacitor, are greater than the applied voltage, V. At first, this may seem to be incorrect, since it would not be possible in a d.c. circuit. In a d.c. circuit, the applied voltage equals the *algebraic* sum of the p.d.s. In the a.c. circuit, the applied voltage is the *phasor* sum of the p.d.s. Thus, it is perfectly possible for the p.d. across the capacitor, or the coil, or both, to be greater than the applied voltage. This is because the total vertical component of voltage is the *difference* between V_L and V_C. This condition is therefore possible only when *all three* components are in series.

7.10 Power in the a.c. Circuit

It has been shown that power is dissipated only by the *resistance* in the circuit. Pure inductance and capacitance do not dissipate any net power. Considering a resistor, we know that the p.d. across it is in phase with the current flowing through it ($\phi = 0°$). For both the inductor and the capacitor, $\phi = 90°$. Thus, we can conclude that only the components of voltage and current, that are in phase with each other, dissipate power. Consider the phasor diagram for an *R-L* series circuit, as shown in Fig. 7.42.

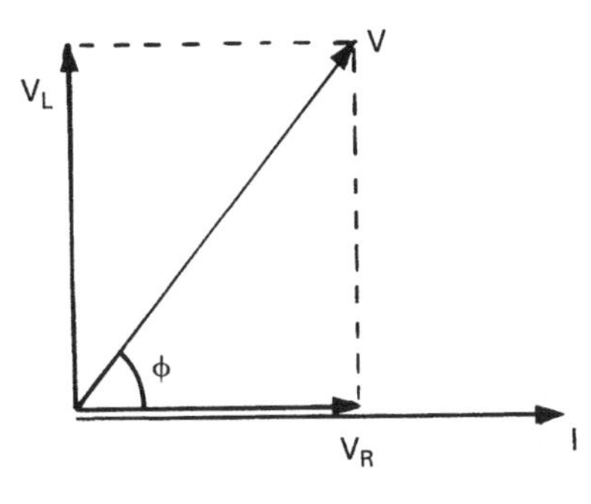

Fig. 7.42

From this diagram, it can be seen that the only two components of current and voltage that are in phase with each other are I and V_R.

Thus, $P = V_R I$ watt

but, $V_R = V\cos\phi$ volt

therefore, $P = VI\cos\phi$ watt (7.11)

7.11 Power Factor

This is defined as the ratio of the true power, to the product of the r.m.s. values of applied voltage and current (apparent power).

$$\text{Thus, power factor (p.f.)} = \frac{VI\cos\phi}{VI}$$

$$\text{hence, p.f.} = \cos\phi \qquad (7.12)$$

To illustrate the concepts of true power and apparent power, consider the *R-L* circuit, shown in Fig. 7.43, where the components are inside a box, and therefore not visible. Let the applied voltage V and the current I be measured, using a voltmeter and ammeter respectively. Taking these two meter readings, it would *appear* that the circuit within the box consumes a power of VI watt.

However, if a wattmeter were connected, this would indicate the true power. This wattmeter reading would be found to be less than the product of the ammeter and voltmeter readings. Since the true and apparent powers are related by the power factor, $\cos\phi$, then they must form two sides of a right-angled triangle.

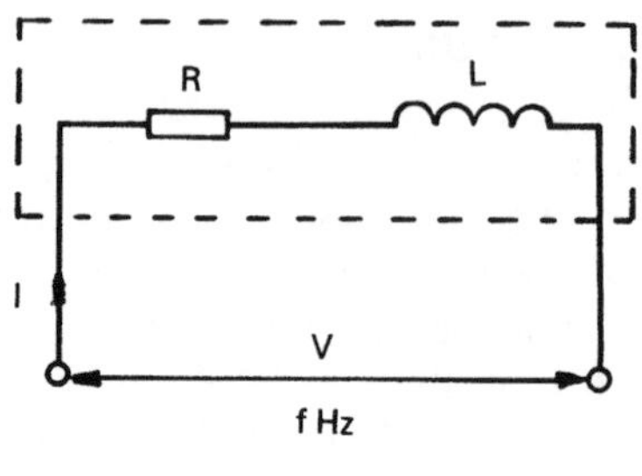

Fig. 7.43

7.12 Power Triangle

The true power is measured in watts (W). The apparent power is measured in volt-ampere (VA). The third side of the triangle is known as the reactive component of power, and is measured in reactive volt-ampere (VAr). The quantity symbols are P, S and Q respectively. These are shown in Fig. 7.44.

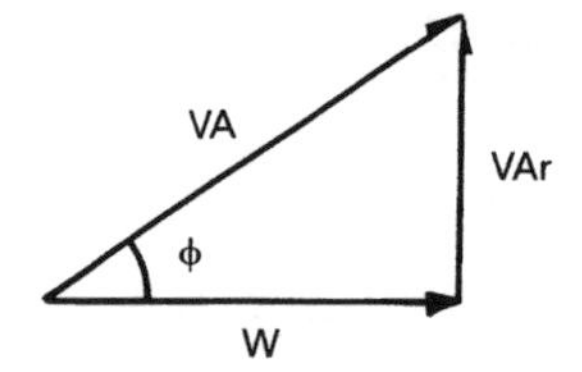

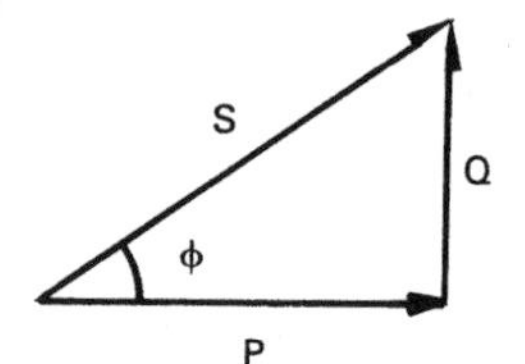

Fig. 7.44

Worked Example 7.13

Question

An R-C circuit consisting of a 4.7 μF capacitor in series with a 200 Ω resistor, is connected to a 250 V, 50 Hz supply. Determine, (a) the current, (b) the power factor, and (c) the values for true, apparent and reactive powers.

Answer

$C = 4.7 \times 10^{-6}$ F; $R = 200\,\Omega$; $V = 250$ V; $f = 50$ Hz

(a) $$X_C = \frac{1}{2\pi fC} \text{ ohm} = \frac{1}{2 \times \pi \times 50 \times 4.7 \times 10^{-6}}$$

so, $X_C = 677.26\,\Omega$

$$Z = \sqrt{R^2 + X_C^2} \text{ ohm} = \sqrt{200^2 + 677.26^2}$$

therefore, $Z = 706.2\,\Omega$

$$I = \frac{V}{Z} \text{ amp} = \frac{250}{706.2}$$

hence $I = 0.354$ A **Ans**

(b) $$\text{p.f.} = \cos\phi = \frac{R}{Z} = \frac{200}{706.2}$$

therefore, p.f. = 0.283 lagging **Ans**

(c) $P = VI\cos\phi$ watt $= 250 \times 0.354 \times 0.283$

therefore, $P = 25.06\,\text{W}$ **Ans**

From the power triangle (Fig. 7.45)

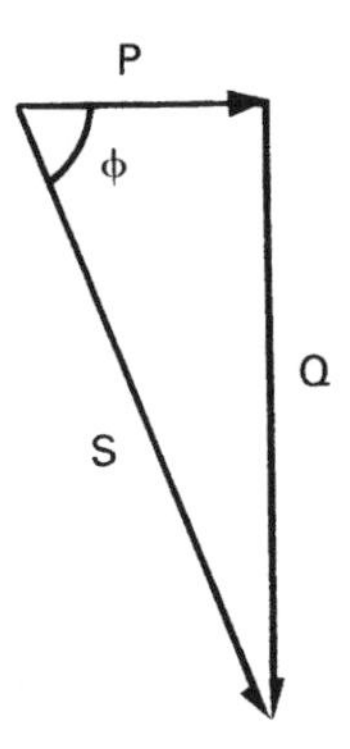

Fig. 7.45

$S = VI$ volt-ampere $= 250 \times 0.354$

therefore, $S = 88.5\,\text{VA}$ **Ans**

$Q = S\sin\phi$ reactive volt-ampere

$= 88.5 \times 0.9591$

hence $Q = 84.88$ VAr **Ans**

Worked Example 7.14

Question

The stator winding of a single-phase a.c. motor consumes 250 W, at a leading power factor of 0.537, when connected to a 240 V, 50 Hz supply. Calculate (a) the current drawn, and (b) the resistance and inductance of the winding.

Answer

$P = 250$ W; $\cos\phi = +0.537$; $V = 240$ V; $f = 50$ Hz

Continued on p. 326

Worked Example 7.14 (*Continued*)

(a)

$$P = VI\cos\phi \text{ watt}$$

$$\text{so, } I = \frac{P}{V\cos\phi} \text{ amp} = \frac{250}{240 \times 0.537}$$

therefore, $I = 1.94$ A **Ans**

(b)

$$Z = \frac{V}{I} \text{ ohm} = \frac{240}{1.94}$$

so $Z = 123.72\,\Omega$

$$\phi = \cos^{-1} 0.537 = 57.52°; \text{ so } \sin^{-\phi} = 0.8436$$

The impedance triangle is shown in Fig. 7.46. From this:

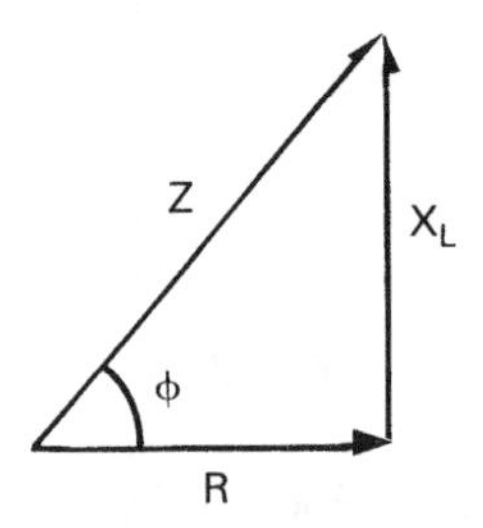

Fig. 7.46

$$R = Z\cos\phi = 123.72 \times 0.537$$

therefore, $R = 66.44\,\Omega$ **Ans**

$$X_L = Z\sin\phi = 123.72 \times 0.8436$$

so $X_L = 104.37\,\Omega$

$$L = \frac{X_L}{2\pi f} \text{ henry} = \frac{104.37}{2 \times \pi \times 50}$$

therefore, $L = 0.332$ H **Ans**

7.13 Series Resonance

Consider a series *R-L-C* circuit, connected to a constant amplitude, variable frequency supply, as in Fig. 7.47. Let the frequency be initially set at a low value. In this case, the capacitive reactance will be relatively large. The inductive reactance will be relatively small. The resistance will remain constant throughout. The relevant phasor diagram is shown in Fig. 7.48(a).

Let the frequency of the supply now be increased. As this is done, so X_C will decrease, and X_L will increase. At one particular frequency, the values of X_C and X_L

will be equal. This is known as the resonant frequency, f_0, and the general condition is known as resonance. This effect is illustrated in Fig. 7.48(b).

If the frequency continues to be increased, then X_L will continue to increase, whilst X_C continues to decrease. This is shown in Fig. 7.48(c).

Considering the circuit phase angle, it can be seen that it has, in turn, changed from lagging to zero (at resonance) to leading. This sequence, in terms of the voltage phasor diagram, is shown in Fig. 7.49(a), (b) and (c).

At the resonant frequency, the applied voltage is in phase with the circuit current. Additionally, under this condition, $V = V_R$. Thus, from the point of view of the supply,

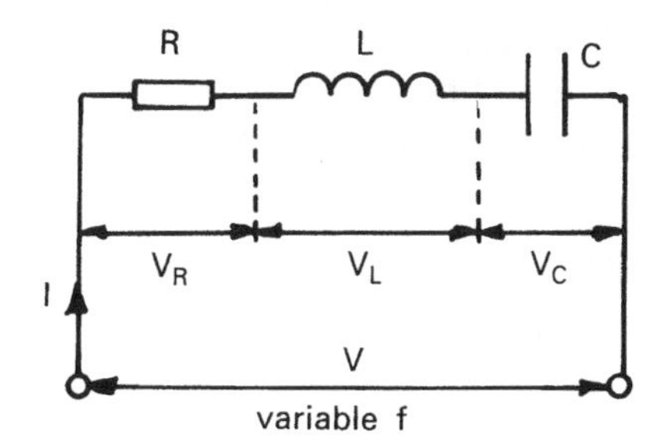

Fig. 7.47

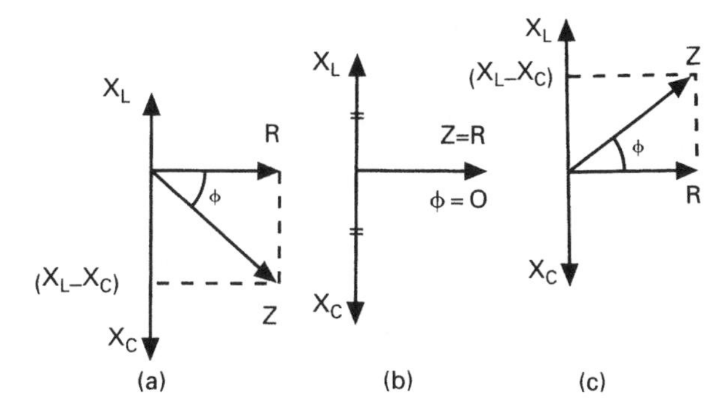

Fig. 7.48

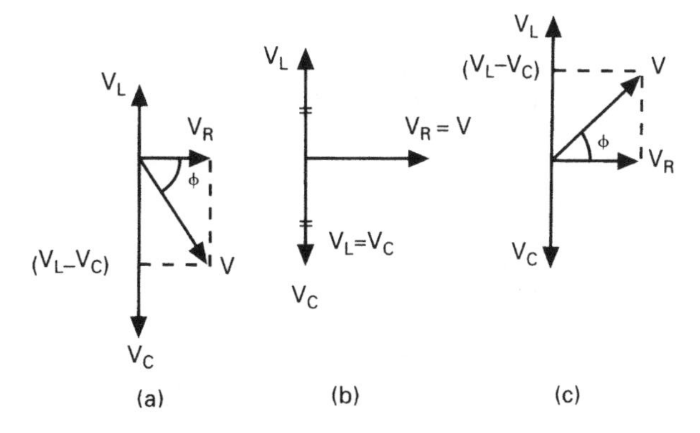

Fig. 7.49

the circuit behaves as if it consists only of the resistor. This is because, at f_0, $X_L = X_C$. These reactances therefore effectively 'cancel' each other. This is illustrated below:

$$Z = \sqrt{R^2 + (X_L - X_C)^2} \text{ ohm}$$

$$\text{and, at } f_0,\ X_L = X_C$$

$$\text{therefore, } Z = \sqrt{R^2} \quad (7.13)$$

$$\text{hence, at resonance } Z = R \text{ ohm}$$

It should be apparent that this is the minimum possible value for the circuit impedance. So, at resonance, the circuit current must reach its maximum possible value. The variation of the reactances, and hence the impedance, is illustrated in Fig. 7.50. The variation of the circuit current, with frequency, is shown in Fig. 7.51.

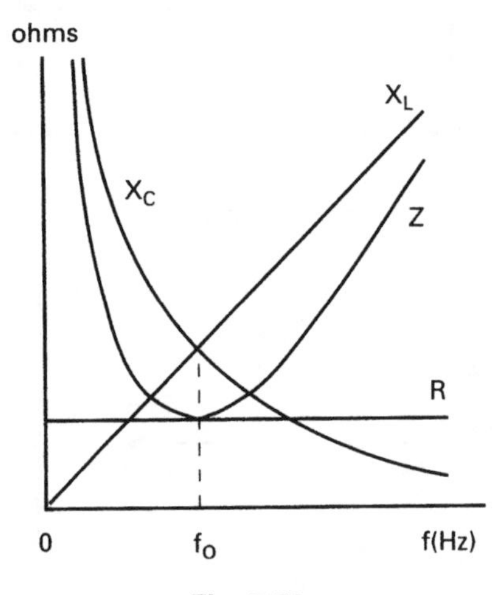

Fig. 7.50

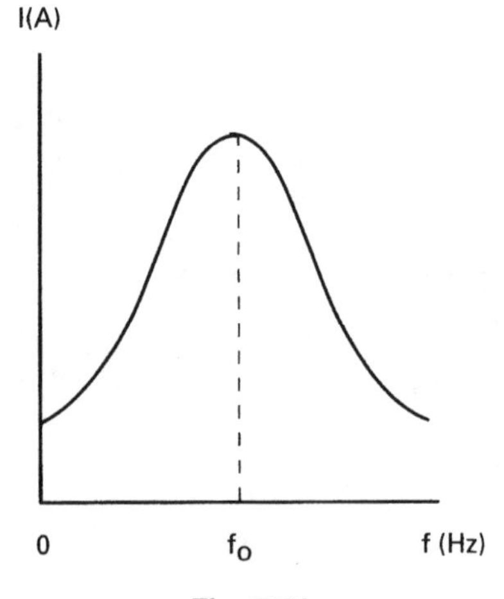

Fig. 7.51

The value for the resonant frequency, for a given circuit, can be determined as follows:

$$\text{At } f_0,\ \omega_0 L = \frac{1}{\omega_0 C} \text{ ohm}$$

$$\text{therefore, } \omega_0^2 LC = 1$$

$$\text{and, } \omega_0^2 = \frac{1}{LC}$$

$$\text{so, } \omega_0 = \frac{1}{\sqrt{LC}} \text{ rad/s}$$

$$\text{hence, } f_0 = \frac{1}{2\pi\sqrt{LC}} \text{ hertz} \qquad (7.14)$$

Note: Under resonant conditions, the p.d.s developed across the capacitor and inductor can be many times greater than the supply voltage. This could have serious consequences for the components used. The dielectric strength of the capacitor may well be exceeded. The insulation on the coil could also prove to be insufficient. For these reasons, the resonant condition is normally avoided, when dealing with 'power' circuits (i.e. 50 Hz supplies). At higher frequencies (e.g. radio frequencies), the applied voltages are normally fairly small. At these frequencies, resonance does not pose the same problem. Indeed, at these frequencies, the resonance effect can be put to good use. This aspect is dealt with in Volume 2 (*Further Electrical and Electronic Principles*).

Worked Example 7.15

Question

A coil of resistance 10 Ω, and inductance 1.013 H, is connected in series with a 10 μF capacitor. Calculate (a) the resonant frequency, (b) the circuit current, when connected to a 240 V, 50 Hz supply, and (c) the p.d. developed across the capacitor.

Answer

$R = 10\,\Omega$; $L = 1.013\,\text{H}$; $C = 10 \times 10^{-6}\,\text{F}$; $V = 240\,\text{V}$

(a) $$f_0 = \frac{1}{2\pi\sqrt{LC}} \text{ hertz} = \frac{1}{2\pi\sqrt{1.013 \times 10^{-5}}}$$

thus, $f_0 = 50$ Hz **Ans**

(b) Since the supply frequency is 50 Hz, then the circuit will resonate. This means that the only opposition to current flow is the *resistance* of the coil, *R* ohm.

$$\text{Therefore, } I = \frac{V}{R} \text{ amp} = \frac{240}{10}$$

hence, $I = 24$ A **Ans**

Continued on p. 330

Worked Example 7.15 (*Continued*)

(c) $$X_C = \frac{1}{2\pi fC} \text{ ohm} = \frac{1}{2 \times \pi \times 50 \times 10^{-5}}$$

so, $X_C = 318.31\ \Omega$

$$V_C = IX_C \text{ volt} = 24 \times 318.31$$

therefore, $V_C = 7639$ V **Ans**

The above example clearly illustrates the problem of resonance at power frequency. For example, suppose the capacitor dielectric has a thickness of 1 mm. The peak voltage between the plates will be $\sqrt{2} \times 7639\ \text{V} = 10\,803\ \text{V}$. The *minimum* dielectric strength required would therefore be $10\,803/10^{-3} = 10.8\ \text{MV/m}$.

If the frequency was increased to 200 Hz, the applied voltage remaining at 240 V, the current would be reduced to 0.518 A, and the p.d. across the capacitor to 46.25 V.

Assignment Questions

1. Calculate the reactance of a perfect capacitor, of 0.47 μF, at the following frequencies. (a) 50 Hz, (b) 200 Hz, (c) 5 kHz, and (d) 300 kHz.
2. Determine the reactance of a perfect inductor, of 0.5 H, at the following frequencies. (a) 40 Hz, (b) 600 Hz, (c) 4 kHz, and (d) 1 MHz.
3. A pure inductor has a reactance of 125 Ω, when connected to a 5 kHz supply. Calculate the inductance.
4. A perfect capacitor has a reactance of 100 Ω, when connected to a 60 Hz supply. Determine its capacitance.
5. A 250 V, 50 Hz supply is connected to a pure inductor. The resulting current is 1.5 A. Calculate the value of the inductance.
6. The current drawn by a perfect capacitor is 250 mA, when connected to a 25 V, 1.5 kHz supply. Determine the capacitance value.
7. Determine the frequency at which a 40 mH inductor will have a reactance of 60 Ω.
8. At what frequency will a 5000 pF capacitor have a reactance of 2 kΩ?
9. A 24 V, 400 Hz a.c. supply is connected, in turn, to the following components. (a) a 560 Ω resistor, (b) a pure inductor, of 20 mH, and (c) a pure capacitor of 220 pF. For each case, calculate the current flow and power dissipation.
10. Determine the impedance of a coil, which has a resistance of 15 Ω and a reactance of 20 Ω.
11. A coil of inductance 85 mH, and resistance 75 Ω, is connected to a 120 V, 200 Hz supply. Calculate (a) the impedance, (b) the circuit current, (c) the circuit phase angle and power factor, and (d) the power dissipated.

12 For the circuit shown in Fig. 7.52, determine (a) the supply voltage, and (b) the circuit phase angle.

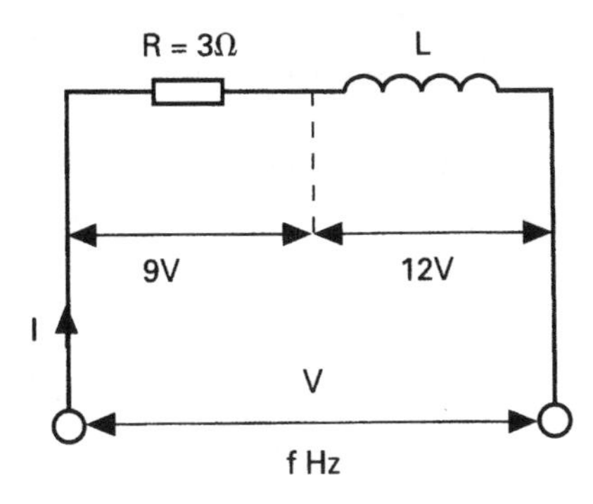

Fig. 7.52

13 For the circuit of Fig. 7.52, assuming a supply frequency of 1.5 kHz, calculate (a) the current, (b) the inductance and (c) the power dissipated.

14 A coil takes a current of 4.5 A, when connected to a 24 V d.c. supply. When connected to a 24 V, 50 Hz supply, it takes a current of 2.6 A. Explain why the current in the second case is less than in the first. Hence, calculate (a) the resistance of the coil and (b) its inductance.

15 A voltage of 45 V, 80 Hz, is applied across a circuit consisting of a capacitor in series with a resistor. Determine the p.d. across the capacitor, if the p.d. across the resistor is 20 V.

16 A 75 Ω resistor is connected in series with a 15 μF capacitor. If this circuit is supplied at 180 V, 100 Hz, calculate (a) the impedance, (b) the current, (c) the power factor and (d) the power dissipated.

17 For the circuit shown in Fig. 7.53, the reactance of the capacitor is 1600 Ω. Determine (a) the supply voltage, (b) the circuit current, (c) the supply frequency and (d) the value of the resistance.

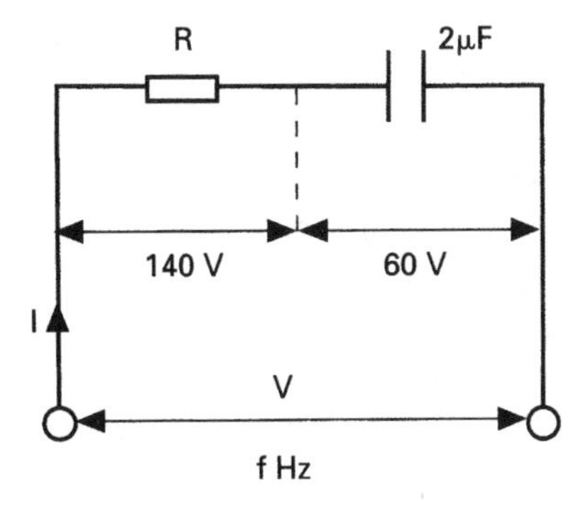

Fig. 7.53

18 An alternating voltage of $v = 180 \sin 628.4t$ volt is applied to a circuit containing a 39 Ω resistor, connected in series with a 47 μF capacitor. Calculate (a) the circuit

current, (b) the p.d.s across the resistor and capacitor, (c) the power dissipated, and (d) the power factor.

19 For the circuit shown in Fig. 7.54, determine (a) the impedance, (b) the current, (c) the phase angle, (d) the p.d. across each component.

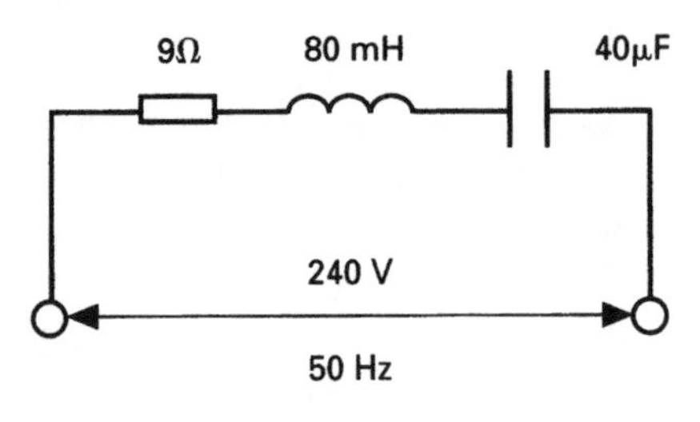

Fig. 7.54

20 The circuit shown in Fig. 7.55 dissipates 162.5 W. Calculate (a) the circuit current, (b) the p.d. across the coil, (c) the p.d. across the capacitor, and (d) the phase angle of the circuit.

21 Determine the resonant frequency for the circuit shown in Fig. 7.55. What would be the current flowing under this condition? Explain why this condition would be avoided at power frequencies.

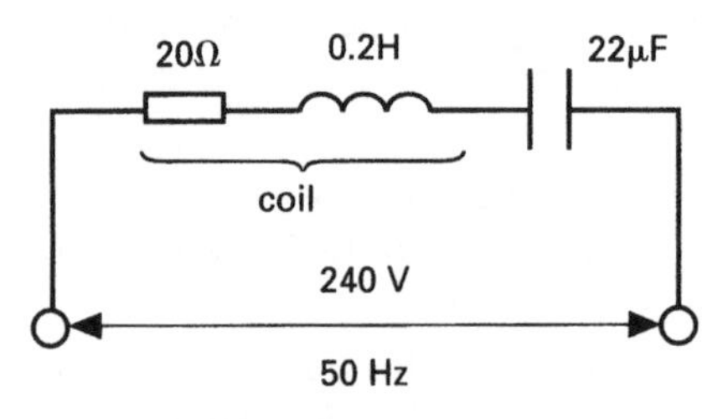

Fig. 7.55

22 An *R-L-C* circuit resonates at a frequency of 10 kHz. If the capacitor has a value of 400 pF, calculate the value of the inductor.

23 A transformer is rated at 15 kVA, at a lagging power factor of 0.75. Determine the rated power output, in kW.

24 A load dissipates 40 kW, at a power factor of 0.6 lagging. Calculate the apparent power, and its reactive component.

25 The power consumed by an *R-C* circuit, when connected to a 100 V, 2 kHz supply, is 800 W. If the circuit current is 12 A, calculate (a) the impedance, (b) the resistance, (c) the reactance, (d) the capacitance, (e) the power factor, and (f) the circuit phase angle.

26 An a.c. motor draws a current of 10 A, when connected to a 240 V, 50 Hz supply. If the power factor is 0.6 lagging, calculate (a) the true power, apparent power, and reactive component of power, (b) the resistance and inductance of the motor winding.

Suggested Practical Assignments

Assignment 1

To determine the variation of capacitive reactance, with variation of frequency.

Apparatus:

$1 \times 1\,\mu F$ capacitor
1 × signal generator
1 × ammeter
1 × voltmeter

Method:

1 Connect the capacitor and ammeter in series, across the output terminals of the signal generator.
2 Set the output voltage to 10 V (as measured across the capacitor), and the frequency to 300 Hz. Measure the current flow. Maintain this voltage value throughout.
3 Alter the frequency, in 100 Hz steps, up to 1 kHz. Record the ammeter and voltmeter readings at each step.
4 Alter the frequency, in 500 Hz steps, up to 4 kHz. Record the meter readings at each step.
5 From your tabulated values of I and V, calculate and tabulate the values for X_C.
6 Plot two graphs. One graph to be X_C versus frequency; the other to be X_C versus reciprocal of frequency $(1/f)$.
7 Write an assignment report, and from your results, state the relationship between capacitive reactance and frequency.

Assignment 2

To investigate the reactance and impedance of a coil.

Apparatus:

1 × coil
1 × signal generator
1 × d.c. supply
1 × ammeter
1 × voltmeter

Method:

1 Connect the coil to the d.c. supply.
2 Increase the voltage applied to the coil, in steps, and note the current at each step.

3 Tabulate these meter readings, and plot a graph of V versus I. From this graph, determine the resistance of the coil.
4 Connect the coil to the signal generator, set to a frequency of 1 kHz.
5 Vary the output voltage of the signal generator, in steps, and record the voltage and current readings at each step.
6 From your tabulated values, plot a graph of V versus I.
7 From this graph, determine the impedance of the coil. Hence, determine the reactance of the coil at 1 kHz, and its inductance.
8 Complete an assignment report.

Assignment 3

To investigate the phase relationship between current and voltage in a capacitor circuit.

Apparatus:

1 × double-beam oscilloscope
1 × signal generator
1 × 0.22 μF capacitor
1 × 1 kΩ resistor

Method:

1 Connect the resistor and capacitor, in series, across the output terminals of the signal generator.
2 Connect the oscilloscope, such that one channel monitors the p.d. across the capacitor, and the other the p.d. across the resistor.
3 Set the frequency to 1 kHz, and sketch the two waveforms observed.
4 Make a careful note of the time interval between the two waveforms, and hence calculate the phase difference between them.

Note: The current through a resistor is in phase with the p.d. across it. Thus, the waveform for the resistor p.d., may be taken to represent the circuit current.

Assignment 4

To investigate the phase relationship, between current and voltage, for an inductor.

Apparatus:

1 × low resistance coil
1 × 5 kΩ resistor

1 × signal generator
1 × double beam oscilloscope

Method:

1 Connect the resistor and coil, in series, across the output terminals of the signal generator. Set the frequency to 10 kHz.
2 Using the two channels of the oscilloscope, monitor the p.d.s across the resistor and coil.
3 Carefully note the time interval between the two waveforms.
4 Calculate the phase angle between the two waveforms, and sketch the waveforms.

Note: The waveform of p.d. across the resistor, may again be taken to represent the circuit current. Provided that the coil resistance is very low compared to the reactance, then the coil may be considered to be almost ideal.

Assignment 5

To investigate the effect of series resonance.

Apparatus:

1 × signal generator
1 × inductor (of known value)
1 × capacitor (of known value)
1 × voltmeter
1 × ammeter

Method:

1 Connect the capacitor, inductor and ammeter in series, across the terminals of the signal generator.
2 Calculate the theoretical resonant frequency of the circuit.
3 Set the frequency to $f_0/10$. Record the circuit current, and the capacitor p.d. Ensure that the output voltage of the signal generator is maintained constant throughout.
4 Increase the frequency, in discrete steps, up to $f_0 \times 10$. Record the current and voltage readings at each step.

Note: Take 'extra' readings at, and on either side of the resonant frequency f_0 Hz.

5 Plot a graph of circuit current versus frequency. Compare the actual resonant frequency to the calculated value.
6 Complete an assignment report.

Supplementary Worked Examples

Example 1: A 2 kΩ resistor, a perfect 0.5 H inductor and a perfect 2.2 μF capacitor are connected, in turn, across a 5 V, 1 kHz supply. For each case calculate the resulting current flow and sketch the relevant phasor diagram.

Solution: $R = 2000\,\Omega$; $L = 0.5$ H; $C = 2.2 \times 10^{-6}$ F; $V = 5$ V; $f = 10^3$ Hz

Resistor: $I = \dfrac{V}{R}$ volt $= \dfrac{5}{2000}$

so, $I = 2.5$ mA **Ans**

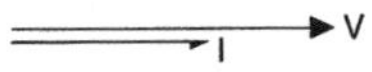

Fig. 7.56

Inductor: Since this is a pure inductor, the only opposition to the flow of current will be the inductive reactance, X_L.

$$X_L = 2\pi f L \text{ ohm} = 2\pi \times 10^3 \times 0.5$$

$$X_L = 3.142\,\text{k}\Omega$$

$$I = \frac{V}{X_L} \text{ amp} = \frac{5}{3142}$$

so, $I = 1.59$ mA **Ans**

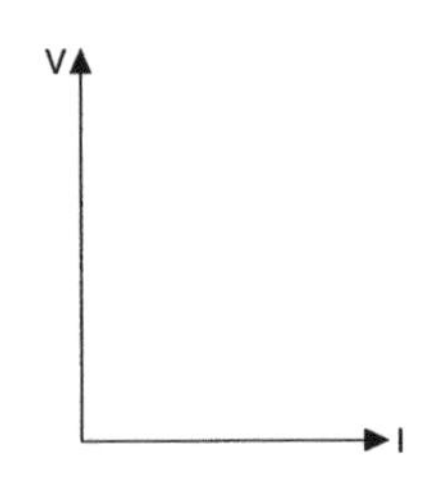

Fig. 7.57

Capacitor: Similarly, since it is a perfect capacitor, then the only opposition to current flow is the capacitive reactance, X_C.

$$X_C = \frac{1}{2\pi f C} \text{ ohm} = \frac{1}{2\pi \times 10^3 \times 2.2 \times 10^{-6}}$$

$$X_C = 73.34\,\text{k}\Omega$$

$$I = \frac{V}{X_C} \text{ amp} = \frac{5}{73\,430}$$

so, $I = 69.1\,\mu$A **Ans**

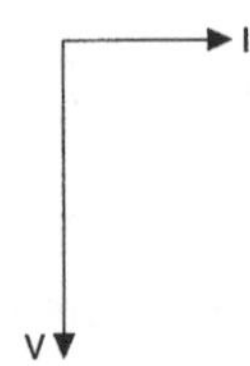

Fig. 7.58

Example 2: A pure inductor is connected across a 10 V, 200 Hz supply, and the current flowing through it is measured as 0.4 A. Determine the value of its inductance.

Solution: $V = 10\,\text{V}; f = 200\,\text{Hz}; I = 0.4\,\text{A}$

$$X_L = \frac{V}{I}\ \text{ohm} = \frac{10}{0.4}$$

so, $X_L = 25\,\Omega$

and, $X_L = 2\pi fL\ \text{ohm}$

$$\text{so, } L = \frac{X_L}{2\pi f}\ \text{henry} = \frac{25}{2\pi \times 200}$$

and, $L = 19.9\,\text{mH}$ **Ans**

Example 3: A perfect capacitor is connected across a 6 V, 5 kHz supply, and the resulting current flow is 88.6 mA. Calculate the capacitance value.

Solution: $V = 6\,\text{V}; f = 5000\,\text{Hz}; I = 88.6 \times 10^{-3}\,\text{A}$

$$X_C = \frac{V}{I}\ \text{ohm} = \frac{6}{88.6 \times 10^{-3}}$$

so, $X_C = 67.72\,\Omega$

$$\text{and, } X_C = \frac{1}{2\pi fC}\ \text{ohm}$$

$$\text{so, } C = \frac{1}{2\pi f X_C}\ \text{farad} = \frac{1}{2\pi \times 5000 \times 67.72}$$

and $C = 4.7 \times 10^{-7} = 0.47\,\mu\text{F}$ **Ans**

Example 4: A coil of wire is tested by connecting it, in turn, to a d.c. supply and then an a.c. supply. The results from these two tests are as follows:
d.c. supply of 10 V; resulting current flow 50 mA
a.c. supply of 10 V, 100 Hz; resulting current flow 32 mA
Using the results of these two tests, determine the resistance and inductance values for the coil.

Solution: d.c. test: $V = 10\,\text{V}; I = 50 \times 10^{-3}\,\text{A}$

Since the d.c. current is a *steady* current then the only opposition to the current will be the resistance of the coil.

$$R = \frac{V}{I}\ \text{ohm} = \frac{10}{50 \times 10^{-3}}$$

so, $R = 200\,\Omega$ **Ans**

a.c. test: $V = 10\,\text{V}; I = 32 \times 10^{-3}\,\text{A}; f = 100\,\text{Hz}$

In this case the opposition to the flow of *alternating* current will be the combined effect of its resistance and its inductive reactance, i.e. the total opposition is the coil impedance, Z.

$$Z = \frac{V}{I} \text{ ohm} = \frac{10}{32 \times 10^{-3}}$$

so, $Z = 312.5\,\Omega$

Now, $Z = \sqrt{R^2 + X_L^2}$ ohm or, $Z^2 = R^2 + X_L^2$

so, $X_L^2 = Z^2 - R^2$

and, $X_L = \sqrt{Z^2 - R^2}$ ohm $= \sqrt{312.5^2 - 200^2}$

$= \sqrt{57\,656}$

so, $X_L = 240\,\Omega$

$$L = \frac{X_L}{2\pi f} \text{ henry} = \frac{240}{2\pi \times 100}$$

hence, $L = 0.382$ H **Ans**

Example 5: A coil of resistance 25 Ω and inductance 40 mH is connected to a 50 Hz a.c. supply, and the current which then flows is 5.36 A. Calculate (a) the supply voltage, (b) the circuit phase angle, and (c) the power dissipated.

Solution: $R = 25\,\Omega$; $L = 0.04$ H; $I = 5.36$ A; $f = 50$ Hz

(a) $X_L = 2\pi fL$ ohm $= 2\pi \times 50 \times 0.04$

so, $X_L = 12.57\,\Omega$

$$Z = \sqrt{R^2 + X_L^2} \text{ ohm} = \sqrt{25^2 + 12.57^2}$$

and, $Z = 28\,\Omega$

$V = IZ$ volt $= 5.36 \times 28$

and, $V = 150$ V **Ans**

(b) The impedance triangle for the coil is shown in Fig. 7.59

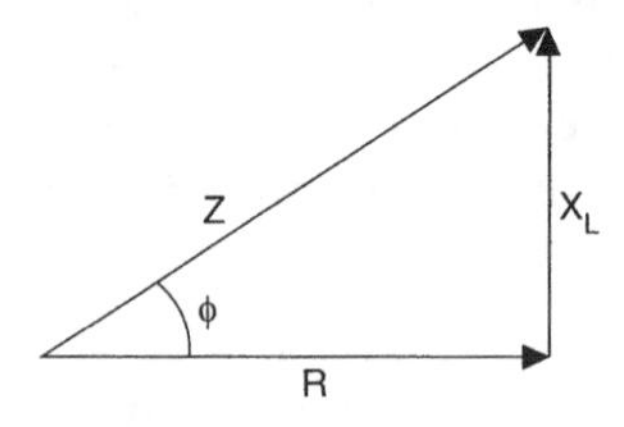

Fig. 7.59

$$\phi = \tan^{-1}\frac{X_L}{R} = \cos^{-1}\frac{R}{Z} = \sin^{-1}\frac{X_L}{Z}$$

In order to minimise possible errors the last of the above equations will be avoided, since it involves the use of two previously calculated values. So, the first equation has been chosen.

$$\phi = \tan^{-1}\frac{X_L}{R} = \tan^{-1}\frac{12.57}{25} = \tan^{-1} 0.5028$$

and, $\phi = 26.7^\circ$ or 0.466 rad **Ans**

(c) $P = VI \cos \phi$ watt $= 150 \times 5.36 \times \cos 26.7°$

so, $P = 718.3$ W **Ans**

Alternatively, since only resistive components dissipate power, then

$P = I^2R$ watt $= 5.36^2 \times 25 = 718.2$ W

Note: In this case the power *cannot* be calculated from $P = VI$ watt. This may be verified by considering the circuit and phasor diagrams as shown in Fig. 7.60. From the circuit diagram it can be seen that the p.d. across the resistive component is V_R and *NOT V* volt. This point illustrates the value of sketching the circuit and phasor diagrams *before* proceeding with the calculations.

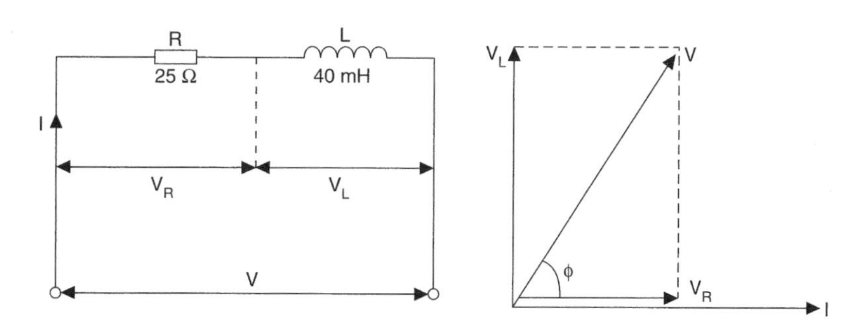

Fig. 7.60

$V_R = IR$ volt $= 5.36 \times 25 = 134$ V

$P = V_R I$ watt $= 134 \times 5.36$

and, $P = 718.2$ W, which verifies the previous calculated answer.

Example 6: A 10 μF capacitor is connected in series with a 270 Ω resistor across a 20 V, 50 Hz supply. Calculate (a) the current flowing, (b) the p.d.s across the resistor and the capacitor, and (c) the circuit power factor.

Solution: $R = 270\ \Omega$; $C = 10^{-5}$ F; $V = 20$ V; $f = 50$ Hz

The relevant circuit and phasor diagrams are shown in Fig. 7.61.

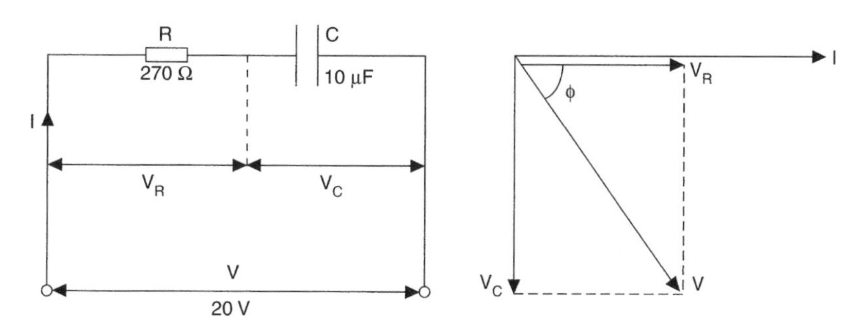

Fig. 7.61

(a) $$X_C = \frac{1}{2\pi fC} \text{ ohm} = \frac{1}{2\pi \times 50 \times 10^{-5}}$$

so, $X_C = 318.3\,\Omega$

$$Z = \sqrt{R^2 + X_C^2} \text{ ohm} = \sqrt{270^2 + 318.3^2}$$

and, $Z = 417.4\,\Omega$

$$I = \frac{V}{Z} \text{ amp} = \frac{20}{417.4}$$

hence, $I = 47.92\,\text{mA}$ **Ans**

(b) $V_R = IR$ volt $= 47.92 \times 10^{-3} \times 270$

$V_R = 12.94\,\text{V}$ **Ans**

$V_C = IX_C$ volt $= 47.92 \times 10^{-3} \times 318.3$

$V_C = 15.25\,\text{V}$ **Ans**

(c) $$\text{p.f.} = \cos\phi = \frac{R}{Z}$$

$$= \frac{270}{417.4}$$

so, p.f. $= 0.347$ lagging **Ans**

Example 7: A circuit draws a power of 2.5 kW at a leading power factor of 0.866. Calculate the apparent power and the reactive component of power.

Solution: $P = 2500$ W; $\cos\phi = 0.866$

Phase angle, $\phi = \cos^{-1} 0.866 = 30°$, and the power triangle will be as shown in Fig. 7.62.

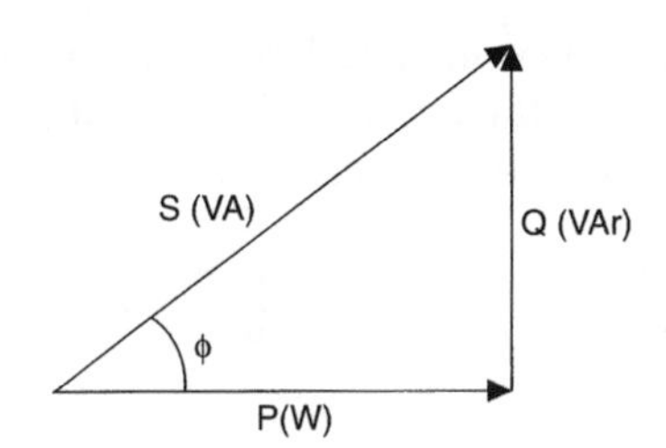

Fig. 7.62

$$\cos\phi = \frac{P}{S}$$

so apparent power, $S = \frac{P}{\cos\phi}$ volt-ampere

$$= \frac{2500}{0.866}$$

hence, $S = 2.89\,\text{kVA}$ **Ans**

$$\tan\phi = \frac{Q}{P}$$

so, reactive component, $Q = P \tan\phi$ volt-ampere reactive

$$= 2.5 \times \tan 30^\circ \text{ kVAr}$$

therefore, $Q = 1.44$ kVAr **Ans**

Example 8: A coil of resistance 330 Ω and inductance 0.25 H is connected in series with a 10 μF capacitor. This circuit is connected across a 100 V, 80 Hz supply. Calculate (a) the circuit current, (b) the p.d.s. across the coil and the capacitor, (c) the circuit phase angle and power factor, and (d) the power dissipated.

Solution: $R = 330\ \Omega$; $L = 0.25$ H; $C = 10^{-5}$ F; $V = 100$ V; $f = 80$ Hz

Note that we are dealing with a *practical* coil, which possesses both resistance and inductance. In order to simplify the calculations, such a coil is always considered as comprising a perfect resistor in series with a perfect inductor, as shown in the circuit diagram of Fig. 7.63.

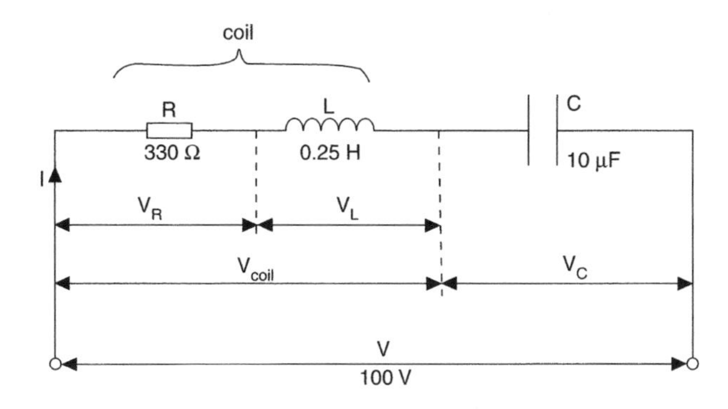

Fig. 7.63

(a)

$$X_L = 2\pi f L \text{ ohm} = 2\pi \times 80 \times 0.25$$

$$X_L = 125.66\ \Omega$$

$$X_C = \frac{1}{2\pi f C} \text{ ohm} = \frac{1}{2\pi \times 80 \times 10^{-5}}$$

$$X_C = 198.94\ \Omega$$

$$Z = \sqrt{R^2 + (X_L - X_C)^2} \text{ ohm} = \sqrt{330^2 + (125.66 - 198.94)^2}$$

and, $Z = 338\ \Omega$

$$I = \frac{V}{Z} \text{ amp} = \frac{100}{338}$$

hence, $I = 0.296$ A **Ans**

(b) $V_R = IR$ volt $= 0.296 \times 330 = 97.68$ V

$V_L = IX_L$ volt $= 0.296 \times 125.66 = 37.2$ V

$$V_{coil} = \sqrt{V_R^2 + V_L^2} \text{ volt} = \sqrt{97.68^2 + 37.2^2}$$

so, $V_{coil} = 104.5\,\text{V}$ **Ans**

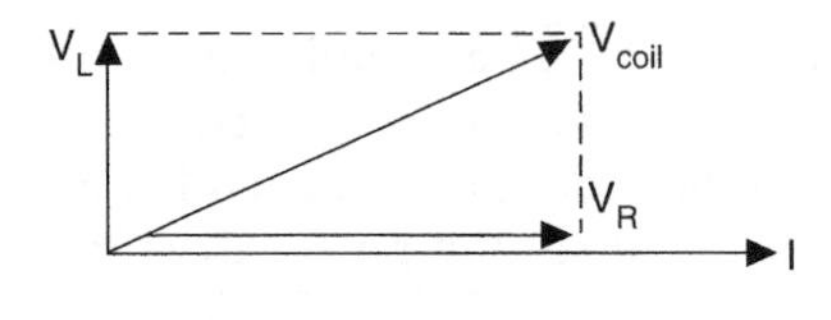

Fig. 7.64

Alternatively: $Z_{coil} = \sqrt{R^2 + X_L^2}$ ohm

$$= \sqrt{330^2 + 125.66^2}$$

$$Z_{coil} = 353.1\,\Omega$$

$$V_{coil} = IZ_{coil} \text{ volt} = 0.296 \times 353.1$$

so, $V_{coil} = 104.5\,\text{V}$ **Ans**

$$V_C = IX_C \text{ volt} = 0.296 \times 198.94$$

so, $V_C = 58.9\,\text{V}$ **Ans**

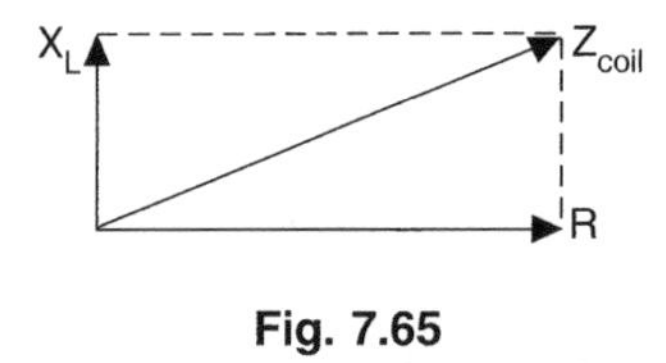

Fig. 7.65

(c) The complete phasor diagram is shown in Fig. 7.66.

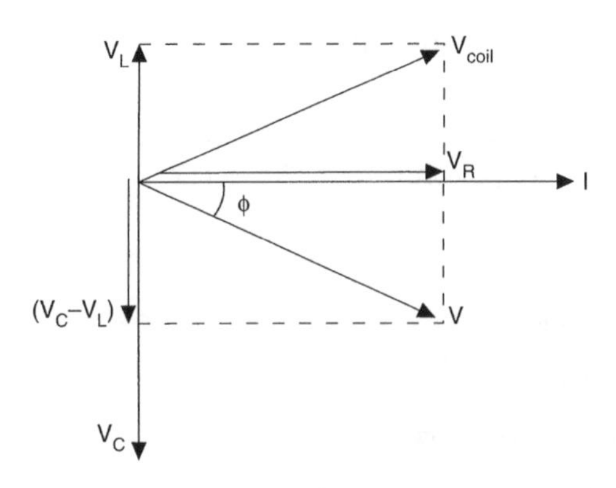

Fig. 7.66

$$\text{p.f.} = \cos\phi = \frac{V_R}{V} = \frac{97.68}{10}$$

hence, p.f. $= 0.977$ lagging **Ans**

phase angle, $\phi = \cos^{-1}\ 0.977$

so, $\phi = 12.5°$ lagging **Ans**

(d) $P = VI\ \cos\phi$ watt or $P = I^2R$ watt

$= 100 \times 0.296 \times 0.977$ $= 0296^2 \times 330$

$P = 28.9\,\text{W}$ **Ans** $P = 28.9\,\text{W}$ **Ans**

Example 9: A coil of resistance 500 Ω and inductance 0.2 H is connected in series with a 20 nF capacitor across a 10 V, variable frequency supply. Determine (a) the frequency at which the circuit current will be at its maximum value, (b) the value of this maximum current, and (c) the p.d.s across both the coil and the capacitor at this frequency.

Solution: $R = 500\,\Omega$; $L = 0.2\,\text{H}$; $C = 20 \times 10^{-9}$ F; $V = 10\,\text{V}$

For the current to be at its maximum value, the circuit must be supplied at its resonant frequency, f_o Hz. This condition is shown by the phasor diagram Fig. 7.67.

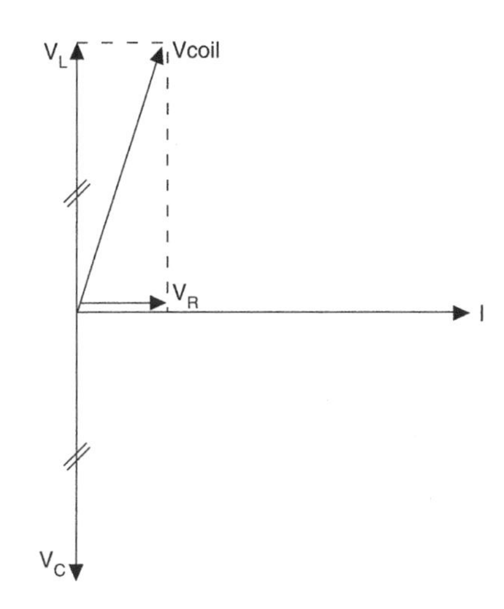

Fig. 7.67

(a) $$f_o = \frac{1}{2\pi\sqrt{LC}}\,\text{Hz} = \frac{1}{2\pi\sqrt{0.2 \times 20 \times 10^{-9}}}$$

hence, $f_o = 2.516$ kHz **Ans**

(b) At resonance, $V_L = V_C$, so $X_L = X_C$
so they 'cancel' each other

$$\text{and } I = \frac{V}{R} \text{ amp} = \frac{10}{500}$$

so, $I = 20\,\text{mA}$ **Ans**

(c)
$$X_C = X_L = 2\pi f_o L \text{ ohm}$$
$$= 2\pi \times 2516 \times 0.2$$
$$\text{so, } X_C = X_L = 3.162\,\text{k}\Omega$$
$$V_C = IX_C \text{ ohm} = 0.02 \times 3162$$

hence, $V_C = 63.23\,\text{V}$ **Ans**

$$V_{\text{coil}} = IZ_{\text{coil}} \text{ volt, where } Z_{\text{coil}} = \sqrt{R^2 + X_L^2} \text{ ohm}$$
$$= \sqrt{500^2 + 3162^2}$$
$$\text{and, } Z_{\text{coil}} = 3201\,\Omega$$

hence, $V_{\text{coil}} = 0.02 \times 3201 = 64\,\text{V}$ **Ans**

Summary of Equations

Resistance: $I = \dfrac{V}{R}$ amp; $P = I^2R$ watt; $\phi = 0°$

Inductance: Reactance, $X_L = 2\pi fL$ ohm; $I = \dfrac{V}{X_L}$ amp; $P = 0\,\text{W}$; $\phi = +90°$ or $+\dfrac{\pi}{2}$ rad

Capacitance: Reactance, $X_C = \dfrac{1}{2f\pi C}$ ohm; $I = \dfrac{V}{X_C}$ amp; $P = 0$ W; $\phi = -90°$ or $-\dfrac{\pi}{2}$ rad

Impedance: $Z = \sqrt{R^2 + (X_L - X_C)^2}$ ohm; $\cos\phi = \dfrac{R}{Z}$; $P = VI\cos\phi = I^2R$ watt

$I = \dfrac{V}{Z}$ amp

Power triangle: Power factor $= \cos\phi$
True power in W; apparent power in VA; reactive 'power' in VAr.

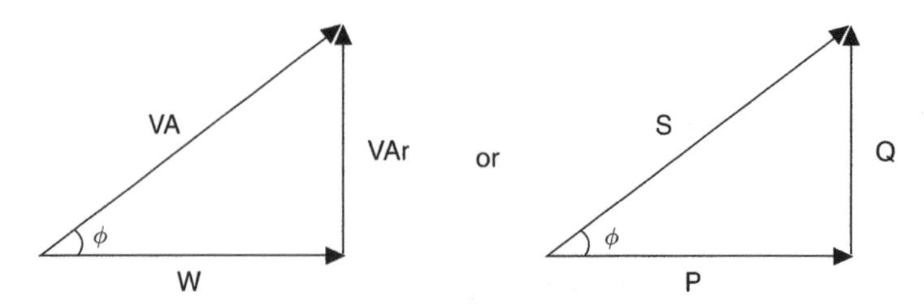

Series resonance: Resonant frequency, $f_o = \dfrac{1}{2\pi\sqrt{LC}}$ hertz

$X_L = X_C$ ohm; $Z = R$ ohm; $I = \dfrac{V}{R}$ amp (max. possible value)

8 Control Principles

This chapter introduces the basic principles and concepts of control systems. On completion, you should be able to:

1. Compare and differentiate between simple open-loop and closed-loop control systems.
2. Compare the relative advantages and disadvantages of these systems.
3. Describe examples of such systems, in both engineering and non-engineering situations.
4. Derive the transfer function for simple control systems.
5. Appreciate the causes of instability in closed-loop control systems.
6. Describe the response of a control system, when subjected to a step input.

8.1 Introduction

The subject of automatic control is a fairly complex one. This chapter, therefore, is designed to give you an insight and 'feel' for the underlying principles involved. It is therefore mainly descriptive, and the mathematical content is minimal.

When the term 'automatic control' is used, most people would take this as applying to some form of engineering system. However, the control principles used in engineering systems apply equally to many other situations. Despite the very wide range of control systems, they may all be analysed by using the same techniques. In this regard, the use of analogies is very useful.

8.2 Non-engineering Applications

A few examples of non-engineering applications of control systems are as follows:

Budgetary control	Management	Educational systems
Economic systems	Marketing	Medical systems
Public transport	Commerce	Human behaviour

It should be apparent that, in all of the above areas, a formalised control is essential. Ideally, the control would be perfect. Unfortunately, in the real world most systems are far from perfect. The main problem with the above examples is that, in each case, there are many variables to take into account. Hence, the system of control needs to be very complex. Thus, if only one of the variables goes out of control, this tends to have a 'knock-on' effect—often with drastic results. Most engineering control systems deal with only one or two variables at a time. They are therefore capable of performing with much greater precision.

8.3 Engineering Applications

Some common examples of engineering control systems are:

Machine tool operation
Temperature control
Generator output control
Liquid and bulk level control
Automatic pilots for aircraft and ships
Radio telescope positioning
Video recorder drives
Motor vehicle power steering and antilock braking
Assembly of electronic components

The variables most commonly requiring control are: Linear and angular position; velocity; pressure; voltage and current; flow rate.

8.4 Classification of Control Systems

Any control system consists of separate components or elements, which are interconnected in order to regulate or control a variable quantity or process. The behaviour of each individual element will affect the behaviour of the whole system. All systems will have a demand or reference (input) **signal**, and a corresponding output **signal**.

> In this context the term **signal** is used in its widest sense. It refers to the quantity involved; thus it may be a mechanical position, a pressure, a voltage, or some other physical quantity

Control systems may be classified into one of two main groups. These are open-loop and closed-loop systems. The essential difference is that, in closed-loop control some form of feedback of information is provided, between output and input. This use of feedback enables *automatic* corrective action, if the output deviates from the demanded value. Open-loop systems do not have any provision for such feedback. Thus, once an input demand has been made the system is left to respond, in the hope that it will respond correctly. A simple example of an open-loop system is that of a golfer making a tee shot. Let us assume that the club strikes the ball at exactly the correct angle, and with exactly the right force. Once the ball leaves the club face, the golfer has no further control over it. Thus, if there is a sudden gust of wind, or the ball hits a small bump on the fairway, it will not come to rest at the intended spot.

8.5 Open-loop Systems

These are the simplest form of control system. They are not necessarily the best solution, but for some applications, may be sufficient. Consider the following examples:

1 A simple electric fire is used to heat a room. When the supply is turned on, the room temperature increases. However, without any further intervention the room will probably become too warm. The only way to correct this would be to turn off the supply. This system is illustrated in the **block diagram** (Fig. 8.1), which also shows the various signals involved.

> A **block diagram** is a means of showing the interconnections in a system. Each element in the system is represented by a box, which is used to identify its function. It allows the overall function of the system to be readily seen, without showing every single detail of each element. It also enables the signal paths to be clearly seen

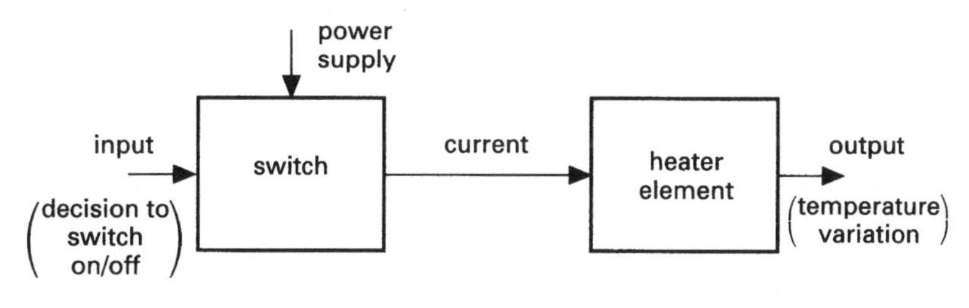

Fig. 8.1

2 A generator is driven by a steam turbine. The voltage output of the generator depends upon its speed. Its speed depends upon the steam pressure into the turbine. Thus, to achieve a particular output voltage, the steam pressure is adjusted accordingly. If the electrical load on the generator is suddenly increased, its terminal voltage will fall. This condition is not therefore automatically corrected. The block diagram is shown in Fig. 8.2.

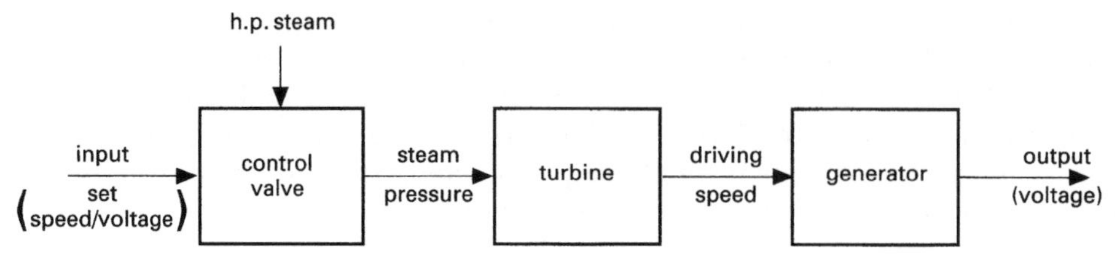

Fig. 8.2

8.6 Closed-loop Systems

A closed-loop system must have some means of comparing the value of its output with the desired value. This involves feeding back information about the output, and making the comparison. Thus, a feedback loop is required, which will form one input to a device known as a *comparator*. The other input to the comparator will be the demand or reference signal. If there is any difference between the desired and actual values of the output, the comparator will produce the appropriate output signal. This signal is known as the *error signal*. Hence, all closed-loop systems are said to be *error actuated*.

Consider the two previous open-loop examples. Each of these could be converted to closed-loop, as follows:

1 By installing a room thermostat, the desired temperature may be set. In addition, the actual temperature of the room can be compared with the desired value. The comparison element in the thermostat is a bimetallic strip. The two metals involved have different coefficients of expansion. Thus, as the temperature varies, so the strip bends. One end of the strip is fixed, and the other end moves the position of an electrical contact. The temperature setting dial alters the effective spacing between a second electrical contact and that operated by the bimetallic strip. By this means, the electrical supply to the heater can be switched. The block diagram is shown in Fig. 8.3.

Note: In this example, there is not a direct physical feedback path. However, the temperature of the air in the room completes the feedback loop.

This form of control is known as ON/OFF closed-loop control. The graph of the variations in room temperature are shown in Fig. 8.4. From this, it may be seen that the temperature varies about the desired value. The resulting **dead-band** is typical of any ON/OFF closed-loop control system. The reason for the dead-band in this example, is the response time of the heater element. When the desired temperature is achieved, the

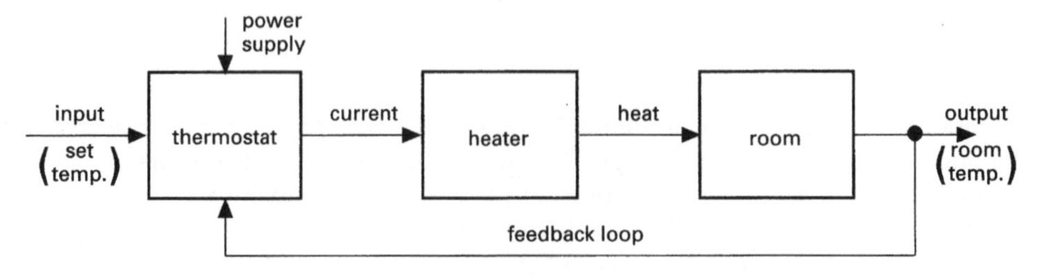

Fig. 8.3

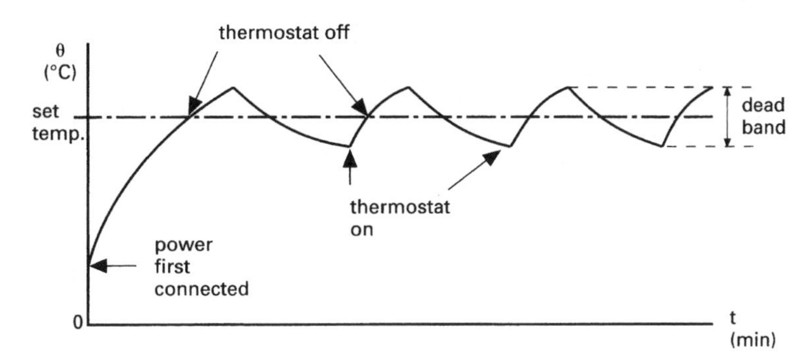

Fig. 8.4

thermostat disconnects the heater. However, the heater element will take a certain amount of time to cool. The room temperature will therefore continue to increase slightly. Similarly, when the room starts to cool, there will be a small time lag in the response of the bimetallic strip. The heater element will also take some time to reach its operating temperature. These two delays together account for the room temperature falling below the desired value.

Dead-band is the range of input change to which the system does not respond

2 For the generator system, the loop could be closed in a number of ways. Firstly, let us consider the use of a human operator. The voltage output of the generator could be displayed on a voltmeter, and the operator could then observe any variation of output. When such a change occurred, the operator could adjust the control valve accordingly. Although, technically, this would constitute a closed-loop system, it would be a very unsatisfactory one. The operator would at some stage start to daydream, fall asleep, or need to take a break. This would then break the feedback loop. Even if the operator could stay fully alert all the time, his/her response in applying the necessary corrections would be relatively slow. In addition, it is quite likely that over/under corrections would result, particularly if the load variations were quite rapid. In order to make the system truly automatic, with a fast reaction time, and not subject to fatigue etc., the human operator must be replaced.

One method of achieving this would be as follows. A **tachogenerator** would be attached to the generator drive shaft. This would provide a feedback signal (voltage) that *represents* the output of the generator. The steam control valve would be a motorised valve. The comparator would be in the form of either a **differential amplifier** or a **summing amplifier**. The demand signal input to the comparator would be derived from a potentiometer. The corresponding block diagram is shown in Fig. 8.5.

A **tachogenerator** (or 'tacho') is a small d.c. generator which will produce an output voltage that is directly proportional to its speed

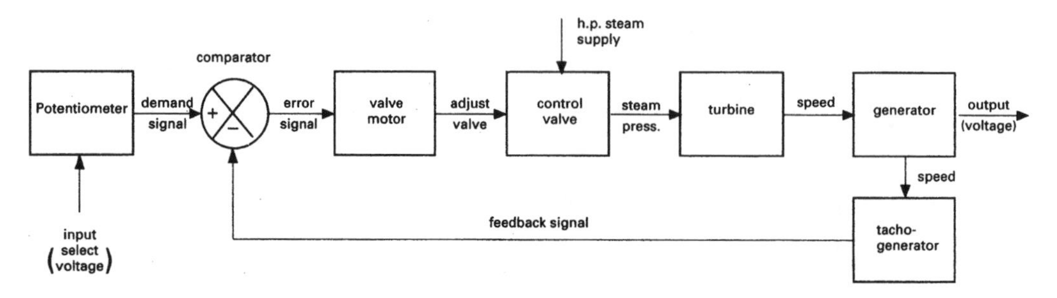

Fig. 8.5

> A **differential amplifier** is one which has two input terminals. The output of the amplifier is equal to the *difference* between its inputs.
> A **summing amplifier** is one which also has two (or more) input terminals. Its output is the sum of the inputs. However, if one input is of the opposite polarity to the other, then the output will be equal to the *difference* between its inputs

The system operates as follows. The desired output voltage is dialled up on the rotary potentiometer at the input. The movable contact on this potentiometer therefore provides the demand signal to the comparator. Assuming that the generator is stationary at this time, a large error signal is then fed to the steam valve motor. This, in turn, will open the valve. The generator now starts to build up speed, and consequently starts to generate an output voltage. The tachogenerator now provides a feedback signal to the comparator. The resulting error voltage therefore decreases. The valve motor continues to open the valve, but at a slower rate. When the generator achieves the correct speed to produce the demanded output voltage, the tachogenerator feedback voltage will be equal to the demand signal voltage from the potentiometer. The error signal to the valve motor will now be zero. This motor stops, and the generator continues to run at the set speed. If, at some later stage, the generator load varies, the slight change in its speed will cause a change in the feedback signal from the tachogenerator. This will produce an error signal from the comparator. The steam valve motor will then automatically adjust the speed, and rapidly return the output voltage to the demanded value.

Note: The circular symbol, with a 'cross' inside, is the general symbol for a comparator. It should also be noted that the demand input to this element is marked with a plus sign, and the feedback signal with a minus sign. This is to indicate that the two signals are in opposition to each other. It also indicates that *negative* feedback is employed.

This form of closed-loop control is known as continuous control. The system is, of course, error actuated, as is the ON/OFF temperature control system described. In the latter case, this resulted in a comparatively large dead-band, and a continuous variation of the temperature, about the demanded value. In the continuous control system these large fluctuations of output do not usually occur. This is because, when the output starts to vary from the desired value, the corrective action is almost instantaneous. Thus, any deviation from demanded value is very brief, and consequently very small. In addition, if the load on the generator remains constant, there will be no deviation from the desired output.

8.7 Transfer Functions and Block Diagrams

The transfer function of an element is the ratio of its output quantity to its input quantity. This ratio should not be confused with efficiency. The efficiency of a device is the ratio of the output *power* to the input *power*. It is usual for the output of an element to be a different physical quantity to the input. For example, consider a rotary potentiometer. The input is in the form of a mechanical displacement of the movable contact. The output is the voltage tapped off from the potentiometer track. Figure 8.6 shows the electrical arrangement, and Fig. 8.7 the system block diagram. When a control system is shown in block diagram form, it is normal practice to write the appropriate transfer function in each block.

Considering Fig. 8.7, the transfer function, Y for the potentiometer is:

$$Y = \frac{V_o}{\theta_i} \text{ volt/rad} \tag{8.1}$$

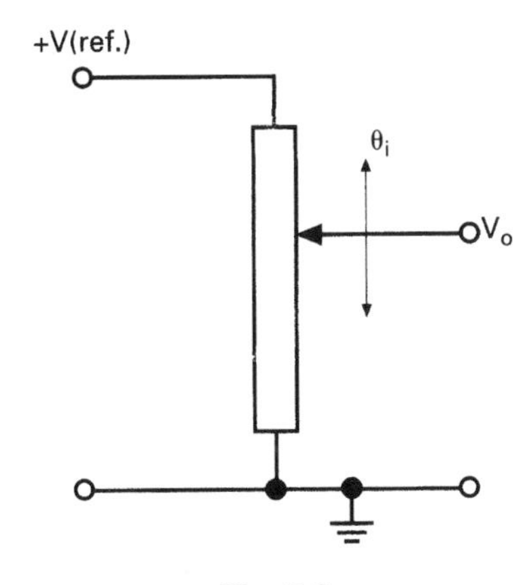

Fig. 8.6

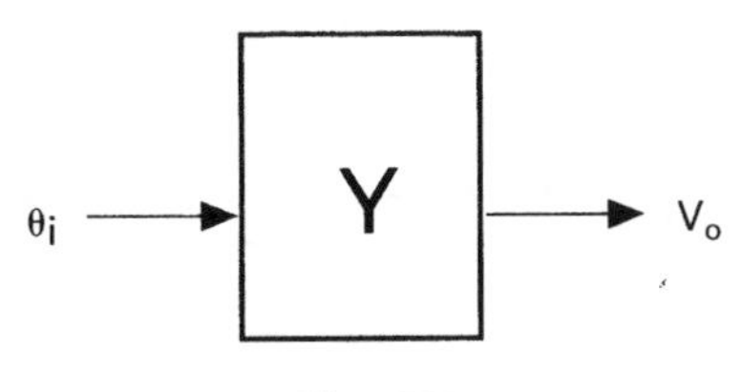

Fig. 8.7

or, in terms of the output of the block:

$$V_o = Y\theta_i \text{ volt} \tag{8.2}$$

When dealing with a complete control system, the overall transfer function for the system is a combination of the individual transfer functions. For an open-loop system, the overall transfer function is simply the product of the individual transfer functions. Consider the generalised open-loop system shown in Fig. 8.8.

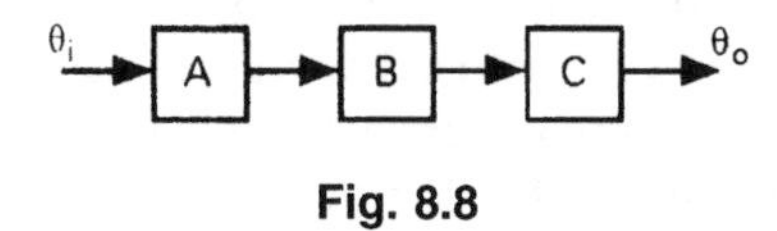

Fig. 8.8

The output of the first element will be A × its input. Thus the output of the first block is $A\theta_i$. This then forms the input to the next block, the output of which will be $AB\theta_i$. Following this sequence through to the output, we have

$$\theta_o = ABC\theta_i$$

hence, the overall transfer function is:

$$\frac{\theta_o}{\theta_i} = ABC \tag{8.3}$$

A generalised closed-loop control system is shown in Fig. 8.9. The overall transfer function for this system can be derived in a similar manner, as follows

$$\text{input to first block} = \theta_i - \theta_o$$

$$\text{input to second block} = A(\theta_i - \theta_o)$$

$$\text{input to third block} = AM(\theta_i - \theta_o)$$

$$\text{therefore, output, } \theta_o = AML(\theta_i - \theta_o)$$

$$\text{so, } \theta_o = AML\theta_i - AML\theta_o$$

$$\text{hence, } \theta_o(1 + AML) = AML\theta_o$$

$$\text{and } \frac{\theta_o}{\theta_1} = \frac{AML}{1 + AML} \tag{8.4}$$

Note: The symbol (θ) used above may refer to *any* physical property. It does not imply that both the input and output quantities are angles.

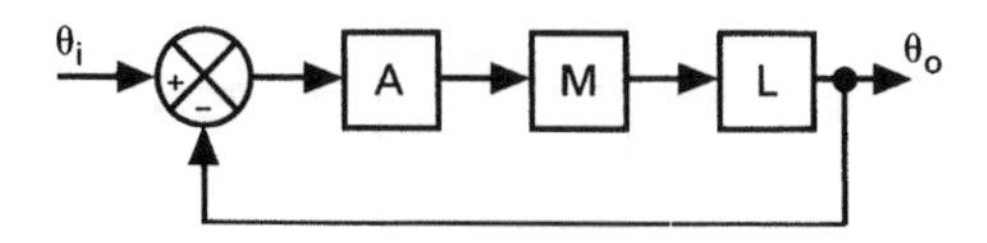

Fig. 8.9

Worked Example 8.1

Question

A simple open-loop speed control system consists of an electronic amplifier and a motor. The block diagram is shown in Fig. 8.10, where V_i is the input demand signal. The transfer function, A, for the amplifier is 10 V/V, and that for the motor, M, is 12 (rad/s)/V. Determine the system transfer function, and hence calculate the demand signal required to produce an output speed of 120 rad/s.

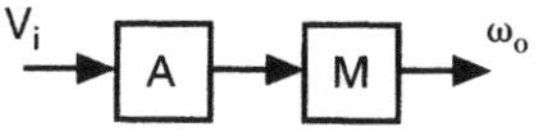

Fig. 8.10

Answer

A = 10 volt/volt (i.e. amplifier gain = 10 times)

$B = 12$(rad/s)/volt; $\omega_o = 120$ rads

$$\text{transfer function} = \frac{\omega_o}{V_i} = \text{AM (rad/s)/V}$$

$$= 10 \times 12 = 120 \text{ (rad/s)/V } \textbf{Ans}$$

$$\text{hence, } V_i = \frac{\omega_o}{\text{AM}} \text{ volt} = \frac{120}{10 \times 12}$$

$$\text{so, } V_i = 1 \text{ V } \textbf{Ans}$$

Worked Example 8.2

Question

The control system above is converted to a closed-loop system, by the inclusion of a tachogenerator. The transfer function for the tacho, T, is 0.1 V/(rad/s). Sketch the system block diagram. Determine the system transfer function, and calculate the input voltage now required to produce an output speed of 120 rad/s.

Answer

A = 10 V/V; M = 12 (rad/s)/V; T = 0.01 V/(rad/s)

The block diagram for the system is shown in Fig. 8.11.

$$\text{input to amplifier} = V_i - \text{T}\omega_o \text{ volt}$$

$$\text{input to motor} = \text{A}(V_i - \text{T}\omega_o) \text{ volt}$$

Continued on p. 354

Worked Example 8.2 (*Continued*)

output of motor, $\omega_o = \text{AM}(V_i - \text{T}\omega o)$ rad/s

therefore, $\omega_o(1 + \text{AMT}) = \text{AM}\,V_i \ldots\ldots\ldots\ldots [1]$

$$\text{and, } \frac{\omega_o}{V_i} = \frac{\text{AM}}{1 + \text{AMT}} \textbf{ Ans}$$

$$\text{From equation [1], } V_i = \frac{\omega_o(1 + \text{AMT})}{\text{AM}}$$

$$= \frac{120 \times (1 + \{10 \times 12 \times 0.01\})}{10 \times 12}$$

hence, $V_i = 2.2$ V **Ans**

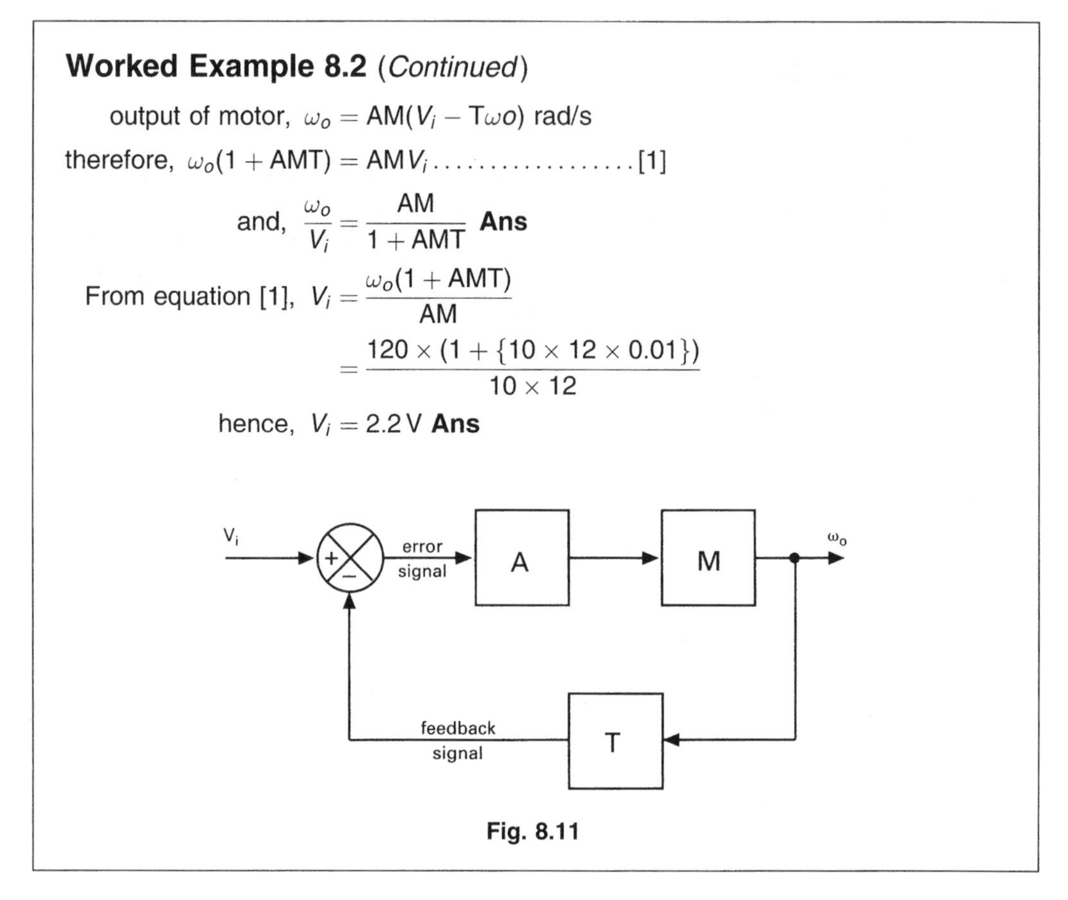

Fig. 8.11

Note: Let us compare the answers for V_i from these two examples. It is clear that, when negative feedback is employed, a larger input voltage is required in order to produce the same output. Thus, the overall 'gain' of the system is reduced. This effect you will study in more detail in the electronics module at level 3.

Considering the second example above, it may well occur to you to ask, how the motor can rotate at some constant speed, when the error voltage must then be zero? The answer is that it cannot! This confirms that any closed-loop system must be error actuated. For the system considered, the motor will actually rotate at a speed slightly less than that demanded. There will therefore be a small error signal into the amplifier. After amplification, this will be just sufficient to keep the motor speed constant. This difference, between the desired output and the actual output under constant output conditions, is known as the steady state error. This concept will be dealt with in more detail in Volume 2 (*Further Electrical and Electronic Principles*).

8.8 A Positional Control System

Let us consider the problem of accurately positioning a heavy load, such as the dish antenna of a large radio telescope. Figure 8.12 shows a simple solution to the problem.

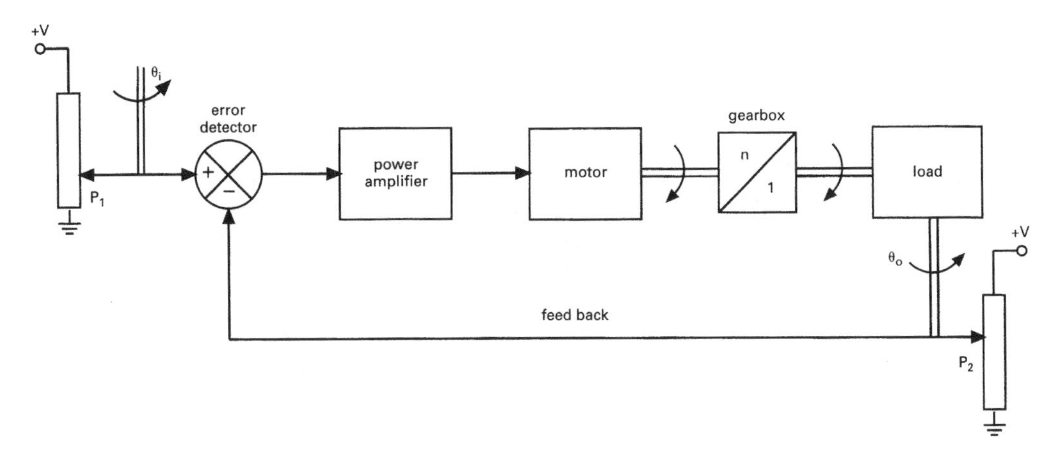

Fig. 8.12

The desired antenna position is applied to the system by rotating the wiper of potentiometer P_1. The voltage tapped off from P_1 provides the demand signal into the error detector (comparator). The wiper of P_2 is driven by the load, so its voltage output represents the load position. This voltage forms the feedback signal. Both potentiometers are supplied from the same reference supply. If θ_o is different to θ_i, the comparator will produce an error signal. The power amplifier amplifies this error signal in order to drive the motor. The gearbox reduces the amount of torque that the motor has to produce, in order to position the load. When the load arrives at the desired position, θ_o will equal θ_i. The error will then be zero, and the antenna will no longer be driven by the motor.

If the system is disturbed, for example by a strong gust of wind, then the antenna would tend to be moved from its correct position. As soon as this happens, a new error signal is created. This will then cause the motor to correct the antenna position, thus reducing the error once more to zero.

8.9 System Response and Stability

The solution proposed for the positional control system just described would appear to be ideal. Unfortunately this is not the case. It will tend to suffer from problems of steady state error and instability. Let us first consider the problem of steady state error. Very large masses have to be moved and positioned. As the load approaches the desired position, the error signal becomes progressively smaller. A point will be reached when the amplified error signal is insufficient to keep the motor driving. Thus the load will come to rest at a position other than that desired. Since the system will now be at rest, there will be a steady state error, θ_{ss}.

The obvious solution to this would be to increase the gain of the amplifier(s) in the system. This would have the effect of keeping the motor running longer. Hence, the load will come to rest closer to the desired position. Thus, the steady state error would be reduced.

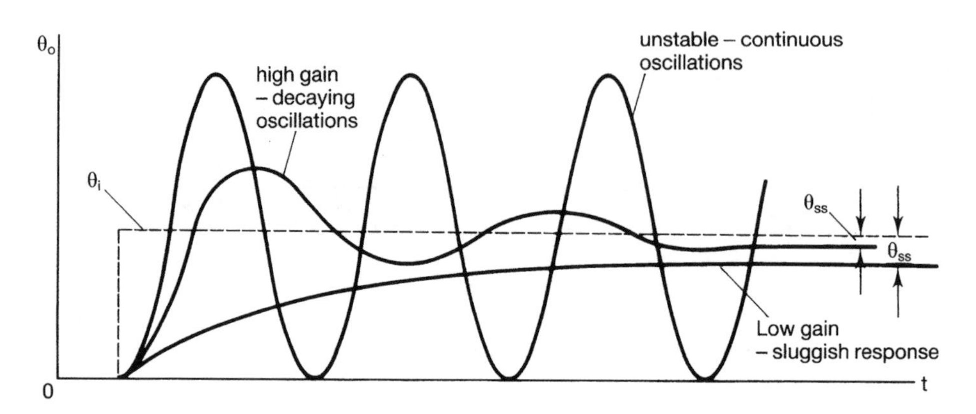

Fig. 8.13

The increased gain in the system will now aggravate the problem of instability. Since the masses involved are very large, they will possess considerable inertia. In addition, the system would be designed to have as little friction as possible. The effect of friction is to slow down, or damp, the movement. Due to the inertia in the system, the antenna will tend to keep rotating, even when the motor input signal has fallen to zero. Thus, the antenna will tend to overshoot the desired position. This will create an error signal to reverse the motor. The load is then driven back towards the desired position. Once more, the inertia effect will cause the load to overshoot, and the above sequence is repeated.

The load will therefore tend to oscillate about the desired position, before finally coming to rest. The greater the gain of the system, the larger the amplitude of these oscillations, but the smaller will be the final steady state error. These effects are illustrated in Fig. 8.13.

Violent oscillations of the antenna are obviously undesirable. With the large masses involved, severe structural damage could result. At the same time, the steady state error needs to be as small as possible. Thus we have conflicting requirements. The full solution to these problems is dealt with in Volume 2.

High system gain is not the only cause of instability in the system. It has been pointed out that negative feedback is employed in control systems. Negative feedback promotes stability, because the system reacts so as to reduce the error signal to zero. In other words, the feedback signal is in opposition to the demand signal. However, it is possible to have time lags and phase shifts around the closed loop. The effect of these is to make the feedback signal reinforce the demand signal. In this situation we have positive feedback. The result is that the oscillations will be maintained, or in the worst case increased. The system then becomes completely unstable. This situation may be compared to pushing a child's swing. To keep the swing in motion, pushes are applied as the swing starts to move away from you. The natural movement of the swing is thus maintained. This is the application of positive feedback. However, if the pushes are timed to coincide with the backward movement, then it will slow down and stop. This is the application of negative feedback.

In Fig. 8.13, the input θ_i, is shown as a sudden change from one value to a new one. This sudden change in the demand is known as a step input.

Assignment Questions

1 State the essential difference between open-loop and closed-loop control systems. Illustrate your answer by describing (include block diagrams) two examples of each type (other than those already described in this book).

2 Explain the following control systems terms: (a) demand or reference signal, (b) error signal, (c) error detector, (d) feedback.

3 Explain the difference between continuous control and ON/OFF (discontinuous) control. Give an example of each, together with relevant block diagrams.

4 What is meant by the term transfer function? With the aid of diagrams, show how the overall transfer function can be obtained for (a) an open-loop system, and (b) a closed-loop system.

5 Explain the effect of steady state error. This form of error is reduced by increasing the system gain. What adverse effect can this course of action have?

6 Explain the difference between positive and negative feedback. Why is positive feedback avoided in control systems?

7 For the control system shown in Fig. 8.14, derive the overall system transfer function.

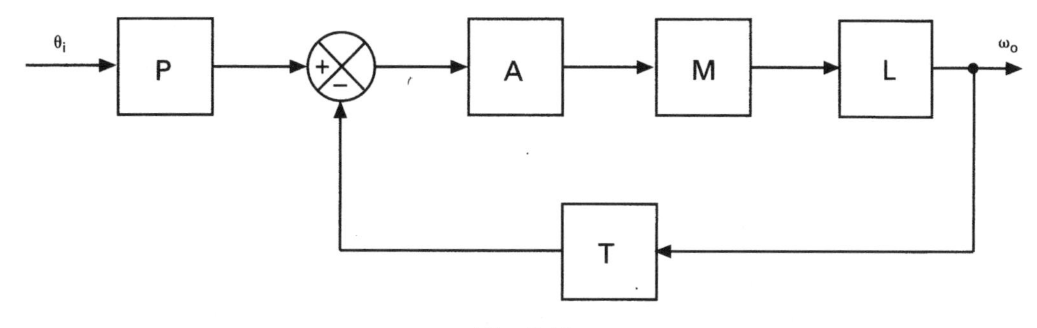

Fig. 8.14

9 Semiconductor Theory and Devices

This chapter explains the behaviour of semiconductors and the way in which they are employed in diodes and transistors.

On completion of this chapter you should be able to:

1 Understand the way in which conduction takes place in semiconductor materials.
2 Understand how these materials are employed to form devices such as diodes and transistors.
3 Know of the different ways in which a transistor may be connected and the effects these will have on the characteristics and parameters of the transistor.
4 Obtain values for the basic transistor parameters from the appropriate transistor characteristics.

9.1 Atomic Structure

In Chapter 1 it was stated that an atom consists of a central nucleus containing positively charged protons, and neutrons, the latter being electrically neutral, surrounded by negatively charged electrons orbiting in layers or shells. Electrons in the inner orbits or shells have the least energy and are tightly bound into their orbits due to the electrostatic force of attraction between them and the nucleus. Electrons in the outermost shell experience a much weaker binding force, and are known as valence electrons.

In conductors it is these valence electrons that can gain sufficient energy to break free from their parent atoms. They thus become 'free' electrons which are available to drift through the material under the influence of an emf and hence are mobile charge carriers which produce current flow.

The shells are identified by letters of the alphabet, beginning with the letter K for the innermost shell, L for the next and so on. Each shell represents a certain energy level, and each shell can contain only up to a certain maximum number of electrons. This maximum possible number of electrons contained in a given shell is governed by the relationship $2n^2$, where n is the number of the shell. Thus the *maximum* number of electrons in the first four shells will be as shown in Table 9.1.

Table 9.1

Shell	*n*	*Max. No. of electrons*
K	1	$2 \times 1^2 = 2$
L	2	$2 \times 2^2 = 8$
M	3	$2 \times 3^2 = 18$
N	4	$2 \times 4^2 = 32$

All things in nature tend to stabilise at their lowest possible energy level, and atoms and electrons are no exception. This results in the lowest energy levels (shells) being filled first until all the electrons belonging to that atom are accommodated. Another feature of the system is that if the outermost shell of an atom is completely full (contains its maximum permitted number of electrons) then the binding force on these valence electrons is very strong and the atom is very stable. To illustrate this consider the inert gas neon. The term inert is used because it is very difficult to make it react to external influences. A neon atom has a total of 10 electrons, two of which are in the K shell and the remaining eight completely fill the L shell. Having a full valence shell is the reason why neon, krypton and xenon are inert gases. In contrast, a hydrogen atom has only one electron, so its valence shell is almost empty and it is a highly reactive element. One further point to bear in mind is that the electrons in the shells (from L onwards) may exist at slightly different energy levels known as subshells. These subshells may also contain only up to a certain maximum number of electrons. This is shown, for the L, M and N shells, in Table 9.2.

Table 9.2

Shell	*L*		*M*			*N*			
Subshell	2s	2p	3s	3p	3d	4s	4p	4d	4f
Max. No.	2	6	2	6	10	2	6	10	14
Total	8		18			32			

9.2 Intrinsic (Pure) Semiconductors

Semiconductors are group 4 elements, which means they have four valence electrons. For this reason they are also known as tetravalent elements. Among this group of elements are carbon (C), silicon (Si), germanium (Ge) and tin (Sn). Of these only silicon and germanium are used as intrinsic semiconductors, with silicon being the most commonly used. Carbon is not normally considered as a semconductor because it can exist in many different forms, from diamond to graphite. Similarly, tin is not used because at normal ambient temperatures it acts as a good conductor. The following descriptions of the behaviour of semiconductor materials will be confined to silicon although the general properties and behaviour of germanium are the same. The arrangement of electrons in the shells and subshells of silicon are shown in Table 9.3.

Table 9.3

K	L		M		
1s	2s	2p	3s	3p	3d
2	2	6	2	2	—

From Table 9.3 it may be seen that the four valence electrons are contained in the M shell, where the 3s subshell is full but the 3p subshell contains only two electrons. However, from Table 9.2 it can be seen that a 3p subshell is capable of containing up to a maximum of six electrons before it is full, so in the silicon atom there is space for a further four electrons to be accommodated in this outermost shell.

Silicon has an atomic bonding system known as covalent bonding whereby each of the valence electrons orbits not only its 'parent' atom, but also orbits its closest neighbouring atom. This effect is illustrated in Fig. 9.1, where the five large circles represent the nucleus and shells K and L of five adjacent atoms (identified as A, B, C, D and E) and the small circles represent their valence electrons, where the letters a, b, c etc. identify their 'parent' atoms.

Concentrating on the immediate space surrounding atom A, it may be seen that there are actually eight valence electrons orbiting this atom; four of its own plus one from each of its four nearest neighbours. This figure is only two-dimensional and is centred on atom A. However, the same arrangement would be found if the picture was centred

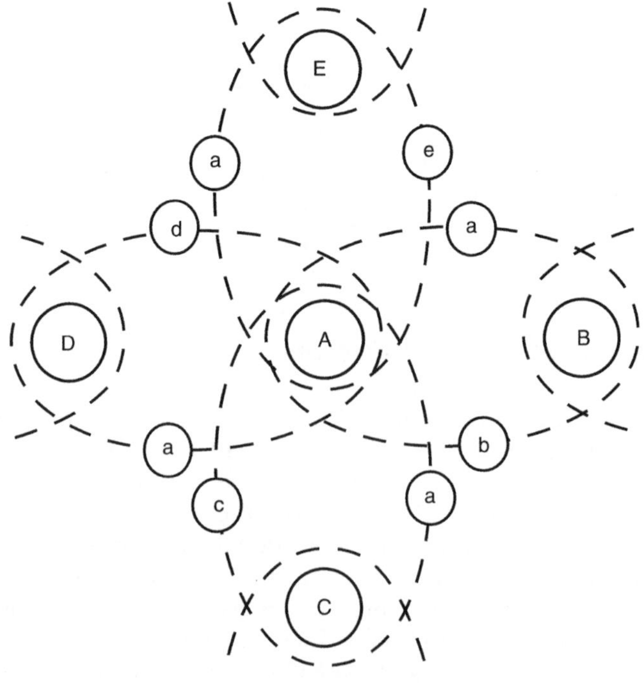

Fig. 9.1

on any given atom in the crystal lattice. In addition, the actual lattice is of course three dimensional. In this case imagine atom A being located at the centre of an imaginary cube with the other four neighbouring atoms being at four of the corners of the cube. Each of these 'corner' atoms is in turn at the centre of another imaginary cube, and so on throughout the whole crystal lattice. The result is what is known as the diamond crystal lattice.

From the above description it may be seen that each silicon atom has an *apparent* valency of 8, which is the same as for the inert gases such as neon. The covalent bonding system is a very strong one so the valence electrons are quite tightly bound into it. It is for this reason that intrinsic silicon is a relatively poor conductor of electrical current, and is called a semiconductor.

9.3 Electron-Hole Pair Generation and Recombination

Although the covalent bond is strong, it is not perfect. Thus, when a sample of silicon is at normal ambient temperature, a few valence electrons will gain sufficient energy to break free from the bond and so become free electrons available as mobile charge carriers. Whenever such an electron breaks free and drifts away from its parent atom it leaves behind a space in the covalent bond, and this space is referred to as a hole. Thus, whenever a bond is broken an electron-hole pair is generated. This effect is illustrated in Fig. 9.2, where the short straight lines represent electrons and the small circle represents a corresponding hole. The large circles again represent the silicon atoms complete with their inner shells of electrons.

The atom which now has a hole in its valence band is effectively a positive ion because it has lost an electron which would normally occupy that space. On the atomic scale, the ion is very massive, is locked into the crystal lattice, and so cannot move. However, electron-hole pair generation will be taking place in a random manner throughout the crystal lattice, and a generated free electron will at some stage drift

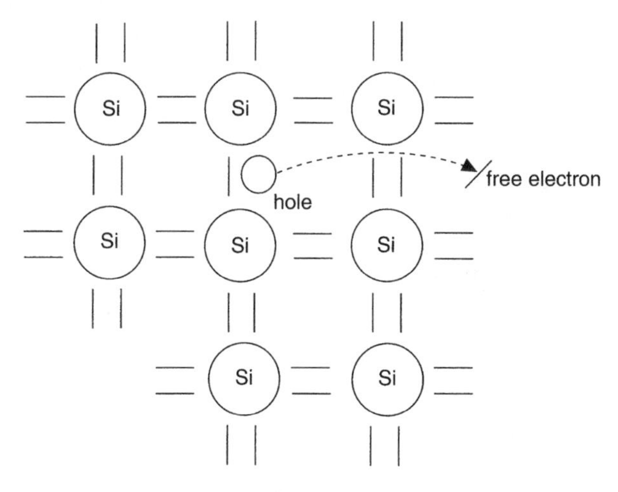

Fig. 9.2

into the vicinity of one of these positive ions, and be captured, i.e. the hole will once more be filled by an electron. This process is known as recombination, and when it occurs the normal charge balance of that atom is restored.

The hole-pair generation and recombination processes occur continuously, and since heat is a form of energy, will increase as the temperature increases. This results in more mobile charge carriers being available, and accounts for the fact that semiconductors have a negative temperature coefficient of resistance, i.e. as they get hotter they conduct more easily. It must be borne in mind that although these thermally generated mobile charge carriers are being produced, the sample of material *as a whole* still remains electrically neutral. In other words, if a 'head count' of all the positve and negative charged particles could be made, there would still be a balance between positive and negative, i.e. for every free electron there will be a corresponding hole.

The concept of the drift of free electrons through the material may be readily understood, but the concept of hole mobility is more difficult to appreciate. In fact the holes themselves cannot move — they are merely generated and filled. However, when a bond breaks down the electron that drifts away will at some point fill a hole elsewhere in the lattice. Thus the hole that has been filled is replaced elsewhere by the newly generated hole, and will *appear* to have drifted to a new location. In order to simplify the description of conduction in a semiconductor, the holes are considered to be mobile positive charge carriers whilst the free electrons are of course mobile negative charge carriers.

9.4 Conduction in Intrinsic Semiconductors

Figure 9.3 illustrates the effect when a source of emf is connected across a sample of pure silicon. The electric field produced by the battery will attract free electrons towards the positive plate and the corresponding holes towards the negative plate. Since the external circuit is completed by conductors, and holes exist only in semiconductors, then how does current actually flow around the circuit without producing an excess of positive charge (the holes) at the left-hand end of the silicon? The answer is quite simple. For every electron that leaves the right-hand end and travels to the positive plate of the battery, another is released from the negative plate and enters the silicon at the left-hand end, where a recombination can occur. This recombination will be balanced by fresh electron-hole pair generation. Thus, within the silicon there

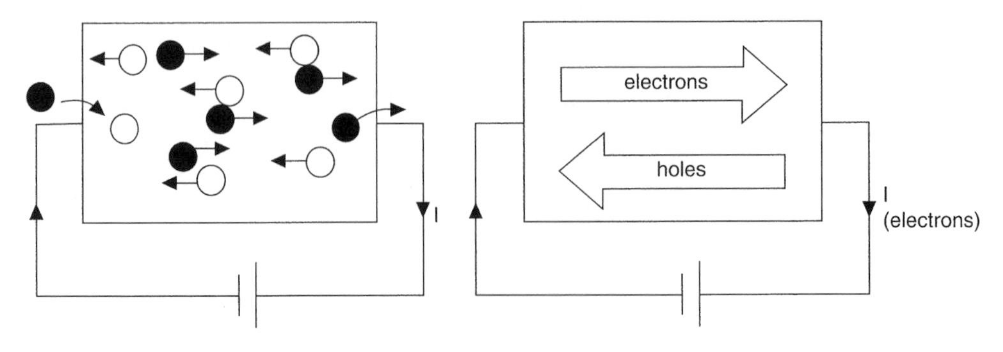

Fig. 9.3

will be a continuous drift of electrons in one direction with a drift of a corresponding number of holes in the opposite direction. In the external circuit the current flow is of course due only to the drift of electrons.

9.5 Extrinsic (Impure) Semiconductors

Although pure silicon and germanium will conduct, as explained in the previous section, their characteristics are still closer to insulators than to conductors. In order to improve their conduction very small quantities (in the order of 1 part in 10^8) of certain other elements are added. This process is known as doping. The impurity elements that are added are either pentavalent (have five valence electrons) or are trivalent (have three valence electrons) atoms. Depending upon which type is used in the doping process determines which one of the two types of extrinsic semiconductor is produced.

9.6 n-type Semiconductor

To produce this type of semiconductor, pentavalent impurities are employed. The most commonly used are arsenic (As), phosphorus (P), and antimony (Sb). When atoms of such an element are added to the silicon a bonding process takes place such that each impurity atom joins the covalent bonding system of the silicon. However, since each impurity atom has five valence electrons, one of these cannot find a place in a covalent bond. These 'extra' electrons then tend to drift away from their parent atoms and become additional free electrons in the lattice. Since these impurities donate an extra free electron to the material they are also known as donor impurities.

As a consequence of each donor atom losing one of its valence electrons, they become positive ions locked into the crystal lattice. Note that free electrons introduced by this process *do not* leave a corresponding hole, although thermally generated electron-hole pairs will still be created in the silicon. The effect of the doping process is illustrated in Fig. 9.4.

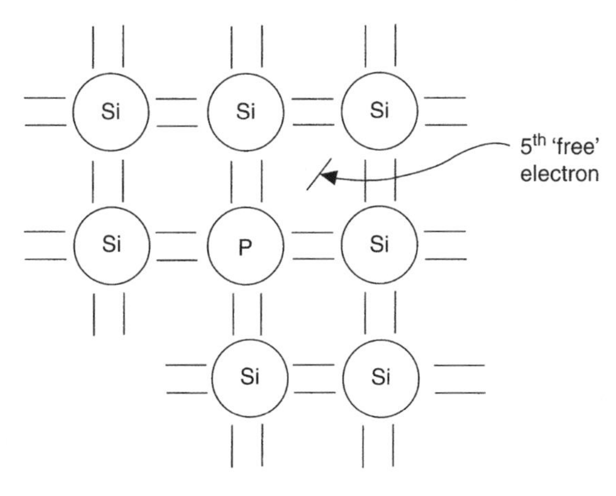

Fig. 9.4

Since the extra charge carriers introduced by the impurity atom are negatively charged electrons, and these will be in addition to the electron-hole pairs, then there will be more mobile negative charge carriers than positive, which is why the material is known as n-type semiconductor. In this case the electrons are the majority charge carriers and the holes are the minority charge carriers. It should again be noted that the material as a whole still remains electrically neutral since for every extra donated free electron there will be a fixed positive ion in the lattice. Thus a sample of n-type semiconductor may be represented as consisting of a number of fixed positive ions with a corresponding number of free electrons, in addition to the thermally generated electron-hole pairs. This is shown in Fig. 9.5.

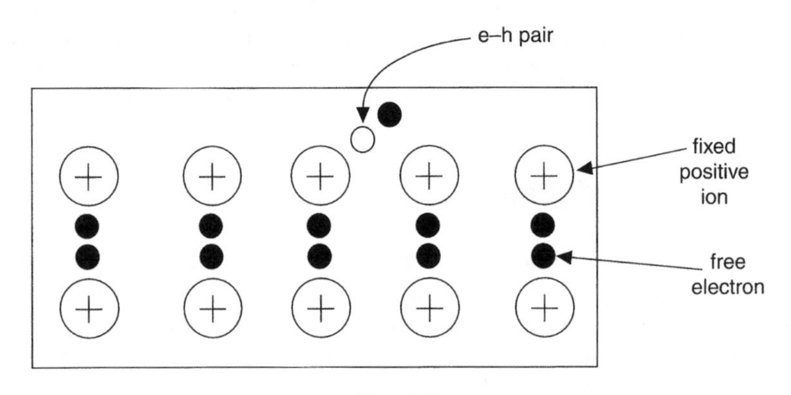

Fig. 9.5

The circuit action when a battery is connected across the material is illustrated in Fig. 9.6. Once more, only electrons flow around the external circuit, whilst within the semiconductor there will be movement of majority carriers in one direction and minority carriers in the opposite direction.

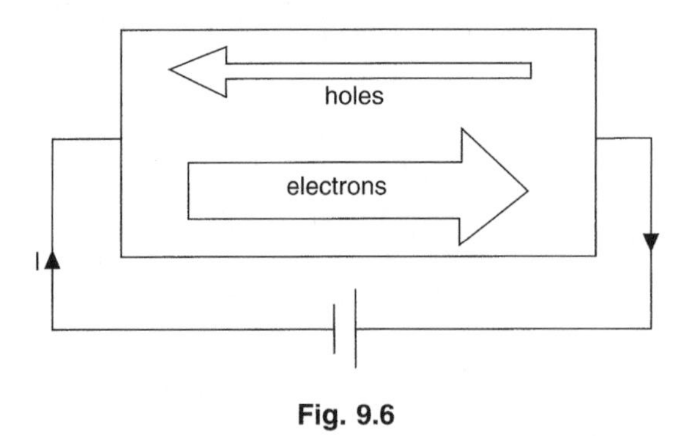

Fig. 9.6

9.7 p-type Semiconductor

In this case a trivalent impurity such as aluminium (Al), gallium (Ga), or indium (In) is introduced. These impurity atoms also join the covalent bonding system, but since they have only three valence electrons there will be a gap or hole in the bond where an electron would normally be required. Due to electron-hole pair generation in the lattice, this hole will soon become filled, and hence the hole will have effectively drifted off elsewhere in the lattice. Since each impurity atom will have accepted an extra electron into its valence band they are known as acceptor impurities, and become fixed negative ions. The result of the doping process is illustrated in Fig. 9.7.

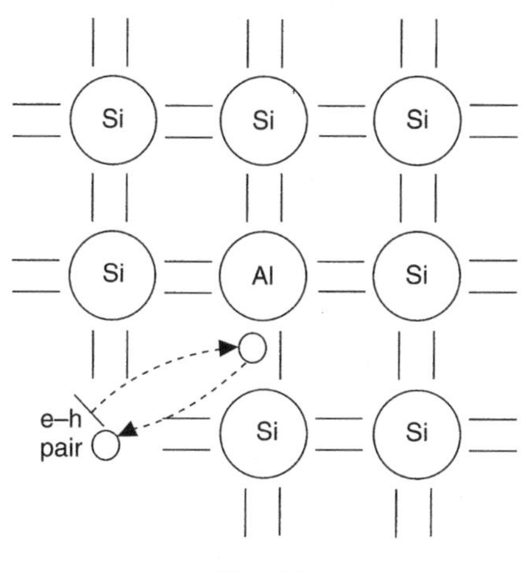

Fig. 9.7

We now have the situation whereby there will be more mobile holes than there are free electrons. Since holes are positive charge carriers, and they will be in the majority, the doped material is called p-type semiconductor, and it may be considered as consisting of a number of fixed negative ions and a corresponding number of mobile holes as shown in Fig. 9.8.

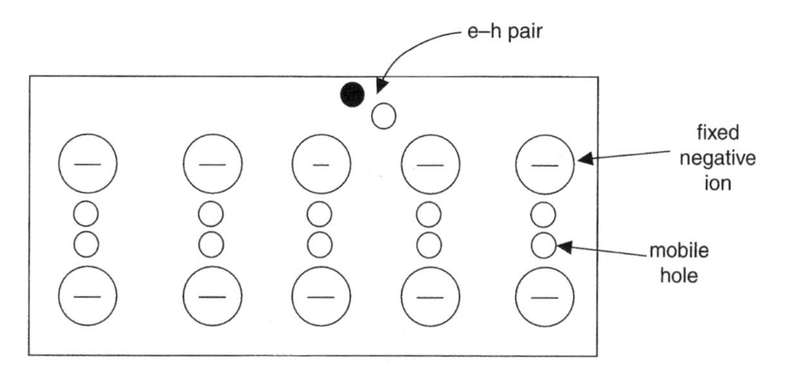

Fig. 9.8

The circuit action when a battery is connected across the material is shown in Fig. 9.9. As the holes approach the left-hand end they are filled by incoming electrons from the battery. At the same time, fresh electron-hole pairs are generated; the electrons being swept to and out of the right-hand end, and the holes drift to the left-hand end to be filled. Once more, the current flow in the semiconductor is due to the movement of

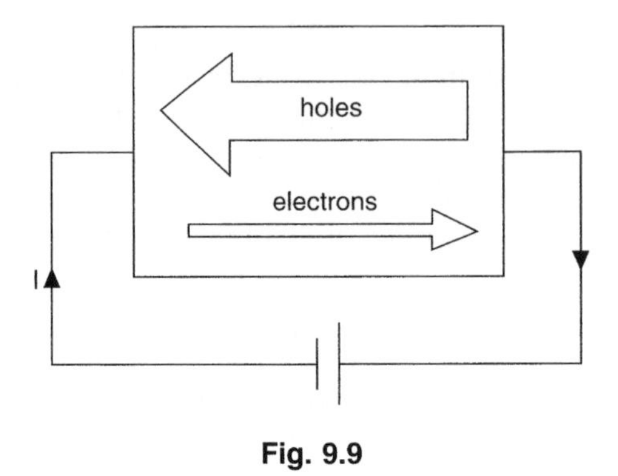

Fig. 9.9

holes and electrons in opposite directions, and only electrons in the external circuit. As with the n-type material, p-type is also electrically neutral.

9.8 The p-n Junction

When a sample of silicon is doped with both donor and acceptor impurities so as to form a region of p-type and a second region of n-type material in the *same crystal lattice*, the boundary where the two regions meet is called a p-n junction. This is illustrated in Fig. 9.10.

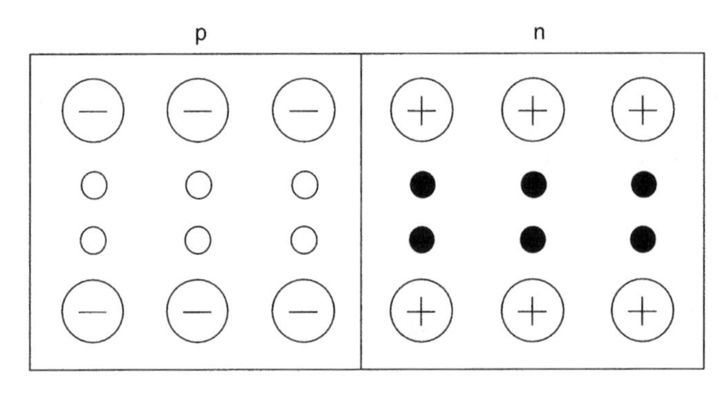

Fig. 9.10

Due to their random movement some of the electrons will diffuse across the junction into the p-type, and similarly some of the holes will diffuse across into the n-type. This effect is illustrated in Fig. 9.11, and from this figure it may be seen that region x acquires a net negative charge whilst region y acquires an equal but positive net charge.

The region between the dotted lines is only about 1 μm wide, and the negative charge on x prevents further diffusion on electrons from the n-type. Similarly the positive charge on y prevents further diffusion of holes from the p-type. This redistribution of charge results in a potential barrier across the junction. In the case of silicon this barrier potential will be in the order of 0.6 to 0.7 V, and for germanium about 0.2 to 0.3 V. Once again note that although there has been some redistribution of charge, the sample of material *as a whole* is still electrically neutral (count up the numbers of positive and negative charges shown in Fig. 9.11).

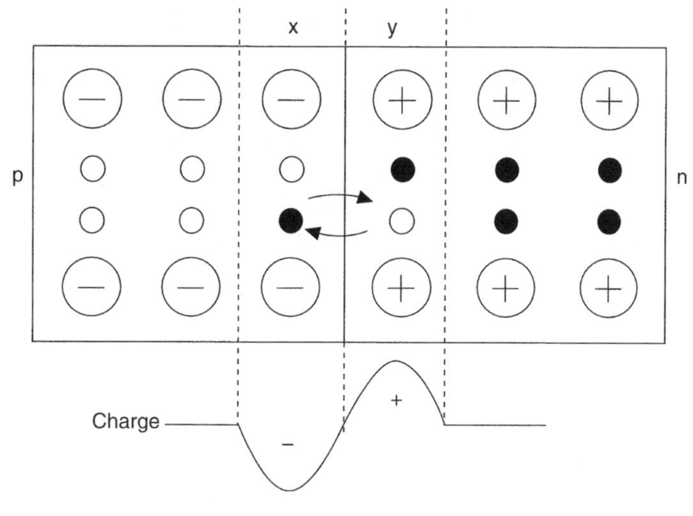

Fig. 9.11

9.9 The p-n Junction Diode

A diode is so called because it has two terminals: the anode, which is the positive terminal, and the cathode, which is the negative terminal. In the case of a p-n junction diode the anode is the p-type and the cathode is the n-type. In Chapter 6 it was stated that a diode will conduct in one direction but not in the other. This behaviour is explained as follows.

9.10 Forward-biased Diode

Figure 9.12 shows a battery connected across a diode such that the positive terminal is connected to the anode and the negative terminal to the cathode.

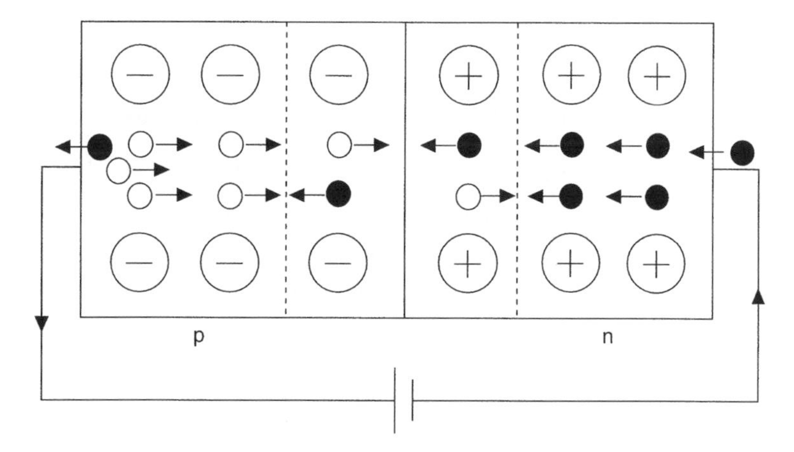

Fig. 9.12

The electric field produced by the battery will cause holes and electrons to be swept toward the junction, where recombinations will take place. For each of these an electron from the battery will enter the cathode. This would have the effect of disturbing the charge balance within the semicoductor, so to counterbalance this a fresh electron-hole pair will be created in the p-type. This newly freed electron will then be attracted to the positive plate of the battery, whilst the hole will be swept towards the junction. Thus the circuit is complete, with electrons moving through the external circuit, and a movement of holes and electrons in the semiconductor. Hence, when the anode of the diode is made positive with respect to the cathode it will conduct, and it is said to be forward biased.

9.11 Reverse-biased Diode

Consider what now happens when the battery connections are reversed (Fig. 9.13). The electric field of the battery will now sweep all the mobile holes into the p-type and all the free electrons into the n-type. This leaves a region on either side of the junction which has been depleted of all of its mobile charge carriers. This layer thus acts as an insulator, and is called the depletion layer. There has been a redistribution of charge within the semiconductor, but since the circuit has an insulating layer in it, current cannot flow. The diode is said to be in its blocking mode.

However, there is no such thing as a perfect insulator, and the depletion layer is no exception. Although all the mobile charge carriers provided by the doping process have been swept to opposite ends of the semiconductor, there will still be some thermally generated electron-hole pairs. If such a pair is generated in the p-type region, the electron will be swept across the junction by the electric field of the battery. Similarly, if the pair is generated in the n-type, the hole will be swept across the junction. Thus a very small reverse current (in the order of microamps) will flow, and is known as the reverse leakage current. Since this leakge current is the result of thermally generated electron-hole pairs, then as the temperature is increased so too will the leakage current.

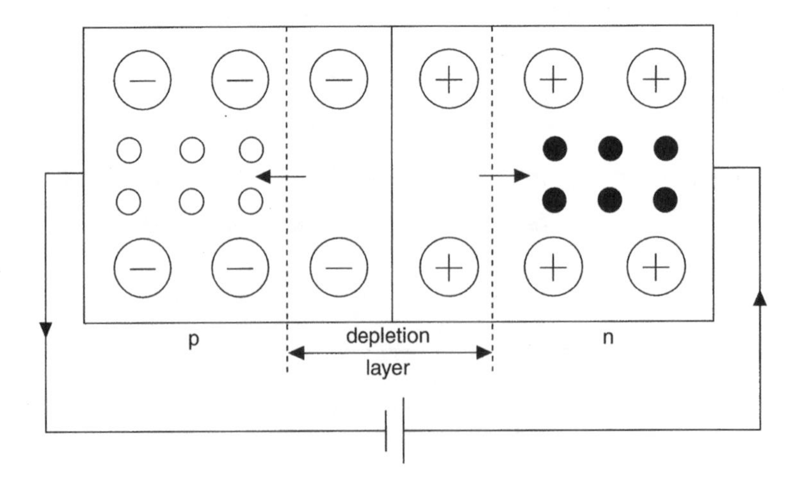

Fig. 9.13

9.12 Diode Characteristics

The characteristics of a device such as a diode can be best illustrated by means of a graph (or graphs) of the current flow through it versus applied voltage. Circuits for determining both the forward and reverse characteristics are shown in Fig. 9.14.

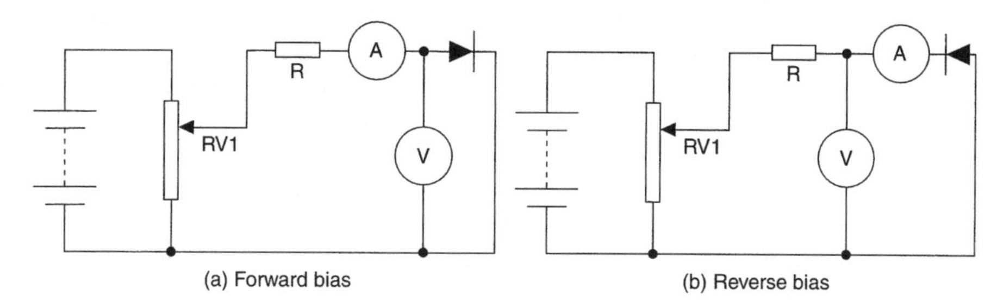

Fig. 9.14

Note the change of position of the voltmeter for the two different tests. In (a) the voltmeter measures only the small p.d. across the diode itself, and not any p.d. across the ammeter. In (b) the ammeter measures only the leakage current of the diode, and does not include any current drawn by the voltmeter. Ideally the voltmeter should not draw any current at all, so it is recommended that a DVM is used rather than a moving coil instrument such as an AVOmeter. The procedure in each case is to vary the applied voltage, in steps, by means of RV1 and record the corresponding current values. When these resuts are plotted, for both silicon and germanium diodes, the graphs will typically be as shown in Fig. 9.15. The very different scales for both current and voltage for the forward and reverse bias conditions should be noted. Also, the actual values shown for the forward current scale and the reverse voltage scale can vary considerably

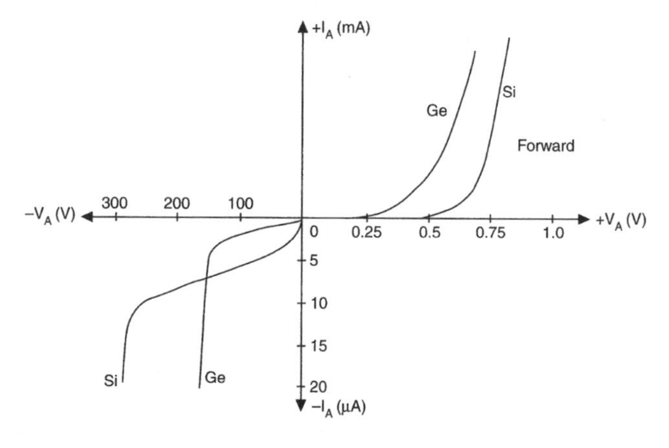

Fig. 9.15

from those shown, depending upon the type of diode being tested, i.e. whether it be a small signal diode or a power rectifying diode. In the case of the latter, the forward current would usually be in amperes rather than milliamps.

The sudden increase in reverse current occurs at a reverse voltage known as the reverse breakdown voltage. The effect occurs because the intensity of the applied electric field causes an increase in electron-hole pair generation. These electron-hole pairs are not due to temperature, but the result of electrons being torn from bonds by the electric field. This same field will rapidly accelerate the resulting charge carriers and as they cross the junction they will collide with atoms. These collisions will free more charge carriers, and the whole process builds up very rapidly. For this reason the effect is known as avalanche breakdown, and will usually result in the destruction of the diode.

When the impurity doping of the semiconductor is heavier than 'normal', the depletion layer produced is much thinner. In this case, when breakdown occurs, the charge carriers can pass through the depletion with very little chance of collisions taking place. This type of breakdown is known as zener breakdown and such diodes are called zener diodes. These devices are always used in their reverse bias mode, and can do so without suffering permanent damage. Zener diodes can be manufactured to have precisely defined breakdown voltages in the range of 1 V up to about 100 V.

9.13 The Bipolar Junction Transistor (BJT)

This is a three-layer device where the semiconductor is doped so as to produce a 'sandwich' of either two n-type layers with a p-type layer between them, or two p-type layers with an n-type layer between them. The former is known as an npn transistor and the latter as a pnp transistor. The term bipolar refers to the fact that conduction within the device is due to the movement of both positive and negative charge carriers (holes and electrons). Both forms of BJT are illustrated in Fig. 9.16.

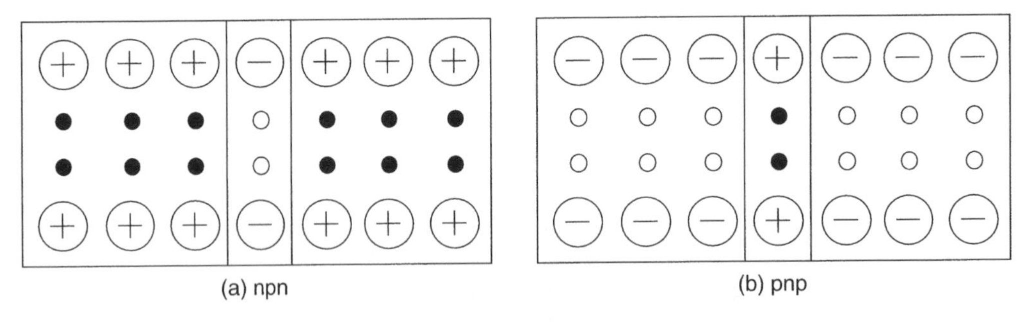

Fig. 9.16

From Fig. 9.16 it may be seen that the transistor has two p-n junctions, each of which will have the same characteristics as a p-n junction diode. The middle layer of the 'sandwich' is much narrower and more lightly doped than the two outer layers.

Since the device has three layers then it is possible to make three electrical connections to it, i.e it is a three-terminal device. In addition, it is possible to connect external sources of emf so as to either forward or reverse bias the two internal p-n junctions. The three layers are named the emitter, base and collector, and the action of an npn transistor is illustrated in Fig. 9.17.

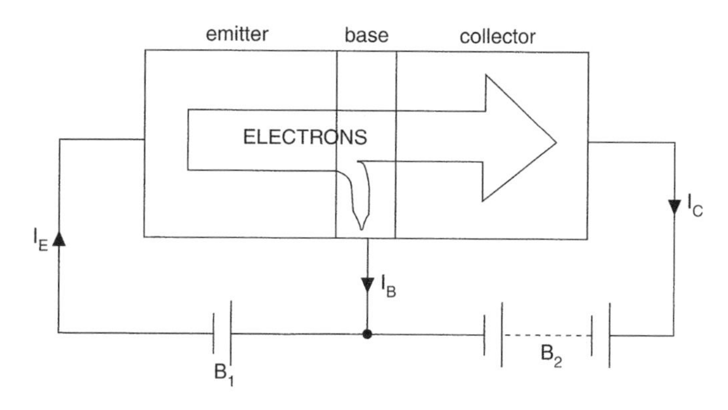

Fig. 9.17

Battery B_1 forward biases the emitter-base junction (positive to 'p' and negative to 'n'), whilst the collector-base junction is reversed biased by battery B_2. The effect of B_1 is to accelerate free electrons from the emitter towards the base. Since the base region is relatively narrow and lightly doped, the vast majority (over 95%) of these electrons will pass straight through the base and cross the second junction into the collector region. Once they reach the collector they will be attracted by the positive potential provided by battery B_2, and flow to this battery. The small percentage of the emitter electrons that recombine with holes in the base will be replaced by electron-hole pairs, the electrons from which will leave the base and flow out to the positive plate of B_1. Applying Kirchhoff's current law to the circuit it may be seen that

$$I_E = I_C + I_B \text{ amp} \tag{9.1}$$

This equation will always hold true regardless of the way in which the transistor is connected into a circuit, and could be said to describe normal transistor action. It may also be seen that in Fig. 9.17 the connection to the base is a common point in the external circuit, and when the transistor is connected in this way it is said to be connected in the common base configuration.

For obvious reasons, the depiction of the transistor in the form of Fig. 9.17 is not convenient when drawing circuit diagrams, and the circuit symbol for an npn transistor and its use for the circuit of Fig. 9.17 are shown in Fig. 9.18.

In the circuit symbol for a transistor the arrowhead indicating the emitter always points in the direction of *conventional* current flow. This is the same convention as used with the junction diode symbol. It should also be noted that in Figs. 9.17 and 9.18(b) the current arrows are actually indicating the direction of electron flow, but if all of these arrows were reversed, the relationship $I_E = I_C + I_B$ would still be true. In Fig. 9.18(b) the current arrows indicate conventional current flow.

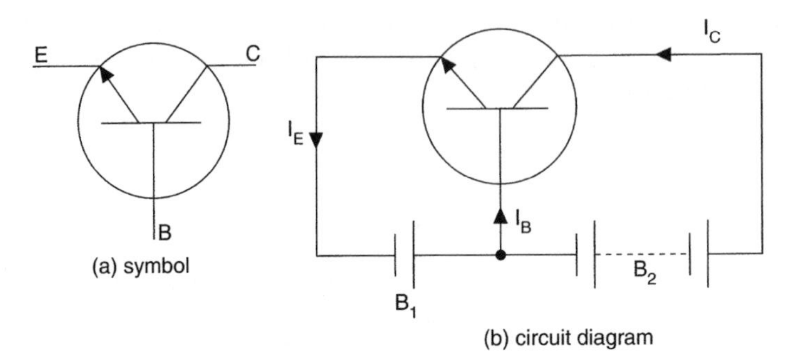

Fig. 9.18

The action of a pnp transistor would be similar to that just described except that electron flow within the transistor would be replaced by holes, and the polarities of the two batteries would have to be reversed to maintain the original biasing conditions for the two p-n junctions. In addition, the circuit symbol for a pnp transistor would have the emitter arrow pointing in the opposite direction to that for an npn transistor as shown in Fig. 9.19.

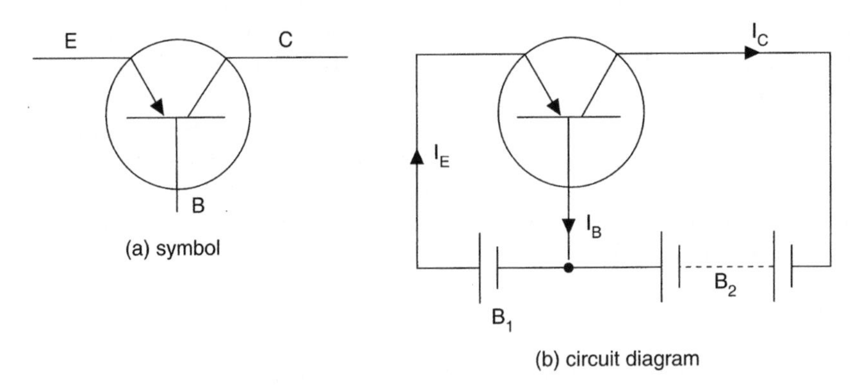

Fig. 9.19

9.14 Transistor Circuit Configurations

Since the transistor is a three-terminal device there are three possible ways of connecting it into a circuit. These are known as common base, common emitter and common collector configurations, and are shown in Fig. 9.20

The characteristics and behaviour of the transistor vary depending upon which configuration is used. The most commonly used configuration is common emitter. The common collector configuration is used for specialised applications and a transistor connected in this way is more normally referred to as an emitter follower.

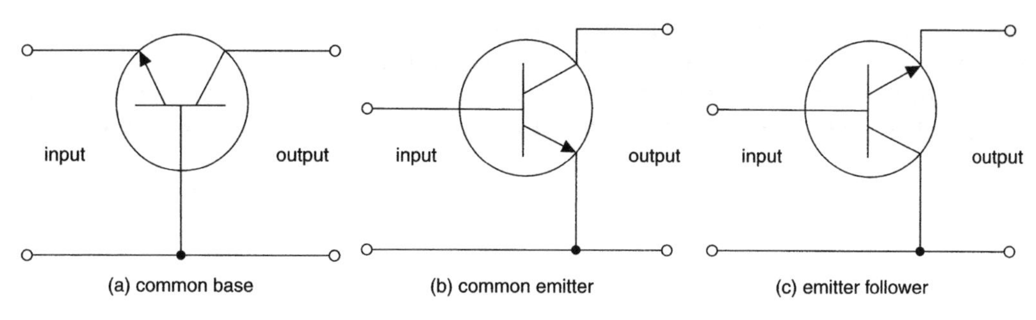

Fig. 9.20

9.15 BJT Common Emitter Characteristics

A circuit suitable for obtaining the various characteristics for a common emitter connected transistor is shown in Fig. 9.21.

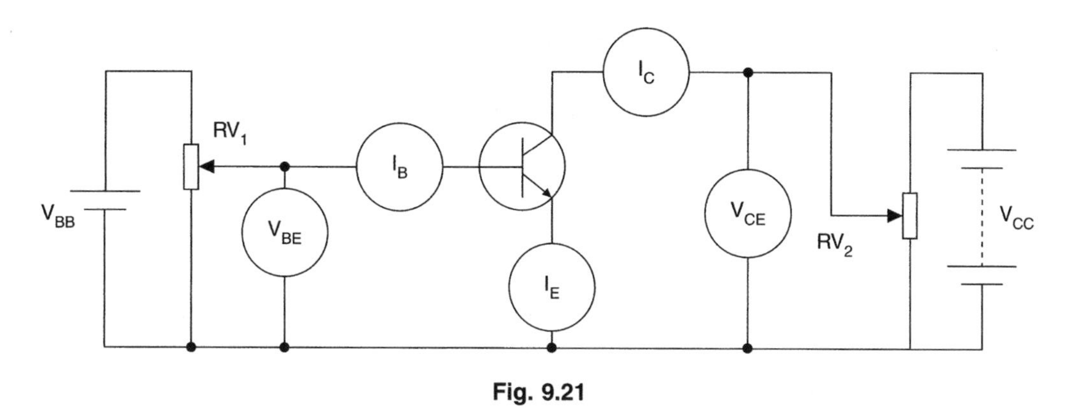

Fig. 9.21

Battery V_{BB} supplies the forward bias for the base-emitter junction, whilst V_{CC} reverse biases the collector-emitter junction. The different characteristics are obtained as follows.

Input Characteristic The collector-emitter voltage, V_{CE}, is set to a predetermined value by means of RV_2. The input base-emitter voltage, V_{BE}, is then varied in steps and the corresponding values of base current, I_B, noted. Since we are dealing with a forward-biased p-n junction it is no surprise that the input characteristic is the same as for a forward-biased junction diode, as shown in Fig. 9.22.

The input resistance of the diode, R_{IN}, is defined as

$$R_{IN} = \frac{\delta V_{BE}}{\delta I_B} \text{ ohm, with } V_{CE} \text{ constant} \tag{9.2}$$

A typical value for R_{IN} would be between 1000 and 1500 ohm.

Output Characteristics For this test I_B is set to a fixed value using RV_1; V_{CE} is varied in steps by means of RV_2 and the corresponding values of I_C noted. This procedure is repeated for a number of different fixed values of I_B. From the results obtained a family of output characteristics can be plotted as shown in Fig. 9.23.

Note that, for the sake of clarity, the height of the characteristic for $I_B = 0$ has been exaggerated. Since there is zero input current the only current that can flow is the

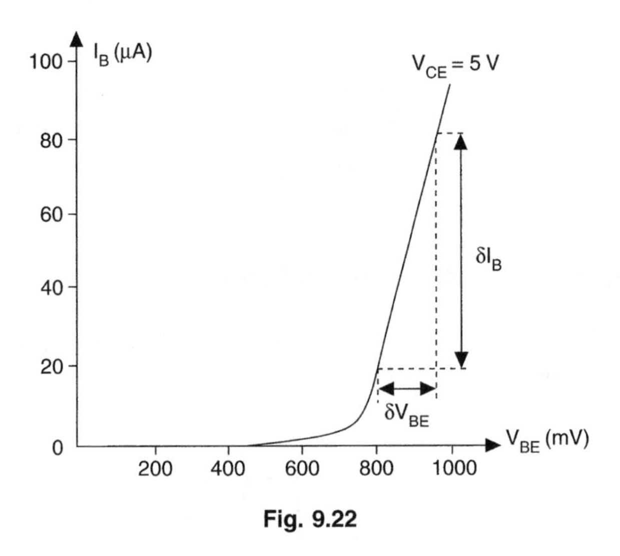

Fig. 9.22

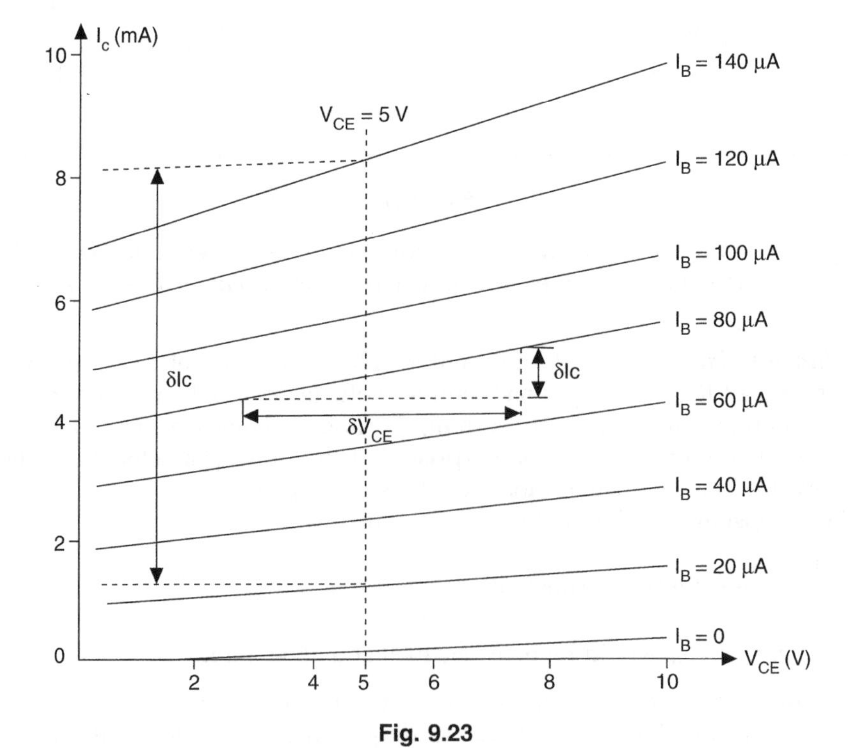

Fig. 9.23

reverse leakage current across the collector-emitter junction. Since this current will be in the order of a few microamps, if this characteristic was drawn to the same scale as the axis for I_C, it would virtually merge with the horizontal axis. This current is referred to as I_{CEO}.

The output characteristics of the transistor are the most useful, since they may be used not only to determine certain parameters of the transistor, but can also be used to predict the behaviour of an amplifier circuit. The output resistance, R_{OUT}, of the transistor is defined as

$$R_{OUT} = \frac{\delta V_{CE}}{\delta I_C} \text{ ohm, with } I_B \text{ constant} \tag{9.3}$$

It is essential to specify the value of base current in this case, since it may be seen that the slope of the characteristics changes with base current. Typical values for R_{OUT} range from 5 kΩ to 100 kΩ.

The large signal or d.c. current gain of the transistor, h_{FE}, is defined as

$$h_{FE} = \frac{\delta I_C}{\delta I_B}, \text{ with } V_{CE} \text{ constant} \tag{9.4}$$

This parameter of the transistor may also be obtained from the output characteristics. In Fig. 9.23, the vertical dotted line represents V_{CE} constant at 5 V. From the intersections of this line with the graphs for $I_B = 20\,\mu$A and $I_B = 100$ A we can determine the total change in collector current, δI_C, for the corresponding change in base current, δI_B, which in this case is 80 A. This relationship between I_C and I_B may also be obtained by plotting the transfer or mutual characteristic as follows.

Transfer Characteristic For this test the collector voltage, V_{CE}, is maintained constant. The base current is varied in steps, and the corresponding values of collector current noted. These results yield the graph shown in Fig. 9.24.

Typical values for h_{FE} can range from 10 to 1000.

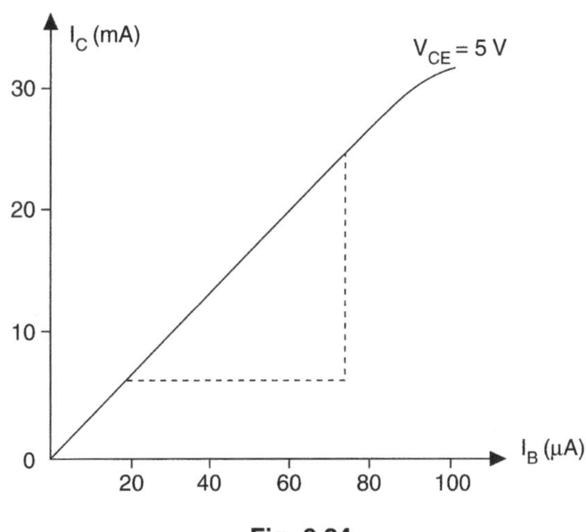

Fig. 9.24

Worked Example 9.1

Question

The input and output characteristics for a certain transistor are as shown in Fig. 9.25.

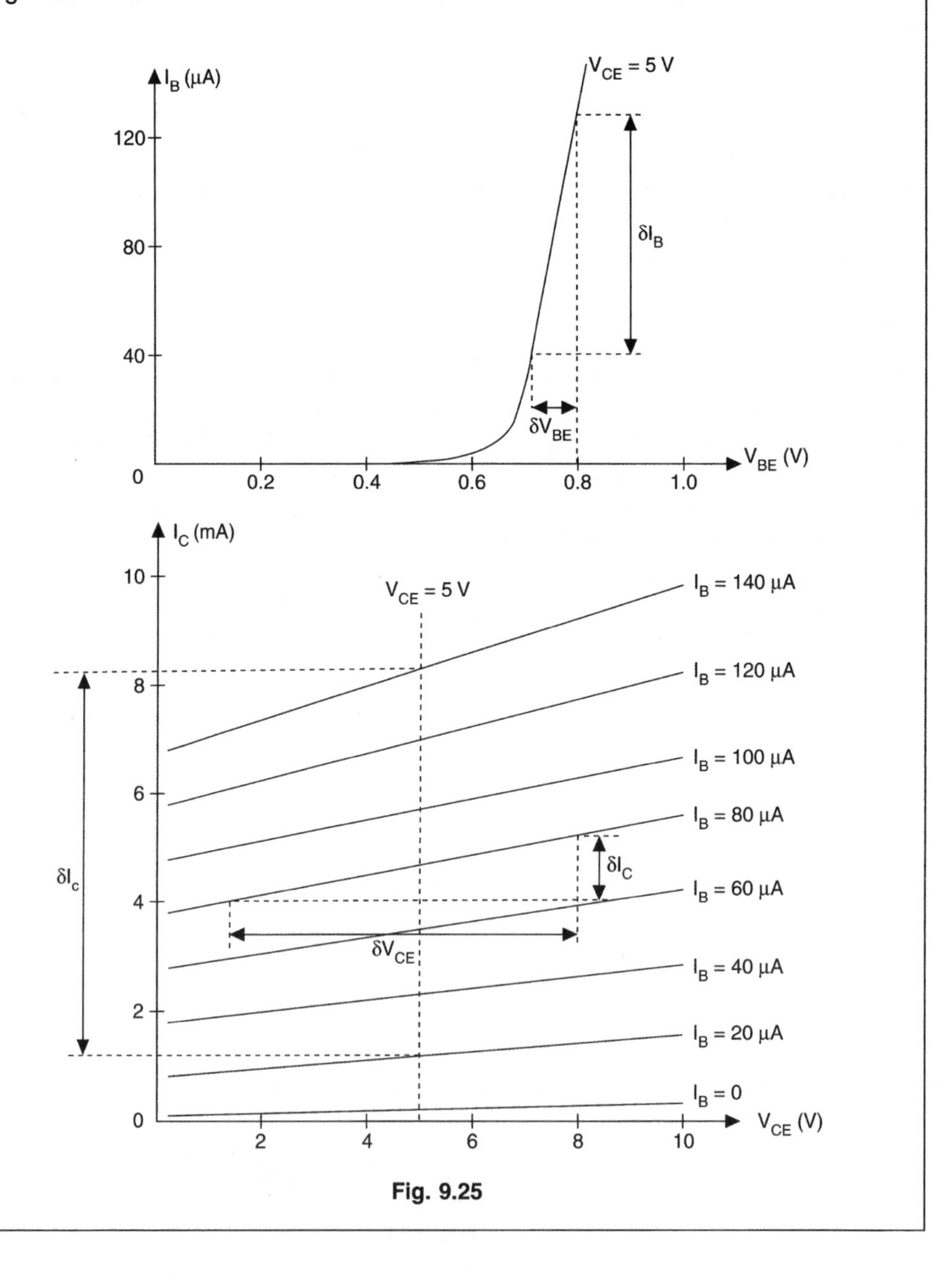

Fig. 9.25

Using these characteristics determine the following transistor parameters (a) input resistance, (b) output resistance at a base current of 80 μA, and (c) large signal current gain with V_{CE} constant at 5 V.

Answer

(a) Using the linear portion of the input characteristic, for a change in V_{BE} of 85 mV the corresponding change of I_B is 88 μA. Hence,

$$R_{IN} = \frac{\delta V_{BE}}{\delta I_B} \text{ ohm} = \frac{85 \times 10^{-3}}{88 \times 10^{-6}}$$

and $R_{IN} = 966\,\Omega$ **Ans**

(b) From the output characteristic for $I_B = 80\,\mu$A, a change of V_{CE} from 8 V to 1.4 V results in a corresponding change in I_C from 5.2 mA to 4 mA. Thus,

$$\delta V_{CE} = 6.6\,V; \text{ and } \delta/I_C = 1.2\,mA$$

$$R_{OUT} = \frac{\delta V_{CE}}{\delta I_C} \text{ ohm} = \frac{6.6}{1.2 \times 10^{-3}}$$

and $R_{OUT} = 5.5\,\text{k}\Omega$ **Ans**

(c) A vertical line at $V_{CE} = 5$ V (i.e. V_{CE} constant at this value) intersects the graphs for $I_B = 140\,\mu$A and $I_B = 20\,\mu$A at $I_C = 8.3$ mA and $I_C = 1.2$ mA respectively. Thus, $\delta I_C = (8.3 - 1.2)$ mA $= 7.1$ mA; and $\delta I_B = (140 - 20)\,\mu$A $=$ 120 μA

$$h_{FE} = \frac{\delta I_C}{\delta I_B} = \frac{7.1 \times 10^{-3}}{120 \times 10^{-6}}$$

and $h_{FE} = 59$ **Ans**

9.16 BJT Common Base Characteristics

Using the circuit of Fig. 9.26, and adopting a similar procedure to that for the common emitter circuit in the previous section (in this case, wherever emitter was specified you now substitute base), the characteristics for this configuration may be obtained.

Input Characteristic This will be of the same shape as that obtained for the common emitter configuration since it is the same reversed-biased junction. The essential difference is that the input current is now the emitter current instead of the base current. This will have a considerable effect on the value of the transistor input resistance, R_{IN}, which is defined as

$$R_{IN} = \frac{\delta V_{EB}}{\delta I_E} \text{ with } V_{CB} \text{ constant} \qquad (9.5)$$

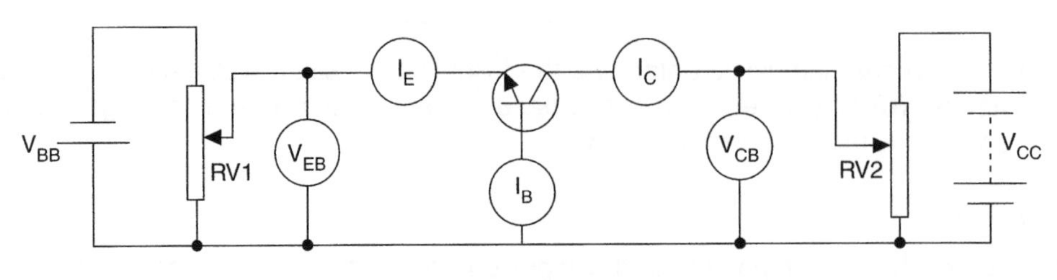

Fig. 9.26

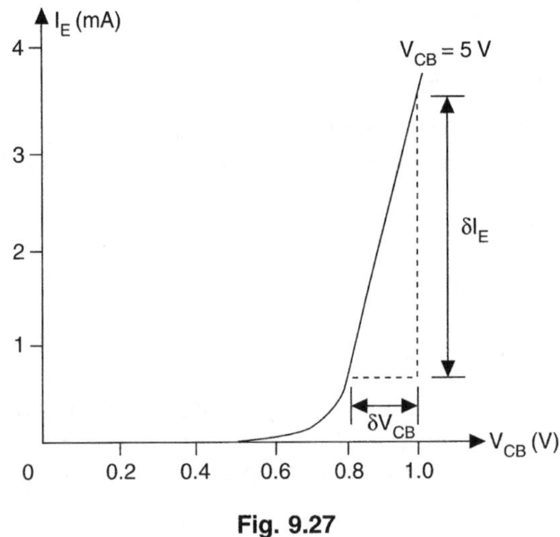

Fig. 9.27

The input characteristic is shown in Fig. 9.27.

Output Characteristics A typical family of output characteristics for a common base connected transistor is shown in Fig. 9.28.

It may be seen that in this case the graphs are almost horizontal, so that the output resistance will be of a high value, typically in megohms, where

$$R_{OUT} = \frac{\delta V_{CB}}{\delta I_C} \text{ ohm, with } I_E \text{ constant} \tag{9.6}$$

The transistor current gain, h_{FB}, is defined as

$$h_{FB} = \frac{\delta I_C}{\delta I_E} \quad \text{with } V_{CB} \text{ constant} \tag{9.7}$$

This parameter may also be obtained from the output characteristics as in the case of the common emitter connection, or from the transfer characteristic as follows.

Transfer Characteristic A plot of collector current versus emitter current is shown in Fig. 9.29. Since both current scales are the same, and the slope of the graph is just less than 45°, then it can be appreciated that the current gain must be less than unity. Typical values for h_{FB} are in the range 0.95 to 0.996.

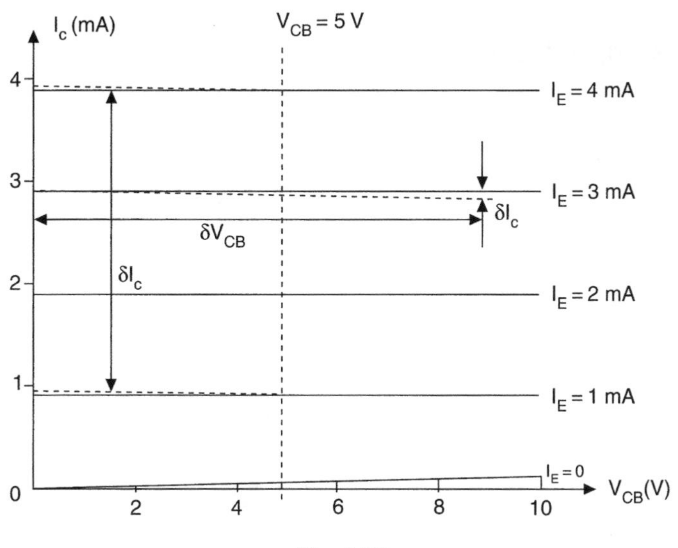

Fig. 9.28

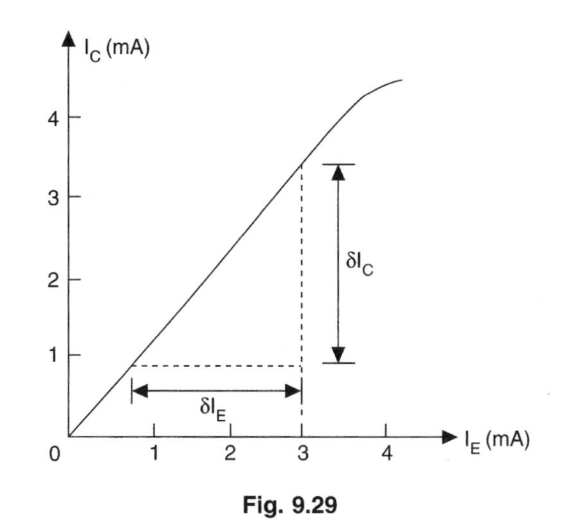

Fig. 9.29

Worked Example 9.2

Question

The characteristics for a transistor connected in common base configuration are as shown in Fig. 9.30. Using these characteristics determine (a) the transistor current gain for a collector voltage of 5 V, (b) its output resistance, and (c) its input resistance.

Continued on p. 380

Worked Example 9.2 (*Continued*)

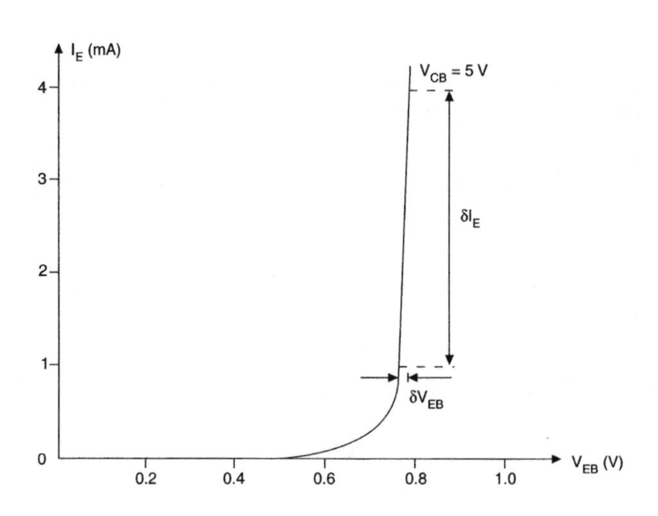

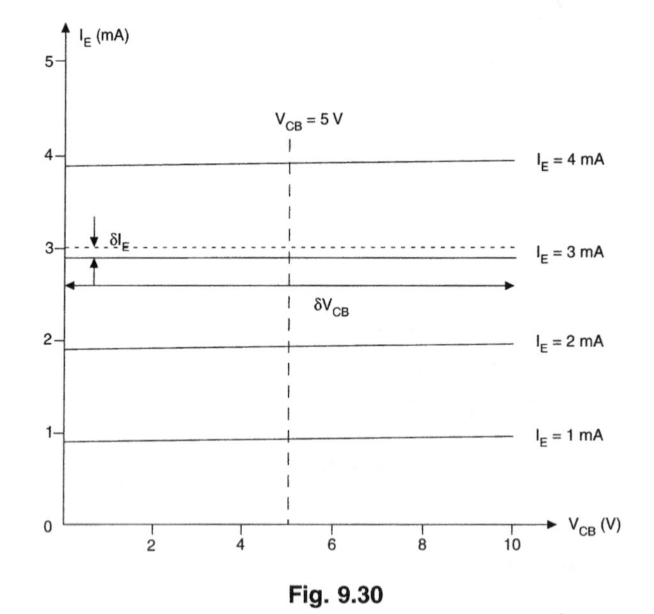

Fig. 9.30

Answer

(a) On the output characteristics, a vertical line is drawn at $V_{CB} = 5\,\text{V}$, and where this line intersects the graphs for $I_E = 4\,\text{mA}$ and $I_E = 1\,\text{mA}$, the corresponding values for I_C are 3.93 mA and 0.95 mA respectively.

Thus $\delta I_C = 2.98\,\text{mA}$; and $\delta I_E = 3\,\text{mA}$

$$\text{and } h_{FB} = \frac{\delta I_C}{\delta I_E} = \frac{2.98}{3.0}$$

$$h_{FB} = 0.993 \text{ } \mathbf{Ans}$$

(b) Since all of the output characteristics have virtually the same slope it does not matter which one is used to determine the output resistance. Thus, from the graph where $I_E = 3\,\text{mA}$, for the change of $\delta V_{CB} = 10\,\text{V}$, the corresponding change $\delta I_E = 0.1\,\text{mA}$.

$$R_{OUT} = \frac{\delta V_{CB}}{\delta I_E} \text{ ohm} = \frac{10}{0.1 \times 10^{-3}}$$

$$R_{OUT} = 100\,\text{k}\Omega \text{ } \mathbf{Ans}$$

(c) From the input characteristic, a change $\delta V_{EB} = 0.2\,\text{V}$ results in a corresponding change $\delta I_E = 3\,\text{mA}$.

$$R_{IN} = \frac{\delta V_{EB}}{\delta I_E} \text{ ohm} = \frac{0.2}{3 \times 10^{-3}}$$

$$R_{IN} = 66.7\,\Omega \text{ } \mathbf{Ans}$$

9.17 Relationship between h_{FE} and h_{FB}

From equation (9.1) we know that regardless of which configuration is used the relationship $I_E = I_C + I_B$. This equation may be rewritten in terms of the corresponding changes in these three currents, as follows

$$\delta I_E = \delta I_C + \delta I_B$$

$$\text{so, } \frac{\delta I_E}{\delta I_C} = \frac{\delta I_C}{\delta I_C} + \frac{\delta I_B}{\delta I_C} = 1 + \frac{\delta I_B}{\delta I_C} \qquad [1]$$

Now, from equation (9.7) we can say that $\frac{\delta I_E}{\delta I_C} = \frac{1}{h_{FB}}$

and, from equation (9.4) we can say that $\frac{\delta I_B}{\delta I_C} = \frac{1}{h_{FE}}$

substituting these into [1] above yields

$$\frac{1}{h_{FB}} = \frac{1}{h_{FE}} + 1 = \frac{1 + h_{FE}}{h_{FE}}$$

$$\text{so, } h_{FB} = \frac{h_{FE}}{1 + h_{FE}} \qquad (9.8)$$

$$\text{similarly, } h_{FE} = \frac{h_{FB}}{1 - h_{FB}} \qquad (9.9)$$

Provided that we know, from manufacturer's data, the value of either h_{FE} or h_{FB} then we can calculate the transistor current gain in either configuration.

The main parameters of a BJT connected in the three different configurations are summarised in Table 9.4.

Table 9.4

Configuration	R_{IN}	R_{OUT}	h_F
common emitter	Medium	Medium	10–1000
common base	Low	High	<1
emitter follower	High	Low	<1

9.18 The Unipolar Junction Transistor (UJT)

The conduction in this device relies solely on the movement of either electrons or holes, and *not* the combination of both; hence the name unipolar. In addition, the current flow through it is controlled by the application of an electric field, so it is more commonly referred to as a field effect transistor or FET.

There are two main types of FET: the junction gate (JUGFET) which is usually simply called a FET, and the insulated gate (IGFET) which is more often called a MOSFET or simply a MOST.

9.19 The JUGFET

This device consists of a bar of either n- or p-type silicon into which is diffused two regions of the opposite type of semiconductor. Since electrons are more mobile than holes an n-type bar is more common. This is called an n-channel JUGFET, and is illustrated in Fig. 9.31.

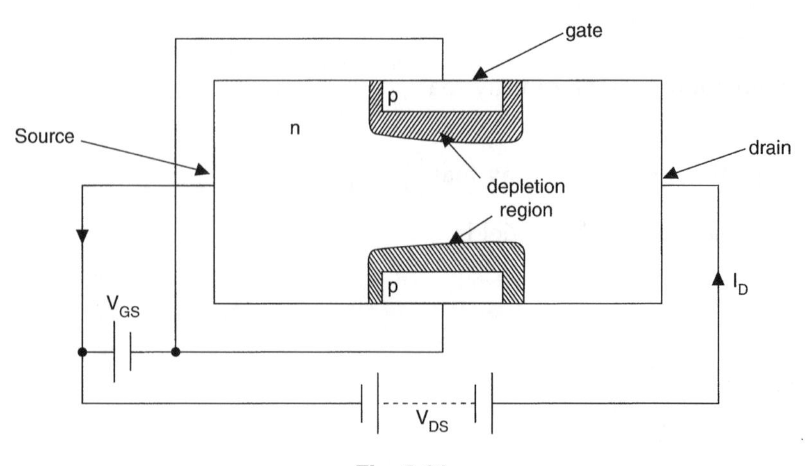

Fig. 9.31

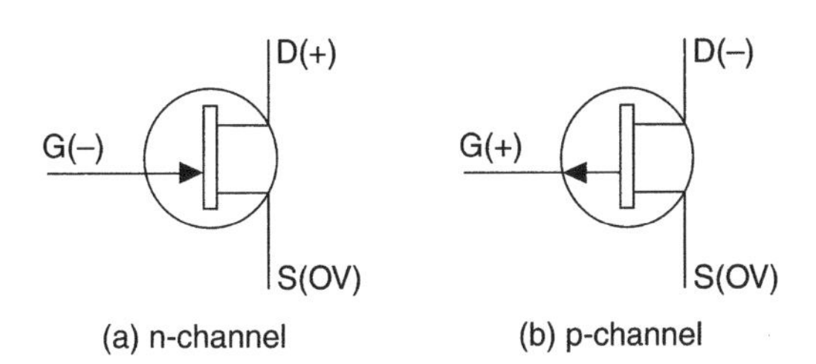

(a) n-channel (b) p-channel

Fig. 9.32

The n-type bar is known as the channel (through which current can flow). Considering Fig. 9.31, the effect of V_{DS} is to cause electrons to flow from left to right through the channel. For this reason the connection at the left-hand end is called the source, and at the other end is the drain. The two p regions are connected together, and the connection here is called the gate. The gate-source junction is reversed biased by V_{GS}, which produces a depletion region in the channel. The shape and size of this depletion region depends upon two factors. The value of V_{GS} determines the extent to which the depletion region extends into the channel. In addition, the electric field through the channel due to V_{DS} will be strongest at the source end. This will have the effect of increasing the width of the depletion regions towards this end of the channel, resulting in the wedge shape shown in Fig. 9.31.

The current flow through the FET is therefore determined by the effective length and cross-section of the conducting channel between the depletion regions. This may be compared to controlling the flow of water through a hosepipe by squeezing it. The circuit symbols for both n-channel and p-channel FETs are shown in Fig. 9.32.

9.20 n-channel JUGFET Characteristics

These characteristics may be obtained in a similar manner to that described for the BJT. However, in the case of the FET, since the gate-source junction is a reverse-biased p-n junction the gate draws negligible current (leakage current only) and the resistance of this junction will be in the order of tens of megohm. For this reason an input characteristic is not relevant. Typical output and mutual transfer characteristics are shown in Fig. 9.33.

From the output characteristics it may be seen that variation of V_{DS} from zero to a value equal to V_p volts causes considerable variation of drain current, I_D. The value V_p is called the pinch-off voltage because when V_{DS} reaches this value the depletion regions at the drain end of the channel almost meet each other. This section of the characteristics is known as the ohmic or resistive region. As V_{DS} is increased beyond this value, I_D becomes almost independent of V_{DS}. I_{DSS} is the drain-source saturation current which is defined later. The dotted portions at high values of drain-source voltage represent the dramatic increase of current due to avalanche breakdown in the p-n junction. For obvious reasons this condition is to be avoided in practice. The FET is normally operated with $V_{DS} > V_p$, and the bias voltage V_{GS} is normally less than or equal to zero.

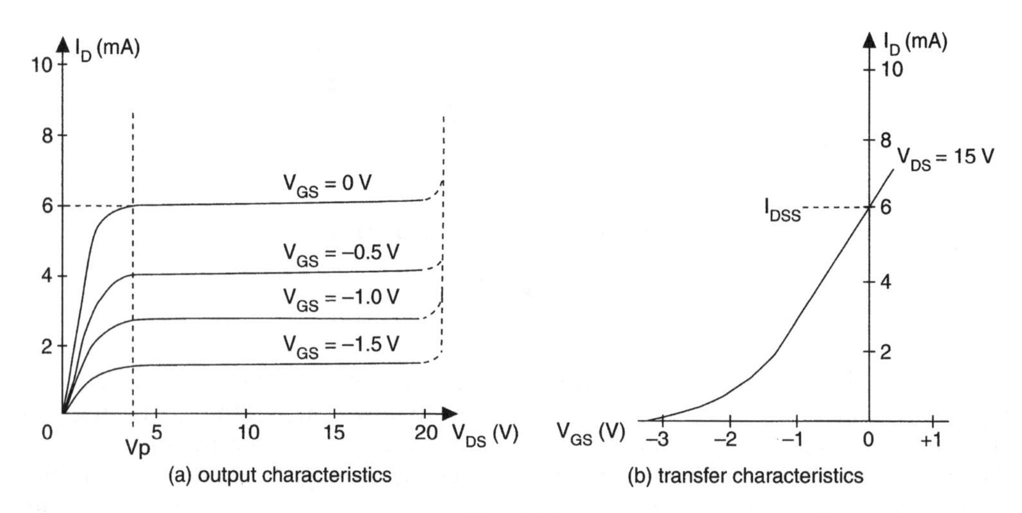

Fig. 9.33

The reasons for the shape of the output characteristics are illustrated in Figs. 9.34(a) to (c) for the condition where $V_{GS} = 0$.

From Fig. 9.34 it may be seen that the dimensions of the conducting channel for $V_{DS} > V_p$ remain virtually constant, hence the current also stays sensibly constant. Although not shown in this series of diagrams, the increase of the negative potential to the gate electrode will also contribute to the narrowing of the conducting channel, and this is clearly demonstrated by the family of output characteristics. The application of a negative potential to the gate therefore reduces (or depletes) the drain current below I_{DSS}, and the FET is said to be operating in depletion mode. This is the normal way in which the device is used. It is not normally operated with the gate potential greater than about 0.5 V since this will forward bias the p-n junction, which will then draw significant current into the gate electrode.

9.21 JUGFET Parameters

There are three main parameters associated with the FET.

Drain-source Saturation Current **(I_{DSS})** This is defined as the drain current that will flow when $V_{GS} = 0\,\text{V}$ and $V_{DS} > V_P$ volt. This will normally be the highest value of drain current flowing.

Drain-source Resistance **(r_{DS})** This parameter is defined as the ratio of drain-source voltage to drain-source current for a given value (often 0 V) of gate-source voltage. Thus

$$r_{DS} = \frac{\delta V_{DS}}{\delta I_{DS}} \text{ ohm, with } V_{GS} \text{ constant}$$

This parameter may be obtained from the output characteristics, and typical values would be in the low to mid kilohm range.

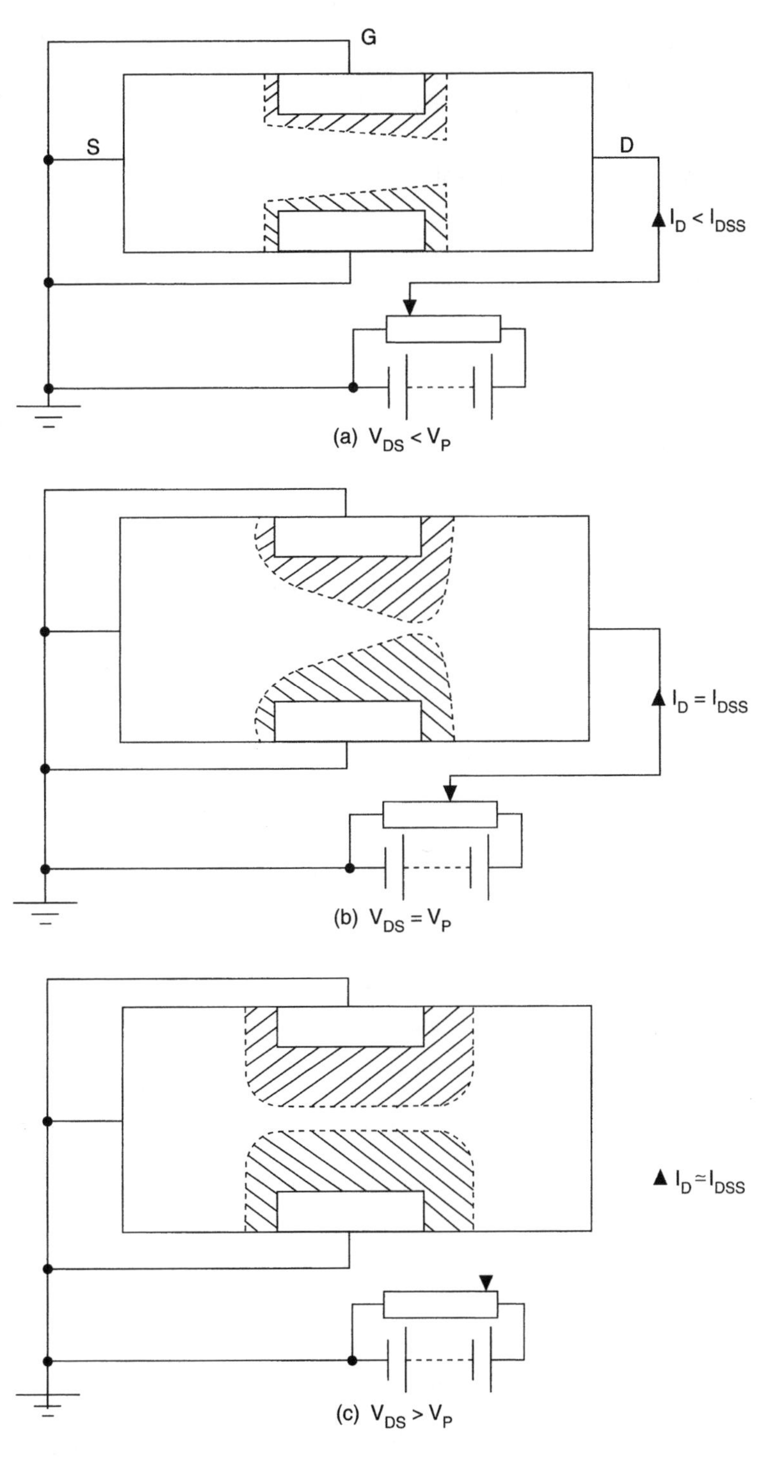

Fig. 9.34

Mutual Conductance or Transconductance (**g_m**) This parameter may be obtained from the transfer characteristic of Fig. 9.33(b), where it may be seen that the slope of the graph at any point is given by the value of drain current at that point divided by the corresponding value of gate-source voltage. Thus

$$g_m = \frac{\delta I_D}{\delta V_{GS}} \text{ siemen, with } V_{DS} \text{ constant}$$

Typical values for g_m range from 0.05 mS to 10 mS. A value for g_m may also be obtained from the output characteristics in a similar manner to that used to obtain h_F for the BJT.

Worked Example 9.3

Question

The output characteristics for a FET are shown in Fig. 9.35. Assuming $V_{DS} = 15\,V$ and $V_{GS} = -0.4\,V$, determine the values for (a) r_{DS} and (b) g_m.

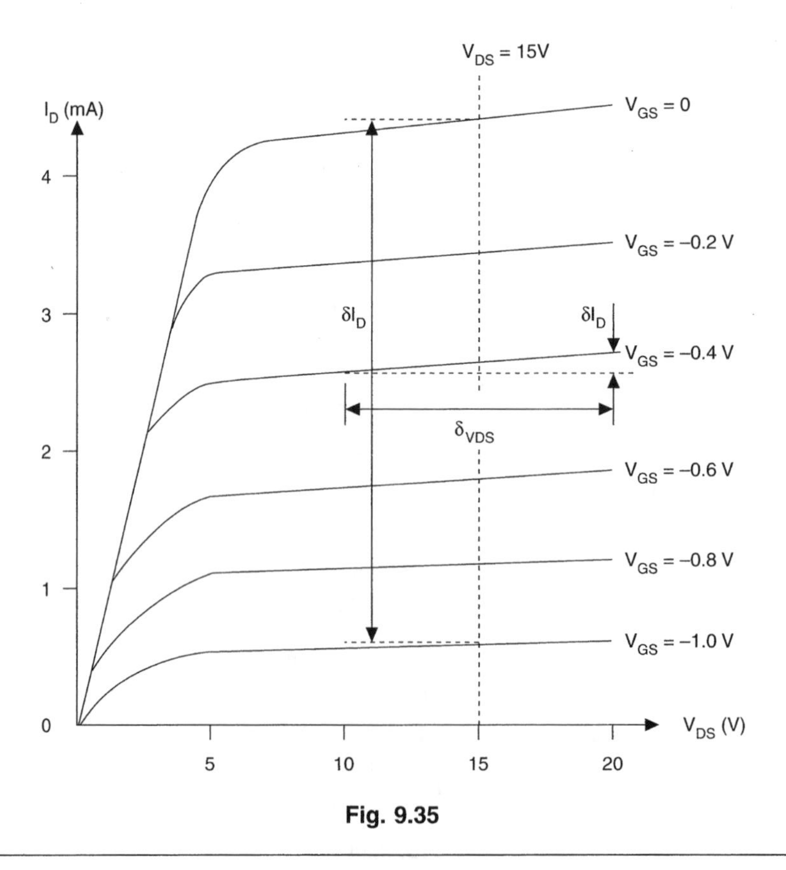

Fig. 9.35

Answer

(a) With $V_{GS} = -0.4\,\text{V}$ and $\delta V_{DS} = (20 - 10) = 10\,\text{V}$ the corresponding change in drain current, $\delta I_D = (2.7 - 2.55)\,\text{mA} = 0.15\,\text{mA}$

$$r_{DS} = \frac{\delta V_{DS}}{\delta I_D}\ \text{ohm} = \frac{10}{0.15}$$

$$r_{DS} = 66.7\,\text{k}\Omega\ \textbf{Ans}$$

(b) For $V_{DS} = 15\,\text{V}$ and $\delta V_{GS} = 0 - (-1) = 1\,\text{V}$ the corresponding change in drain current, $\delta I_D = (4.4 - 0.6)\,\text{mA} = 3.8\,\text{mA}$

$$g_m = \frac{\delta I_D}{\delta V_{GS}}\ \text{siemen} = 3.8 \times 10^{-3}$$

$$g_m = 3.8\,\text{mS}\ \textbf{Ans}$$

9.22 The Metal-Oxide-Semiconductor Transistor (MOSFET)

This device is an insulated gate FET (IGFET) as previously stated. It consists of a lightly doped p-type substrate into which are diffused two heavily doped n-type regions which form the source and drain. On the upper surface is deposited a very thin layer of silicon oxide which acts as an insulating layer. External electrical connections to the source and drain regions are via holes or windows left in the oxide layer. A cross-section of the transistor is shown in Fig. 9.36.

When a positive potential is applied to the gate terminal the resulting electric field will attract electrons (from thermally generated electron-hole pairs) towards the top surface of the substrate, and repel holes away from it. As the gate potential is increased more electrons are attracted towards the upper surface until, at a voltage known as the threshold voltage, V_T, a conducting or inversion channel of n-type semiconductor is induced between source and drain. For this reason this device is called an n-channel

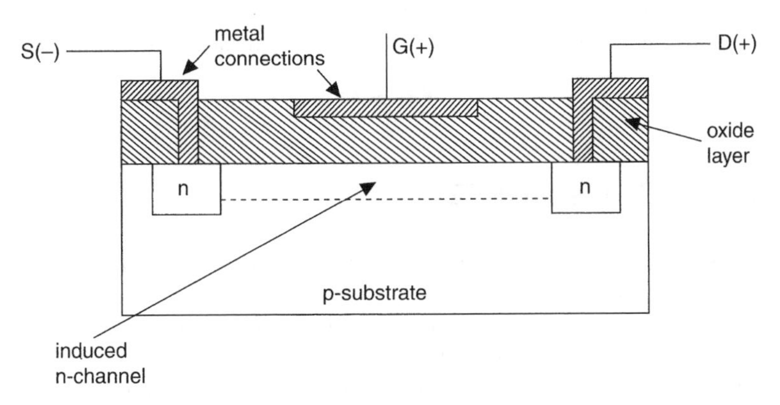

Fig. 9.36

MOSFET. A typical value for V_T is between 2 V and 4 V. When the gate potential is increased beyond V_T even more electrons are attracted into the conducting channel. This will have the effect of increasing or enhancing the drain current, and this mode of operation is known as enhancement mode. If an n-type substrate is used with p-type drain and source, and the polarities at the three electrodes are reversed, then a p-type channel is induced, and the device is called a p-channel MOSFET.

The circuit symbols for both types are shown in Fig. 9.37. Note that the channel between source and drain is shown as a broken line, because until $V_{GS} > V_T$ the transistor will not conduct. In addition, although there is never any electrical connection made to the substrate, it is always shown on the circuit symbol, and the arrowhead indicates which type of channel is employed.

Typical characteristics for an n-channel enhancement mode MOSFET are shown in Fig. 9.38.

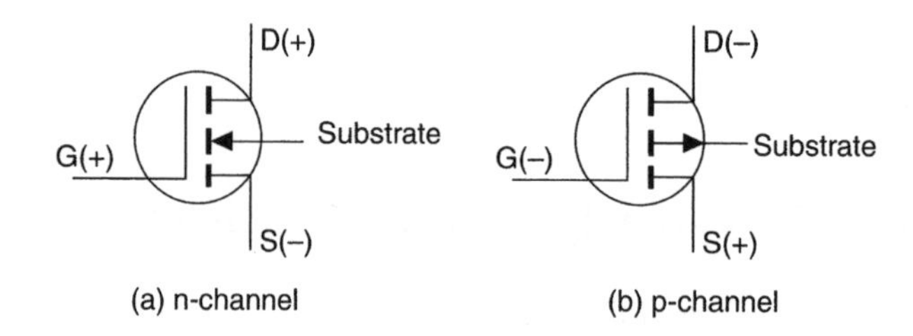

Fig. 9.37

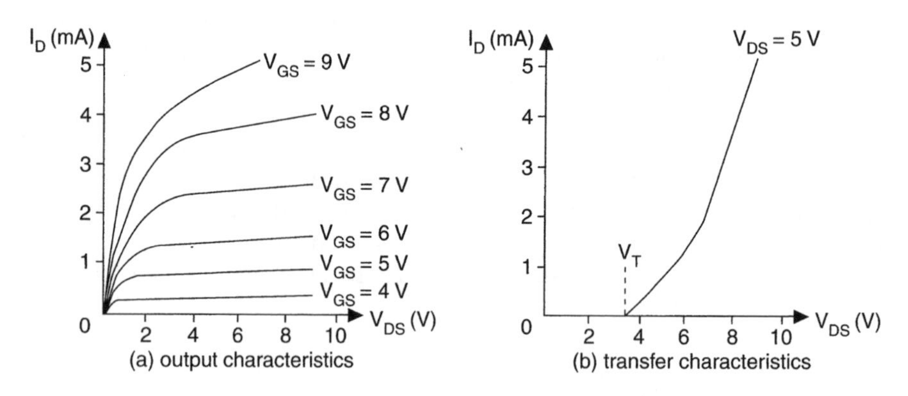

Fig. 9.38

9.23 Depletion Mode MOSFET

The construction of this device is very similar to that for the enhancement mode device but with the exception that, during the manufacturing process, a conducting channel is diffused between source and drain. This is referred to as the initial channel, and the physical arrangement is illustrated in Fig. 9.39(a).

Figure 9.39(b) is used to describe the operation, as follows. Since the source and drain are joined by a conducting channel then drain current can flow even when V_{GS} is zero.

When the gate potential is made negative, the electric field will attract holes from the substrate into the n-channel, where recombinations will occur. Thus the number of free electrons in the n-channel is reduced or depleted, and the drain current will be reduced. It should be noted that if the gate is made positive then more electrons will be attracted into the channel and drain current will increase. Thus this device may be used in either depletion or enhancement mode. The circuit symbols for both n- and p-channel devices are shown in Fig. 9.40. In these cases the conducting channel being a permanent feature, it is shown by a continuous line.

Typical output and transfer characteristics for the device are shown in Fig. 9.41.

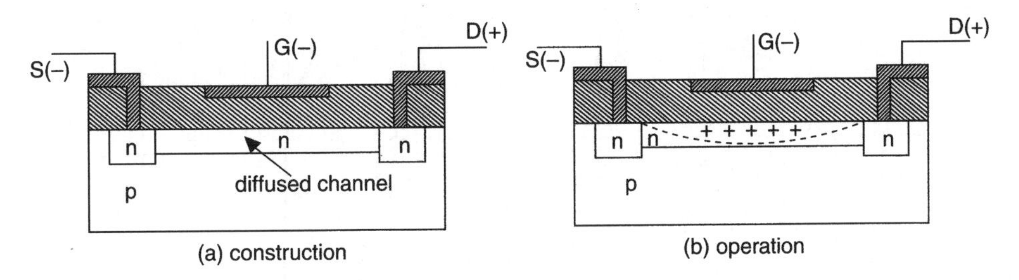

Fig. 9.39

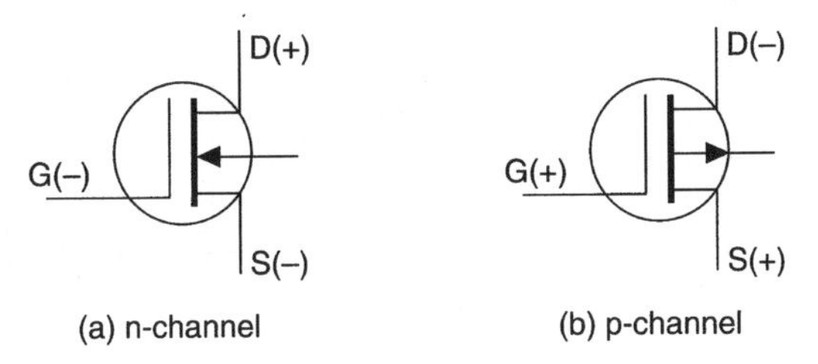

Fig. 9.40

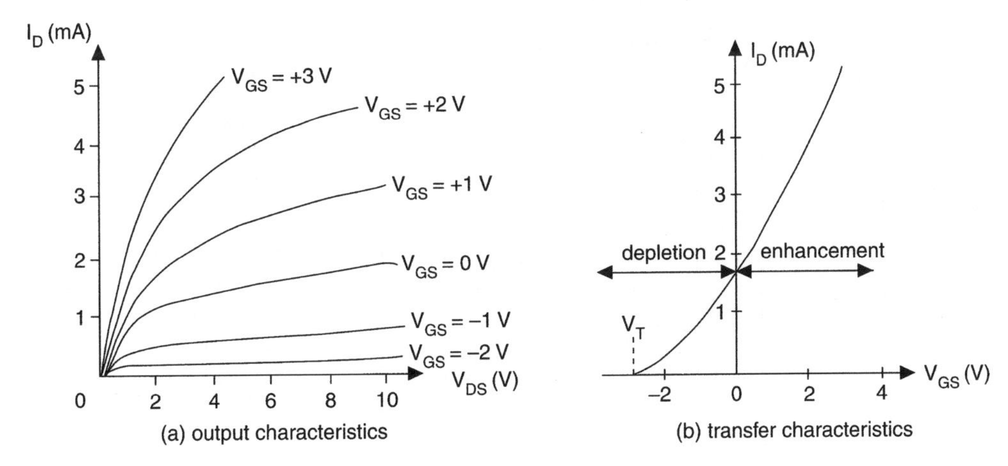

Fig. 9.41

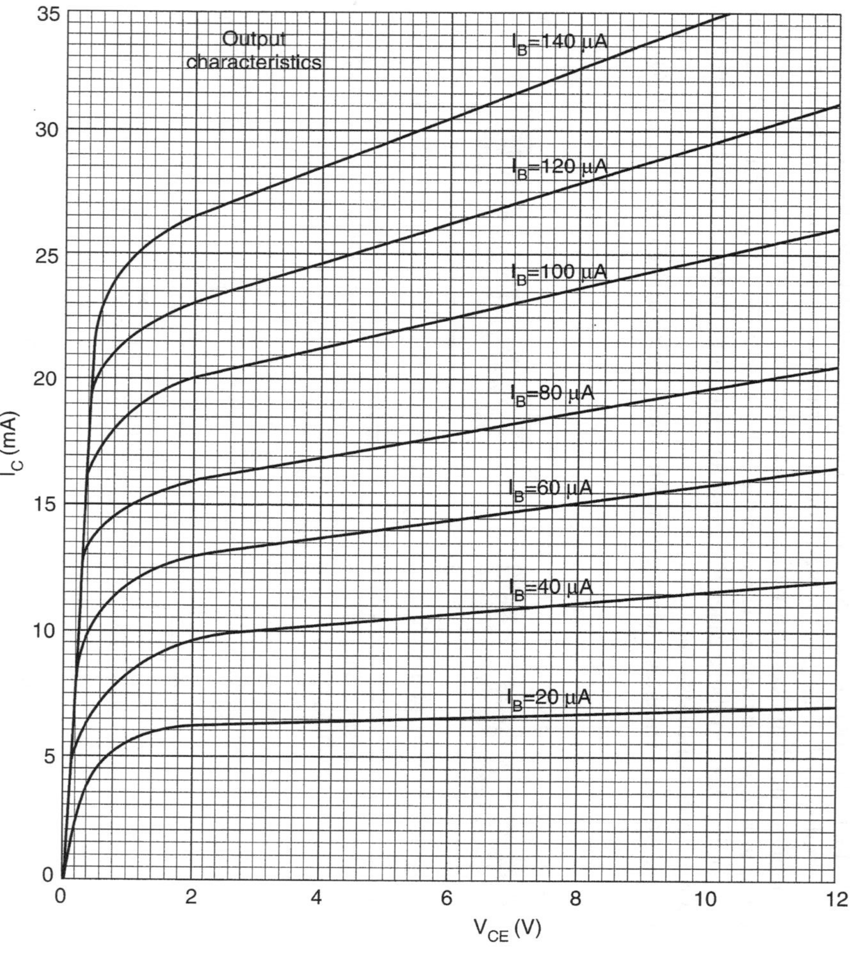
I_B (μA)
Input characteristics
140
120
100
80
60
40
20
0
0.2
0.4
0.6
0.8
V_BE (V)
Output characteristics
I_C (mA)
35
30
25
20
15
10
5
0
2
4
6
8
10
12
V_CE (V)
I_B=140 μA
I_B=120 μA
I_B=100 μA
I_B=80 μA
I_B=60 μA
I_B=40 μA
I_B=20 μA

Fig. 9.42

Typical values for the MOSFET parameters are:

Drain-source resistance, $r_D = 1$ to $50\,\mathrm{k\Omega}$
Gate input resistance, $r_G > 10\,\mathrm{G\Omega}$
Mutual conductance, $g_m = 0.1$ to $25\,\mathrm{mS}$

9.24 Comparison of FETs with BJTs

FETs differ from BJTs in a number of important ways.

1 Conduction in FETs consists only of majority carriers. Hence the general title of unipolar junction transistors.
2 BJTs are current-operated devices whereas conduction in FETs is controlled by an electric field.
3 As a consequence of 2 above, FETs have an extremely high input resistance, and (theoretically at least) draw no current at the input, i.e there is negligible loading effect on any input source.
4 FETs are simpler to manufacture and occupy less physical space. This can be particularly important for integrated circuit (IC) chips, where the size and hence possible density of devices accommodated on the chip can be vital.

Assignment Questions

1 A common emitter connected transistor has characteristics as shown in Fig. 9.42. Using these graphs determine (a) transistor output resistance for a base current of $80\,\mu\mathrm{A}$, (b) transistor input resistance, and (c) transistor current gain for $V_{CE} = 6\,\mathrm{V}$.
2 For the common base characteristics shown in Fig. 9.43 determine (a) transistor current gain, (b) output resistance, and (c) the corresponding value of transistor current gain if the transistor was connected in common emitter configuration.
3 A FET has characteristics as given in the table below. Plot these characteristics and hence determine the parameters (a) drain-source resistance, and (b) mutual conductance for $V_{GS} = 12\,\mathrm{V}$.

V_{DS}	I_{DS} (mA)				
(V)	$V_{GS} = -2.5\,\mathrm{V}$	$V_{GS} = -2.0\,\mathrm{V}$	$V_{GS} = -1.5\,\mathrm{V}$	$V_{GS} = -1.0\,\mathrm{V}$	$V_{GS} = -0.5\,\mathrm{V}$
4	0.6	1.4	2.0	3.0	4.0
16	1.0	3.1	3.6	4.8	6.4
24	1.4	3.7	4.6	6.0	8.0

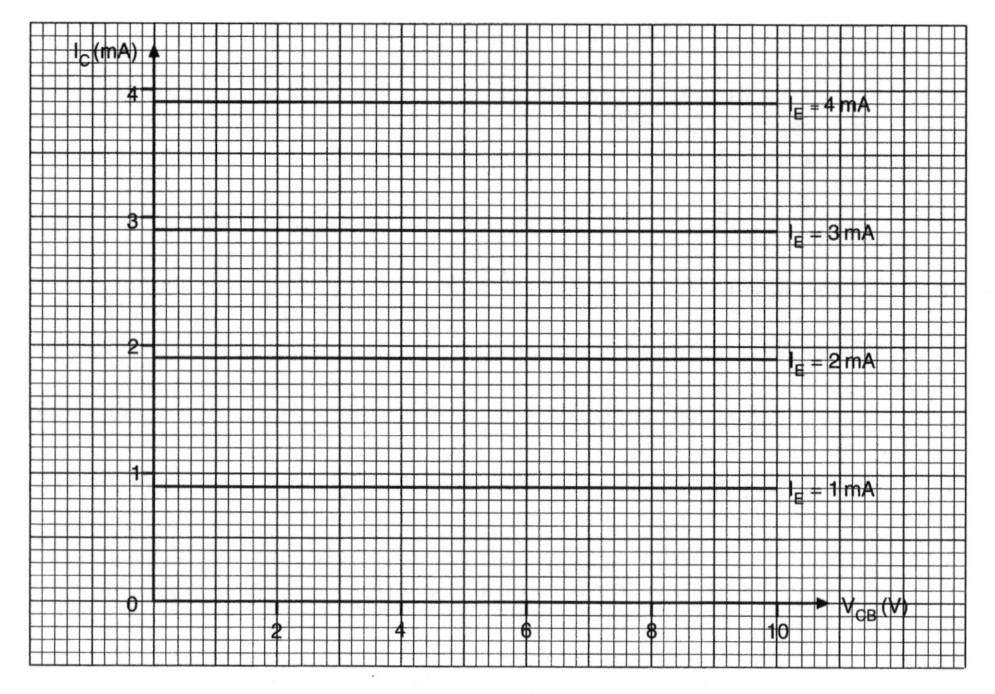

Fig. 9.43

Suggested Practical Assignments

For the following assignments it is left to the student (under the guidance of the teacher) to select appropriate apparatus. The circuit diagrams and method to follow are to be found in the relevant sections (9.12, 9.15, 9.16 and 9.20). For each assignment all results should be tabulated and the corresponding graphs plotted.

Assignment 1

To obtain the forward and reverse characteristics for silicon and germanium p-n junction diodes. An investgation into the effects of increased temperature on the reverse leakage current could also be undertaken.

Assignment 2

To obtain the output and input characteristics for a BJT when connected (a) in common base, and (b) in common emitter configuration. From the plotted graphs determine the transistor parameters in each configuration, and compare the relationship between transistor gain of the two configurations with the theoretical relationship.

Assignment 3

To obtain the output and transfer characteristics for a FET, and from the plotted graphs determine the transistor parameters.

Summary of Equations

BJT – normal transistor action: $I_E = I_C + I_B$ amp or $I_e = I_c + I_b$ amp

Common emitter BJT: $R_{IN} = \dfrac{\delta V_{BE}}{\delta I_B}$ ohm, with V_{CE} constant

$$R_{OUT} = \frac{\delta V_{CE}}{\delta I_C} \text{ ohm, with } I_B \text{ constant}$$

$$h_{FE} = \frac{\delta I_C}{\delta I_B} \text{ with, } V_{CE} \text{ constant}$$

Common base BJT: $R_{IN} = \dfrac{\delta V_{EB}}{\delta I_E}$ with, V_{CB} constant

$$R_{OUT} = \frac{\delta V_{CB}}{\delta I_C} \text{ with, } I_E \text{ constant}$$

$$h_{FB} = \frac{\delta I_C}{\delta I_B} \text{ with, } V_{CB} \text{ constant}$$

Relationship between h_{FE} and h_{FB}: $h_{FE} = \dfrac{h_{FB}}{1 - h_{FB}}$

$$h_{FB} = \frac{h_{FE}}{1 + h_{FE}}$$

FET parameters: $r_{DS} = \dfrac{\delta V_{DS}}{\delta I_D}$ ohm, with V_{GS} constant

$$g_m = \frac{\delta I_D}{\delta V_{GS}} \text{ siemen, with } V_{DS} \text{ constant}$$

10 Semiconductor Circuits

This chapter deals with a variety of circuits involving semiconductor devices. These will include the zener diode, bias and stabilisation for transistors, and small-signal a.c. amplifier circuits using both BJTs and FETs. The use of both of these devices as an electronic switch is also considered.

On completion of this chapter you should be able to:

1 Understand the action of a zener diode and perform basic calculations involving a simple regulator circuit.
2 Understand the need for correct biasing for a transistor, and perform calculations to obtain suitable circuit components to achieve this effect.
3 Understand the operation of small-signal amplifiers and carry out calculations to select suitable circuit components, and to predict the amplifier gain figure(s).
4 Understand how a transistor may be used as an electronic switch, and carry out simple calculations for this type of circuit.

10.1 The Zener Diode

The main feature of the zener diode is its ability to operate in the reverse breakdown mode without sustaining permanent damage. In addition, during manufacture, the precise breakdown voltage (zener voltage) for a given diode can be predetermined. For this reason they are also known as voltage reference diodes. The major application for these devices is to limit or stabilise a voltage between two points in a circuit. Diodes are available with zener voltages from 2.6 V to about 200 V. The circuit symbol for a zener diode is shown in Fig. 10.1.

The forward characteristic for a zener diode will be the same as for any other p-n junction diode, and also, since the device is always used in its reverse bias mode, only its reverse characteristic need be considered. Such a characteristic is shown in Fig. 10.2.

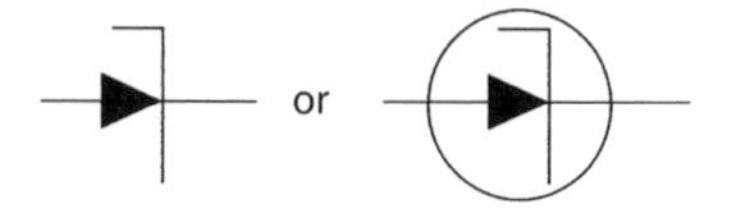

Fig. 10.1

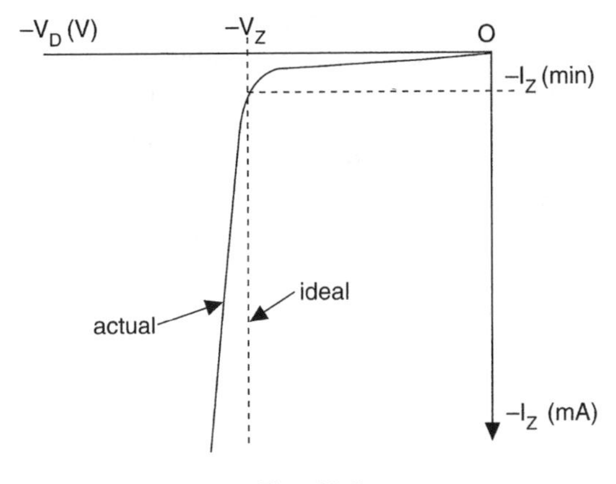

Fig. 10.2

In Fig. 10.2, V_Z represents the zener breakdown voltage, and if it were an ideal device, this p.d. across it would remain constant, regardless of the value of current, I_Z, flowing through it. In practice the graph will have a fairly steep slope as shown. The inverse of the slope of the graph is defined as the diode slope resistance, r_Z, as follows

$$r_Z = \frac{\delta V_D}{\delta I_Z} \text{ ohm} \tag{10.1}$$

Typical values for r_Z range from $0.5\,\Omega$ to about $150\,\Omega$. For satisfactory operation the current through the diode must be at least equal to $I_{Z(min)}$. Due to the diode slope resistance, the p.d. across the diode will vary by a small amount from the ideal of V_Z volt as the diode current changes. For example, if $r_Z = 1\,\Omega$ and $V_Z = -15\,\text{V}$, a change in diode current of 30 mA would cause only a 0.02% change in the diode p.d. This figure may be verified by applying equation (10.1).

The value of current that may be allowed to flow through the device must be limited so as not to exceed the diode power rating. This power rating is always quoted by the manufacturer, and zener diodes are available with power ratings up to about 75 W.

Consider now the application of a zener diode to provide simple voltage stabilisation to a load. A circuit is shown in Fig. 10.3.

In order for satisfactory operation the supply voltage, V_S, needs to be considerably greater than the voltage required at the load. The purpose of the series resistor R_S is to

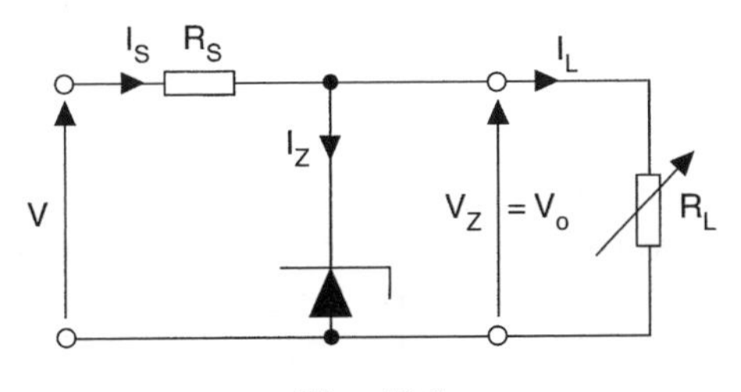

Fig. 10.3

limit the maximum diode current to a safe value, bearing in mind the diode's power rating. Considering Fig. 10.3, the diode current will be at its maximum when the load is disconnected, because under this condition all of the current from the supply will flow through the diode, i.e. $I_Z = I_S$. When the load is connected it will draw a current I_L, and since $I_Z = I_S - I_L$, then under this condition the diode current will decrease, since it must divert current to the load. The output voltage, however, will remain virtually unchanged. Knowing the diode power rating a suitable value for R_S may be calculated as shown in the following worked example. This example also demonstrates the stabilising action of the circuit.

Worked Example 10.1

Question

A 9.1 V, 500 mW zener diode is used in the circuit of Fig. 10.3 to supply a 2.5 kΩ load. The diode has a slope resistance of 1.5 Ω, and the input supply has a nominal value of 12 V.

(a) Calculate a suitable value for the series resistor R_S.
(b) Calculate the value of diode current when the load resistor is connected to the circuit.
(c) If the input supply voltage decreases by 10%, calculate the percentage change in the p.d. across the load.

Answer

$V_Z = 9.1\,\text{V};\ P_Z = 0.5\,\text{W};\ r_Z = 1.5\,\Omega;\ V = 12\,\text{V};\ R_L = 2500\,\Omega$

(a) $P_Z = V_Z I_Z$ watt

$$I_Z = \frac{P_Z}{V_Z}\ \text{amp} = \frac{0.5}{9.1}$$

so, $I_Z = I_S = 54.95\,\text{mA}$

($I_S = I_Z$ because for this condition the load is disconnected)

$$V_S = V - V_Z\ \text{volt} = 12 - 9.1$$

$$V_S = 2.9\,\text{V}$$

$$R_S = \frac{V_S}{I_S}\ \text{ohm} = \frac{2.9}{54.95 \times 10^{-3}}$$

$$R_S = 52.78\,\Omega$$

A resistor of this precise value would not be readily available, so the nearest preferred value resistor would be chosen. However, to ensure that the diode power rating cannot be exceeded, the nearest preferred value *greater than* 52.78 Ω would be chosen. Thus a 56 Ω resistor would be chosen.

In order to protect the resistor, its own power rating must be taken into account. In this circuit, the maximum power dissipated by R_S is:

$$P_{max} = I_S^2 R_S \text{ watt} = (54.95 \times 10^{-3})^2 \times 56$$

$P_{max} = 0.169\,\text{W}$, so a 0.25 W resistor would be chosen, and the complete answer to part (a) is:

R_S should be a 56 Ω, 0.25 W resistor **Ans**

(b) With $R_L = 2500\,\Omega$ and $V_o = 9.1\,\text{V}$

$$I_L = \frac{V_o}{R_L} \text{ amp} = \frac{9.1}{2.5 \times 10^3}$$

$$I_L = 3.64\,\text{mA}$$

$$I_Z = I_S - I_L \text{ amp} = (55 - 3.64)\,\text{mA}$$

$I_Z = 51.36\,\text{mA}$ **Ans**

(c) When V falls by 10% from its nominal value, then

$$V = 12 - (0.1 \times 12) = 12 - 1.2$$

hence, $V = 10.8\,\text{V}$

$$I_S = \frac{V - V_Z}{R_S} \text{ amp} = \frac{10.8 - 9.1}{56}$$

$$I_S = 30.36\,\text{mA}$$

The current for the load must still be diverted from the diode, so

$$I_Z = I_S - I_L \text{ amp} = (30.36 - 3.64)\,\text{mA} = 26.72\,\text{mA}$$

therefore, $\delta I_Z = (51.36 - 26.72)\,\text{mA} = 24.64\,\text{mA}$, and from equation (10.1):

$$\delta V_Z = \delta I_Z r_Z \text{ volt} = 24.64 \times 10^{-3} \times 1.5$$

$$\delta V_Z = 0.037\,\text{V}$$

Thus the voltage applied to the load changes by 0.037 V, which expressed as a percentage change is:

$$\text{change} = \frac{0.037}{9.1} \times 100$$

change $= 0.41\%$ **Ans** (compared with a 10% change in supply)

10.2 Transistor Bias

In order to use a transistor as an amplifying element it needs to be biased correctly. Although d.c. signals may be amplified, the amplification of a.c. signals is more common. However, the bias is provided by d.c. conditions. Consider a common emitter connected BJT and its input characteristic as illustrated in Fig. 10.4. The inclusion of resistor R_C is not required at this stage, but would be present in any practical amplifier circuit, so is shown merely for completeness. This resistor is called the collector load resistor.

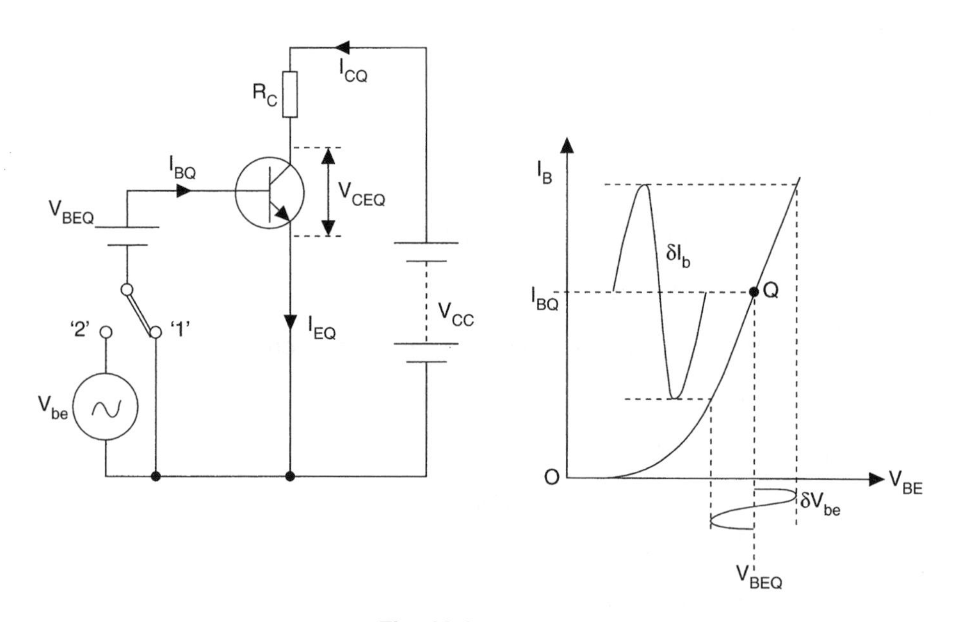

Fig. 10.4

With the switch in position '1' the value of forward bias V_{BEQ} has been chosen such that it coincides with the centre of the linear portion of the input characteristic. This point on the characteristic is identified by the letter Q, because, without any a.c. input signal connected, the transistor is said to be in its quiescent state. Under this condition the base current will be I_{BQ} with the corresponding values of collector and emitter currents being I_{CQ} and I_{EQ} respectively. The use of capital letters in the subscripts indicates d.c. quantities, and the letter Q that they are quiescent values.

Consider now what happens when the switch is moved to position '2', thus connecting the a.c. signal generator between base and emitter. The a.c. signal V_{be} (note the use of lower case letters to indicate an a.c. quantity) from the generator would normally vary about 0 V, but it will now be superimposed on the quiescent d.c. bias level V_{BEQ}. Thus the effective bias voltage will vary in sympathy with the input signal, causing a corresponding variation of the base current about its quiescent d.c. level as shown. Due to transitor action there will be corresponding variations of both collector and emitter currents about their quiescent values.

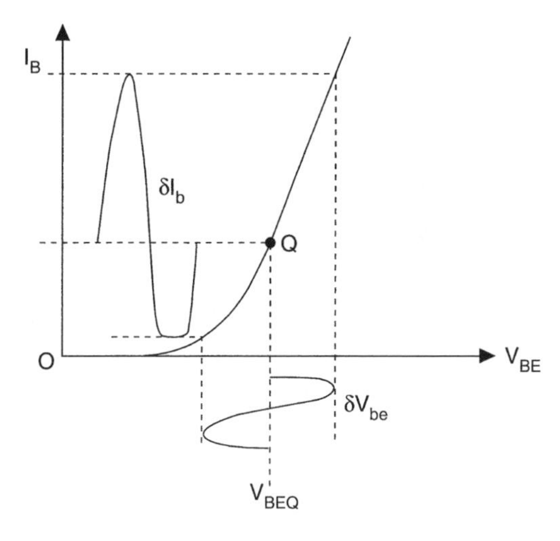

Fig. 10.5

Figure 10.5 shows the effect when a different d.c. bias voltage, and hence different Q point, is selected.

The sinusoidal variation of V_{be} will again cause variation of I_b, but due to the curvature at the bottom of the characteristic the resulting base current will be distorted as shown. This same distortion will be reflected in the collector and emitter currents also. This distortion needs to be avoided, so the bias voltage and hence Q point must be carefully chosen.

The provision of V_{CC} by means of a battery or other d.c. source will always be required, but the provision of the forward bias voltage V_{BEQ} by a second battery or other d.c. source is inconvenient. There are a number of ways of providing the required bias conditions without the need for a second source of emf.

10.3 A Simple Bias Circuit

In the circuit of Fig. 10.6 the required quiescent base current I_{BQ} is provided from V_{CC} via resistor R_B.

The current flowing through R_B depends upon the p.d. across it, which will be $(V_{CC} - V_{BEQ})$ volt. Thus if we know the required values for I_{BQ} and V_{BEQ} by studying the input characteristic, the value for R_B may be calculated as follows.

$$R_B = \frac{(V_{CC} - V_{BEQ})}{I_{BQ}} \text{ ohm} \tag{10.2}$$

In practice, $V_{CC} \gg V_{BEQ}$, so the above equation is often simplified to

$$R_B = \frac{V_{CC}}{I_{BQ}} \text{ ohm} \tag{10.3}$$

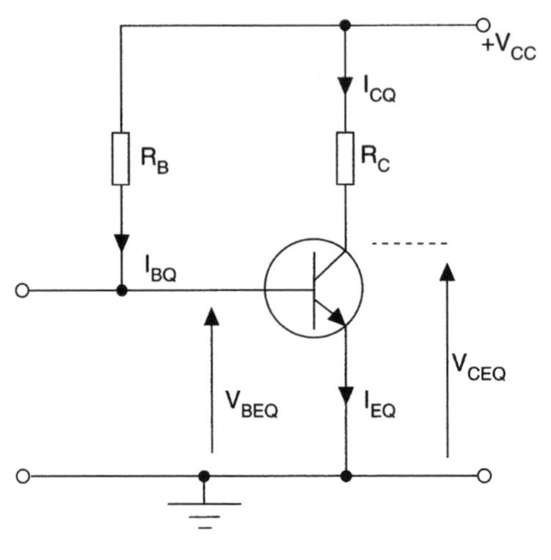

Fig. 10.6

This approximation is valid since when selecting a value for R_B the nearest preferred value would have to be used. This is demonstrated in the following example.

Worked Example 10.2

Question

A simply biased transistor circuit is shown in Fig. 10.7. The required quiescent values for base current and base-emitter voltage are 60 μA and 0.8 V respectively. Determine a suitable value for resistor R_B.

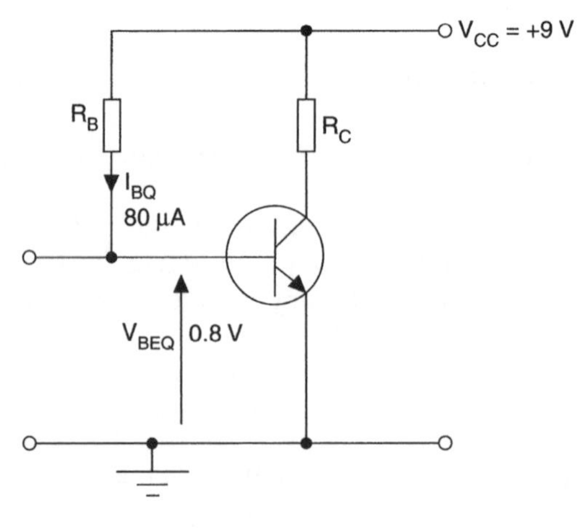

Fig. 10.7

Answer

$V_{CC} = 9\,\text{V}$; $I_{BQ} = 60 \times 10^{-6}\,\text{A}$; $V_{BEQ} = 0.8\,\text{V}$
Using equation (10.2) we have

$$R_B = \frac{(V_{CC} - V_{BEQ})}{I_{BQ}}\ \text{ohm} = \frac{(9 - 0.8)}{60 \times 10^{-6}}$$

$R_B = 137\,\text{k}\Omega$ **Ans**

Using the approximation of equation (10.3) gives

$$R_B = \frac{V_{CC}}{I_{BQ}}\ \text{ohm} = \frac{9}{0.8}$$

$R_B = 150\,\text{k}\Omega$ **Ans**

Now, the nearest preferred value would be 150 kΩ, so the use of equation (10.3) is justified.

Although this bias circuit has the advantage of simplicity, it cannot overcome the problem of the bias and hence Q point varying with temperature change, and the even more serious problem of thermal runaway.

10.4 Thermal Runaway

When the transistor is operating in a circuit, the collector current will tend to cause a temperature increase, with a consequent increase of transistor current gain h_{FB}. For example, assume that h_{FB} increases from 0.98 to 0.99 under this condition. This is an increase of only about 1%, and so will result in only a very small increase of collector current. For this reason, if the transistor is connected in the common base configuration it will be relatively stable, and the simple bias circuit described will usually be acceptable.

Consider now the effect of the same increase of temperature when this same transistor is connected in the common emitter configuration. h_{FB} will increase from 0.98 to 0.99 as before, but the effect on h_{FE} is more dramatic as shown below.

$$\text{When } h_{FB} = 0.98;\ h_{FE} = \frac{h_{FB}}{1 - h_{FB}} = \frac{0.98}{0.02} = 49$$

$$\text{When } h_{FB} = 0.99;\ h_{FE} = \frac{0.99}{0.01} = 99$$

Thus for the same temperature rise the transistor current gain has more than doubled, with a corresponding increase in collector current. This will result in a further increase in temperature, a further small increase in h_{FB}, and a further massive increase in h_{FE}, etc.

The process is a cumulative one known as thermal runaway which will result in the rapid destruction of the transistor. Since the runaway condition is the result of a rapid increase of collector current, then some means of preventing this needs to be employed.

10.5 Bias with Thermal Stabilisation

One simple method of providing a degree of thermal stability is to connect the bias resistor directly to the collector, as shown in Fig. 10.8.

In this circuit the p.d. across R_B is dependent upon the potential at the collector. With a temperature rise the collector current will tend to rise. This increase will cause increased voltage drop across resistor R_C, resulting in a lower potential at the collector, which in turn will tend to reduce base current I_B. Any tendency for base current to fall will be reflected by the same tendency in collector current. Thus we have gone full circle, whereby the original tendency for I_C to increase has been counteracted.

Although this circuit will give good stability for a transistor connected in common base, it may not be entirely satisfactory for the common emitter mode, particularly when the operating conditions elevate the temperature still further. For this reason a more effective method is more often employed, as described in the next section.

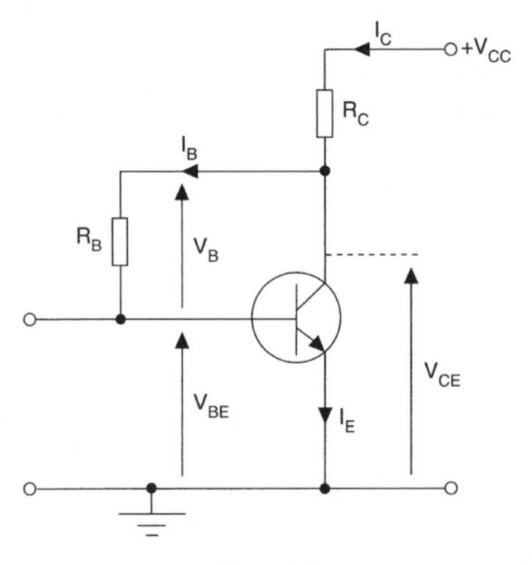

Fig. 10.8

10.6 Three-resistor Bias and Stabilisation

A circuit showing this method of biasing is shown in Fig. 10.9, and this is by far the most common biasing technique employed in practice.

The potential divider chain formed by R_1 and R_2 provides a fixed potential V_2 to the base of the transistor. Resistor R_E is involved in both providing the bias and the thermal stability. The potential at the emitter is V_E, where $V_E = I_E R_E$ volt.

The emitter-base bias voltage will be the difference between the potentials at these two electrodes, so $V_{BE} = (V_2 - V_E)$ volt. Thermal stability is achieved as follows.

As a result of temperature rise, I_C will tend to rise. This will be mirrored by a corresponding increase in emitter current I_E. However, the base potential V_2 is fixed, so the overall bias V_{BE} will tend to decrease, which in turn will tend to reduce I_C. Once more the system has gone full circle such that the original tendency for collector current

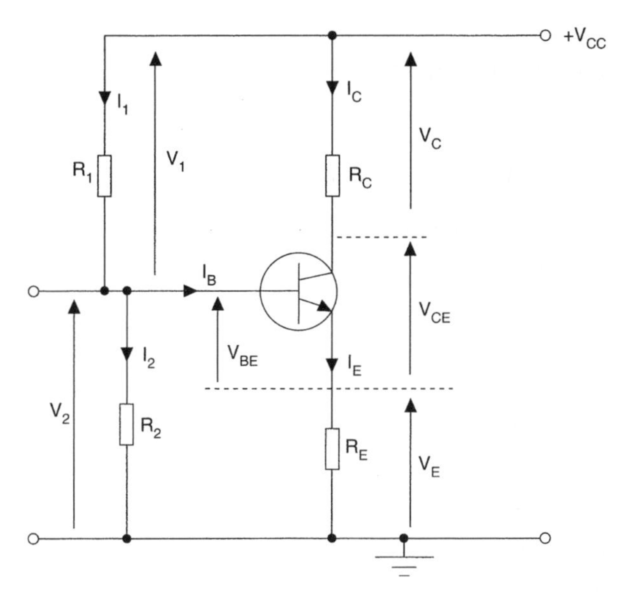

Fig. 10.9

to increase has been counteracted. This effect is illustrated below, where the arrows show the tendencies for the various quantities to increase or decrease.

$$\begin{array}{l} \uparrow \; \uparrow \; \uparrow \quad \text{(bias)} \\ I_C, I_E, V_E, (V_2 - V_E), I_B, I_C \\ \qquad\qquad\quad \downarrow \quad\; \downarrow \; \downarrow \end{array}$$

When calculating suitable values for the bias components the following points should be borne in mind.

(a) For good thermal stability the p.d. across R_E should be about $V_{CC}/10$ volt.

(b) The value of R_2 should be at least $10 \times R_E$ ohm.

(c) The current I_1 through R_1 should be approximately 10 to $20 \times I_{BQ}$.

(d) The collector-emitter voltage V_{CE} should be approximately $V_{CC}/2$ volt.

These points are best illustrated by means of the following example.

Worked Example 10.3

Question

For the circuit of Fig. 10.9 the transistor has a current gain, $h_{FE} = 80$ and the collector supply voltage, $V_{CC} = +10\,\text{V}$. The required bias conditions are $V_{BE} = 0.7\,\text{V}$, and $I_C = 1\,\text{mA}$. Determine suitable values for resistors R_1, R_2, R_E and R_C.

Continued on p. 404

Worked Example 10.3 (*Continued*)

Answer

$V_{CC} = 10\,V$; $I_C = 10^{-3}\,A$; $V_{BE} = 0.7\,V$; $h_{FE} = 80$

$$\text{Since } I_E \approx I_C, \text{ then } I_E = 1\,\text{mA}$$

$$\text{and if } V_E = V_{CC}/10, \text{ then } V_E = 1\,V$$

$$\text{hence, } R_E = \frac{V_E}{I_E}\ \text{ohm} = \frac{1}{10^{-3}}$$

$$R_E = 1\,\text{k}\Omega\ \textbf{Ans}$$

$$\text{if } R_2 = 10\ R_E, \text{ then}$$

$$R_2 = 10\,\text{k}\Omega\ \textbf{Ans}$$

$$V_{BE} = (V_2 - V_E)\ \text{volt}$$

$$\text{so, } V_2 = V_{BE} + V_E = 0.7 + 1 = 1.7\,V$$

$$I_2 = \frac{V_2}{R_2}\ \text{amp} = \frac{1.7}{10^4}$$

$$\text{and, } I_2 = 0.17\,\text{mA}$$

$$I_B = \frac{I_C}{h_{FE}} = \frac{10^{-3}}{80}$$

$$I_B = 12.5\,\mu\text{A or } 0.0125\,\text{mA}$$

$$I_1 = I_2 + I_B = 0.17 + 0.0125 = 0.1825\,\text{mA}$$

$$V_1 = V_{CC} - V_2\ \text{volt} = 10 - 1.7 = 8.3\,V$$

$$R_1 = \frac{V_1}{I_1}\ \text{ohm} = \frac{8.3}{0.1825 \times 10^{-3}}$$

$$R_1 = 46\,\text{k}\Omega, \text{ and the neareset preferred value is } 47\,\text{k}\Omega\ \textbf{Ans}$$

$$V_C = V_{CC} - (V_E + V_{CE})\ \text{volt} = 10 - (1 + 5) \text{ assuming that } V_{CE} = V_{CC}/2$$

$$V_C = 4\,V$$

$$R_C = \frac{V_C}{I_C}\ \text{ohm} = \frac{4}{10^{-3}}$$

$$R_C = 4\,\text{k}\Omega, \text{ and the nearest preferred value is } 3.9\,\text{k}\Omega\ \textbf{Ans}$$

10.7 Biasing Circuits for FETs

In general, FETs are much simpler devices than BJTs, and since the gate draws negligible current the bias arrangements can also be simpler. FETs also have the advantage that they tend not to suffer thermal runaway, though change in temperature can cause drift of the Q point.

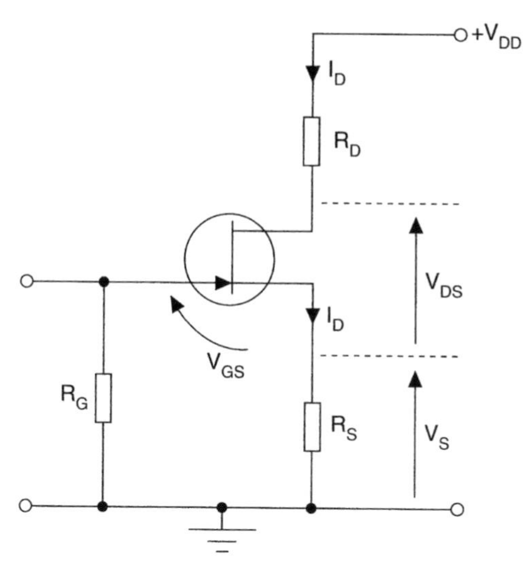

Fig. 10.10

The simplest arrangement for a common source circuit is shown in Fig. 10.10. The resistor R_D is equivalent to the collector resistor R_C in the BJT circuit.

Essentially, the gate-source bias voltage is provided by R_S. Resistor R_G will be a high value resistor, typically in megohms, which is used to connect the 0 V rail to the gate. Now, you may well ask the question, how can a large value resistor act as a direct connection between the 0 V rail and the gate? The answer is very simple. Since the gate draws negligible current there will be negligible current through R_G, hence negligible voltage drop across it, so both ends of the resistor must be at the same potential, i.e. 0 V. The FET itself has a very high input impedance, and therefore will have almost no loading effect on a signal source connected between gate and source. In order not to compromise this effect, resistor R_G is chosen to have a high value.

Drain current I_D flowing through R_S results in voltage V_S being developed across it. Thus the potential at the source will be positive with respect to the gate, which is the required bias condition.

If full bias stabilisation is required, then a three-resistor method similar to that used for the BJT may be employed, and such a circuit is shown in Fig. 10.11.

In this circuit, the bias voltage is provided by R_2 and R_S, since $V_{GS} = (V_2 - V_S)$ volt. Resistor R_G performs a similar function to that in the previous circuit, i.e. to connect potential V_2 to the gate. The stabilising action will be very similar to that for the BJT circuit, in that increase of I_D due to temperature will increase V_S. This has the effect of making the gate more negative with respect to the source, i.e. the forward bias is reduced and I_D is reduced.

10.8 Small Signal a.c. Amplifiers

A simply biased common emitter amplifier circuit is shown in Fig. 10.12. On this diagram both d.c. (biasing conditions) and a.c. (signal conditions) quantities have been

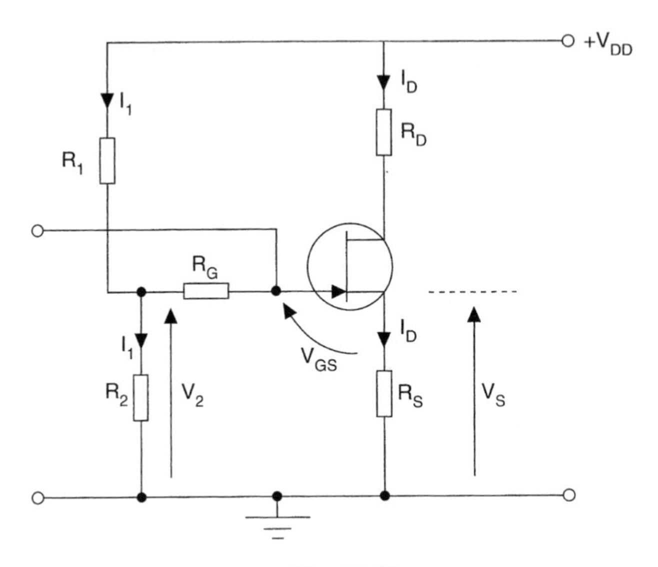

Fig. 10.11

identified. The use of upper case letters, such as V_{CE}, identify the d.c. quantities, and the lower case letters, such as i_b, identify the a.c. quantities.

The purpose of the collector load resistor R_C is to develop an a.c. output voltage when the a.c. component of collector current, i_c, flows through it. Capacitors C_1 and C_2 are known as coupling capacitors. C_1 allows the a.c. input signal to be connected to the base, but will block any d.c. level (normally 0 V) from the signal generator affecting the d.c. bias conditions of the transistor. Similarly, C_2 couples the a.c. signal developed

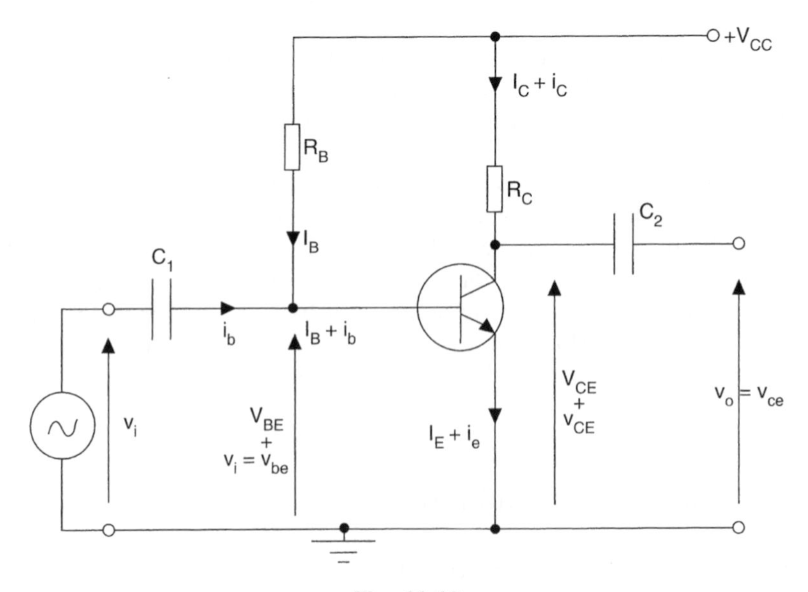

Fig. 10.12

across R_C to the output terminals, but prevents the quiescent collector voltage from affecting the output waveform. The values of these capacitors are chosen so that they will have very low reactance at the signal frequency, and therefore are virtually 'transparent' to the a.c. signal, i.e. they will have imposed negligible voltage drop. The effect of the capacitors on the various currents and voltages is illustrated in Fig. 10.13.

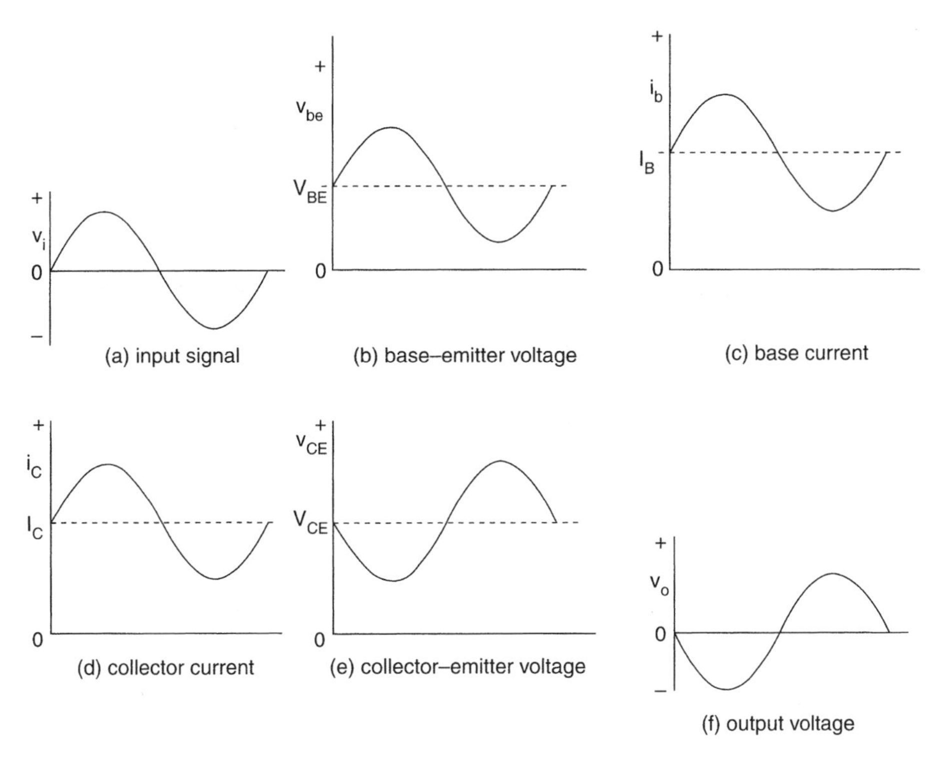

Fig. 10.13

Notes:

1 The amplitude of $V_{be} = V_i$, but capacitor C_1 has prevented the 0 V level of the signal generator from altering the transistor bias V_{BE}.

2 Observe that v_{ce} and hence output voltage v_o is phase-inverted (is antiphase to) the input signal v_i.

3 The amplitude of the output $v_o = v_{ce}$, and capacitor C_2 has prevented the quiescent d.c. level V_{CE} from being connected to the output terminals.

With reference to Note 2 above, the reason for the phase inversion is explained as follows. The potential at the collector depends upon the value of current flowing through R_C. From Fig. 10.13(d) it can be seen that I_C goes more positive in the first half cycle, in response to the input signal. This will result in a larger p.d. across R_C, but since the upper end of this resistor is tied to $+V_{CC}$, then the potential at the collector end must fall. Thus during this half cycle v_{ce} will decraese from its quiescent value. In the next half cycle i_c decreases, so the potential at the collector rises and we have phase inversion.

With R_C in the amplifier circuit an output voltage has been developed. Consider now how we can utilise this so as to analyse the performance of the amplifier. Firstly let us apply Kirchhoff's voltage law between the power supply rails (V_{CC} and 0 V), taking the route through R_C and the transistor, for the circuit of Fig. 10.12.

$$V_{CC} = I_C R_C + V_{CE} \text{ volt}$$

$$\text{when } I_C = 0; \text{ then } V_{CE} = V_{CC} \text{ volt} \ldots\ldots [1]$$

$$\text{when } V_{CC} = 0; \text{ then } I_C = \frac{V_{CC}}{R_C} \text{ amp} \ldots\ldots [2]$$

The conditions where $I_C = 0$ and $V_{CC} = 0$ are two points on the axes of the transistor's output characteristics, and the straight line joining these two points is known as the d.c. or static load line for resistor R_C. A typical set of characteristics with the d.c. load line drawn on it is shown in Fig. 10.14. The slope of the load line is $-1/R_C$ mA/V. The minus sign is due to the negative slope.

Where the load line intersects the characteristic for I_{BQ} gives the Q point for the transistor. Indeed this gives an alternative method for determining a suitable Q point

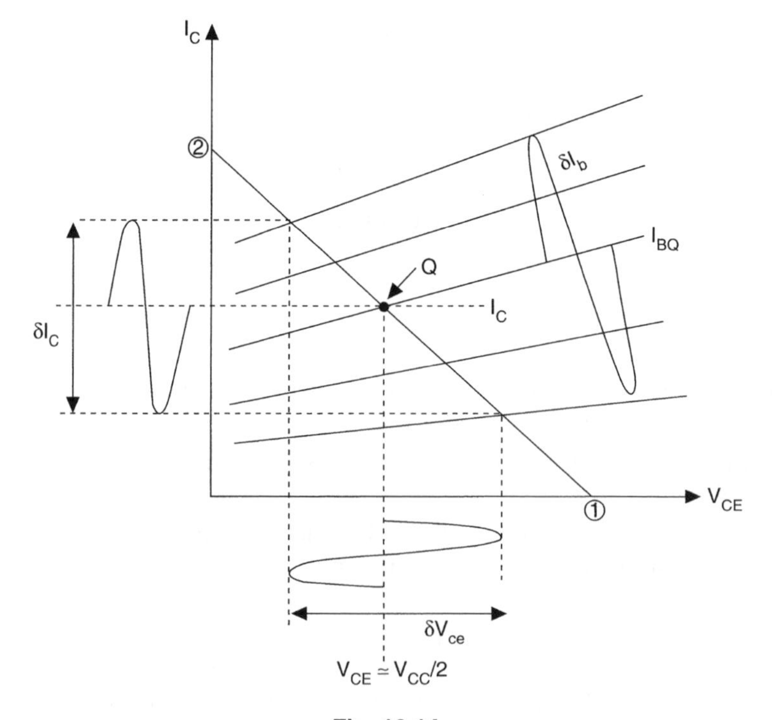

Fig. 10.14

without having to refer to the input characteristic. Bearing in mind that V_{CEQ} needs to be $\approx V_{CC}/2$ volt, then the nearest intersection that corresponds to this requirement will be a suitable choice.

The intersections of the load line with the characteristics shows the limitations imposed on the excursions of collector current (δI_c) and voltage (δV_{ce}) for a given excursion of base current (δI_b). From this we can determine the *amplifier* current and voltage gains, which are defined as follows.

$$\text{Amplifier current gain, } A_i = \frac{\delta I_c}{\delta I_b} \tag{10.4}$$

$$\text{Amplifier voltage gain, } A_v = \frac{\delta V_{ce}}{\delta V_{be}} \tag{10.5}$$

$$\text{and amplifier power gain, } A_p = A_i \times A_v \tag{10.6}$$

The variation δV_{be} corresponds to the a.c. input signal excursion δV_i. Note that the *amplifier* current gain will *always* be less than the transistor current gain h_{FE}, because the latter is defined with V_{CE} constant. Also note that the amplifier power gain refers to the ratio of the *signal* output power to the *signal* input power. When the comparatively large power input from the V_{CC} power supply is taken into account, the overall efficiency of such an amplifier circuit in terms of power output to the total power input will be found to be less than 25%.

The a.c. Equivalent Circuit The circuit of Fig. 10.12 may be redrawn in terms of what the a.c. signals will 'see'. In this case the coupling capacitors C_1 and C_2 are transparent to a.c. and therefore act as short circuits to a.c. signals. The battery supplying V_{CC} will also behave as if it were a large value capacitor, and therefore act as a short-circuit to a.c. signals. Considering a battery in this way is quite valid when you consider what it is comprised of—a set of large oppositely charged plates separated by an electrolyte (insulator), i.e. the same construction as a capacitor! Any other form of d.c. supply used to provide V_{CC} will have the same effect. The result as far as a.c. signals are concerned is that the + V_{CC} rail is directly connected to the 0 V rail. Thus the upper ends of both R_B and R_C are effectively connected to the common 0 V rail. The a.c. equivalent circuit will therefore be as shown in Fig. 10.15.

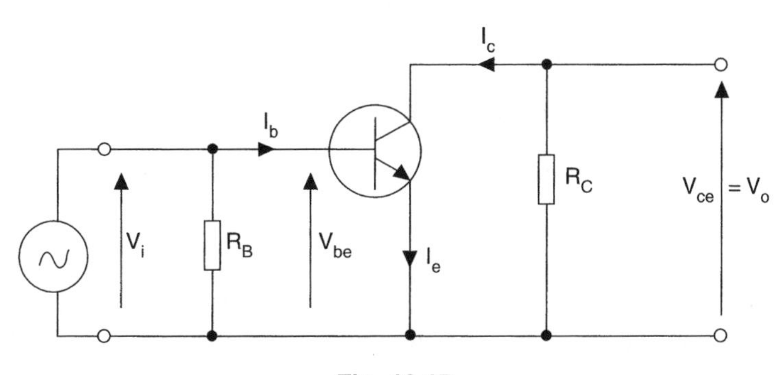

Fig. 10.15

Worked Example 10.4

Question

For the amplifier circuit of Fig. 10.16, an input signal of 300 mV pk-pk causes the base current to vary by 80 μA pk-pk about its quiescent value of 60 μA. The output characteristics for the transistor are given in Fig. 10.17.
Draw the load line on these characteristics and hence determine
(a) the transistor current gain, h_{FE},
(b) the amplifier current gain, A_i,
(c) the amplifier voltage and power gains, A_v and A_p, and
(d) sketch the a.c. equivalent circuit.

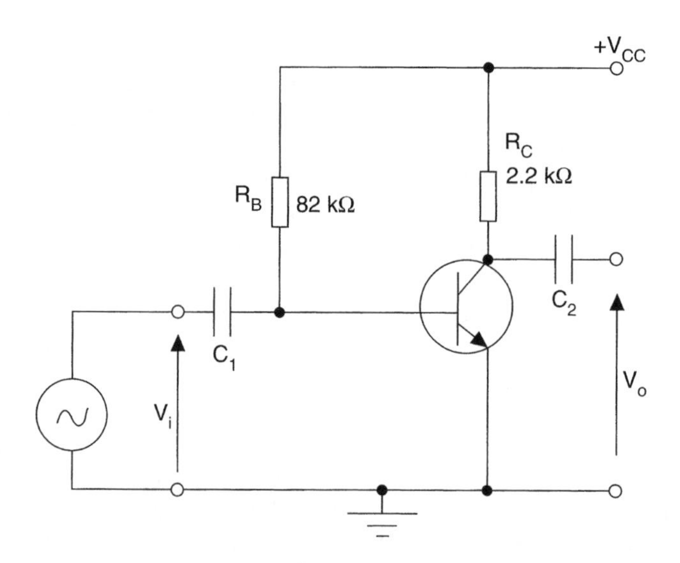

Fig. 10.16

Answer

$\delta V_i = 0.3\,\text{V}$ pk-pk; $\delta I_b = 80 \times 10^{-6}\,\text{A}$ pk-pk; $I_{BQ} = 60 \times 10^{-6}\,\text{A}$; $V_{CC} = 9\,\text{V}$;

$R_C = 2.2 \times 10^3\,\Omega$; $R_B = 82 \times 10^3\,\Omega$

For the load line: $V_{CC} = I_C R_C + V_{CE}$ volt

when $I_C = 0$; $V_{CE} = V_{CC} = 9\,\text{V}$ ....[1]

when $V_{CE} = 0$; $I_C = \dfrac{V_{CC}}{R_C}$ amp $= \dfrac{9}{2.2} = 4.1\,\text{mA}$ ....[2]

Joining these two points by a straight line gives the load line as shown in Fig. 10.17 and its intersection with $I_B = 60\,\mu\text{A}$ gives the Q point, with $V_{CEQ} = 4.6\,\text{V}$.

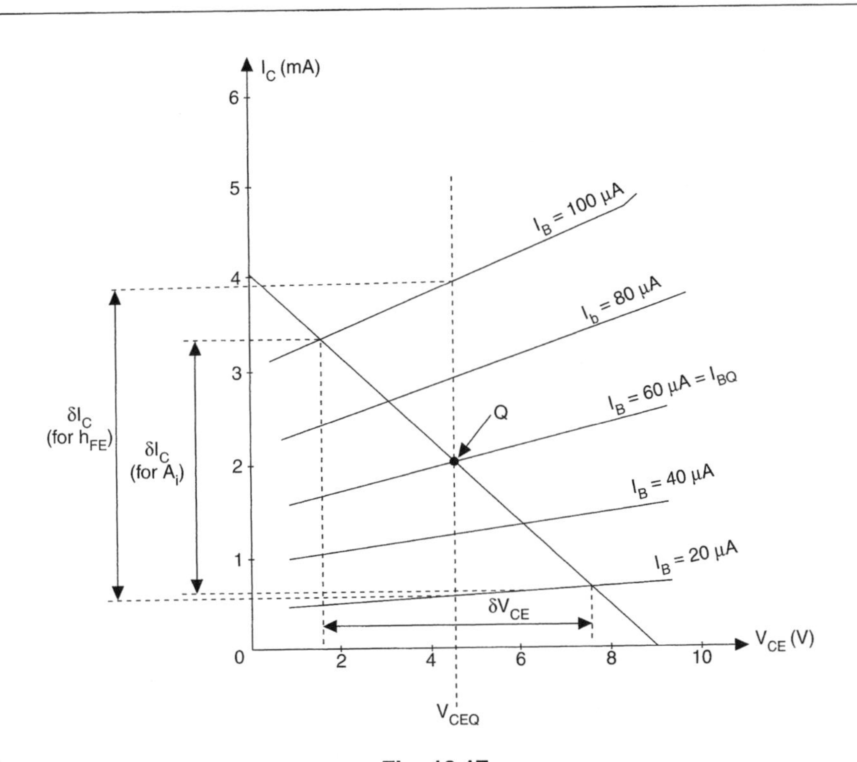

Fig. 10.17

(a) $h_{FE} = \frac{\delta I_C}{\delta I_B}$ (with $V_{CE} = 4.6\,\text{V}$) $= \frac{(3.95 - 0.55) \times 10^{-3}}{(100 - 20) \times 10^{-6}}$

$h_{FE} = 42.5$ **Ans**

(b) For amplifier gains, the excursions of I_c and V_{ce} are determined by the intercepts with the load line.

$$\delta I_c = (3.35 - 0.62)\,\text{mA} = 2.73\,\text{mA pk-pk}$$

$$\delta I_b = 80\,\mu\text{A pk-pk}$$

$$A_i = \frac{\delta I_c}{\delta I_b} = \frac{2.73 \times 10^{-3}}{80 \times 10^{-6}}$$

$A_i = 34.1$ **Ans** Note that this figure is $< h_{FE}$.

(c) $\delta V_{ce} = (7.6 - 1.6)\,\text{V} = 7\,\text{V pk-pk}$

$$\delta V_{be} = 0.3\,\text{V pk-pk}$$

$$A_v = \frac{\delta V_{ce}}{\delta V_{be}} = \frac{7}{0.3}$$

Continued on p. 412

Worked Example 10.4 (*Continued*)

$A_v = 23.3$ **Ans**

$A_p = A_i \times A_v = 34.1 \times 23.3$

$A_p = 796$ **Ans**

(d) The a.c. equivalent circuit will be as shown in Fig. 10.18.

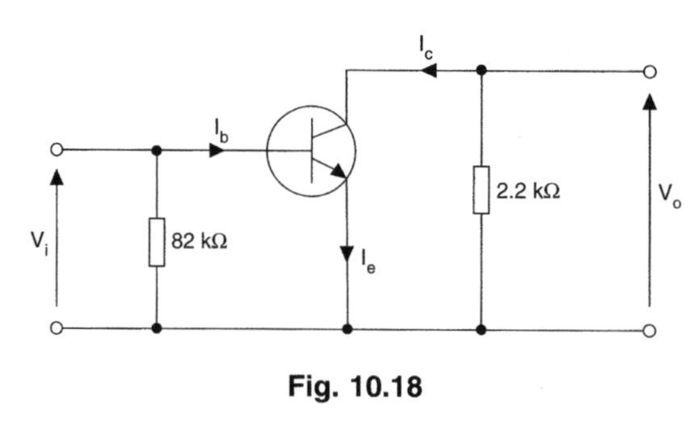

Fig. 10.18

Affect of External load In the previous example it has been assumed that no current was drawn from the circuit at the output terminals, i.e. any external load connected there has an infinite input impedance. The result was that both the d.c. quiescent collector current and the a.c. component of this current both flowed through R_C only, and the d.c. load line limited the excursions of the a.c. components of current and voltage. In practice, any load connected to the output terminals will have a finite impedance, and will therefore offer another path for the a.c. signal via C_2. Due to the d.c. blocking action of this capacitor, the d.c. conditions will not be affected.

Considering the circuit for the previous example, where a 10 kΩ external load, R_L, is now connected between the output terminals. The a.c. equivalent will now be as shown in Fig. 10.19.

The effective a.c. load, R_e, will be the parallel combination of R_C and R_L.

$R_e = \dfrac{R_C R_L}{R_C + R_L}$ ohm and the a.c. load line will have a slope of $-1/R_e$ amp/volt passing through the Q point already established by the d.c. load line. The following example shows how this will affect the analysis of the amplifier circuit.

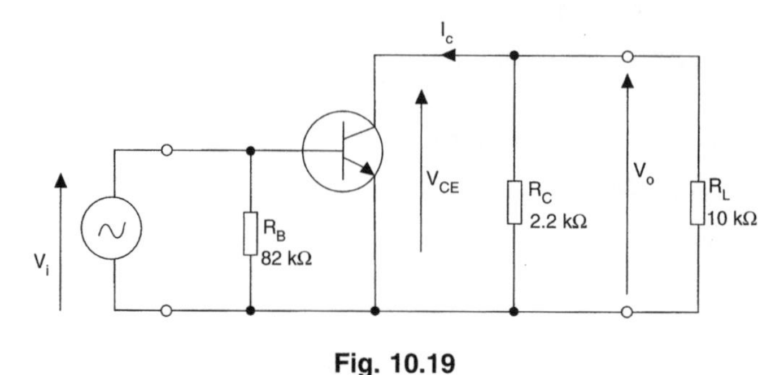

Fig. 10.19

Worked Example 10.5

Question

Using the same set of output characteristics as for Example 10.4, and adding a 10 kΩ external load to the circuit of Fig. 10.16, calculate the amplifier current, voltage and power gains.

Answer

The d.c. load line would be determined and plotted in exactly the same way as before, and this is shown in Fig. 10.20.

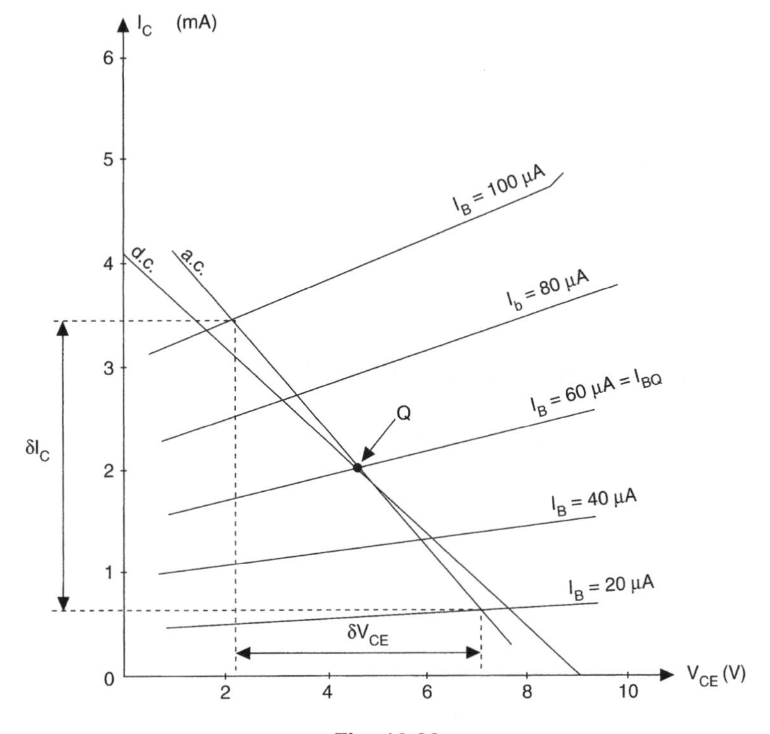

Fig. 10.20

$$\text{a.c. load line: } R_e = \frac{R_C R_L}{R_C + R_L} \text{ ohm} = \frac{10 \times 2.2}{10 + 2.2} \text{ k}\Omega$$

$$R_e = 1.8 \text{ k}\Omega \text{ and the slope} = \frac{-1}{1.8} \text{ mA/V} = -0.56 \text{ mA/V}$$

Now we know that the a.c. load line will pass through the Q point; so starting at this point, if we drop vertically by 0.56 mA and then move to the right by 1 V, we will

Continued on p. 414

Worked Example 10.5 (*Continued*)

have a second point through which the a.c. load line passes. However, this second point is very close to the Q point, so to improve the accuracy, double up the above figures, i.e. starting at the Q point drop down 1.12 mA and move right 2 V. The resulting a.c. load line will then be plotted as in Fig. 10.20, and the excursions of I_c and V_{ce} determined from the intersections with it, as follows.

$$\delta I_c = (3.45 - 0.6)\,\text{mA} = 2.85\,\text{mA pk-pk}$$

$$A_i = \frac{\delta I_c}{\delta I_b} = \frac{2.85 \times 10^{-3}}{80 \times 10^{-6}}$$

$$A_i = 35.6\ \textbf{Ans}$$

$$\delta V_{ce} = (7.1 - 2.2)\,\text{V} = 4.9\,\text{V pk-pk}$$

$$A_v = \frac{\delta V_{ce}}{\delta V_{be}} = \frac{4.9}{0.3}$$

$$A_v = 16.3\ \textbf{Ans}$$

$$A_p = A_i \times A_v = 35.6 \times 16.3$$

$$A_p = 580\ \textbf{Ans}$$

10.9 Three-resistor-biased Amplifier Circuit

When this circuit is employed there will be implications for both the d.c. and the a.c. load lines. A typical circuit is shown in Fig. 10.21, where an external load resistor is also included.

d.c. load line: applying Kirchhoff's voltage law as before:

$$V_{CC} = I_C R_C + V_{CE} + I_E R_E \text{ volt}$$

but $I_C \approx I_E$, so

$$V_{CC} = I_C R_C + V_{CE} + I_C R_E$$

and, $V_{CC} = I_C(R_C + R_E) + V_{CE}$

when $I_C = 0$; $V_{CE} = V_{CC}$ (as before) .... [1]

when $V_{CE} = 0$; $I_C = \dfrac{V_{CC}}{(R_C + R_E)}$ amp .... [2]

The d.c. load line will now have a slope of $-1/(R_C + R_E)$ volt/amp, and when plotted on the output characteristics will establish the Q point.

a.c. load line: the effective a.c. load, $R_e = \dfrac{R_C R_L}{R_C + R_E}$ ohm

and the a.c. load line will have a slope $= -1/R_E$ amp/volt, passing through the Q point, as in the previous example.

The a.c. equivalent circuit is shown in Fig. 10.22.

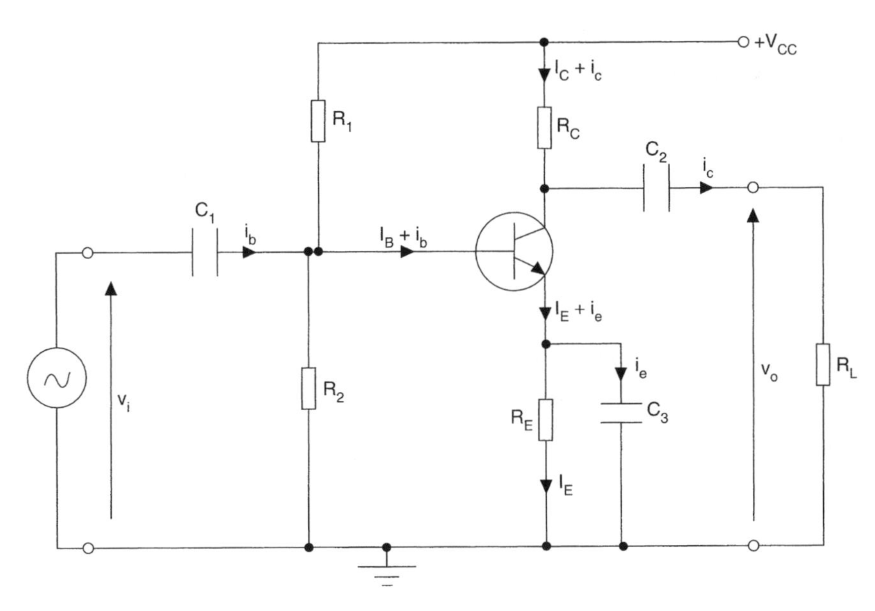

Fig. 10.21

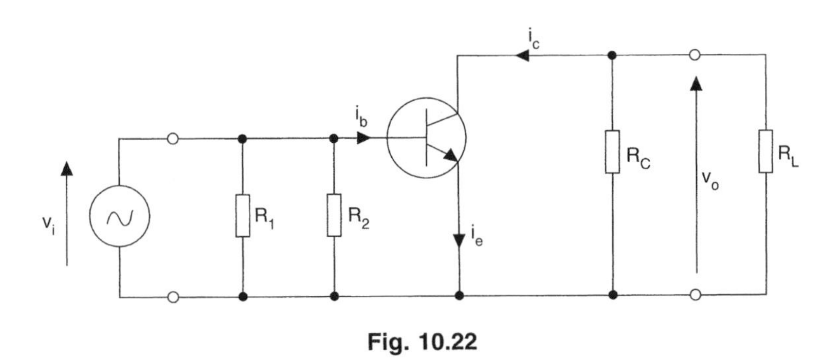

Fig. 10.22

Worked Example 10.6

Question

For the circuit of Fig. 10.21 let the component values be $R_C = 3.3\,\text{k}\Omega$; $R_E = 560\,\Omega$; $R_L = 10\,\text{k}\Omega$ and $V_{CC} = +10\,\text{V}$. You may assume that the capacitors have negligible reactance at the signal frequency. The transistor output characteristics are as in Fig. 10.23, and the input signal generator causes the base current to vary by $80\,\mu\text{A}$ pk-kp about its quiescent value.

(a) Plot the d.c. load line and hence determine a suitable Q point.

(b) Plot the a.c. load line and hence determine the amplifier current gain.

Continued on p. 416

Worked Example 10.6 (*Continued*)

Answer

$R_C = 3.3 \times 10^3\,\Omega$; $R_E = 560\,\Omega$; $R_L = 10^4\,\Omega$; $V_{CC} = 10\,\text{V}$; $\delta I_b = 80 \times 10^{-6}\,\text{A}$

(a) $V_{CC} = I_C(R_C + R_E) + V_{CE}$ volt

when $I_C = 0$; $V_{CE} = V_{CC} = 10\,\text{V}$ ....[1]

when $V_{CC} = 0$; $I_C = \dfrac{V_{CC}}{R_C + R_E}$ amp $= \dfrac{10}{3.86}$ mA ....[2]

$I_C = 2.6\,\text{mA}$

This load line is then plotted as shown in Fig. 10.23, and since, for practical purposes, V_{CEQ} should be about $V_{CC}/2$ volt, the Q point is chosen where the d.c. load line intersects with the $I_B = 60\,\mu\text{A}$ graph. This gives $V_{CEQ} = 4.7\,\text{V}$.

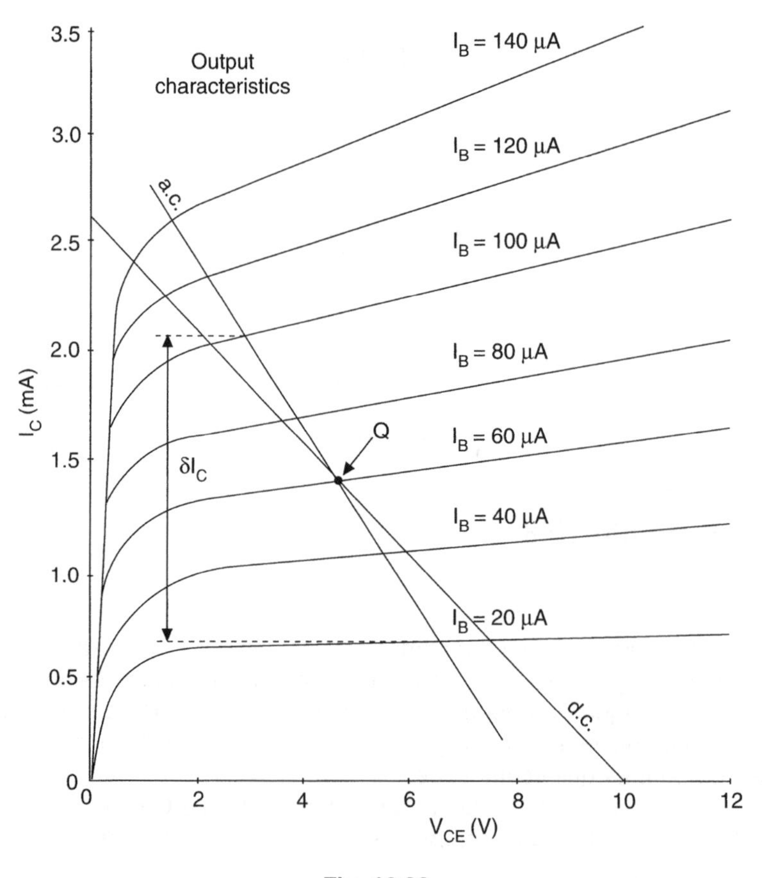

Fig. 10.23

(b) a.c. load line: $R_e = \frac{R_C R_L}{R_C + R_L}$ ohm $= \frac{33}{13.6}$ kΩ

$R_e = 2.48$ kΩ

$$\text{slope} = \frac{-1}{2.48}\ \text{mA/V} = -0.4\ \text{mA/V} \quad (\text{use} - 1.2\ \text{mA/3V})$$

and the load line is plotted with this slope passing through the Q point as shown.

$\delta I_b = 80\ \mu\text{A}$ pk-pk about $I_B = 60\ \mu\text{A}$

so, $\delta I_c = (2.05 - 0.65)\ \text{mA} = 1.4\ \text{mA}$ pk-pk

$$A_i = \frac{\delta I_c}{\delta I_b} = \frac{1.4 \times 10^{-3}}{80 \times 10^{-6}}$$

$A_i = 17.5$ **Ans**

10.10 FET Small-signal Amplifier

A typical FET signal amplifier circuit is shown in Fig. 10.24.

The amplifier voltage gain, A_v, may be determined from load lines plotted on the output characteristics in a similar manner to that for the BJT amplifier.

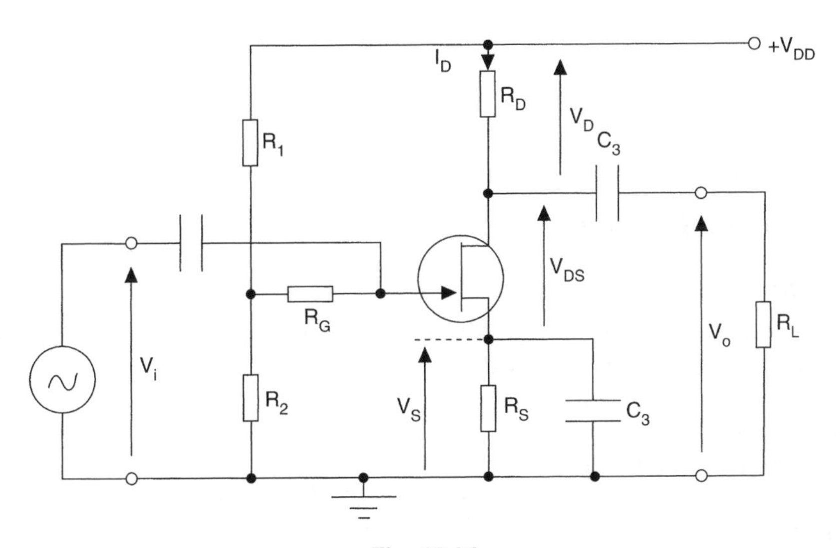

Fig. 10.24

d.c. load line: $V_{DD} = V_D + V_{DS} + V_S$ volt

where $V_D = I_D R_D$ and $V_S = I_D R_S$ volt

$V_{DD} = I_D(R_D + R_S) + V_{DS}$ volt

when $I_D = 0$; $V_{DS} = V_{DD}$ .... [1]

when $V_{DS} = 0$; $I_{DS} = \dfrac{V_{DD}}{R_D + R_S}$ amp .... [2]

Thus the d.c. load line may be plotted and a suitable Q point selected.

a.c. load line: the effective a.c. load, $R_e = \dfrac{R_D R_L}{R_D + R_L}$ ohm

and this load line will have a slope of $-1/R_e$ amp/volt.
Both load lines are shown plotted on Fig. 10.25.

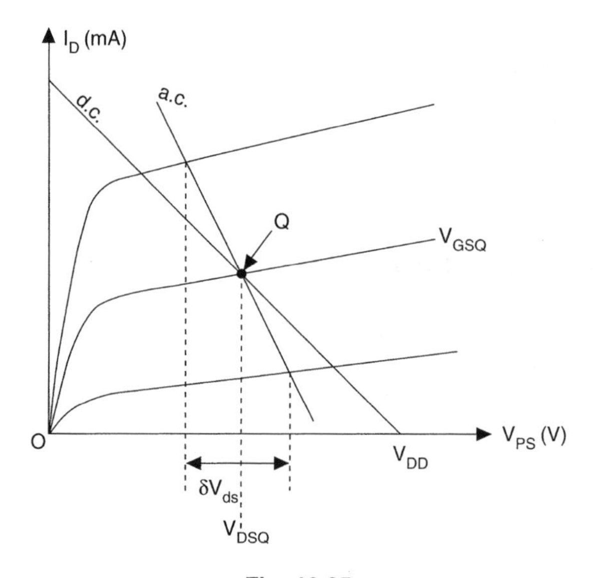

Fig. 10.25

Worked Example 10.7

Question

The FET used in the circuit of Fig. 10.26 has characteristics as shown in Fig. 10.27.

(a) From the characteristics, determine:
  (i) the mutual conductance, g_m, when $V_{DS} = 12$ V, and
  (ii) the drain-source resistance, r_{DS} for $V_{GS} = -3.5$ V.

(b) Plot the d.c. load line and select a suitable Q point.

(c) Plot the a.c. load line and determine the amplifier voltage gain when the input signal causes the gate-source voltage to vary by 2 V pk-pk.

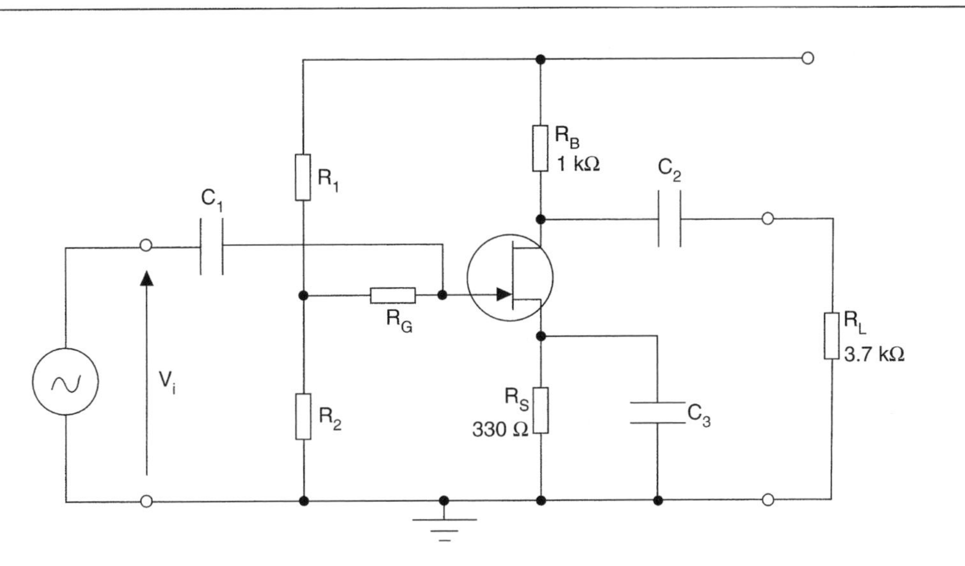

Fig. 10.26

Answer

(a) (i) $g_m = \dfrac{\delta I_D}{\delta V_{GS}}$ with $V_{DS} = 12\,\text{V}$

and from the graphs: $\delta I_D = (13.1 - 6.2)\,\text{mA} = 6.9\,\text{mA}$

and $\delta V_{GS} = (4.5 - 2.5)\,\text{V} = 2\,\text{V}$

$$g_m = \frac{6.9 \times 10^{-3}}{2}$$

$g_m = 3.45\,\text{mS}$ **Ans**

(ii) $r_{DS} = \dfrac{\delta V_{DS}}{\delta I_D}$ with $V_{GS} = -3.5\,\text{V}$

$$\delta V_{DS} = (18 - 2) = 16\,\text{V}$$

$$\delta I_D = (11.2 - 7.2)\,\text{mA} = 4\,\text{mA}$$

$$r_{DS} = \frac{16}{4 \times 10^{-3}}$$

$r_{DS} = 4\,\text{k}\Omega$ **Ans**

(b) Effective a.c. load, $R_e = \dfrac{R_D R_L}{R_D + R_L}$ ohm $= \dfrac{1 \times 3.7}{4.7}$ kΩ

$R_e = 787\,\Omega$ **Ans**

Continued on p. 420

Worked Example 10.7 (*Continued*)

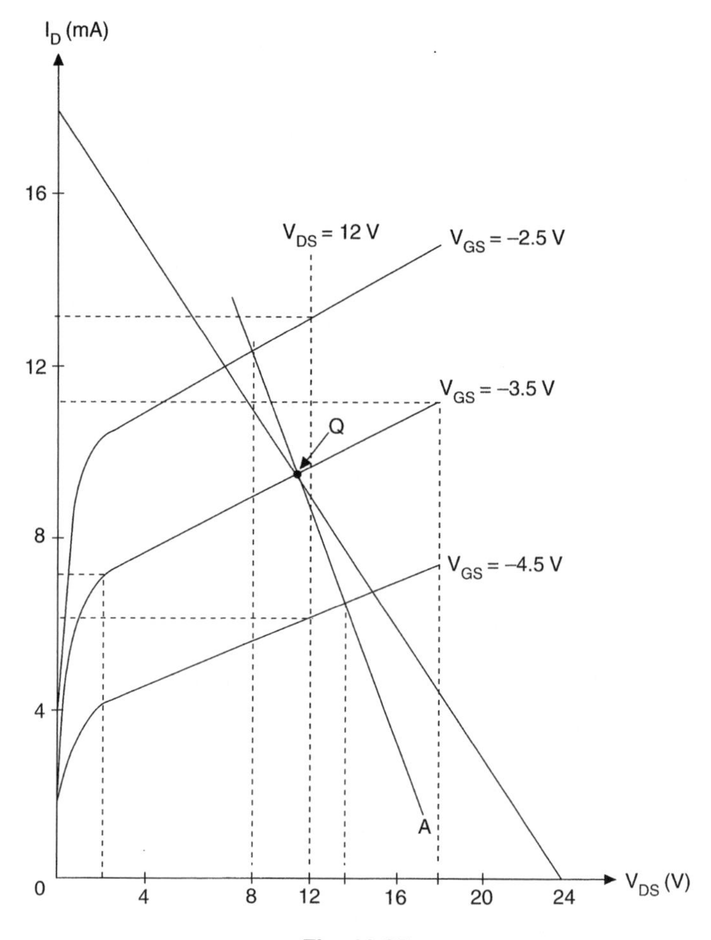

Fig. 10.27

$$\text{slope} = -1/R_e \text{ amp/volt} = -1/787$$
$$\text{slope} = -1.27\,\text{mA/V} \quad (\text{use } 7.62\,\text{mA}/6\,\text{V})$$

and the load line is plotted as shown.

$$\delta V_{ds} = (13.6 - 9.2)\,\text{V} = 4.4\,\text{V pk-pk}$$

$$\delta V_{gs} = 2\,\text{V pk-pk}$$

$$A_v = \frac{\delta V_{ds}}{\delta V_{gs}} = \frac{4.4}{2}$$

$A_v = 2.2$ **Ans**

10.11 The Transistor as a Switch

The important properties of any switch are the switching time, and the ON and OFF resistances. An ideal switch would have zero ON resistance, infinite OFF resistance and take zero time to change between these two states. A mechanical switch comes very close to meeting the resistance requirements, but is slow in operation. It also has the disadvantages of comparatively large physical size and cost, and also suffers from contact 'bounce'. The latter means that as the contacts close they tend to bounce up and down rapidly before settling. This can cause major problems in digital circuits, so another solution is required. A transistor does not have zero ON resistance or infinite OFF resistance, but it is small, cheap, has a fast response time and does not have the problem of contact bounce.

A common emitter connected transistor employed as an electronic switch, together with its characteristics, is shown in Fig. 10.28.

For amplification purposes the transistor is operated over the linear portions of its characteristics, between points X and Y shown on the load line for R_C. For switching purposes the saturation and cut-off regions are used.

When the input voltage is at 0 V then $I_B = 0$ and the transistor is said to be cut off. This corresponds to the OFF condition for the switch. In this case the circuit resistance is certainly not infinite, but will be very high since only the reverse collector current, I_{CEO}, will be flowing (a few microamps). When the input is at $+V_i$ volt the base input current will be sufficient to drive the transistor into saturation at point X, when the collector saturation current, $I_{C(sat)}$, will be flowing. This corresponds to the ON condition for the switch, and the only voltage drop across the transistor will be the small voltage (about 0.1 V) due to $V_{CE(sat)}$.

$$I_{C(sat)} = \frac{V_{CC} - V_{CE(sat)}}{R_C} \text{ amp} \tag{10.7}$$

and the least value required for I_B to produce saturation, $I_{B(sat)}$, is

$$I_{B(sat)} = \frac{I_{C(sat)}}{h_{FE}} \text{ amp} \tag{10.8}$$

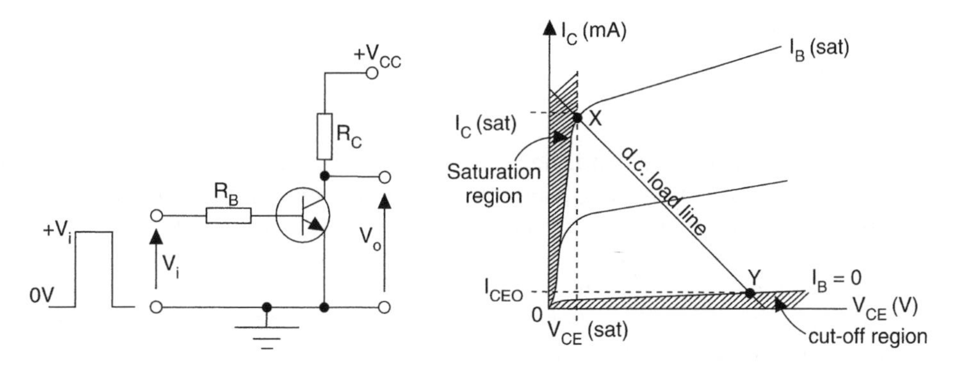

Fig. 10.28

To ensure saturation the base current is arranged to be about three times the figure obtained from equation (10.8).

The required value for R_C may be obtained from equation (10.7), and the required value for R_B is obtained as follows.

$$R_B = \frac{V_i - V_{BE}}{3 \times I_{B(sat)}} \text{ ohm} \qquad (10.9)$$

Worked Example 10.8

Question

The transistor used in the circuit of Fig. 10.28 has a current gain, $h_{FE} = 90$, base-emitter voltage, $V_{BE} = 0.6\,\text{V}$, $V_{CE(sat)} = 0.1\,\text{V}$ and $I_{CE(sat)} = 4\,\text{mA}$. The collector supply voltage, $V_{CC} = +10\,\text{V}$, and the input signal level changes from 0 V to +5 V. Calculate suitable values for R_C and R_B.

Answer

$$I_{CE(sat)} = \frac{V_{CC} - V_{CE(sat)}}{R_C} \text{ amp}$$

$$\text{so,} \quad R_C = \frac{V_{CC} - V_{CE(sat)}}{I_{CE(sat)}} \text{ ohm} = \frac{10 - 0.1}{4 \times 10^{-3}}$$

$R_C = 2.475\,\text{k}\Omega$ and the nearest preferred value is $2.2\,\text{k}\Omega$ **Ans**

Note that in this case the nearest *lower* preferred value is chosen so as to ensure saturation is achieved.

$$I_{B(sat)} = \frac{I_{CE(sat)}}{h_{FE}} \text{ amp} = \frac{4 \times 10^{-3}}{90}$$

$$I_{B(sat)} = 44.4\,\mu\text{A}$$

$$R_B = \frac{V_i - V_{BE}}{3 \times I_{B(sat)}} \text{ ohm} = \frac{5 - 0.6}{133 \times 10^{-6}}$$

$R_B = 33\,\text{k}\Omega$ **Ans**

As with the common emitter amplifier circuit, inversion occurs between input and output of the transistor switch circuit shown. Thus, when the input is HIGH (saturation conditions), the output is LOW ($V_{CE(sat)} \approx 0\,\text{V}$), and when the input is LOW (cut-off conditions), the output is HIGH ($V_{CE} \approx V_{CC}$). If it is required for the output HIGH and LOW conditions to mimic the input conditions then the load resistor can be relocated to the emitter circuit as shown in Fig. 10.29. This circuit is in fact a common collector or emitter follower circuit. This name is used because the output follows the high and low states of the input.

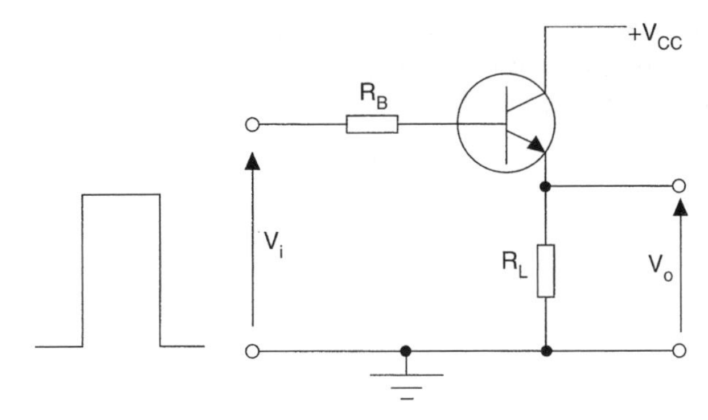

Fig. 10.29

If the load being supplied by the electronic switch is in the form of a power device, such as the operating coil of a relay which requires significantly higher current, then a power transistor would be employed. These can be capable of carrying currents measured in amps rather than milliamps.

A FET can also be used as an electronic switch. Being a voltage operated device it has the advantage that there is no requirement for the equivalent of R_B in the BJT circuit. The switching time for a FET is also faster than that for a BJT. For this reason FET devices are used extensively in digital switching applications, where speed of operation is paramount. On the other hand, FETs cannot handle as much power as BJTs.

Assignment Questions

1 A simple voltage stabiliser circuit is shown in Fig. 10.30. The zener diode is a 7.5 V, 500 mW device and the supply voltage is 12 V. Calculate (a) a suitable value for R_S, and (b) the value of zener current when $R_L = 470\,\Omega$.

2 Using the circuit of Fig. 10.30 with a 10 V, 1.3 W zener diode having a slope resistance of 2.5 Ω, and a supply voltage of 24 V, calculate (a) a suitable value for R_S, (b) the value of zener current when on no-load, and (c) the variation of zener voltage when R_L is changed from 500 Ω to 200 Ω.

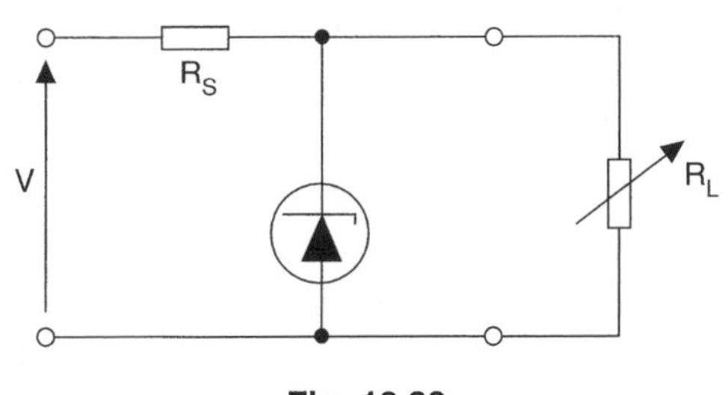

Fig. 10.30

3 For the simply biased transistor circuit of Fig. 10.31, the required quiescent values for base current and base-emitter voltage are 100 μA and 0.75 V respectively. Calculate suitable values for R_B and R_C.

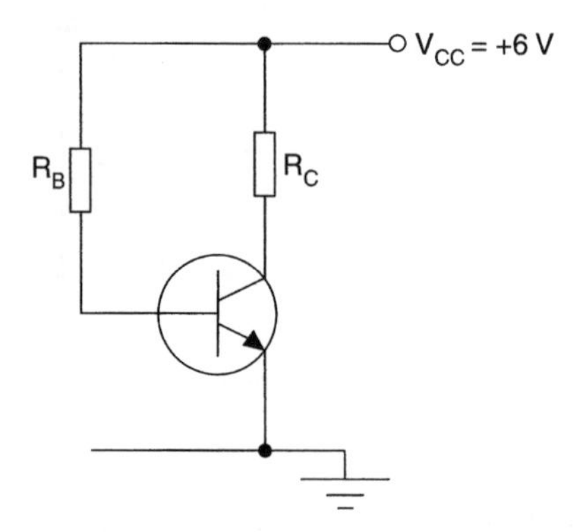

Fig. 10.31

4 The transistor in Fig. 10.32 has a current gain of 40 and the required bias conditions are $V_{BE} = 0.7\,\text{V}$ and $I_C = 2\,\text{mA}$. Calculate suitable values for the four resistors.

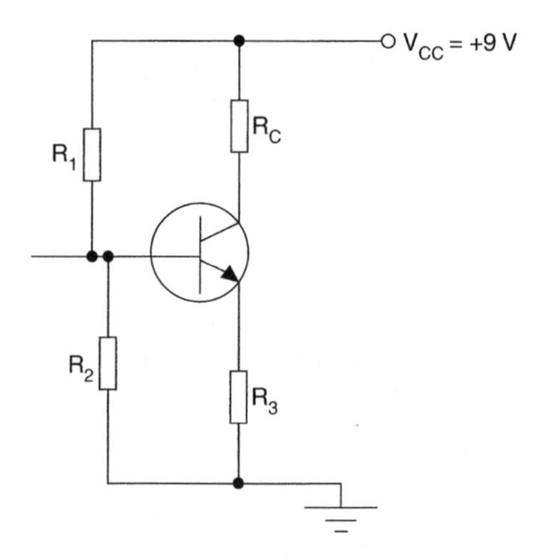

Fig. 10.32

5 The FET in Fig. 10.33 requires a quiescent gate-source voltage of $-2\,\text{V}$ with a drain current of 5 mA. Determine suitable values for R_S and R_D.

6 The characteristics for the FET in Fig. 10.34 are as in Fig. 10.35. Using these characteristics

(a) Determine the mutual conductance and drain-source resistance for $V_DS = 6\,\text{V}$.

(b) Plot the d.c. load line and hence select a suitable Q point.

(c) Plot the a.c. load line and determine the amplifier voltage gain when the input signal causes the gate-source voltage to vary by 2 V pk-pk about its quiescent value.

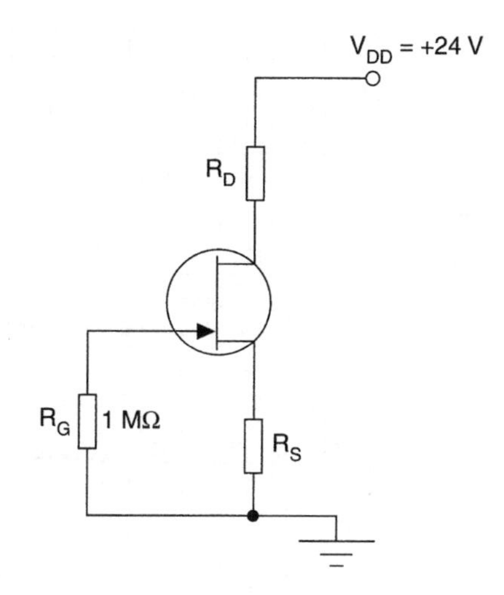

Fig. 10.33

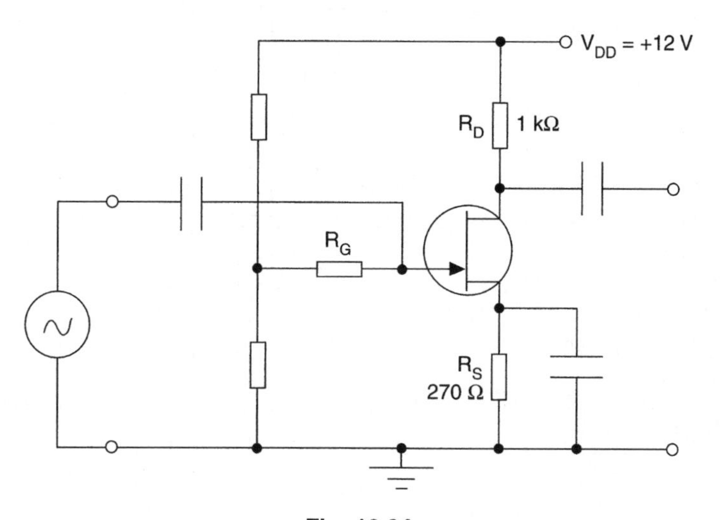

Fig. 10.34

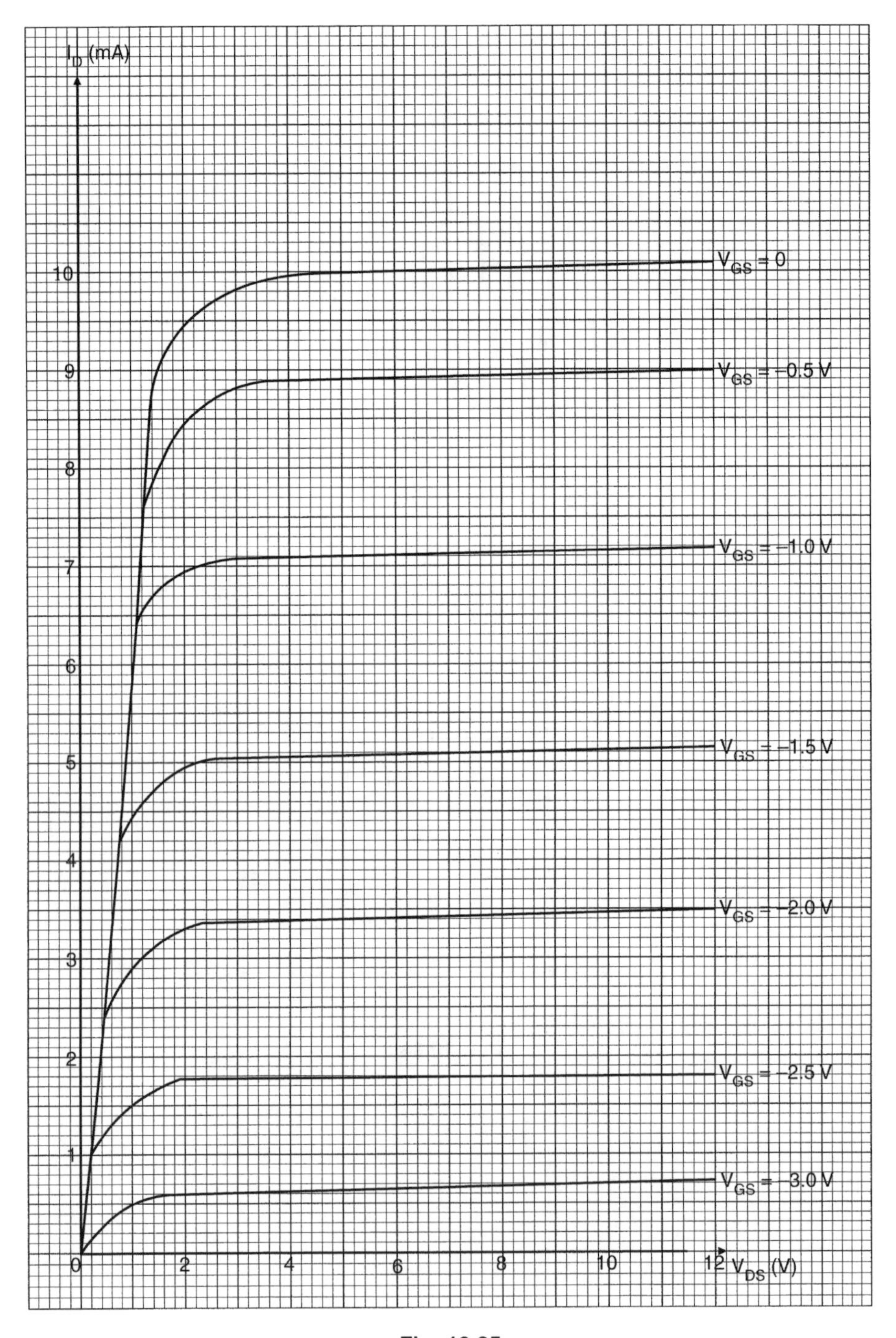

I_D (mA)
10
9
8
7
6
5
4
3
2
1
0
2
4
6
8
10
12 V_{DS} (V)
V_{GS} = 0
V_{GS} = −0.5 V
V_{GS} = −1.0 V
V_{GS} = −1.5 V
V_{GS} = −2.0 V
V_{GS} = −2.5 V
V_{GS} = −3.0 V

Fig. 10.35

Suggested Practical Assignments

Assignment 1

To investigate the operation of the zener diode.

Apparatus:

1 × 5.6 V, 400 mW zener diode
1 × 9.1 V, 400 mW zener diode
1 × 470 Ω resistor
1 × variable d.c. power supply unit (psu)
1 × voltmeter (DVM)
1 × ammeter

Method:

1 Connect the circuit of Fig. 10.36 using the 5.6 V diode.
2 Vary the input voltage in 1 V steps from 0 V to +15 V, and note the corresponding values of V_Z and I.

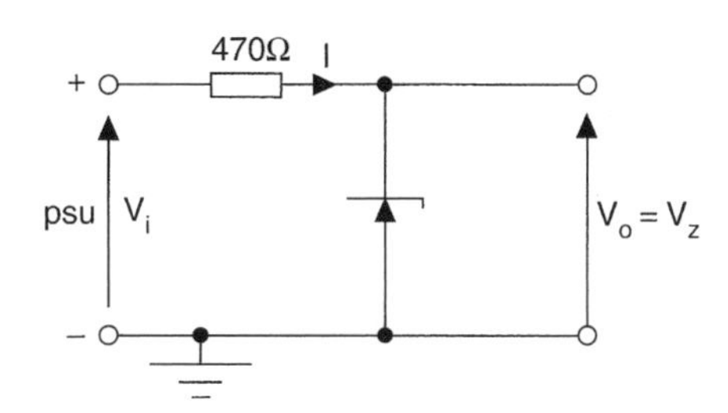

Fig. 10.36

3 Tabulate your results and plot the reverse characteristic for the diode.
4 Repeat steps 1 to 3 above for the 9.1 V diode.
5 From the plotted characteristics, determine the diode slope resistance in each case.

Assignment 2

To investigate the use of a transistor as a switch.

Apparatus:

1 × general purpose BJT (e.g. a BC108)
1 × 10 kΩ resistor

1 × 100 kΩ resistor
1 × single pole switch
1 × d.c. psu
1 × voltmeter

Method:

1 Connect the circuit of Fig. 10.37, and set the psu output to 9 V.
2 With the switch in position '1', note the values of V_i and V_o.
3 Move the switch to position '2' and again note the values of V_i and V_o.
4 Switch off the psu.
5 Transfer the 1 kΩ resistor from the collector circuit into the emitter circuit, and connect the collector directly to the V_{CC} rail.
6 Switch on the psu and repeat steps 2 and 3 above.

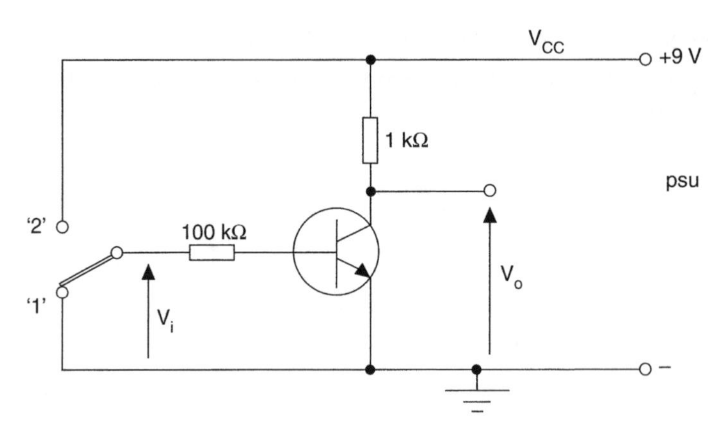

Fig. 10.37

Assignment 3

To investigate a small-signal BJT amplifier.

Apparatus:

1 × BC 108 transistor
4 × resistors: 4.7 kΩ, 1 kΩ, 56 kΩ, and 12 kΩ
1 × signal generator (and possibly an attenuator box)
2 × 10 μF capacitors
1 × 100 μF capacitor
1 × double beam oscilloscope
1 × DVM
1 × d.c. psu

Method:

1 Connect the circuit of Fig. 10.38, taking care to observe the correct polarity for the electrolytic capacitors.

2 Set the signal generator to a sinewave output at a frequency of 5 kHz and zero volts, but do not connect to the amplifier input terminals yet.

3 Adjust the d.c. psu output to 12 V.

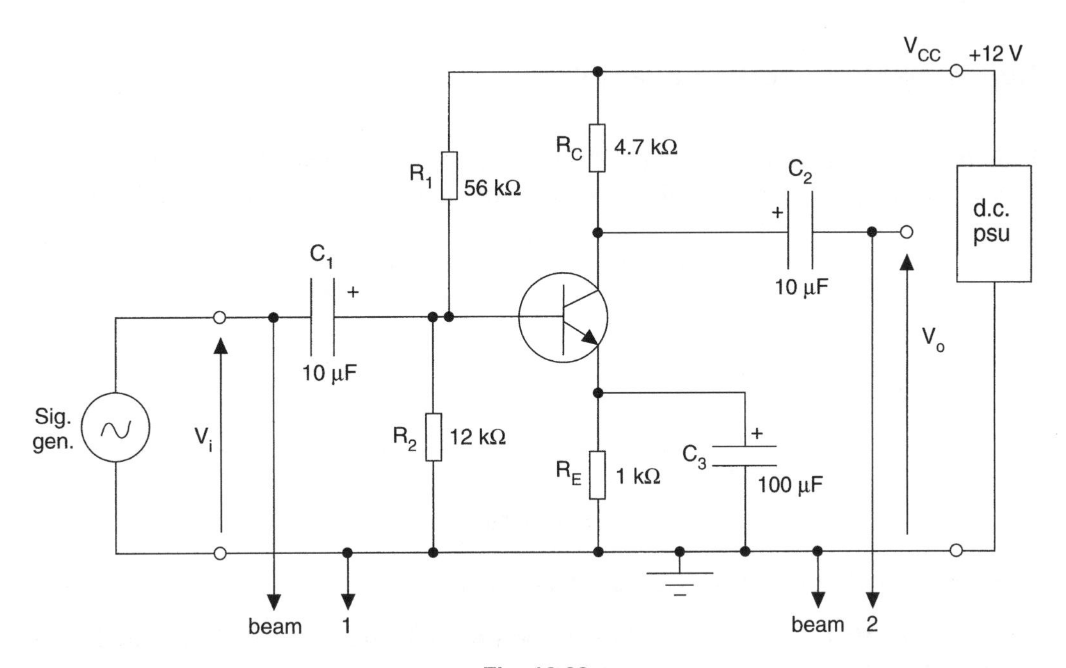

Fig. 10.38

4 Using the DVM measure and note the quiescent d.c. voltages listed below:

(a) between base and emitter;

(b) across R_C;

(c) between collector and emitter;

(d) across R_E.

5 Connect the signal generator to the input terminals and very carefully increase its voltage so that the amplifier output voltage, V_o, is an undistorted sinewave, as shown by beam 2 of the oscilloscope. It may be necessary to connect the signal generator via an attenuator in order to provide a sufficiently small input signal to the amplifier.

6 Using the oscilloscope measure the amplitudes of V_i and V_o and compare their phase relationship.

7 Monitor the waveforms at the following points in the circuit, and by means of the DC/AC switch on the oscilloscope, determine the d.c. level about which each waveform varies:

(a) between base and emitter;

(b) between collector and emitter;

(c) between collector and the 0 V rail;

(d) across R_C.

8 Determine the amplifier voltage gain.

Supplementary Worked Examples

Example 1: A d.c. voltage of 15 V±5% is required to be supplied from a 24 V unstabilised source. This is to be achieved by the simple regulator circuit of Fig. 10.39.

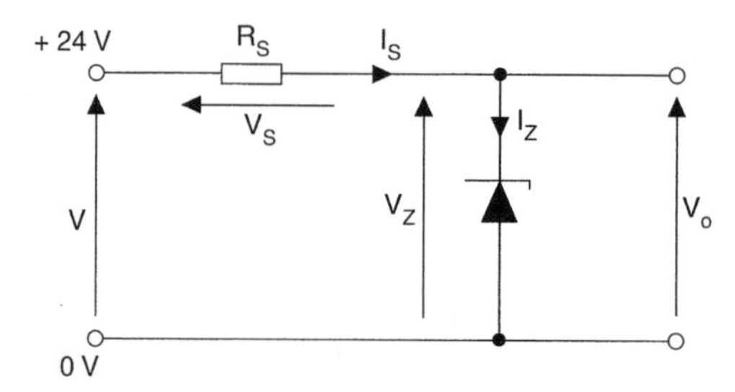

Fig. 10.39

The available diodes and resistors are listed below.

(a) For each diode listed determine the appropriate resistor required and hence determine the total unit cost for each circuit.

(b) In order to satisfy the specified output voltage tolerance of ±5%, determine which of the three circuits will meet the specification at lowest cost.

Diode No.	V_Z (V)	*Slope resistance* (Ω)	*Max power* (W)	*Unit cost* (£)
1	15	30	0.5	0.07
2	15	15	1.3	0.20
3	15	2.5	5.0	0.67

Resistors are available in the following values and unit costs
18, 27, 56, 100, 120, 150, 220, 270, and 330 Ω

0.25 W	**£0.026**
0.5 W	**£0.038**
1.0 W	**£0.055**
2.5 W	**£0.260**
7.5 W	**£0.280**

Solution:

(a) For all three diodes:

$$V_S = (V - V_Z)\,\text{volt} = 24 - 15 = 9\,\text{V}$$

Diode 1: $I_Z = \dfrac{P_Z}{V_Z}\,\text{amp} = \dfrac{0.5}{15}$

$$I_Z = 33.3\,\text{mA}$$

$$R_S = \frac{V_S}{I_Z}\,\text{ohm} = \frac{9}{33.3 \times 10^{-3}}$$

$$R_S = 270\,\Omega$$

$$P_S = \frac{V_S^2}{R_S}\,\text{watt} = \frac{81}{270}$$

$$P_S = 0.3\,\text{W}$$

and, $R_S = 270\,\Omega$, 0.5 W **Ans**

total unit cost = £(0.07 + 0.038) = £0.045 **Ans**

Diode 2: $I_Z = \dfrac{1.3}{15}\,\text{amp} = 86.7\,\text{mA}$

$$R_S = \frac{9}{86.7 \times 10^{-3}} = 103.85\,\Omega\text{, so choose the } 120\,\Omega \text{ resistor}$$

$$P_S = \frac{81}{120} = 0.675\,\text{W, so choose } 1.0\,\text{W rating}$$

hence, $R_S = 120\,\Omega$, 1.0 W **Ans**

total unit cost = £(0.20 + 0.055) = £0.26 **Ans**

Diode 3: $I_Z = \dfrac{5}{15} = 333.3\,\text{mA}$

$$R_S = \frac{9}{333.3 \times 10^{-3}} = 27\,\Omega$$

$$P_S = \frac{81}{27} = 3\,\text{W so choose } 7.5\,\text{W resistor}$$

hence, $R_S = 27\,\Omega$, 7.5 W **Ans**

total unit cost = £(0.67 + 0.28) = £0.95 **Ans**

(b) Allowable $\delta V_Z = \dfrac{5 \times 15}{100} = 0.75\,\text{V}$

Diode 1: $\delta V_Z = \delta I_{zrz}$ volt $= 30 \times 10^{-3} \times 30 = 0.9\,\text{V}$, which is unacceptable

Diode 2: $\delta V_Z = 30 \times 10^{-3} \times 15 = 0.45\,\text{V}$, which is acceptable

Diode 3: $\delta V_Z = 30 \times 10^{-3} \times 2.5 = 0.075\,\text{V}$, which is acceptable

Thus to meet the specification at lowest cost, circuit 2 would be adopted **Ans**

Example 2: The FET used in the amplifier circuit of Fig. 10.40 has characteristics as given in Table 10.1. Using this data:

(a) Plot the characteristics.

(b) Plot the load line to determine a suitable Q point and hence the corresponding quiescent values for V_{GS}, I_D and V_{DS}.

(c) Determine the values for R_{ds} and g_m under quiescent conditions.

(d) Plot the a.c. load line and hence obtain the amplifier voltage gain when V_{GS} varies about its quiescent value by 2 V pk-pk.

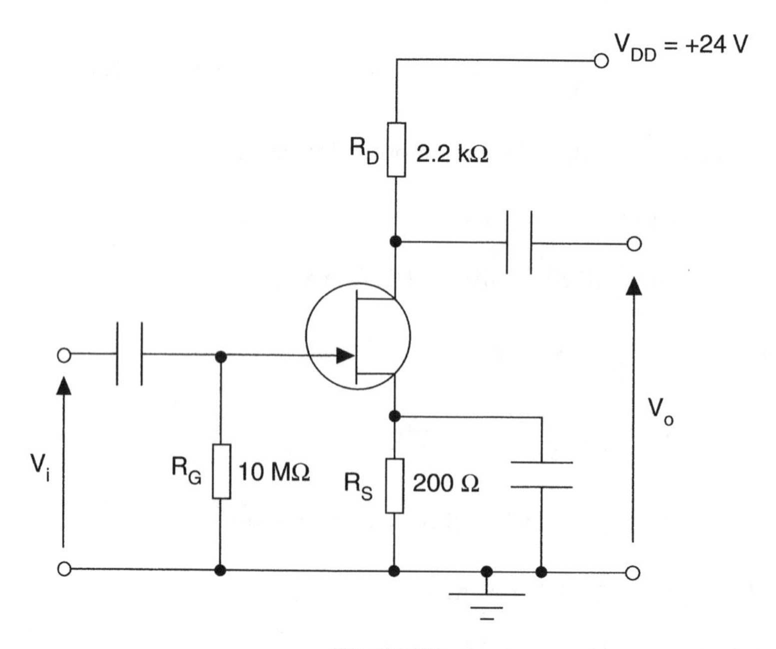

Fig. 10.40

Table 10.1

V_{DS} (V)	$I_D(mA)$				
	$V_{GS} = -2.5\,\text{V}$	$V_{GS} = -2.0\,\text{V}$	$V_{GS} = -1.5\,\text{V}$	$V_{GS} = -1.0\,\text{V}$	$V_{GS} = -0.5\,\text{V}$
4	1.4	2.8	4.4	6.1	8.0
16	1.6	3.0	4.6	6.3	8.2
24	1.8	3.1	4.7	6.4	8.4

Solution:

(a) The plotted characteristics are shown in Fig. 10.41.

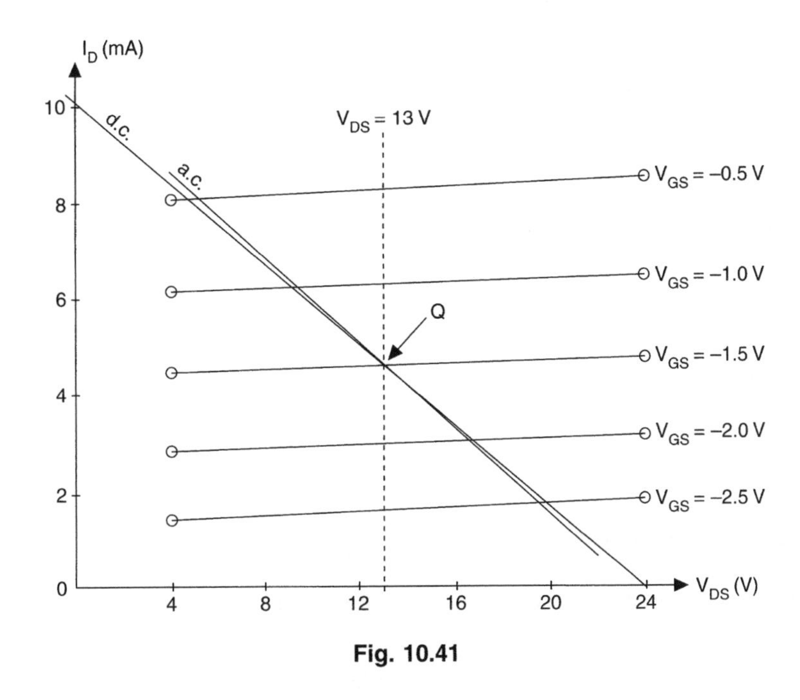

Fig. 10.41

(b) ***d.c. load line:*** $V_{CC} = I_D(R_D + R_S) + V_{DS}$ volt

when $I_D = 0$; $V_{DS} = V_{CC} = 24\,\text{V} \ldots\ldots\ldots\ldots\ldots\ldots [1]$

when $V_{DS} = 0$; $I_D = \dfrac{V_{CC}}{R_D + R_S}$ ohm $= \dfrac{24}{2400} \ldots\ldots\ldots [2]$

$I_D = 10\,\text{mA}$

The Q point is chosen where this load line intersects the $V_{GS} = -1.5\,\text{V}$ graph, and the resulting quiescent values are:

$V_{DSQ} = 13\,\text{V}$; $I_{DQ} = 4.58\,\text{mA}$ **Ans**

(c) $R_{ds} = \dfrac{\delta V_{DS}}{\delta I_D}$ ohm, with $V_{GS} = -1.5\,\text{V}$

$\delta V_{DS} = 24 - 4 = 20\,\text{V}$; and $\delta I_D = (4.7 - 4.4)\,\text{mA} = 0.3\,\text{mA}$

$$R_{ds} = \frac{20}{0.3 \times 10^{-3}}\ \text{ohm}$$

$R_{ds} = 66.7\,\text{k}\Omega$ **Ans**

$g_m = \dfrac{\delta I_D}{\delta V_{GS}}$ siemen, with $V_{DS} = V_{DSQ} = 13\,\text{V}$

$\delta V_{GS} = 2\,\text{V}$ pk-pk; and $\delta I_D = (8.2 - 1.6)\,\text{mA} = 6.6\,\text{mA}$

$g_m = \dfrac{6.6 \times 10^{-3}}{2} = 3.3\,\text{mS}$ **Ans**

(d) ***a.c. load line:*** effective a.c. load $= R_D = 2.2\,\text{k}\Omega$

$$\text{slope} = \frac{-1}{R_D}\ \text{A/V} = \frac{-1}{2.2}\ \text{mA/V}$$

slope $= -0.455\,\text{mA/V}$ (use $-4.55\,\text{mA}/10\,\text{V}$), and the plotted load line is as shown on Fig. 10.41.

$$A_v = \frac{\delta V_{DS}}{\delta V_{GS}}$$

$\delta V_{GS} = 2\,\text{V}$ pk-pk; and $\delta V_{DS} = 19.6 - 5.2 = 14.4\,\text{V}$ pk-pk

$A_v = \dfrac{14.4}{2} = 7.2$ **Ans**

Summary of Equations

Small-signal common emitter amplifier: Current gain, $A_i = \dfrac{\delta I_c}{\delta I_b}$

Voltage gain, $A_v = \dfrac{\delta V_{ce}}{\delta V_{be}}$

Power gain, $A_p = A_i \times A_v$

Small-signal FET amplifier: Voltage gain, $A_v = \dfrac{\delta V_{ds}}{\delta V_{gs}}$

BJT as a switch: $I_{C(sat)} = \dfrac{V_{CC} - V_{CE(sat)}}{R_C}$ amp

$$I_{B(sat)} = \frac{I_{C(sat)}}{h_{FE}}\ \text{amp}$$

$$R_B = \frac{V_i - V_{BE}}{3 \times I_{B(sat)}}\ \text{ohm}$$

Answers to Assignment Questions

Chapter 1

1.1 (a) 4.563×10^2
(b) 9.023×10^5
(c) 2.85×10^{-4}
(d) 8×10^3
(e) 4.712×10^{-2}
(f) 1.8×10^{-4} A
(g) 3.8×10^{-2} V
(h) 8×10^{10} N
(i) 2×10^{-3} F

1.2 (a) 1.5 kΩ
(b) 3.3 mΩ
(c) 25 μA
(d) 750 V or 0.75 kV
(e) 800 kV
(f) 47 nF

1.3 750 C

1.4 0.104 s

1.5 0.917 A

1.6 (a) 2.25 kV
(b) 18.75 V

1.7 (a) 2.273 A
(b) 61 mA
(c) 18.52 μA
(d) 0.512 mA

1.8 54.14 MJ or 15.04 kWh

1.9 £55.88

1.10 (a) 0.5 A
(b) 280 V
(c) 140 W
(d) 42 kJ

1.11 (a) 49.64 V
(b) 27.58 Ω

1.12 (a) 0.15 Ω
(b) 2.25 Ω

1.13 225 W; 67.5 kJ

1.14 (a) 2 A
(b) 8 V
(c) 2.4 kC

1.15 $2\,m/s^2$; $2.67\,m/s^2$ $0.5\,m/s^2$

1.16 8 μA

1.17 19.2 Ω; 12.5 A

1.18 15.65 kW

1.19 71.72°C

1.20 33.6 Ω

Chapter 2

2.1 0.375 A

2.2 (a) 29.1 Ω
(b) 24.75 A

2.3 0.45 A

2.4 14 Ω

2.5 (a) 15.67 Ω
(b) 0.766 A
(c) 0.426 A

2.6 (a) 2.371 A; 1.779 A; 1.804 A; 1.443 A; 0.902 A

(b) 35.37 V (15 Ω and 20 Ω); 14.43 V (8 Ω, 10 Ω, 16 Ω)
(c) 63.26 W

2.7 2.5 V

2.8 (a) 12.63 V
(b) 2.4 A
(c) 0.947 A

2.9 (a) 3.68 Ω
(b) 5.44 A
(c) 9.32 V
(d) 2.33 A

2.10 12.8 Ω

2.11 (a) 1.57 Ω
(b) 127.33 Ω

2.12 40 Ω

2.13 80 Ω

2.14 (a) 1.304 A
(b) 5.217 V
(c) 0.87 A

2.15 (a) 6.67 A
(b) 26.67 A
(c) 13.33 A

2.16 (a) 2 V
(b) 1.933 V

2.17 0.024 Ω

2.18 (a) 9.6 A
(b) 43.2 V
(c) 4.32 A; 2.16 A; 1.08 A

2.19 3.094 V

2.20 6.82 V

2.21 (a) 20 Ω
(b) 10 A
(c) 60 V
(d) 2 kW
(e) 300 W

2.22 (a) 1 A
(b) 30 V; 36 V
(c) 54 kJ
(d) 180 C

2.23 2.368 A; −0.263 A; 2.105 A; 10.526 V

2.24 1 A (discharge); 0.5 A (discharge); 1.5 A; 9 W

2.25 (a) −0.183 A (charge); 5.725 A (discharge); 5.542 A (discharge)
(b) 108.55 V

2.26 (a) 2.273 A (discharge); −0.455 A (charge);
(b) 1.32 A
(c) 11.363 V

2.27 0.376 A; 0.388 A; 0.764 A; 61.14 V

2.28 (b) 3 Ω
(c) 0.133 A (10 Ω and 5 Ω); 0.222 A (6 Ω and 3 Ω); 1.33 V (10 Ω and 6 Ω); 0.67 V (5 Ω and 3 Ω)

2.29 3.498 Ω; 16.85 kΩ; 22.5 Ω

2.30 1.194 V

2.31 1.314 A

2.32 0.116 A from B to A

2.33 5.11 V, A positive

2.34 0.635 A

2.35 1.314 A

2.36 0.116 A from B to A

2.37 1.2 V

2.38 (a) $E_o = 13.24$ V; $R_o = 4.71\ \Omega$
(b) $I_{sc} = 1.25$ A; $R_o = 4.71\ \Omega$ (c) 4.71 Ω

2.39 8.45 V; 2.17 A

2.40 (a) 5.19 V (B positive); 1.037 A (B to A)
(b) 0.4 Ω
(c) 16.8 W

2.41 2:1

2.42 (a) 14 dB
(b) −4.4 dB
(c) 24.1 dB
(d) 60 dB
(e) −6.6 dB
(f) 19.3 dB

2.43 1.052 V

2.44 0.28 V

2.45 (a) 20 dBm
(b) 4 dBm
(c) −8.2 dBm
(d) −6 dBm
(e) 40 dBm

2.46 5.13 mW

2.47 +15 dB

Chapter 3

3.1 (a) $0.2\,\mu C$
(b) $2.29\mu C/m^2$
3.2 $16.94\,kV/m$
3.3 $50\,kV/m$
3.4 $60\,mC/m^2$
3.5 $125\,kV/m$
3.6 $40\,mC$
3.7 $165.96\,V$
3.8 $300\,V$
3.9 500 pF or 0.5 nF
3.10 240 pF or 0.24 nF
3.11 3
3.12 5.53 nF
3.13 (a) 442.7 pF
(b) $0.177\,\mu C$
(c) $400\,kV/m$
3.14 0.089 mm
3.15 188 nF
3.16 1.36
3.17 25
3.18 (a) 4.8 mm
(b) $1.213 \times 10^{-3}\ m^2$
3.19 (a) $14\,\mu F$
(b) $2.86\,\mu F$
3.20 (i) $1.463\,\mu F$; $17\,\mu F$
(ii) $0.013\,\mu F$; $0.29\,\mu F$
(iii) 19.18 pF; 490 pF
(iv) 215 pF; 10.2 nF
3.21 8 nF
3.22 (a) $5\,\mu F$
(b) 200 V
(c) 3 mC
3.23 (a) $24\,\mu F$
(b) $480\,\mu C$
(c) 1.44 mC
3.24 (a) 13.85 V (4 nF); 6.15 V
(b) 18.46 nC
3.25 $C_2 = 4.57\mu F$; $C_3 = 3.56\,\mu F$
3.26 200 V; 200 V; 1.2 mC; 2 mC; 3.2 mC;
3.27 $V_1 = 360\,V$; $V_2 = 240\,V$; $C_3 = 40\mu F$
3.28 200 V
3.29 $80.67\,cm^2$
3.30 (a) 48 pF
(b) 267 pC
(c) 40 V
3.31 625 mJ
3.32 200 V
3.33 $5\,\mu F$
3.34 (a) 40 nF
(b) 0.8 mJ
(c) $400\,kV/m$
3.35 (a) 1.6 mC; 0.32 J
(b) 266.7 V; 0.213 J
3.36 (a) 0.6 mC; 150 V; 100 V
(b) 120 V; 0.48 mC; 0.72 mC
3.37 (a) 1.5 mm
(b) $52.94\ cm^2$
(c) 75 nC; $28\,\mu J$
(d) $14.2\,\mu C/m^2$
3.38 $0.5\,\mu m$

Chapter 4

4.1 0.417 T
4.2 1.98 mWb
4.3 $40\ cm^2$
4.4 21.25 At
4.5 0.8 A
4.6 270
4.7 5633 At/m
4.8 112.5 At
4.9 2 A
4.10 (a) 900 At
(b) 1.11 T
(c) 5000 At/m
4.11 64
4.12 278
4.13 (a) 1200 At
(b) 5457 At/m

(c) 1.37 T
(d) 549 μWb

4.14 (a) 0.206 A
(b) 1777

4.15 2.95 A

4.16 (a) 360 At/m; 1830 At/m
(b) 582

4.17 176.7

4.18 5.14 A

4.19 1.975 A

4.20 1.54 T

4.21 11.57 μWb

Chapter 5

5.1 352.9 V

5.2 0.571 ms

5.3 37.5 V

5.4 (a) 32 V
(b) 5.33 V

5.7 1 Wb/s

5.8 0.05 T

5.9 (a) 0.5 V
(b) 0.433 V
(c) 0.354 V

5.11 1.125 N

5.12 2.5 A

5.13 0.143 m

5.14 3 N

5.15 0.94 μNm

5.16 (a) 27 μN
(b) 0.405 μNm

5.17 12.5 mN

5.18 1750 A

5.19 (a) 0.641 Ω
(b) 6.25 mΩ

5.20 11.975 kΩ; 39.975 kΩ

5.22 1.47 kΩ; 14.97 kΩ;
44.97 kΩ; 149.97 kΩ

5.23 −0.125%

5.24 1.12 H

5.25 4 A

5.26 15 625

5.27 5 H; 800 V

5.28 120

5.29 22.5 A/s

5.30 0.75 mH

5.31 (a) 1364
(b) 33.6 mH
(c) 1.344 V

5.32 0.12 H

5.33 30 V

5.34 (a) 628.3 mH
(b) 56.5 mH
(c) 5.09 V

5.35 (a) 0.3 mH
(b) 1.125 mH
(c) 0.3 V; 1.125 V

5.36 58.67 mH

5.37 (a) 150 V
(b) 5.625 mWb

5.38 12.5 kV

5.40 12 V

5.41 27.5 V; 400 mA; 11 W

5.42 (a) 42.67 V
(b) 3.56 A

5.43 51.2 mJ

5.44 60.4 mH

5.45 2.45 A

Chapter 6

6.1 (a) 150 Hz
(b) 15 Hz
(c) 31.83 Hz

6.2 (a) 100 Hz
(b) 12.5 rev/s

6.3 24

6.4 5 mA; 50 μs; 20 kHz
6.5 (b) $i = 7.5\sin(200\pi t)$ milliamp
6.6 (a) 427.3 V
(b) 50 Hz
(c) 302 V; 272.2 V
6.7 $v = 63.64\sin(3000\pi t)$ volt; 22.31 V
6.8 (a) 250 V; 176.8 V; 75 mA; 53 mA; 20 mWb; 14.14 mWb; 6.8 V; 4.81 V
(b) 25 Hz; 100 Hz; 50 Hz; 1.5 kHz
6.9 $i = 7.07\sin(4000\pi t)$ amp
(a) 6.724 A
(b) 47.86 μs
6.10 353.6 V; 159.25 V
6.11 4.22 mA; 5.97 mA
6.12 1.429
6.13 16 V
6.14 22.2 V
6.18 $v = 17.44\sin(314t + 0.409)$ volt
6.19 $i = 22.26\sin(\omega t - 0.396)$ amp
6.20 $v = 43.06\sin(\omega t - 0.019)$ volt
6.23 3.2 V; 2.26 V; 400 μs; 2.5 kHz

Chapter 7

7.1 (a) 6.772 kΩ
(b) 1.693 kΩ
(c) 67.73 Ω
(d) 1.13 Ω
7.2 (a) 125.7 Ω
(b) 1.885 kΩ
(c) 12.57 kΩ
(d) 3.14 MΩ
7.3 3.98 mH
7.4 26.5 μF
7.5 0.531 H
7.6 1.06 μF
7.7 238.7 Hz
7.8 15.92 kHz
7.9 (a) 42.86 mA; 1.029 W
(b) 0.477 A; zero
(c) 13.27 μA; zero
7.10 25 Ω
7.11 (a) 130.52 Ω
(b) 0.92 A
(c) +54.93 deg; 0.5746
(d) 63.44 W
7.12 (a) 15 V
(b) +53.13 deg
7.13 (a) 3A
(b) 0.424 mH
(c) 27 W
7.14 (a) 5.33 Ω
(b) 0.024 H
7.15 40.3 V
7.16 (a) 130 Ω
(b) 1.385 A
(c) 0.577
(d) 143.9 W
7.17 (a) 152.3 V
(b) 37.5 mA
(c) 49.74 Hz
(d) 3.733 kΩ
7.18 (a) 2.46 A
(b) 96.13 V; 83.46 V
(c) 236.5 W
(d) 0.755
7.19 (a) 55.18 Ω
(b) 4.35 A
(c) −80.6 deg
(d) 39.14 V; 109.3 V; 346 V
7.20 (a) 2.85 A
(b) 187.9 V
(c) 412.4 V
(d) −76.3 deg.
7.21 75.87 Hz; 12 A
7.22 0.633 H
7.23 11.25 kW
7.24 66.67 kVA; 53.33 kVAr
7.25 (a) 8.33 Ω
(b) 5.56 Ω
(c) 6.21 Ω
(d) 12.8 μF
(e) 0.677
(f) −48.2 deg.
7.26 (a) 1.44 kW; 2.4 kVA; 1.92 kVAr
(b) 14.4 Ω; 0.061 H

Chapter 9

9.1 (a) 1.58 kΩ
(b) 360 Ω
(c) 212

9.2 (a) 0.997
(b) 200 kΩ
(c) 332

9.3 (a) 7.7 kΩ
(b) 2.7 mS

Chapter 10

10.1 (a) 67.5 Ω [NPV 68 Ω, 0.5 W]
(b) 50.7 mA

10.2 (a) 108 Ω [NPV 120 Ω, 2 W]
(b) 117 mA
(c) 12 mV

10.3 $R_B = 52.5$ kΩ [NPV 56 kΩ];
$R_c = 2.5$ kΩ [NPV 2.2 kΩ]

10.4 $R_E = 470\,\Omega$; $R_2 = 4.7$ kΩ;
$R_1 = 15$ kΩ; $R_C = 2$ kΩ
[NPV 2.2 kΩ]

10.5 $R_S = 1$ kΩ; $R_D = 2$ kΩ [NPV 2.2 kΩ]

10.6 (a) $g_m = 3.15$ mS; $r_{ds} = 53$ kΩ
(b) $V_{GSQ} = -1.5$ V; $V_{DSQ} = 5.6$ V
(c) $A_V = 2.15$

Appendix A

Physical Quantities with SI and other preferred units

General quantities	*Symbol*	*Units*
Acceleration, linear	a	m/s^2 (metre/second/second)
Area	A	m^2 (square metre)
Energy or work	W	J (joule)
Force	F	N (newton)
Length	l	m (metre)
Mass	m	kg (kilogram)
Power	P	W (watt)
Pressure	p	Pa (pascal)
Temperature value	θ	K or °C (Kelvin or degree Celsius)
Time	t	s (second)
Torque	T	Nm (newton metre)
Velocity, angular	ω	rad/s (radian/second)
Velocity, linear	v or u	m/s (metre/second)
Volume	V	m^3 (cubic metre)
Wavelength	λ	m metre

Electrical quantities	*Symbol*	*Units*
Admittance	Y	Ω (ohm)
Charge (quantity)	Q	C (coulomb)
Conductance	G	S (siemen)
Current	I	A (ampere)
Current density	J	A/m^2 (ampere/square metre)
Electromotive force (emf)	E	V (volts)
Frequency	f	Hz (hertz)
Impedance	Z	Ω (ohm)
Period	T	s (second)
Potential difference (p.d.)	V	V (volt)
Power, active	P	W (watt)
Power, apparent	S	VA (volt ampere)
Power, reactive	Q	VAr (volt ampere reactive)
Reactance	X	Ω (ohm)

Resistance	R	Ω (ohm)
Resistivity	ρ	Ωm (ohm metre)
Time constant	τ	s (second)

Electrostatic quantities	*Symbol*	*Unit*
Capacitance	C	F (farad)
Field strength	$\mathbf{E}$	V/m (volt/metre)
Flux	ψ	C (coulomb)
Flux density	D	C/m^2 (coulomb/square metre)
Permittivity, absolute	ϵ	F/m (farad/metre)

General quantities	*Symbol*	*Unit*
Permittivity, relative	ϵ_r	no units
Permittivity, of free space	ϵ_0	F/m (farad/metre)

Electromagnetic quantities	*Symbol*	*Unit*
Field strength	H	A/m (ampere/metre)(1)
Flux	Φ	Wb (weber)
Flux density	B	T (tesla)
Inductance, mutual	M	H (henry)
Inductance, self	L	H (henry)
Magnetomotive force (mmf)	F	A (ampere)(2)
Permeability, absolute	μ	H/m (henry/metre)
Permeability, relative	μ_r	no units
Permeability, of free space	μ_0	H/m (henry/metre)
Reluctance	S	A/Wb (ampere/weber)(3)

(1) At/m (ampere turn/metre) in this book
(2) (ampere turn) in this book
(3) At/Wb (ampere turn/weber) in this book

Index